Exploring the Multifaceted Roles of Glycosaminoglycans (GAGs) - New Advances and Further Challenges

Exploring the Multifaceted Roles of Glycosaminoglycans (GAGs) - New Advances and Further Challenges

Editors

Dragana Nikitovic
Serge Perez

MDPI • Basel • Beijing • Wuhan • Barcelona • Belgrade • Manchester • Tokyo • Cluj • Tianjin

Editors
Dragana Nikitovic
Medical School
University of Crete
Heraklion
Greece

Serge Perez
CNRS, CERMAV
University Grenoble Alpes
Grenoble
France

Editorial Office
MDPI
St. Alban-Anlage 66
4052 Basel, Switzerland

This is a reprint of articles from the Special Issue published online in the open access journal *Biomolecules* (ISSN 2218-273X) (available at: www.mdpi.com/journal/biomolecules/special_issues/ Multifaceted_Roles_Glycosaminoglycans).

For citation purposes, cite each article independently as indicated on the article page online and as indicated below:

LastName, A.A.; LastName, B.B.; LastName, C.C. Article Title. *Journal Name* **Year**, *Volume Number*, Page Range.

ISBN 978-3-0365-2677-5 (Hbk)
ISBN 978-3-0365-2676-8 (PDF)

© 2021 by the authors. Articles in this book are Open Access and distributed under the Creative Commons Attribution (CC BY) license, which allows users to download, copy and build upon published articles, as long as the author and publisher are properly credited, which ensures maximum dissemination and a wider impact of our publications.

The book as a whole is distributed by MDPI under the terms and conditions of the Creative Commons license CC BY-NC-ND.

Contents

About the Editors . . . . **vii**

Dragana Nikitovic and Serge Pérez
Preface for the Special Issue on the Exploration of the Multifaceted Roles of Glycosaminoglycans: GAGs
Reprinted from: *Biomolecules* **2021**, *11*, 1630, doi:10.3390/biom11111630 . . . . **1**

Elizabeth K. Whitmore, Gabriel Vesenka, Hanna Sihler and Olgun Guvench
Efficient Construction of Atomic-Resolution Models of Non-Sulfated Chondroitin Glycosaminoglycan Using Molecular Dynamics Data
Reprinted from: *Biomolecules* **2020**, *10*, 537, doi:10.3390/biom10040537 . . . . **9**

Serge Pérez, François Bonnardel, Frédérique Lisacek, Anne Imberty, Sylvie Ricard Blum and Olga Makshakova
GAG-DB, the New Interface of the Three-Dimensional Landscape of Glycosaminoglycans
Reprinted from: *Biomolecules* **2020**, *10*, 1660, doi:10.3390/biom10121660 . . . . **33**

Alfonso Maria Ponsiglione, Maria Russo and Enza Torino
Glycosaminoglycans and Contrast Agents: The Role of Hyaluronic Acid as MRI Contrast Enhancer
Reprinted from: *Biomolecules* **2020**, *10*, 1612, doi:10.3390/biom10121612 . . . . **51**

Anthony J Hayes and James Melrose
Aggrecan, the Primary Weight-Bearing Cartilage Proteoglycan, Has Context-Dependent, Cell-Directive Properties in Embryonic Development and Neurogenesis: Aggrecan Glycan Side Chain Modifications Convey Interactive Biodiversity
Reprinted from: *Biomolecules* **2020**, *10*, 1244, doi:10.3390/biom10091244 . . . . **67**

Charlie Colin-Pierre, Valérie Untereiner, Ganesh D. Sockalingum, Nicolas Berthélémy, Louis Danoux, Vincent Bardey, Solène Mine, Christine Jeanmaire, Laurent Ramont and Stéphane Brézillon
Hair Histology and Glycosaminoglycans Distribution Probed by Infrared Spectral Imaging: Focus on Heparan Sulfate Proteoglycan and Glypican-1 during Hair Growth Cycle
Reprinted from: *Biomolecules* **2021**, *11*, 192, doi:10.3390/biom11020192 . . . . **103**

Aikaterini Berdiaki, Monica Neagu, Eirini-Maria Giatagana, Andrey Kuskov, Aristidis M. Tsatsakis, George N. Tzanakakis and Dragana Nikitovic
Glycosaminoglycans: Carriers and Targets for Tailored Anti-Cancer Therapy
Reprinted from: *Biomolecules* **2021**, *11*, 395, doi:10.3390/biom11030395 . . . . **119**

Isabel Faria-Ramos, Juliana Poças, Catarina Marques, João Santos-Antunes, Guilherme Macedo, Celso A. Reis and Ana Magalhães
Heparan Sulfate Glycosaminoglycans: (Un)Expected Allies in Cancer Clinical Management
Reprinted from: *Biomolecules* **2021**, *11*, 136, doi:10.3390/biom11020136 . . . . **151**

Elena Caravà, Paola Moretto, Ilaria Caon, Arianna Parnigoni, Alberto Passi, Evgenia Karousou, Davide Vigetti, Jessica Canino, Ilaria Canobbio and Manuela Viola
HA and HS Changes in Endothelial Inflammatory Activation
Reprinted from: *Biomolecules* **2021**, *11*, 809, doi:10.3390/biom11060809 . . . . **179**

Shravan Morla and Umesh R. Desai
Discovery of Sulfated Small Molecule Inhibitors of Matrix Metalloproteinase-8
Reprinted from: *Biomolecules* **2020**, *10*, 1166, doi:10.3390/biom10081166 193

John Garcia, Helen S. McCarthy, Jan Herman Kuiper, James Melrose and Sally Roberts
Perlecan in the Natural and Cell Therapy Repair of Human Adult Articular Cartilage: Can Modifications in This Proteoglycan Be a Novel Therapeutic Approach?
Reprinted from: *Biomolecules* **2021**, *11*, 92, doi:10.3390/biom11010092 207

Elena Lantero, Jessica Fernandes, Carlos Raúl Aláez-Versón, Joana Gomes, Henrique Silveira, Fatima Nogueira and Xavier Fernàndez-Busquets
Heparin Administered to *Anopheles* in Membrane Feeding Assays Blocks *Plasmodium* Development in the Mosquito
Reprinted from: *Biomolecules* **2020**, *10*, 1136, doi:10.3390/biom10081136 229

Joman Javadi, Katalin Dobra and Anders Hjerpe
Multiplex Soluble Biomarker Analysis from Pleural Effusion
Reprinted from: *Biomolecules* **2020**, *10*, 1113, doi:10.3390/biom10081113 241

Harkanwalpreet Sodhi and Alyssa Panitch
Glycosaminoglycans in Tissue Engineering: A Review
Reprinted from: *Biomolecules* **2020**, *11*, 29, doi:10.3390/biom11010029 255

Riccardo Ladiè, Cesare Cosentino, Irene Tagliaro, Carlo Antonini, Giulio Bianchini and Sabrina Bertini
Supramolecular Structuring of Hyaluronan-Lactose-Modified Chitosan Matrix: Towards High-Performance Biopolymers with Excellent Biodegradation
Reprinted from: *Biomolecules* **2021**, *11*, 389, doi:10.3390/biom11030389 279

Tatiana Guzzo, Fabio Barile, Cecilia Marras, Davide Bellini, Walter Mandaliti, Ridvan Nepravishta, Maurizio Paci and Alessandra Topai
Stability Evaluation and Degradation Studies of DAC® Hyaluronic-Polylactide Based Hydrogel by DOSY NMR Spectroscopy
Reprinted from: *Biomolecules* **2020**, *10*, 1478, doi:10.3390/biom10111478 297

Mélanie Leroux, Julie Michaud, Eric Bayma, Sylvie Armand, Sophie Drouillard and Bernard Priem
Misincorporation of Galactose by Chondroitin Synthase of *Escherichia coli* K4: From Traces to Synthesis of Chondbiuronan, a Novel Chondroitin-Like Polysaccharide
Reprinted from: *Biomolecules* **2020**, *10*, 1667, doi:10.3390/biom10121667 313

About the Editors

Dragana Nikitovic

Dragana Nikitovic is currently an Associate Professor and Head of the Laboratory of Histology-Embryology at the Medical School, University of Crete (UOC). She has earned a Doctor of Medical Sciences degree from the UOC, Greece. She has carried out postgraduate research at the Queen Mary and Westfield College, London, the UK, and Karolinska Institute, Stockholm, Sweden. Her scientific interests are focused explicitly on studying proteoglycans (PGs) and glycosaminoglycans (GAGs) effects in disease progression, including cancer and inflammation and their mechanisms of action. She serves as an associate editor, review editor, and editorial board member for several journals in Life Sciences and is an active member of respective scientific societies. Her research interests have resulted in the authorship of more than 100 original publications and book chapters.

Serge Perez

Serge Perez holds a Doctorate es Sciences from the University of Grenoble, France. He had international exposure throughout several academic and industry positions in research laboratories in the U.S.A., Canada, and France (Centre de Recherches sur les Macromolecules Végétales, CNRS, Grenoble, as the chairperson and as Director of Research at the European Synchrotron Radiation Facility). His research interests span across the whole area of structural glycoscience, emphasizing polysaccharides, glycoconjugates, and protein-carbohydrate interactions. He has a strong interest in the economy of glycoscience, and e-learning for which he created the www.glycopedia.eu, He is actively involved in scientific societies, as President and past-president of the European Carbohydrate Organisation. He is the author of more than 300 research publications, among which several are the subject of a large number of citations and references.

Editorial

Preface for the Special Issue on the Exploration of the Multifaceted Roles of Glycosaminoglycans: GAGs

Dragana Nikitovic [1,*] and Serge Pérez [2]

1. Laboratory of Histology-Embryology, School of Medicine, University of Crete, 71003 Heraklion, Greece
2. University Grenoble Alpes, CNRS, CERMAV, 38000 Grenoble, France; serge.perez@cermav.cnrs.fr
* Correspondence: nikitovic@uoc.gr

Citation: Nikitovic, D.; Pérez, S. Preface for the Special Issue on the Exploration of the Multifaceted Roles of Glycosaminoglycans: GAGs. *Biomolecules* **2021**, *11*, 1630. https://doi.org/10.3390/biom11111630

Received: 22 October 2021
Accepted: 26 October 2021
Published: 4 November 2021

Publisher's Note: MDPI stays neutral with regard to jurisdictional claims in published maps and institutional affiliations.

Copyright: © 2021 by the authors. Licensee MDPI, Basel, Switzerland. This article is an open access article distributed under the terms and conditions of the Creative Commons Attribution (CC BY) license (https://creativecommons.org/licenses/by/4.0/).

Glycosaminoglycans (GAGs) are linear, anionic polysaccharides that consist of repeating disaccharides of hexosamine and hexuronic acid. The exception to this is keratan sulfate, whose building blocks consist of hexosamine and galactose. Differences in the primary disaccharide unit structure regarding uronic acid and hexosamine, the number and position of the sulfate residues, the presence of *N*-acetyl and/or *N*-sulfate groups, and the relative molecular mass are evident. All such differences bestow these biomolecules with impressive complexity and diversity. The fine structure of the disaccharide units defines the types of GAGs. These include chondroitin/dermatan sulfate (CS/DS), heparin/heparan sulfate (Hep/HS), and keratan sulfate (KS), as well as non-sulfated hyaluronan (HA) (Figure 1).

GAGs are ubiquitously localized throughout the extracellular matrix (ECM) and to the cell membranes of cells in all tissues. They are either conjugated to protein cores in the form of proteoglycans, e.g., CS/DS, HS, and KS, or as free GAGs (HA and Hep). Through their interaction with proteins, GAGs can affect the cell-extracellular matrix (ECM) and cell–cell interactions, finely modulating ligand-receptor binding and thus chemokine and cytokine activities as well as growth factor sequestration. Thus, GAGs regulate several biological processes under homeostasis; they also participate in disease progression. Recently, significant advances have been made in the analytic, sequencing, and structural characterization of GAG oligosaccharides as well as in GAG profiling in tissues and cells (GAGomics). Moreover, studies focused on the structure/sequence-function relationships of GAGs have resulted in critical novel insights. Furthermore, advances in the characterization of protein–GAG complexes provide invaluable tools to decipher GAG's roles in the intricate tissue milieu and answer critical questions regarding GAG participation in the molecular basis of disease and embryonic development.

This Special Issue of *Biomolecules*, entitled "Exploring the multifaceted roles of glycosaminoglycans (GAGs)—new advances and further challenges", features original research and review articles. These articles cover several timely topics in structural biology and imaging; morphogenesis, cancer, and other disease therapy and drug developments; tissue engineering; and metabolic engineering. This Special Issue also includes an article illustrating how metabolic engineering can be used to create the novel chondroitin-like polysaccharide.

A prerequisite for communicating in any discipline and across disciplines is familiarity with the appropriate terminology. Several nomenclature rules exist in the field of biochemistry. The historical description of GAGs follows IUPAC and IUB nomenclature. New structural depictions such as the structural nomenclature for glycan [1] and their translation into machine-readable formats [2,3] have opened the route for cross-references with popular bioinformatics resources and further connections with other "omics".

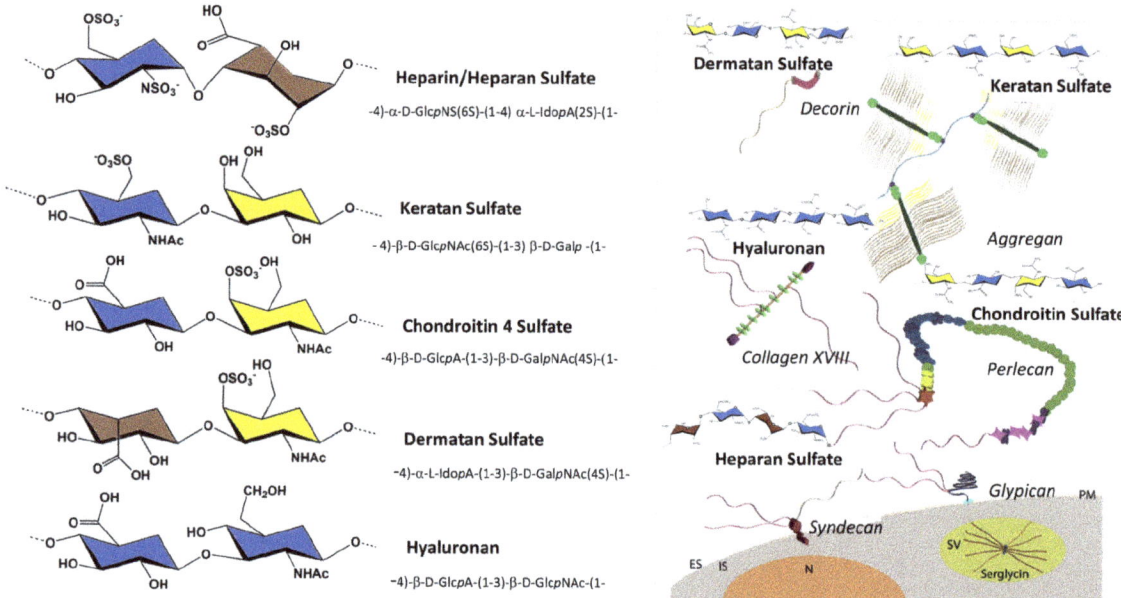

Figure 1. Cartoon representation of the chemical constitution of the five families of GAGs and of six categories of proteoglycans (aggrecan; decorin, perlecan, and collagen; glypican; and syndecan and serglycin). ES, extracellular; IS, intracellular; N, nucleus; SV, secretory vesicle. (Adapted from K. Rodgers, J.D. San Antonio, O. Jacenko. Dev Dyn. 2008, 237, 2622–2642).

1. Structure and Imaging

The structural heterogeneity of GAGs complicates the composition and sequence analysis of GAGs. No less than 200 different monosaccharides have been identified; this has resulted in a very high number of disaccharide segments exhibiting high conformational flexibility. In addition, there have also been intrinsic difficulties in establishing the three-dimensional structures of GAGs. Despite these difficulties and recognizing that the shape of molecules is a fundamental principle in chemistry, physics, and biology, scientists are developing experimental and computational tools to elucidate and understand molecular shapes and molecular motions. Elisabeth Whitmore and collaborators [4] report on the development of efficient atomic resolution models using molecular GAG dynamics. They illustrate the outcome of their application for the case of non-sulfated chondroitin, which may provide insights and arguments for the understanding of disciplines where molecular dynamics play a crucial role.

There is a healthy dialogue between computational and experimental endeavors, a typical example being the structural determination of the oligo of GAGS and polysaccharides and their interactions with proteins. Results have accumulated over time due to X-ray single-crystal diffraction methods, X-ray fiber diffractometry, solution NMR spectroscopy, and scattering data. These data have been curated, annotated, and organized before their structuration into a three-dimensional database containing three-dimensional data on GAGs and GAGs–protein complexes retrieved from the PDB [5]. The database includes protein sequences and the standard nomenclature for GAG composition, sequence, and topology. It provides a family-based classification of GAGs that is cross-referenced with glyco-databases with links to UniProtKB via accession numbers. The 3D visualization of contacts between GAGs and their protein ligands is implemented via the protein–ligand interaction profiler (PLIP). The nature of the structure that GAG polysaccharides can adopt, either solid-state or solution, is also reported. Finally, characterized quaternary structures of the complexes improve our understanding as to if and how GAGs participate in long-range, multivalent binding with potential synergy when several chains are involved in interactions.

Molecular interactions involving GAGs are not restricted to proteins. Many authors consider the large GAG polymeric backbones and their chemical properties to be essential features for the rational design of drug delivery and diagnostic systems. Magnetic resonance imaging is an established diagnostic method for which GAGs, when adequately decorated, offer the benefit of contrast enhancers.

The administration of a paramagnetic contrast agent, such as a metal chelate, such as gadolinium diethylene triamine penta-acetic acid (Gd-DTPA), helps visualize relative GAG distribution in vivo. For example, the negative charge of the contrast agent will distribute itself within articular cartilage in a spatially inverse relationship to the concentration of negatively charged GAG molecules. Alfonso Ponsiglione and collaborators [6] explore the range of advantages that could represent fine control over the combination of GAGs and imaging agents in the formulation of novel multifunctional diagnostic probes.

2. Morphogenesis and Development

GAGs, as essential constituents of the human glycome, play pivotal roles in a multitude of biological processes during embryonic development and in the maintenance of homeostasis. Such roles can be observed throughout the structural mold of aggregan and the diversity of its decoration by GAGs. The development of vertebrates from a single cell to the generation of various cell types and organs is carried out throughout a synchronized developmental program consisting of the spatial and temporal coordination of specific signaling molecules, including morphogens and growth factors. The importance of specific arrangements of GAG chains on aggrecan in all of its forms is also a primary morphogenetic functional determinant. It provides aggrecan with unique tissue-context-dependent regulatory properties. The versatility displayed by aggrecan in biodiverse contexts is a function of its GAG side chains [7].

The article by Colin-Pierre et al. [8] describes the evolution of heparan sulfate proteoglycans in hair follicles. The heparan sulfate proteoglycan distribution in hair follicles has traditionally been done by conventional histology, biochemical analysis, and immunohistochemistry. The authors use the absorption region that is relevant to sulfation as a spectral marker. This is performed using infrared spectral imaging (IRSI), which has been used intensively for cell (spectral cytology) and tissue (spectral histology) characterization. Supported by Western blot and immunohistochemistry analysis, infrared spectral imaging specifically shows the qualitative and/or quantitative evolution of the GAGs expression pattern between the anagen, catagen, and telogen phases. Moreover, this demonstrates that IRSI could be utilized for GAG cytology and tissue characterization.

3. Therapy: GAGs as Targets and Novel Therapy Agents

GAGs are essential ECMs and cell membrane components and are extensively altered under various pathological conditions, including cancer. Indeed, during disease progression, the fine GAG structure and expression change in a manner that is associated with disease evolution. Furthermore, pathological conditions are characterized by the extent of GAG remodeling that is either due to the increased expression of glycosidases or to a chemical reaction with elevated, radical oxygen species. Specific disease-dependent GAG alterations have been identified as druggable entities, with industry and academic research efforts examining their potential in drug development. Berdiaki et al. [9] discuss the up-to-date developments of implementing GAG disease-dependent changes in two directions: (i) utilizing GAGs as the targets of therapeutic strategies and (ii) employing GAG specificity and excellent physicochemical properties for the targeted delivery of cancer therapeutics.

Faria-Ramos et al. [10] specifically focus on the role of HS in carcinogenesis. Due to HS's well-established regulation of critical cellular receptors and respective downstream signaling pathways and the aberrant expression of HS in tumor tissue, these GAGs have been characterized as modulators of malignant features. This review article highlights the significant clinical potential of HS to improve both the diagnosis and prognosis of cancer, either as HS-based biomarkers or as therapeutic targets [10].

GAG functions are implicated in inflammatory processes. Notably, cardiovascular disease propagation and the inflammatory status of tissues are closely correlated. The treatment of endothelial cells with the cytokine TNF-α, which is known to be increased in obese patients and has been reported to induce cardiometabolic diseases, strongly affects the expression patterns of hyaluronan and the HS-containing proteoglycans known as syndecans [11]. These changes seem to facilitate the onset of a pathological state by altering (i) the endothelial barrier properties, (ii) increasing HA in the pericellular coat and the possibility of consequent monocyte recruitment from the blood; or (iii) altering the sulfation pattern of membrane-bound HS, which can cause modifications to the endothelium response to growth factors and cytokines. Therefore, the authors confirm the critical role of ECM components such as GAGs in disease progression.

Matrix metalloproteinases (MMPs) are endopeptidases that are able to cleave both matrix and non-matrix proteins. MMPs activity and the resulting extracellular matrix remodeling are increased in acute and chronic diseases and are correlated with disease pathogenesis. Thus, the enhanced activity of MMP-8 facilitates the progression of various pathologies, including atherosclerosis, pulmonary fibrosis, and sepsis. Since natural GAGs are known to modulate the functions of various MMPs, the synthetic non-sugar mimetics of GAGs have been hypothesized to inhibit MMP-8 activity. The strategy of Moria and Desai [12], upon screening a library of 58 synthetic, sulfated mimetics consisting of a dozen scaffolds, led to the identification of sulfated benzofurans and sulfated quinazolinones as promising inhibitors of MMP-8. Interestingly, this work provides the first proof that the sulfated mimetics of GAGs could lead to potent, selective, and catalytic activity-tunable, small molecular inhibitors of MMP-8.

Due to the lack of blood vessels and the consequently limited bioavailability of oxygen and nutrients, articular cartilage has restricted regenerative capacity, resulting in frequent degenerative disease in older individuals. Therefore, therapeutic strategies limiting or halting the progression of cartilage destruction are an unmet health need. Perlecan, a multifunctional HS proteoglycan, promotes embryonic cartilage development and stabilizes mature tissue. Using immunohistochemistry, Garcia et al. [13] showed a pericellular and diffuse matrix staining pattern for perlecan in both natural and cell-therapy-repaired cartilage. This observation was related to whether the morphology of the newly formed tissue was hyaline cartilage or fibrocartilage. In addition, immunostaining was significantly more enhanced in these repair tissues for perlecan than it was for normal age-matched controls and was sensitive to heparanase treatment. Thus, the modulation of HS could be helpful in the treatment of degenerative cartilage disease.

A novel, interesting therapeutic function of heparin has been shown by Lantero et al. [14]. Indeed, these authors report an antimalarial activity of heparin. Innovative antimalarial strategies are urgently needed as plasmodium parasites continue to express increased resistance to the available drugs that have been developed against plasmodium parasites. Heparin delivered in membrane feeding assays together with *Plasmodium berghei*-infected blood of *Anopheles stephensi* mosquitoes was shown to inhibit the parasite's ookinete–oocyst transition by binding the ookinetes. The inhibition of the parasite life-cycle by heparin might represent a new antimalarial strategy for rapid implementation and is an excellent example of the ubiquitous use of these multifaceted molecules.

4. Biomarkers

Because the expression of GAGs and their fine structure are markedly altered in disease, these features could have an essential clinical implementation. Malignant pleural mesothelioma (MPM) is a highly aggressive and therapy-resistant pleural malignancy with poor prognoses and short patient survival. When patients' pleural infusion was analyzed by a Luminex multiplex assay, syndecan-1 (SDC-1) and MMP-7 levels were significantly lower, whereas mesothelin and galectin-1 levels were significantly higher in malignant mesothelioma effusions compared to in adenocarcinoma. Javadi et al. [15] suggest that MMP-7, shed SDC-1, mesothelin, and galectin-1 can be diagnostic and that VEGF and SDC-1 are prognostic markers in MPM patients. Indeed, this study confirms the vital role of ECM components in malignant disease progression.

5. Tissue Engineering and Biomaterial

GAGs are native components of the ECM that drive cell behavior and control the microenvironment surrounding cells, making them promising therapeutic targets for many diseases. Recent studies have shown that the recapitulation of cell interactions with ECM is critical in tissue engineering, which aims to mimic and regenerate endogenous tissues. Because of this, the incorporation of GAGs to drive stem cell fate and to promote cell proliferation in engineered tissues has gained increasing attention. This review article [16] summarizes the role of glycosaminoglycans in tissue engineering and their recent use in these constructs. In addition, the evaluation of the general research trends in this niche offers insight into future research directions in this field.

Hyaluronan displays such properties as biocompatibility, biodegradability, high viscoelasticity, and immunoneutrality, making it attractive for biomedical and pharmaceutical applications. Furthermore, from the standpoint of physical properties, the polyelectrolyte nature of negatively charged hyaluronan provides a way to create new high-performance complexes. One such complex occurs when hyaluronan self-assembles with a positively charged lactose-modified chitosan. The authors of this investigation [17] show that the complex that is formed has a monodisperse molecular weight distribution and a high viscosity and is susceptible to enzymatic degradation by hyaluronidase and lysozyme. Due to the wide range of applications in biomedicine and biotechnology, the development of such polyelectrolyte complexes is of scientific and biotechnological interest.

The conjunction of the wide range of biological activities and unique physicochemical properties confer a distinctive place as an implantable biomaterial used in orthopedics and traumatology to hyaluronan. Infections related to implanted medical devices depend on the bacterial capability to establish highly structured multilayered biofilms on artificial surfaces. One way to prevent such peri-implant infection is to apply an implanted biomaterial, defensive antibacterial coating (DAC), which can act as a resorbable barrier that delivers local antibiofilm and antibacterial compounds. The copolymer of hyaluronic acid and poly-D, L-lactic acid produces a hydrogel that retains the hydrophobic character of the poly-D, L-lactide sidechains and the hydrophilic character of a hyaluronic acid backbone. The suitability of such a hydrogel depends on the stability and degradation of both the hyaluronan backbone and the polylactic chains over time and temperature. T. Guzzo and her collaborators [18] performed chromatographic analysis and explored the suitability of diffusion-ordered NMR spectroscopy to characterize the outcome of the biomaterial over time in physiological conditions.

6. Metabolic Engineering

Bacterial cells exhibit a wide diversity of capsular polysaccharides that constitute the cell surface of the outer membrane and that mediate interactions with the environment. The capsules are often composed of GAG-like polymers. The structural similarity of microbial capsular polysaccharides to these biomolecules makes these ideal bacteria candidates for non-animal GAG-derived products. It is true that the capsular polysaccharide of *Escherichia coli* K4, the chondroitin synthase polymerase, KfoC, synthesizes a chondroitin-like polysaccharide. While exploring novel methods and conditions to produce chondroitin via metabolic engineering, Leroux and her colleagues expressed KofC in a recombinant strain of *Escherichia coli* deprived of 4-epimerase activity [19]. They realized that KfoC could polymerize a GalNAc-free polysaccharide, giving rise to a novel GAG that they call chondbiuronan.

7. Conclusions

While the authors of these articles discussed the multifaceted features of GAGs, they are all well motivated by specific applications in biology and medicine and will develop appropriate tools that are likely to be important in structuring a large amount of available data and opening the field to cross-disciplinary endeavors. We hope that these articles will not only provide timely case studies but that they will also form the basis of a series of questions aimed at answering the broader question of what remains to be solved about GAGs?

Funding: Mentioned research by S.P. profited from Campus Rhodanien Co-Funds (https://unige-cofunds.ch/alliance-campus-rhodanien, accessed on 11 March 2021) support from ANR PIA Glyco@Alps [ANR-15-IDEX-02]; Alliance), and Labex Arcane/CBH-EUR-GS [ANR-17-EURE-0003]. Innogly COST Action covered the costs of the meeting where the Special issue was inaugurated. D.N. was partially funded by the Research Committee of University of Crete (ELKE), grant number (KA:10028).

Institutional Review Board Statement: Not applicable.

Informed Consent Statement: Not applicable.

Data Availability Statement: Not applicable.

Acknowledgments: This article is part of the Innogly Cost action initiative.

Conflicts of Interest: The authors declare no conflict of interest.

References

1. Varki, A.; Cummings, R.D.; Aebi, M.; Packer, N.; Seeberger, P.H.; Esko, J.D.; Stanley, P.; Hart, G.; Darvill, A.; Kinoshita, T.; et al. Symbol nomenclature for graphical representations of glycans. *Glycobiology* **2015**, *25*, 1323–1324. [CrossRef]
2. Clerc, O.; Mariethoz, J.; Rivet, A.; Lisacek, F.; Pérez, S.; Ricard-Blum, S. A pipeline to translate glycosaminoglycan sequences into 3D models. Application to the exploration of glycosaminoglycan conformational space. *Glycobiology* **2018**, *29*, 36–44. [CrossRef]
3. Clerc, O.; Deniaud, M.; Vallet, S.D.; Naba, A.; Rivet, A.; Perez, S.; Thierry-Mieg, N.; Ricard-Blum, S. MatrixDB: Integration of new data with a focus on glycosaminoglycan interactions. *Nucleic Acids Res.* **2018**, *47*, D376–D381. [CrossRef]
4. Whitmore, E.K.; Vesenka, G.; Sihler, H.; Guvench, O. Efficient construction of atomic-resolution models of non-sulfated chondroitin glycosaminoglycan using molecular dynamics data. *Biomolecules* **2020**, *10*, 537. [CrossRef] [PubMed]
5. Pérez, S.; Bonnardel, F.; Lisacek, F.; Imberty, A.; Blum, S.R.; Makshakova, O. GAG-DB, the new interface of the three-dimensional landscape of glycosaminoglycans. *Biomolecules* **2020**, *10*, 1660. [CrossRef] [PubMed]
6. Ponsiglione, A.M.; Russo, M.; Torino, E. Glycosaminoglycans and contrast agents: The role of hyaluronic acid as MRI contrast enhancer. *Biomolecules* **2020**, *10*, 1612. [CrossRef] [PubMed]
7. Hayes, A.J.; Melrose, J. Aggrecan, the primary weight-bearing cartilage proteoglycan, has context-dependent, cell-directive properties in embryonic development and neurogenesis: Aggrecan glycan side chain modifications convey interactive biodiversity. *Biomolecules* **2020**, *10*, 1244. [CrossRef] [PubMed]
8. Colin-Pierre, C.; Untereiner, V.; Sockalingum, G.; Berthélémy, N.; Danoux, L.; Bardey, V.; Mine, S.; Jeanmaire, C.; Ramont, L.; Brézillon, S. Hair histology and glycosaminoglycans distribution probed by infrared spectral imaging: Focus on heparan sulfate proteoglycan and Glypican-1 during hair growth cycle. *Biomolecules* **2021**, *11*, 192. [CrossRef] [PubMed]
9. Berdiaki, A.; Neagu, M.; Giatagana, E.-M.; Kuskov, A.; Tsatsakis, A.; Tzanakakis, G.; Nikitovic, D. Glycosaminoglycans: Carriers and targets for tailored anti-cancer therapy. *Biomolecules* **2021**, *11*, 395. [CrossRef] [PubMed]
10. Faria-Ramos, I.; Poças, J.; Marques, C.; Santos-Antunes, J.; Macedo, G.; Reis, C.; Magalhães, A. heparan sulfate glycosaminoglycans: (Un)Expected allies in cancer clinical management. *Biomolecules* **2021**, *11*, 136. [CrossRef] [PubMed]
11. Caravà, E.; Moretto, P.; Caon, I.; Parnigoni, A.; Passi, A.; Karousou, E.; Vigetti, D.; Canino, J.; Canobbio, I.; Viola, M. HA and HS changes in endothelial inflammatory activation. *Biomolecules* **2021**, *11*, 809. [CrossRef] [PubMed]
12. Morla, S.; Desai, U.R. Discovery of sulfated small molecule inhibitors of matrix metalloproteinase. *Biomolecules* **2020**, *10*, 1166. [CrossRef] [PubMed]
13. Garcia, J.; McCarthy, H.S.; Kuiper, J.H.; Melrose, J.; Roberts, S. Perlecan in the natural and cell therapy repair of human adult articular cartilage: Can modifications in this proteoglycan be a novel therapeutic approach? *Biomolecules* **2021**, *11*, 92. [CrossRef] [PubMed]
14. Lantero, E.; Fernandes, J.; Aláez-Versón, C.R.; Gomes, J.; Silveira, H.; Nogueira, F.; Fernàndez-Busquets, X. Heparin administered to *Anopheles* in membrane feeding assays blocks *Plasmodium* development in the mosquito. *Biomolecules* **2020**, *10*, 1136. [CrossRef] [PubMed]
15. Javadi, J.; Dobra, K.; Hjerpe, A. Multiplex soluble biomarker analysis from pleural effusion. *Biomolecules* **2020**, *10*, 1113. [CrossRef] [PubMed]
16. Sodhi, H.; Panitch, A. glycosaminoglycans in tissue engineering: A review. *Biomolecules* **2020**, *11*, 29. [CrossRef] [PubMed]

17. Ladiè, R.; Cosentino, C.; Tagliaro, I.; Antonini, C.; Bianchini, G.; Bertini, S. Supramolecular structuring of hyaluronan-lactose-modified chitosan matrix: Towards high-performance biopolymers with excellent biodegradation. *Biomolecules* **2021**, *11*, 389. [CrossRef] [PubMed]
18. Guzzo, T.; Barile, F.; Marras, C.; Bellini, D.; Mandaliti, W.; Nepravishta, R.; Paci, M.; Topai, A. stability evaluation and degradation studies of DAC® hyaluronic-polylactide based hydrogel by DOSY NMR spectroscopy. *Biomolecules* **2020**, *10*, 1478. [CrossRef] [PubMed]
19. Leroux, M.; Michaud, J.; Bayma, E.; Armand, S.; Drouillard, S.; Priem, B. Misincorporation of galactose by chondroitin synthase of escherichia Coli K4: From traces to synthesis of chondbiuronan, a novel chondroitin-like polysaccharide. *Biomolecules* **2020**, *10*, 1667. [CrossRef] [PubMed]

Article

Efficient Construction of Atomic-Resolution Models of Non-Sulfated Chondroitin Glycosaminoglycan Using Molecular Dynamics Data

Elizabeth K. Whitmore [1,2], Gabriel Vesenka [1], Hanna Sihler [1] and Olgun Guvench [1,2,*]

[1] Department of Pharmaceutical Sciences, University of New England College of Pharmacy, 716 Stevens Avenue, Portland, ME 04103, USA; ewhitmore@une.edu (E.K.W.); gvesenka1@une.edu (G.V.); hsihler@une.edu (H.S.)
[2] Graduate School of Biomedical Science and Engineering, University of Maine, 5775 Stodder Hall, Orono, ME 04469, USA
* Correspondence: oguvench@une.edu; Tel.: +1-207-221-4171

Received: 26 February 2020; Accepted: 1 April 2020; Published: 2 April 2020

Abstract: Glycosaminoglycans (GAGs) are linear, structurally diverse, conformationally complex carbohydrate polymers that may contain up to 200 monosaccharides. These characteristics present a challenge for studying GAG conformational thermodynamics at atomic resolution using existing experimental methods. Molecular dynamics (MD) simulations can overcome this challenge but are only feasible for short GAG polymers. To address this problem, we developed an algorithm that applies all conformational parameters contributing to GAG backbone flexibility (i.e., bond lengths, bond angles, and dihedral angles) from unbiased all-atom explicit-solvent MD simulations of short GAG polymers to rapidly construct models of GAGs of arbitrary length. The algorithm was used to generate non-sulfated chondroitin 10- and 20-mer ensembles which were compared to MD-generated ensembles for internal validation. End-to-end distance distributions in constructed and MD-generated ensembles have minimal differences, suggesting that our algorithm produces conformational ensembles that mimic the backbone flexibility seen in simulation. Non-sulfated chondroitin 100- and 200-mer ensembles were constructed within a day, demonstrating the efficiency of the algorithm and reduction in time and computational cost compared to simulation.

Keywords: molecular dynamics; glycosaminoglycan; proteoglycan; chondroitin sulfate; carbohydrate conformation; carbohydrate flexibility; glycosidic linkage; ring pucker; force field; explicit solvent

1. Introduction

The diverse group of protein–carbohydrate conjugates called proteoglycans (PGs) is a fundamental component of tissue structure in animals and can be found in the extracellular matrix (ECM) as well as on and within cells. PGs bind growth factors [1–12], enzymes [2,12], membrane receptors [12], and ECM molecules [2,12,13]. By doing so, they modulate signal transduction [13,14], tissue morphogenesis [2,8–11], and matrix assembly [2,15–17]. PG bioactivity is often dependent on the covalently linked carbohydrate chains called glycosaminoglycans (GAGs), which are linear, highly negatively charged, and structurally diverse carbohydrate polymers. GAGs mediate receptor–ligand complex formation by either forming non-covalent complexes with proteins or inhibiting the formation of complexes with other biomolecules. This makes GAGs key modulators in many diseases, giving them potential therapeutic applications. For example, heparan sulfate (HS) is released during sepsis and induces septic shock [18,19]; the removal of chondroitin sulfate (CS) may enhance memory retention and slow neurodegeneration in patients with Alzheimer's disease [20–22]; and dermatan sulfate (DS) deficiency has been implicated in Ehlers–Danlos syndrome, thus the screening of DS in urine could be used as an early diagnostic tool [23,24].

GAG binding sites on proteins are determined by protein sequence and structure, with requirements for both shape and charge complementarity [12,25]. Thus, GAG function depends on GAG three-dimensional structure and conformation. Even subtle structural differences impact GAG function. For example, while CS and DS have many functional differences, the only structural difference is in the chirality of the uronic acid monosaccharides. While much is known about GAG function, attempting to study GAG conformational thermodynamics at atomic resolution presents a largely unsolved problem for existing experimental methods. This is largely due to the structural and conformational complexities of GAGs. For example, a given GAG consists of a repeating sequence of a particular disaccharide, but conformational complexity is introduced through flexibility in the glycosidic linkages between monosaccharides [26–30] (Figure 1). Additional complexity results from non-template-based synthesis [31] and variable enzymatic sulfation [32], which means a biological sample of a GAG composed of a specific disaccharide repeat will be polydisperse and heterogeneous owing to the variable length and sulfation of the individual polymer molecules. Liquid chromatography–mass spectrometry (LC-MS) [33–35], X-ray crystallography [36–41], and nuclear magnetic resonance (NMR) [42–45] are used to study GAGs but are limited in their ability to account for all of these complexities. Additionally, some studies have used results from LC-MS [45], X-ray crystallography [46], and NMR [46–51] to compare and validate conformational data from molecular dynamics (MD) simulations. This suggests that MD simulations can produce results complementary to experimental analysis methods by providing realistic three-dimensional atomic-resolution molecular models of GAG conformational ensembles [52–56].

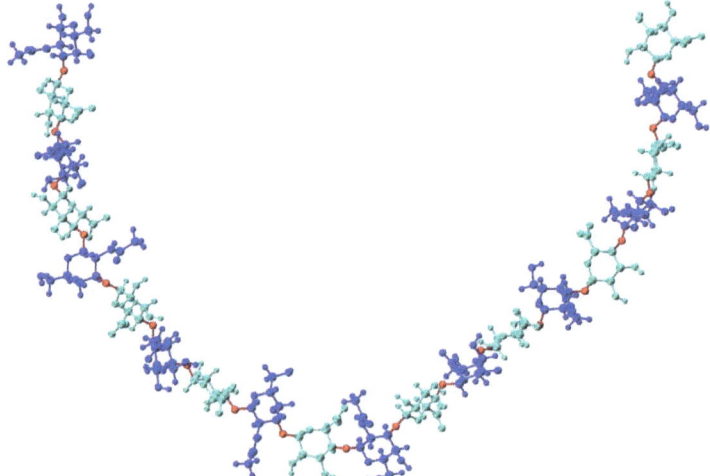

Figure 1. Compact non-sulfated chondroitin 20-mer conformation arising from flexible glycosidic linkages (red) between monosaccharide rings (GalNAc in blue and GlcA in cyan). The molecular graphics throughout are produced with the VMD program [57].

A critical challenge with MD simulations of GAGs is that a single biological GAG polymer chain may contain up to 200 monosaccharide units [9]. When fully solvated, the resulting system will have in excess of 10^6 atoms. It is not feasible to routinely simulate such a system using current graphics processing unit (GPU)-accelerated MD codes with a modern GPU and multi-core CPU. This limits the utility of all-atom explicit-solvent MD as a tool for routine conformational analysis of GAGs of this size.

Coarse-grained (CG) MD simulations are the most feasible current alternative to all-atom explicit-solvent MD as they entail fewer degrees of freedom for the solute [48] and often an implicit (continuum) description of the solvent [58,59]. This can make CG MD two to three orders of magnitude faster, thereby allowing for the handling of large systems [60], such as GAG 200-mers. Indeed, a recent CG model using glycosidic linkage and ring pucker energy functions has provided previously-unseen details of the structure–dynamics relationship of GAGs in the context of PGs [48]. An important insight from that study was that GAGs, in contrast to the unique ordered conformations of folded proteins, need to be considered as existing in conformational ensembles containing a large diversity of three-dimensional conformations.

As an alternative approach to using CG MD to generate such conformational ensembles for GAGs, we propose using glycosidic linkage and monosaccharide ring conformations from unbiased all-atom explicit-solvent MD simulations [56,61–63] of short GAG polymers to rapidly construct conformational ensembles for GAGs of an arbitrary length. Toward this end, we studied a non-sulfated chondroitin 20-mer with the sequence [-4 glucuronate β1-3 N-acetylgalactosamine β1-]$_{10}$ for its simplicity and homogeneity. We first ran microsecond-scale all-atom explicit solvent MD on the 20-mer and used the resulting trajectories to develop a database of conformations. From this database, we randomly selected individual values for the bond lengths, bond angles, and dihedral angles in the glycosidic linkages connecting glucuronate (GlcA) and N-acetylgalactosamine (GalNAc) and in the monosaccharide rings. These values were used to construct a 20-mer conformational ensemble. The comparison of the constructed ensemble with the MD-generated ensemble of 20-mer conformations revealed similar end-to-end distance distributions, with a strong bias toward extended conformations in both cases. Short end-to-end distances associated with more compact conformations were facilitated by the sampling of non-$^4C_1$ ring puckering by GlcA. This change in ring geometry, which occurs rarely on the microsecond timescale, introduced kinks into the polymer, causing it to bend back toward itself. The fact that the MD-generated ensemble had a great deal of variability in both end-to-end distances and radii of gyration demonstrates the inherent flexibility of the chondroitin polymer in aqueous solution. The fact that the constructed ensemble has very similar conformational properties to the MD-generated ensemble suggests that there is little correlation between the individual dihedral angle values that determine the internal geometry of a given conformation. Therefore, on the timescale of the simulations, non-sulfated chondroitin 20-mer does not appear to have any higher-order structure, in contrast to, for example, the secondary and tertiary structure seen in proteins. This lack of higher-order structure was borne out in a comparison of end-to-end distances for constructed vs. MD-generated ensembles of 10-mers, with the constructed ensemble built using the 20-mer database. Finally, we used the methodology to produce conformational ensembles of 100-mers and of 200-mers. The ability to model polymers with biologically-relevant chain lengths (e.g., 100- to 200-mers) will provide insights into GAG binding by other biomolecules. This will be especially useful in understanding the formation of complexes containing multiple biomolecules bound to a single GAG.

Other programs that construct three-dimensional atomic-resolution models of GAG polymers exist, for example, Glycam GAG Builder [64], POLYS Glycan Builder [65], CarbBuilder [66], and MatrixDB GAG Builder [67,68], which allow the user to choose GAG type, length, and sequence and are useful tools for producing an initial structure for MD simulations. Glycam and POLYS Glycan Builder allow the user either to specify particular glycosidic linkage dihedral angle values or use default parameters pulled from their databases. The databases used by Glycam, POLYS Glycan Builder, and Carb Builder include GAG mono- and disaccharide structures determined by molecular mechanics and/or MD. MatrixDB pulls from databases of experimentally determined conformations of GAG disaccharides from crystallized GAG–protein complexes. While the user has the option to choose the GAG length, these tools are intended for shorter GAG polymers. In contrast to these tools, our algorithm pulls from a database of full conformational landscapes of unbound GAG 20-mers. Additionally, our algorithm is intended for modeling long GAG polymers with biologically-relevant chain lengths and can quickly produce large ensembles (e.g., on the order of 10,000 3-D models) of polymer conformations that we would expect to see in simulation. Thus, it eliminates the need for simulation, reducing time and computational cost.

2. Materials and Methods

2.1. Molecular Dynamics

2.1.1. System Construction

Coordinates for all systems were constructed using the CHARMM software [69–71] v. c41b2 with the CHARMM36 (C36) biomolecular force field for carbohydrates [56,61–63]. Of note, it has been shown that MD simulations can reproduce ring puckers observed by NMR [49–51], with one study demonstrating the capacity of the CHARMM36 force field to reproduce NMR data for an iduronate derivative in the context of a heparin analogue [51]. The initial conformation for an MD simulation of non-sulfated chondroitin 20-mer was fully extended, with glycosidic linkage dihedrals $\phi = -83.75°$ and $\psi = -156.25°$ in all GlcAβ1-3GalNAc linkages and $\phi = -63.75°$ and $\psi = 118.75°$ in all GalNAcβ1-4GlcA linkages. These glycosidic linkage dihedral angle values were found to be the most energetically favorable in MD-simulated, non-sulfated chondroitin disaccharides [54]. All other internal coordinates were taken from the force-field files. In this conformation, the 20-mer had an end-to-end distance of 101.8 Å and was solvated in a cubic periodic unit cell with an edge length of 124.3 Å (~63,000 water molecules). The explicit solvent consisted of the TIP3P water model [72,73], neutralizing Na+ counterions, and 140 mM sodium chloride.

2.1.2. Energy Minimization and Heating

The NAMD program [74] v. 2.12 (http://www.ks.uiuc.edu/Research/namd/) was used to minimize the potential energy for 1000 steps using the conjugate gradient method [75,76] then heat the system to the target temperature of 310 K by reassigning velocities from a random distribution at the target temperature every 1000 steps for 20,000 steps with a timestep of 0.002 ps (40 ps). During heating, harmonic positional restraints were placed on non-hydrogen atoms of the solute and constraints [77–79] were applied to maintain equilibrium values for TIP3P geometries and for bond lengths involving hydrogen atoms. The Lennard–Jones (L-J) [80] and electrostatic potential energies had cutoff distances of 10 Å. An energy switching function [81] was applied to L-J interactions between 8 and 10 Å and an isotropic pressure correction accounted for contributions from L-J interactions beyond the cutoff [82]. The particle mesh Ewald (PME) method [83] with fourth order B-spline interpolation for a cubic unit cell and fast Fourier transform (FFT) grid spacing of 1.0 Å along each axis was used to account for electrostatic interactions beyond the cutoff. Consistent with the CHARMM additive force fields for proteins, nucleic acids, lipids, and small molecules [84], carbohydrate 1-4 non-bonded interactions were not scaled (i.e., scaling factor = 1.0). Heating was done under constant pressure with pressure regulated at 1 atm by a Langevin Piston barostat [85]. A 500,000-step (1-ns) unbiased constant particle number/constant pressure/constant temperature (NPT) MD run, followed with a temperature of 310 K maintained by a Langevin thermostat [86] and without positional restraints. The average periodic cell parameters from the last half of this NPT ensemble trajectory (123.7 Å) were used as cell basis vectors for the quadruplicate canonical (NVT) ensemble MD simulations detailed below. These were preceded by minimization and heating as detailed above, with the exception of constant volume (i.e., no Langevin piston barostat) with a box edge length of 123.7 Å.

2.1.3. Production Simulations

Unbiased canonical (NVT) ensemble MD was run using CHARMM software with the OpenMM GPU acceleration interface [87–92] on CUDA platform and GTX 1080 Ti graphics cards (NVIDIA Corp., Santa Clara, USA.). Non-bonded interaction truncations and energy calculations were performed using the same methods from the heating stage, and Ewald summation of Gaussian electrostatic charge density distributions [93,94] with a width of 0.320 was performed. The SHAKE algorithm [77] was used to constrain all water geometries and bonds involving hydrogen atoms using bond distances from the parameter table and the leapfrog Verlet integration algorithm [95] was used for Langevin dynamics with a friction coefficient of 0.1 ps^{-1}, a constant temperature of 310 K, and a 0.002-ps timestep. Prior to production, each of four replicates was equilibrated for 50,000 steps (100 ps). Simulations were run for 250,000,000 steps (500 ns) and atomic coordinates were saved at 25,000-step (50-ps) intervals for analyses (10,000 snapshots per quadruplicate simulation).

2.1.4. Conformational Analysis

Chondroitin conformational properties were quantified by bond length, bond angle, and dihedral angle values. Glycosidic linkage conformational values considered were bond lengths for C_1-O and O-C_n, bond angles defined by O_5-C_1-O and C_1-O-C_n, and ϕ and ψ dihedral angles with IUPAC definitions: $\phi = O_5$-C_1-O-C_n and $\psi = C_1$-O-C_n-$C_{(n-1)}$ (Figure 2). Glycosidic linkage dihedral free energies $\Delta G(\phi, \psi)$ were analyzed to characterize potential conformational patterns. ϕ, ψ dihedral values from the MD-generated 20-mer ensemble are taken to have uniform probabilities. $\Delta G(\phi, \psi)$ is therefore computed by binning these values into 2.5° × 2.5° bins and then using the relationship $\Delta G(\phi_i, \psi_j) = -RT\ln(n_{ij}) - k$, where n_{ij} is the bin count for the bin corresponding to ϕ_i, ψ_j, R is the universal gas constant, T is the temperature of the MD simulations, and k is chosen so that the global minimum is located at $\Delta G = 0$ kcal/mol.

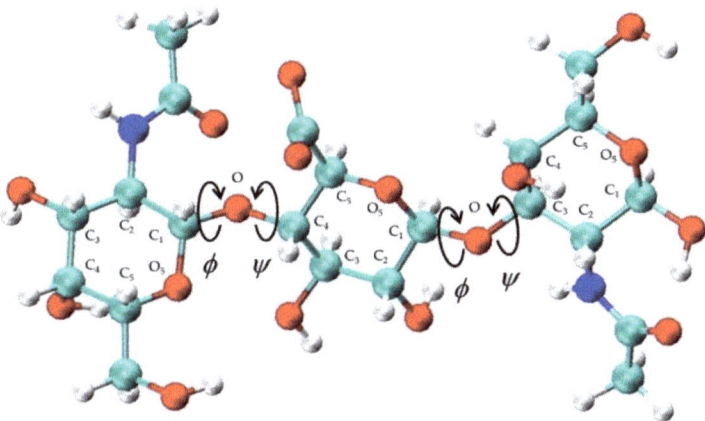

Figure 2. Non-sulfated chondroitin trisaccharide with glycosidic linkage dihedral angles, monosaccharide ring atoms, and linker oxygen atoms labeled; glycosidic linkage parameters used in the construction algorithm include C_1-O and O-C_n bond lengths, O_5-C_1-O and C_1-O-C_n bond angles, and $\phi = O_5$-C_1-O-C_n and $\psi = C_1$-O-C_n-$C_{(n-1)}$ dihedral angles.

Geometric values defining monosaccharide ring conformations included all bond lengths, bond angles, and dihedral angles within the ring and in exocyclic functional groups that are not part of a glycosidic linkage. To characterize potential conformational patterns in monosaccharide rings, Cremer-Pople (C-P) ring-puckering parameters (ϕ, θ, Q) of each monosaccharide ring in the MD-simulated 20-mer ensemble were computed. Conformations of each element (i.e., each GlcA monosaccharide, GalNAc monosaccharide, β1-3 linkage, and β1-4 linkage) were extracted separately from each saved snapshot of the 20-mer MD trajectories. Initially, data were separated out by run and residue/linkage number and aggregated across all snapshots in each run to determine if conformational data were the same in different runs and if individual linkage and ring conformations are dependent upon one another. Subsequently, all individual conformations were aggregated across all snapshots in all runs (e.g., 10,000 snapshots * 4 runs * 10 GlcA monosaccharides = 400,000 samples of GlcA monosaccharide conformations) to yield one set of data for each of: (1) GlcA monosaccharide conformation, (2) GalNAc monosaccharide conformation, (3) β1-3 linkage conformation, and (4) β1-4 linkage conformation.

2.2. Construction Algorithm to Generate GAG Conformational Ensembles

The conformational data described above served as inputs to an algorithm we developed to generate chondroitin polymer conformational ensembles of user-specified length and with a user-specified number of conformations. The algorithm works as follows:

1. In each constructed polymer conformation, each glycosidic linkage and monosaccharide ring is treated independently, and conformational parameters are randomly selected from the database containing the corresponding linkage or ring conformations from the 20-mer MD trajectories;
2. Two CHARMM stream files are written, one to define the sequence and linkages in the polymer and another to perform the following procedure for each frame: (1) All internal geometry conformation values selected by the algorithm are assigned and used to construct atomic coordinates. (2) End-to-end distance (i.e., distance between C_1 of the reducing end and C_4 of the non-reducing end) and radius of gyration are calculated. (3) A 100-step steepest descent (SD) potential energy minimization followed by a 100-step conjugate gradient minimization, each with intramolecular restraints, is performed to relieve bonded strain and steric clashes. The Lennard–Jones potential (E_{L-J}) is calculated on an atom–atom pair (i, j) basis using an energy switching function, as implemented in CHARMM with r_{on} = 7.5 Å and r_{off} = 8.5 Å [69]. As there is no solvent and thus no solvent screening of electrostatic interactions, electrostatics are excluded from energy calculations to prevent the non-physical intramolecular association of charged and polar groups. All glycosidic linkage and endocyclic ring dihedral angles, along with a dihedral angle in each GlcA carboxylate group (C_4-C_5-C_6-O_{61}) and GalNAc N-acetyl group (C-N-C_2-C_3), are restrained to their starting values (i.e., those randomly selected from the database) during minimization so as not to change the conformations observed in simulation. Dihedral restraint energy (E_{rdihe}) is calculated by comparing each restrained dihedral angle's database value (ϕ_0) to its value (ϕ_1) in the current frame of minimization with a force constant (k_{dihe}) of 100.0 kcal/mol/radian/radian (Equation (1))

$$E_{rdihe} = k_{dihe} * \sum (\phi_1 - \phi_0)^2 \tag{1}$$

3. To ensure conformational ensembles do not contain non-physical conformations, a bond potential energy (E_b) cutoff is applied. This cutoff is the sum of a polymer-length-specific cutoff and a constant independent of polymer length. The length-specific component of the cutoff is the bond potential energy after energy minimization, performed using the same restraints and minimization protocol used for each frame of the constructed ensemble (outlined above), of the polymer constructed in a fully-extended conformation (i.e., with the same glycosidic linkage ϕ and ψ angles as the starting conformation for MD simulations). The constant is added as a buffer to account for slight variations in the energies of other extended conformations. As linkage and ring conformations are treated independently and selected at random, it is possible to have a bond piercing another monosaccharide ring that may not be corrected by minimization. To estimate the ring-piercing bond strain energy for each exocyclic bond not participating in a glycosidic linkage, a system containing two non-bonded monosaccharides (i.e., GlcA and GalNAc, GlcA and GlcA, or GalNAc and GalNAc) was constructed such that an exocyclic bond of one monosaccharide pierces the ring of the other. To estimate the bond strain energy for each bond participating in a glycosidic linkage, a system containing one disaccharide unit (i.e., GlcAβ1-3GalNAc or GalNAcβ1-4GlcA) and a single monosaccharide (i.e., GlcA or GalNAc) was constructed such that a linkage bond in the disaccharide pierces the ring of the single monosaccharide. Systems containing interlocking rings (i.e., GlcA-GalNAc, GlcA-GlcA, and GalNAc-GalNAc) were also constructed to estimate the bond strain energy of the bonds piercing

the opposite ring. The same energy minimization protocol used in the algorithm was performed on this conformation, as well as a conformation in which the non-bonded saccharide units are 20 Å apart, and the post-minimization lengths of the bond piercing the ring in the initial conformation were compared. The pierced bond length (x_2), the non-pierced bond length (x_1), and the equilibrium bond length (x_0) and corresponding force-field bond-stretching constant (k_b) from the CHARMM parameter file were used to estimate a lower bound on the energy (ΔE_b) resulting from the bond distortion (Equation (2)).

$$\Delta E_b = k_b * \left[(x_2 - x_0)^2 - (x_1 - x_0)^2\right] \quad (2)$$

Of all conformations that still had a bond piercing a ring after minimization, the smallest ΔE_b = 132.3 kcal/mol. Of the conformations in which ring piercing was corrected during minimization, the maximum $\Delta E_b < 1$ kcal/mol. Thus, a buffer of 100 kcal/mol is added to the post-minimization bond potential of the initial extended conformation for any given polymer length. If the post-minimization bond potential of a given frame is beyond this cutoff, the frame is excluded from the ensemble.

For internal validation of our implementation of the algorithm, bond length probability distributions for each type of bond (i.e., C-C single bond, C-O single bond, C=O double bond, C-O partial double bond of GlcA carboxylate group, C_2-N single bond between GalNAc amide and ring carbon, C-N single bond within GalNAc amide, C-H bond, O-H bond, and N-H bond), free energies $\Delta G(\phi, \psi)$ for β1-3 and β1-4 glycosidic linkages, C-P parameters of GlcA and GalNAc monosaccharide rings, end-to-end distance distributions, and scatterplots of radius of gyration as a function of end-to-end distance from MD-generated ensembles and constructed ensembles both before and after energy minimization were compared. Additionally, bond potential energy distributions from constructed ensembles after energy minimization were plotted to verify that the algorithm calculated an appropriate energy cutoff and gave the expected energy distributions for the given polymer size.

To assess the expediency of application of MD-generated 20-mer conformations to construct chondroitin polymers of variable length, we constructed a non-sulfated chondroitin 10-mer ensemble using the algorithm and compared it to chondroitin 10-mer conformational ensembles generated by MD using the same protocol as the 20-mer simulations. We also constructed conformational ensembles of a non-sulfated chondroitin 100-mer and 200-mer to demonstrate the efficacy and efficiency of our algorithm to construct conformational ensembles of chondroitin polymers with biologically-relevant chain lengths.

3. Results and Discussion

3.1. Glycosidic Linkage Geometries

In non-sulfated chondroitin 20-mer MD simulations, we found that all ϕ, ψ dihedrals sampled in GlcAβ1-3GalNAc linkages were centered about a global free energy minimum (Min I) while GalNAcβ1-4GlcA linkages showed more flexibility. In addition to a global minimum, $\Delta G(\phi, \psi)$ for GalNAcβ1-4GlcA also has two local minima (Min II and Min II') (Figure 3 and Table 1). To validate these observed glycosidic linkage geometries, we looked at the free energy minima of non-sulfated chondroitin glycosidic linkage dihedrals from biased MD simulations of disaccharides (using dihedral definitions $\phi = O_5\text{-}C_1\text{-}O\text{-}C_n$ and $\psi = C_1\text{-}O\text{-}C_n\text{-}C_{(n+1)}$ as opposed to the IUPAC $\psi = C_1\text{-}O\text{-}C_n\text{-}C_{(n-1)}$ used in our study) [54] (Table S1). We found that at each free energy minimum in β1-3 and β1-4 linkages, our ϕ dihedrals differed by no more than +/-2.5° and our ψ dihedrals differed by no more than +/-127.5°, which is in close agreement if we assume $C_1\text{-}O\text{-}C_3\text{-}C_2 = C_1\text{-}O\text{-}C_3\text{-}C_4 + 120°$ and $C_1\text{-}O\text{-}C_4\text{-}C_3 = C_1\text{-}O\text{-}C_4\text{-}C_5 - 120°$. Additionally, our data were mostly in agreement with the most energetically-favorable glycosidic linkage dihedrals (i.e., at global minima) in non-sulfated chondroitin hexasaccharides from MD simulations (using dihedral definitions $\phi = O_5\text{-}C_1\text{-}O\text{-}C_n$ and ψ

= C_1-O-C_n-$C_{(n+1)}$) and validated by NMR [96] (Table S1). The biggest difference was in our β1-3 ψ dihedrals, which differed by about +100° (+120° difference expected). This study restrained pyranose rings to a $^4C_1$ chair and did not use explicit solvent in simulations. Each of these factors may contribute to interactions between neighboring monosaccharides and thus glycosidic linkage conformation, which would explain the variation from our results.

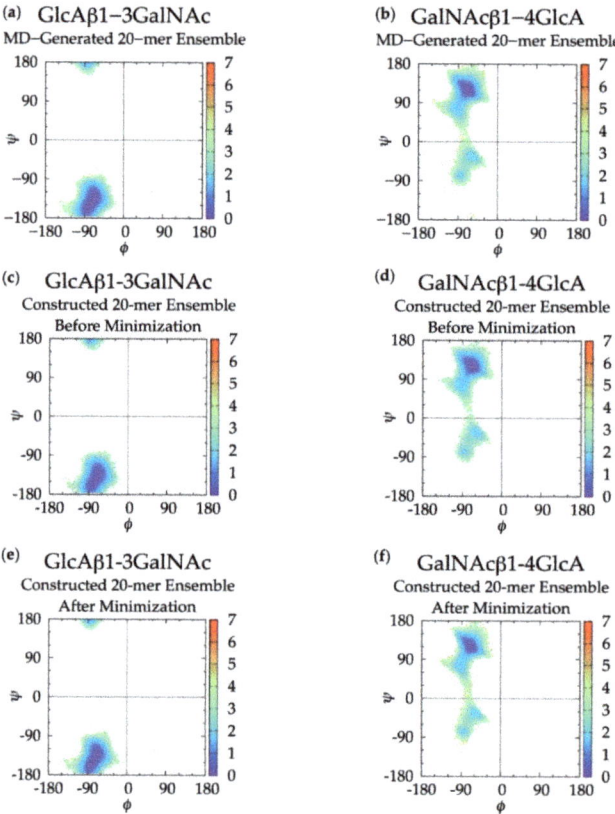

Figure 3. $\Delta G(\phi, \psi)$ in non-sulfated chondroitin 20-mer ensembles for aggregated GlcAβ1-3GalNAc and GalNAcβ1-4GlcA glycosidic linkage data (**a**,**b**) MD-generated ensembles, (**c**,**d**) constructed ensembles before minimization, and (**e**,**f**) constructed ensembles after minimization; contour lines every 1 kcal/mol.

Table 1. Glycosidic Linkage Dihedrals (ϕ, ψ) [1] and Free Energy ($\Delta G(\phi, \psi)$ kcal/mol) Minima.

	MD-Generated 20-mer Ensemble				Constructed 20-mer Ensemble (Before Energy Minimization)				Constructed 20-mer Ensemble (After Energy Minimization)			
	GlcAβ1-3 GalNAc		GalNAcβ1-4 GlcA		GlcAβ1-3 GalNAc		GalNAcβ1-4 GlcA		GlcAβ1-3 GalNAc		GalNAcβ1-4 GlcA	
Min	ϕ, ψ	ΔG	ϕ, ψ	ΔG	ϕ, ψ	ΔG	ϕ, ψ	ΔG	ϕ, ψ	ΔG	ϕ, ψ	ΔG
I	−81.25°, −153.75°	0.00	−66.25°, 116.25°	0.00	−76.25°, −148.75°	0.00	−66.25°, 116.25°	0.00	−81.25°, −153.75°	0.00	−68.75°, 121.25°	0.00
II			−58.75°, −33.75°	1.57			−61.25°, −33.75°	1.54			−58.75°, −33.75°	1.57
II'			−86.25°, −73.75°	1.80			−86.25°, −73.75°	1.71			−86.25°, −78.75°	1.73

[1] ϕ, ψ dihedral angles were sorted into 2.5° bins.

For internal validation of the construction algorithm, we compared glycosidic linkage input and output data. If the algorithm is performing correctly, $\Delta G(\phi, \psi)$ from the MD ensemble and the constructed ensemble *before minimization* will be nearly identical, and $\Delta G(\phi, \psi)$ from the constructed ensemble *after minimization* will not be substantially different. Performing the comparison between the ensemble of 40,000 20-mer conformations from the MD and a constructed ensemble of the same size confirms this to be the case. Figure 3 demonstrates that $\Delta G(\phi, \psi)$ for GlcAβ1-3GalNAc and for GalNAcβ1-4GlcA glycosidic linkages are qualitatively identical when comparing the MD-generated and constructed ensembles. Quantitative analysis (Table 1) shows that the global minima (Min I) for both types of linkages and the secondary local minima (Min II and Min II') for GalNAcβ1-4GlcA linkages are basically identical with 0° to 5° differences between the MD-generated input data and the constructed ensemble output data before minimization. The minimization in the construction algorithm, used to resolve any steric clashes, results in relatively minor changes in the location of the global minima, also ranging from 0° to 5°. Constructed ensemble glycosidic dihedral values after minimization are within 5° of the MD data, and $\Delta G(\phi, \psi)$ values are within 0.1 kcal/mol.

As detailed in Methods and discussed below, the minimization portion of the construction algorithm not only relieves any steric clashes, but also is used to detect bond-strain energies indicative of ring piercing. The close similarity of $\Delta G(\phi, \psi)$ for the constructed ensemble after minimization in comparison to the MD-generated ensemble suggests that few constructed conformations have large steric clashes, resulting in large conformational shifts after minimization. It also suggests that a few constructed conformations have ring-piercing events that necessitate their exclusion from the constructed ensemble altogether. Indeed, this is the case: during the creation of the 40,000-member constructed ensemble, only 18 conformations were excluded because they failed to meet the bond-strain energy criterion.

3.2. GlcA Ring Pucker Effects

An initial implementation of the construction algorithm used default force-field geometries for all GalNAc and GlcA rings. The result was that all GalNAc and GlcA rings in an algorithmically-constructed conformation had the same internal geometries. Ensembles constructed using this version of the algorithm had longer average end-to-end distances than MD-generated ensembles (Figure S1), which meant that, on average, constructed conformations were overly extended. The default force-field ring pucker geometry for both types of monosaccharides was $^4C_1$. With that ring pucker, all βGlcA and all but one βGalNAc exocyclic functional groups are equatorial, and therefore the $^4C_1$ ring pucker is expected to be strongly preferred to other ring pucker geometries. To validate this simple approach to assigning ring pucker geometry, we computed C-P parameters of each monosaccharide ring in the MD-simulated 20-mer ensemble (10 * 40,000 = 400,000 ring conformations for each of the two monosaccharide types). As seen in NMR and force-field studies, the stable $^4C_1$ chair ring pucker was the principal conformer for both GlcA [50,96–99] and GalNAc [46,96,98,100] in the MD simulations, with slight deformations (0° < C-P θ < 30°) (Figure 4a,b). However, a small minority of GlcA ring conformers were skew-boat or boat, namely $^3S_1$, $B_{1,4}$, $^5S_1$, $^{2,5}B$, ^2S_O, $B_{3,O}$, $^1S_3$, $^{1,4}B$, and $^1S_5$ (60° < C-P θ < 120°) (Figure 4b). Studies that performed unbiased MD simulations with other force fields observed skew-boat and boat ring puckers of non-sulfated GlcA monosaccharides on the microsecond timescale, but the occurrences were negligible due to high energy barriers [50,98]. In line with those findings, we observed only occasional GlcA skew-boat and boat pucker transitions in chondroitin 20-mers in our 500-ns unbiased CHARMM simulations. However, the C-P ϕ values in non-$^4C_1$ GlcA conformers in these studies differed from ours. Specifically, one study found ^2S_O, $B_{3,O}$, $^1S_3$, $^{1,4}B$, and $^1S_5$ [98]. Slight differences could be explained by differing ion concentrations which likely impacted pyranose ring puckers [101]. However, it is likely that the differences primarily result from intramolecular interactions. The aforementioned literature data come from simulated GlcA monosaccharides only, whereas our results come from simulated chondroitin 20-mers.

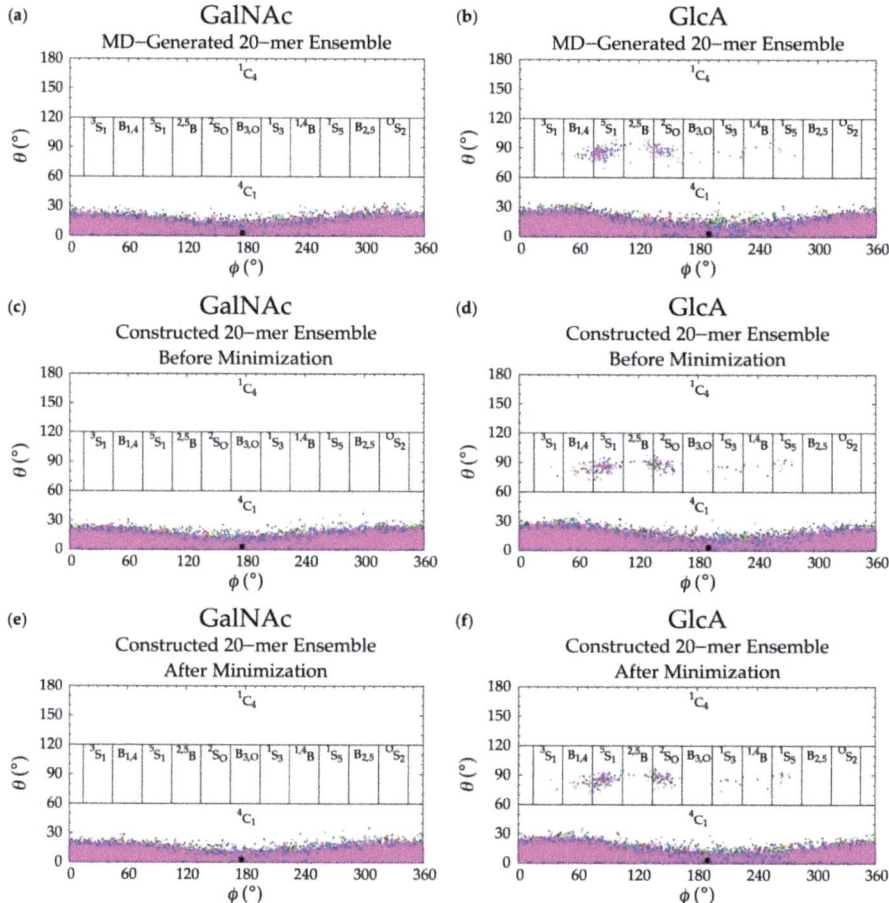

Figure 4. Cremer–Pople data for GalNAc and GlcA in (**a**,**b**) MD-generated ensembles and constructed ensembles (**c**,**d**) before and (**e**,**f**) after energy minimization; geometries from the four sets of each type of ensemble are represented by red, green, blue, and magenta dots, respectively and the force-field geometry is represented by a single large black dot. Cremer–Pople parameters (ϕ, θ) for all rings in every tenth snapshot from each ensemble were plotted (i.e., 10 rings * 1,000 snapshots per run * 4 runs = 40,000 parameter sets). As the algorithm reads all ring conformations sampled in MD, not all datapoints in panels (**c**–**f**) are seen in panels (**a**,**b**) but the full MD-generated dataset contains all datapoints in the constructed ensembles.

The MD-generated 20-mer GlcA ring conformations can be separated into two broad categories: those that do not introduce a kink into the polymer and those that do. With the inclusion of $^4C_1$, the former category encompasses $^4C_1$ and ^2S_O GlcA ring puckers, both of which place the two glycosidic linkage oxygen atoms, located at opposite ends of the ring, in an equatorial conformation (Figure 5a). As such, the O-C_1 and C_4-O bond vectors therein are approximately parallel and promote extended polymer conformations. The latter category encompasses $^3S_1$, $B_{1,4}$, $^5S_1$, $^{2,5}B$, $B_{3,O}$, $^1S_3$, $^{1,4}B$, and $^1S_5$ GlcA ring puckers (Figure 5b). These ring puckers all place one of these glycosidic linkage oxygen atoms in the equatorial position and the other in the axial position. For these ring puckers, the O-C_1 and C_4-O bond vectors are approximately perpendicular, which results in a kink in the polymer chain, and can reduce end-to-end distance even when the remainder of the polymer is fully extended (Figure 5b).

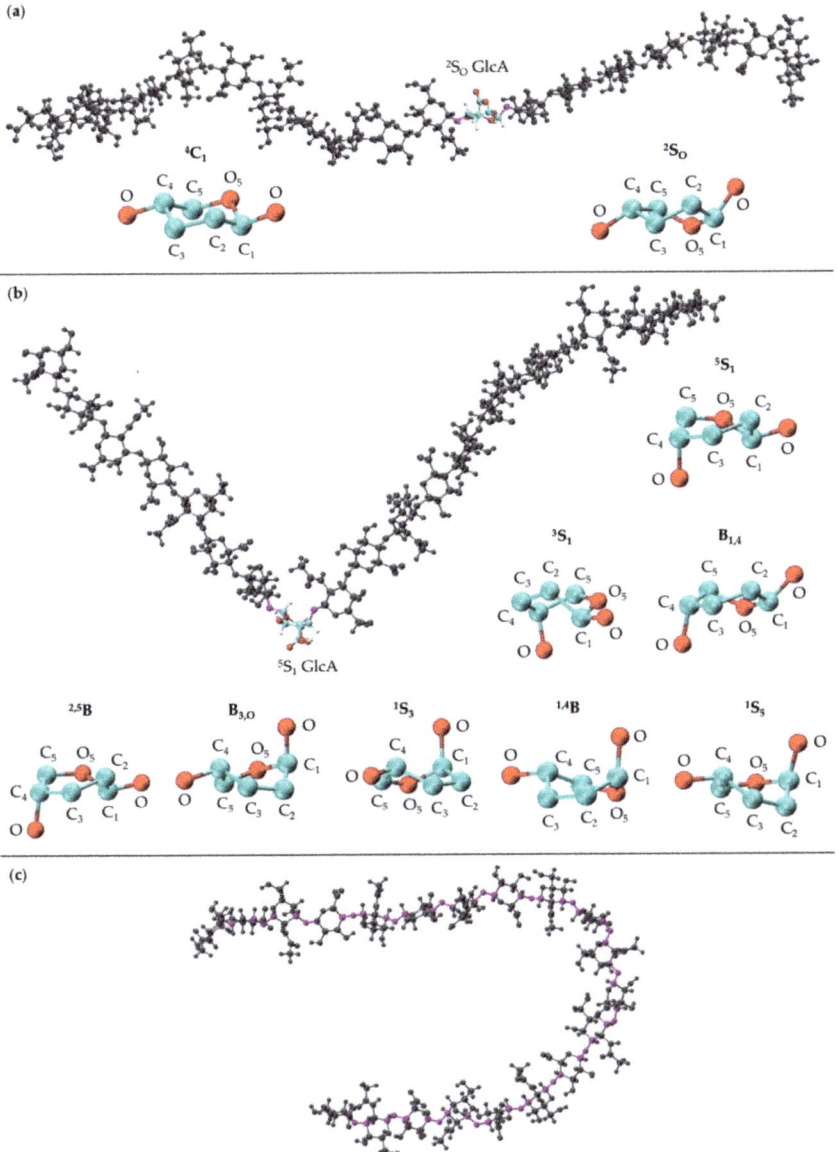

Figure 5. (**a**) 20-mer conformation with a ^2S_O GlcA conformer (colored by atom type with flanking linkage atoms highlighted in purple) and close-ups of GlcA monosaccharide rings in $^4C_1$ and ^2S_O conformations (shows endocyclic ring atoms and linker oxygen atoms only); (**b**) 20-mer conformation with a kink at a $^5S_1$ GlcA conformer (colored by atom type with flanking linkage atoms highlighted in purple) and GlcA monosaccharide rings in $^5S_1$, $^3S_1$, $B_{1,4}$, $^{2,5}B$, $B_{3,O}$, $^1S_3$, $^{1,4}B$, and $^1S_5$ conformations (shows endocyclic ring atoms and linker oxygen atoms only); (**c**) 20-mer conformation with a curve caused by flexible glycosidic linkage geometries (highlighted in purple) and all monosaccharides in $^4C_1$ conformations; all images came from MD-generated ensembles.

As such, the final version of the construction algorithm uses MD-generated ring conformations instead of default force-field topology geometries. As an added benefit, this approach includes not only MD-generated ring dihedral angles but also bond lengths and angles. Using this finalized version of the algorithm, the peak in the end-to-end distance histogram for the constructed ensemble was shifted left compared to that resulting from force-field topology ring geometries (Figure 6 vs. Figure S1) and much more closely matches the reference MD-generated ensemble data. This finding shows the importance of accounting for ring flexibility in constructing chondroitin glycosaminoglycan polymer conformations similar to those sampled in all-atom explicit-solvent MD simulations. Of note, the radius of gyration was also analyzed as a function of end-to-end distance in MD-generated and constructed ensembles after minimization (Figure S2a,b). These results showed that the radius of gyration is highly correlated with end-to-end distance in both MD-generated and constructed ensembles.

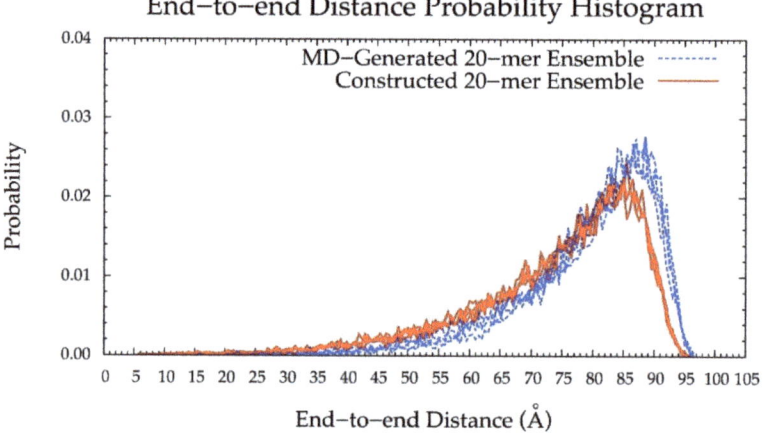

Figure 6. End-to-end distance probability distribution of MD-generated (blue dashed lines) and constructed (red solid lines) 20-mer ensembles; each type of ensemble includes four sets of 10,000 conformations; probabilities were calculated for end-to-end distances sorted into 0.5 Å bins.

While polymer kinks from ring puckering can lead to the shortening of polymer end-to-end distances, they are not required to achieve this. For 20-mers, flexibility in the glycosidic linkages even with $^4C_1$ ring puckering in all constituent monosaccharides can be sufficient to produce compact conformations (Figure 5c). Furthermore, because of the flexibility in the glycosidic linkages flanking non-$^4C_1$ ring puckers, polymer kinks from ring puckering do not always lead to compact conformations. Thus, the leftward shift in the end-to-end distance histogram upon the inclusion of non-$^4C_1$ ring puckers supplements glycosidic linkage flexibility in yielding compact conformations.

3.3. Treating Glycosidic Linkage and Ring Pucker Geometries as Independent Variables

To determine if linkage geometries and ring deformations are interdependent, individual glycosidic linkage $\Delta G(\phi, \psi)$ plots (Figure S3) were created (as opposed to the aggregate data $\Delta G(\phi, \psi)$ plots in Figure 3) and no distinguishing patterns emerged. These per-linkage glycosidic linkage data were also examined in the context of C-P plots of adjacent rings (Figure S4), for which there were also no distinguishing patterns. Additionally, ϕ and ψ values in linkages flanking GlcA rings not in a $^4C_1$ chair conformation were checked. For each linkage type, these conformations were all centered about the global $\Delta G(\phi, \psi)$ minima for the aggregate data (Figure 3), and 99.96% of conformations fell within the basin extending to $\Delta G(\phi, \psi)$ = +2 kcal/mol (Figure 7). Furthermore, different types of non-$^4C_1$ chair conformers did not have unique flanking linkage geometries. As no connection between linkage and ring conformations was observed in this analysis of the MD data, each linkage conformation and ring pucker was treated independently in the construction algorithm.

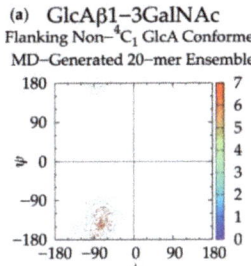
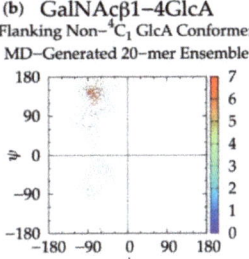

Figure 7. $\Delta G(\phi, \psi)$ plots for glycosidic linkages flanking non-$^4C_1$ GlcA conformers in non-sulfated chondroitin 20-mer MD-generated ensembles: (**a**) GlcAβ1-3GalNAc and (**b**) GalNAcβ1-4GlcA.

3.4. Handling Non-physical Constructed Conformations

To determine a criterion to exclude non-physical conformations from constructed ensembles, energy minimization with restrained endocyclic ring and glycosidic linkage dihedrals was performed on each conformation, and post-minimization bond potential energy and bond length probability distributions were analyzed. Conformations with outlying bond energies and abnormally long bonds may point to ring piercing (Figure 8) that cannot be fixed by minimization. To confirm this possibility, the post-minimization conformations with bond energies greater than that of the fully-extended 20-mer conformation after minimization were visualized. As anticipated, among these conformations, most with outlying total bond energies contained pierced rings which were not resolved by minimization. For each of the 12 20-mer conformations with a pierced ring, the difference between the post-minimization bond energy and that of the fully-extended 20-mer conformation is greater than the predicted energy change caused by bond distortions of that pierced ring (Table S2). The six conformations with outlying bond energies that did not contain pierced rings had kinks that resulted in bond length and bond angle distortions in glycosidic linkages that were nearly overlapping even after minimization. Of note, those conformations with bond energies that were not outlying were fully extended. These findings motivated using a bond potential energy cutoff in the construction algorithm. As stated previously, applying this cutoff resulted in 18 conformations being excluded during creation of the 40,000-member constructed ensemble. The resulting constructed ensemble contains no outlying bond lengths (Figure S5).

Dihedral angles before and after energy minimization were compared by analyzing glycosidic linkage $\Delta G(\phi, \psi)$ (Table 1 and Figure 3c–f), monosaccharide ring C-P parameters (Figure 4c–f), and the change in ϕ and ψ dihedral angles due to minimization. All $\Delta G(\phi, \psi)$ and C-P plots after energy minimization match those from before energy minimization and 99.6% of all ϕ and ψ dihedral angle differences before and after minimization are within 4° (Figure S6). This confirms that dihedrals do not undergo any major changes during minimization. Additionally, differences in end-to-end distance before and after minimization were calculated and the maximum change is 2.13 Å with 99.9% of changes under 0.5 Å, confirming that overall backbone conformation does not change as a result of minimization. According to these results, the selected bond potential energy cutoff and restraint scheme during minimization give conformations with little deviation from the initial constructed conformations before minimization.

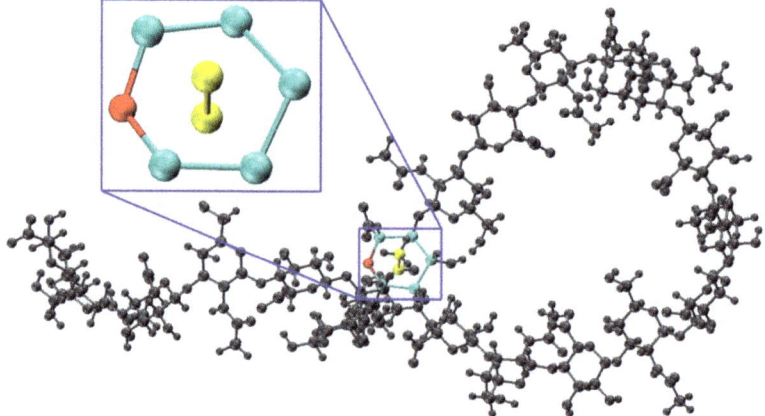

Figure 8. Constructed 20-mer conformation with a GlcA ring pierced by a GalNAc C-CT bond and a close up panel showing atoms involved in the ring pierce; E_b = 787.7 kcal/mol, fully-extended 20-mer post-minimization E_b = 29.6 kcal/mol, ΔE_b = 758.1 kcal/mol (Table S2).

3.5. Internal Validation on 10-mers

To further validate the algorithm and test the use of conformational parameters from MD-generated 20-mer ensembles to construct polymers of variable length, we constructed non-sulfated chondroitin 10-mer ensembles and compared them to MD-generated 10-mer ensembles. All linkage and most ring conformations in MD-generated 10-mer (Figure 9 and Figure S7) and 20-mer ensembles matched (Figure 3a,b and Figure 4a,b), with the exception of non-$^4C_1$ GlcA rings which were not sampled in 10-mer simulations. This finding, combined with the report from NMR and force-field studies that GlcA skew-boat and boat conformations are negligible in non-sulfated chondroitin mono- and oligosaccharides [50,98], suggests that these GlcA conformations may result from intramolecular interactions in longer GAG polymers.

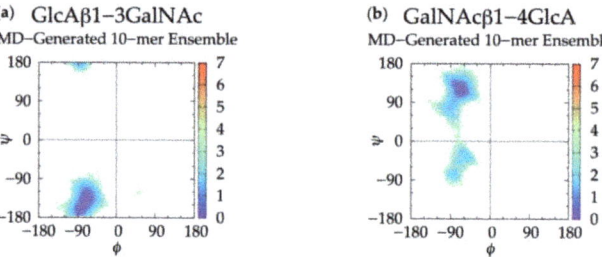

Figure 9. $\Delta G(\phi, \psi)$ plots for each glycosidic linkage in non-sulfated chondroitin 10-mer MD-generated ensembles: (**a**) GlcAβ1-3GalNAc and (**b**) GalNAcβ1-4GlcA.

Additionally, the end-to-end distance distributions of constructed and MD-generated 10-mer ensembles matched with minimal difference in the most probable end-to-end distance (Figure 10 and Table 2). Further, the radius of gyration is highly correlated with end-to-end distance in both MD-generated and constructed ensembles (Figure S2c,d). Of note, the end-to-end distance distributions of MD-generated 20-mer ensembles more closely matched those of 20-mer ensembles constructed using MD-generated 20-mer conformations (Figure 6) than those of 20-mer ensembles constructed using MD-generated disaccharide conformations (Figures S8 and S9). Together, these findings suggest that MD-generated 20-mer conformational parameters are ideal for constructing chondroitin polymers of different lengths.

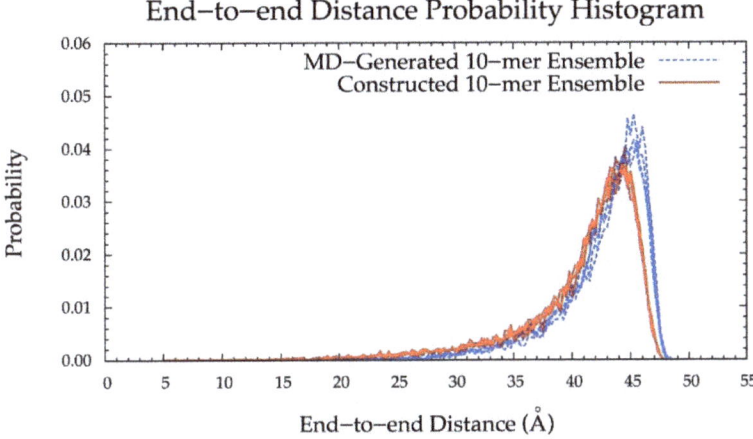

Figure 10. End-to-end distance probability distribution of MD-generated (blue dashed lines) and constructed (red solid lines) 10-mer ensembles; each type of ensemble includes four sets of 10,000 conformations; probabilities were calculated for end-to-end distances sorted into 0.25 Å bins.

Table 2. Most Probable End-to-End Distances (d) in MD-Generated and Constructed Ensembles [1].

	20-mer Ensembles			10-mer Ensembles		
	MD-Generated d (Å)	Constructed d (Å)	% Difference	MD-Generated d (Å)	Constructed d (Å)	% Difference
Run 1	88.5	83.0		45.25	44.50	
Run 2	88.5	85.5		45.25	43.50	
Run 3	86.0	85.0		45.50	44.50	
Run 4	86.5	85.0		45.25	44.25	
All [2]	88.5	85.0	4.03%	45.25	44.50	1.671%

[1] Probabilities were calculated for end-to-end distances sorted into 0.5 Å bins for the 20-mer ensembles and 0.25 Å bins for the 10-mer ensembles. [2] All = end-to-end distance distribution aggregated across all four runs.

3.6. Application to Longer Chondroitin Polymers

To implement the algorithm in the construction of conformational ensembles of non-sulfated chondroitin polymers of biologically-relevant chain lengths, chondroitin 100-mer and 200-mer ensembles were constructed and the end-to-end distance (Figures 11 and 12), radius of gyration (Figure S2e,f), and bond potential energy distributions (Figure S10c,d) were examined. The skewness of the end-to-end distance distributions shifts toward the right with increasing polymer length. This stands to reason, as there is a greater chance of folding with longer chains. This also explains why there are more frames excluded from these ensembles (i.e., 457 and 1407 frames excluded from the 100-mer and 200-mer ensembles, respectively). Bond potential energy distributions have similarly-shaped curves in all polymer lengths (Figure S10) and energy values increase linearly as a function of atom count (Figure S11). These results suggest that 100-mer and 200-mer conformational ensembles constructed using our algorithm (Figures 11 and 12) are reasonable predictions of biological conformations given their high number of degrees of freedom.

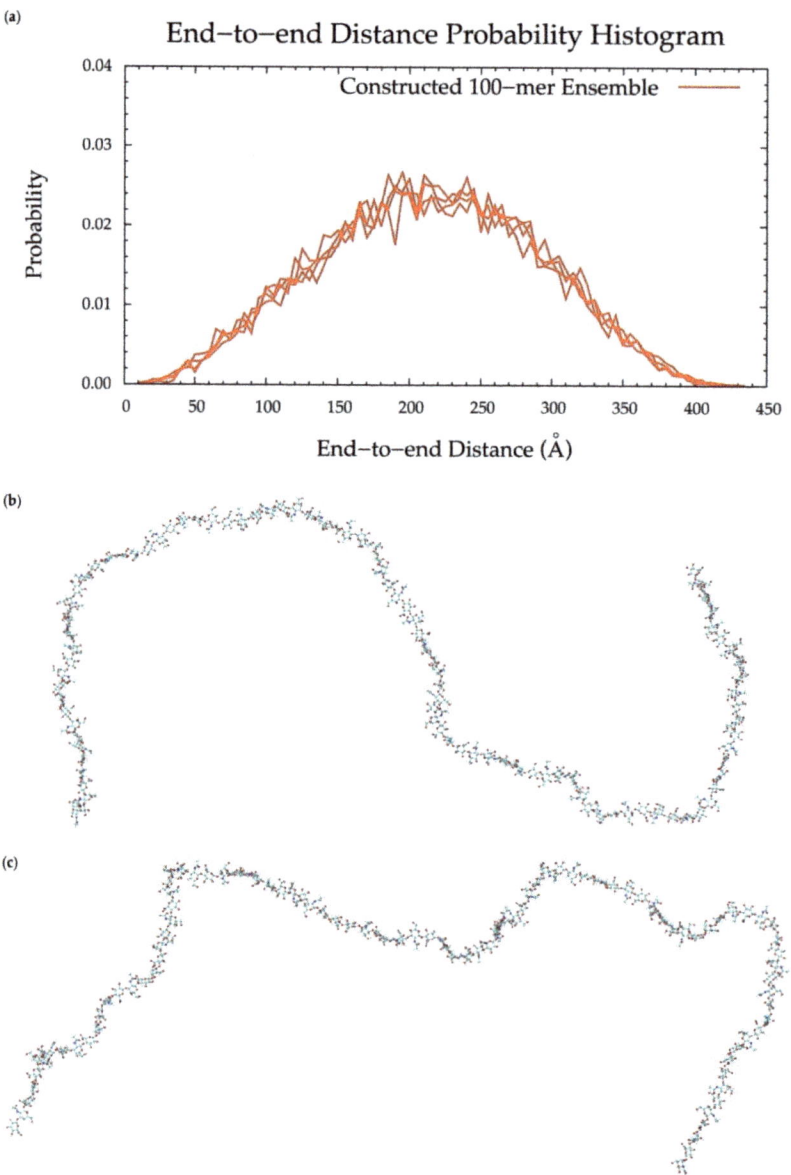

Figure 11. (a) End-to-end distance probability distribution of constructed 100-mer ensemble; includes four sets of 10,000 conformations; probabilities were calculated for end-to-end distances sorted into 5 Å bins. (b,c) Snapshots of the non-sulfated chondroitin 100-mer having the most-probable end-to-end distance (225 Å in both snapshots) from constructed ensembles.

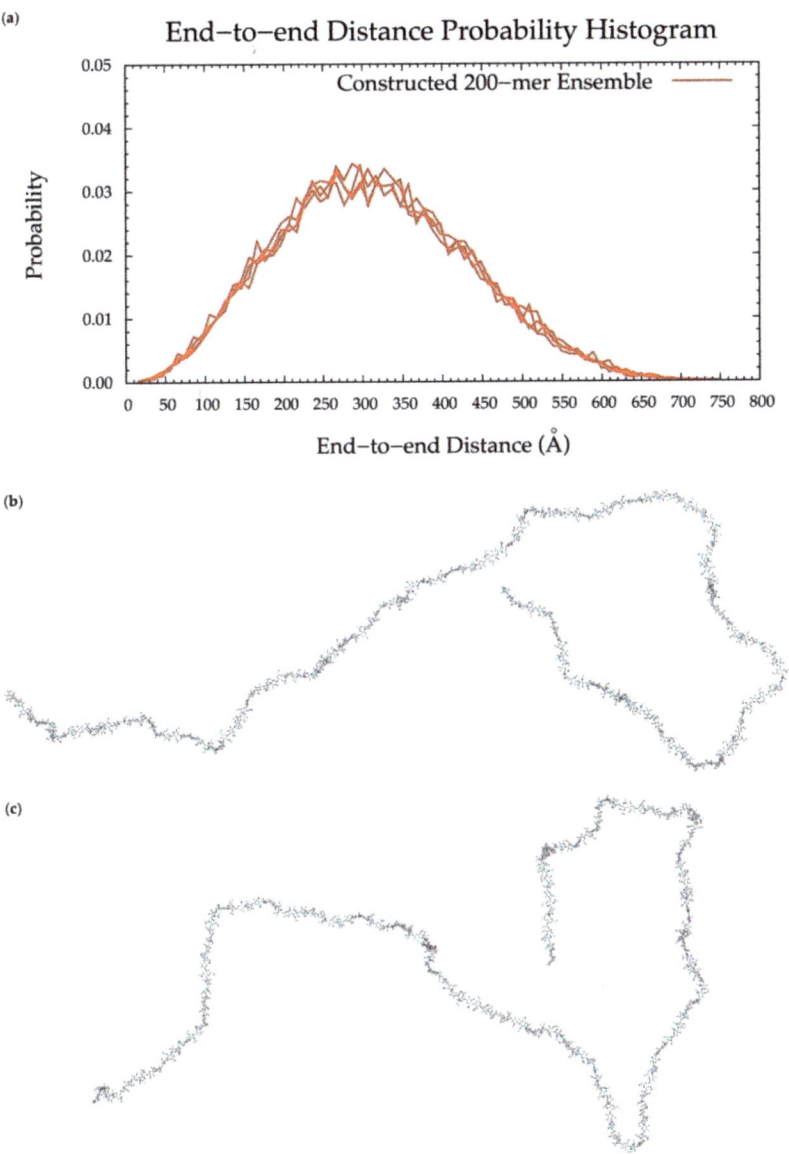

Figure 12. (a) End-to-end distance probability distribution of constructed 200-mer ensemble; includes four sets of 10,000 conformations; probabilities were calculated for end-to-end distances sorted into 10 Å bins. (b,c) Snapshots of the non-sulfated chondroitin 200-mer having the most-probable end-to-end distance (300 Å in both snapshots) from constructed ensembles.

4. Conclusions

With (1), all bond, bond angle, and dihedral angle conformational parameters from MD incorporated into the algorithm, (2) monosaccharide rings and glycosidic linkages treated independently, (3) energy minimization performed on each constructed conformation, and (4) a bond potential energy cutoff applied, end-to-end distance probability distributions from constructed and MD-generated ensembles match with minimal differences in most probable end-to-end distances (Table 2 and Figures 6 and 10–12) suggesting that our algorithm produces conformational ensembles that mimic the backbone flexibility seen in MD simulations of non-sulfated chondroitin polymers.

Our program is also valuable for its efficiency. For example, the fully-solvated chondroitin 20-mer system contains ~191,000 atoms and took about one month to simulate. It took only about 12.5 h to construct the 20-mer conformational ensembles using our algorithm, in which all end-to-end distances, radii of gyration, bond lengths, dihedral angles, monosaccharide ring PDB files for C-P analysis of every tenth frame, and bond and system potential energies before and after minimization, C-P parameters of every GlcA ring in every frame, and PDBs of all conformations with bond energies greater than that of the fully-extended 20-mer (including conformations with pierced rings) are written. The fully-solvated chondroitin 10-mer system contains ~36,000 atoms and took about five days to simulate. The 10-mer ensembles were constructed using our algorithm in about 40 min and energies, dihedral angles, end-to-end distances, and radii of gyration after minimization, C-P parameters of every GlcA ring in every frame, and PDBs of all conformations with bond energies greater than that of the fully-extended 10-mer were written. Fully-solvated chondroitin 100-mer and 200-mer systems would contain ~3,370,000 atoms and ~75,450,000 atoms, respectively. Systems of this magnitude are not feasible to simulate with current computational resources but if they could be simulated, they would take on the order of years to complete. Construction of the chondroitin 100-mer and 200-mer conformational ensembles using our algorithm took about four and nine hours, respectively, and the algorithm produced the same output data types as for the 10-mer (n.b.: these timings are less than the 20-mer timing above because of all of the additional output written to disk for analysis in the case of the 20-mer ensemble construction).

In conclusion, our algorithm, incorporating glycosidic linkage and monosaccharide ring conformations from the MD simulation of non-sulfated chondroitin 20-mers, can be used to efficiently generate conformational ensembles of non-sulfated chondroitin polymers of arbitrary length. We are investigating the applicability of this approach to various sulfo-forms of CS and different types of GAGs, including hyaluronan (HA), dermatan sulfate (DS), heparan sulfate (HS), and keratan sulfate (KS). Given the variability and complexity of GAGs, as well as existing barriers to the experimental characterization of the three-dimensional conformational properties of GAGs of lengths relevant in the context of PGs, there are currently very few efforts to target GAGs. We anticipate that the presented algorithm, combined with experimental data on PG core proteins and conformational analysis of the linker tetrasaccharide [48,52,102–104], may provide a useful means of generating atomic-resolution three-dimensional models of full PGs. The algorithm could also be used to model full GAG–protein complexes, which may provide insights into potential interactions between multiple biomolecules within a single GAG complex. The ability to model these complex biomolecules would be a key step towards improving understanding of GAG bioactivity, assessing the druggability of GAGs, designing agonists or antagonists to treat disease, and developing diagnostic tools. Thus, this methodology may open a new avenue into disease modulation.

Supplementary Materials: The following are available online http://www.mdpi.com/2218-273X/10/4/537/s1, Table S1: Comparison to Observed Literature Values of Glycosidic Linkage Dihedrals (ϕ, ψ) in Non-Sulfated Chondroitin, Table S2: Bond Energies of Constructed Chondroitin 20-mer Conformations with Pierced Rings, Figure S1: End-to-end distance probability distribution of 20-mer ensembles generated by MD and an early version of the construction algorithm which applied glycosidic linkage geometries from MD-generated 20-mer ensembles, Figure S2: Scatterplots of radius of gyration as a function of end-to-end distance in MD-generated and constructed chondroitin ensembles, Figure S3: $\Delta G(\phi,\psi)$ plots for each glycosidic linkage in the chondroitin 20-mer from MD-generated ensembles, Figure S4: Cremer-Pople plots for each monosaccharide ring in the chondroitin 20-mer from MD-generated ensembles, Figure S5: Probability histograms of bond lengths for each type of bond in the chondroitin 20-mer from MD-generated and constructed ensembles, Figure S6: Probability histogram showing changes in glycosidic linkage ϕ and ψ dihedral angles during energy minimization in constructed 20-mer ensembles, Figure S7: Cremer-Pople plots of GalNAc and GlcA in MD-generated chondroitin 10-mer ensembles, Figure S8: End-to-end distance probability distribution of chondroitin 20-mer ensembles generated by MD and an early version of the construction algorithm which applied glycosidic linkage geometries from MD-generated non-sulfated chondroitin disaccharide ensembles, Figure S9: $\Delta G(\phi,\psi)$ plots for GlcAβ1-3GalNAc and GalNAcβ1-4GlcA glycosidic linkages in MD-generated non-sulfated chondroitin disaccharide ensembles and 20-mer ensembles constructed using glycosidic linkage dihedral probabilities from these MD-generated disaccharide ensembles, Figure S10: Bond energy distribution probability histograms from constructed 10-, 20-, 100-, and 200-mer ensembles, Figure S11: Average bond energies as a function of polymer length.

Author Contributions: Conceptualization, E.K.W. and O.G.; methodology, E.K.W., G.V., H.S., and O.G.; software, E.K.W., G.V., H.S., and O.G.; validation E.K.W. and O.G.; formal analysis, E.K.W.; investigation E.K.W. and O.G.; resources, O.G.; data curation, E.K.W. and O.G.; writing—original draft preparation, E.K.W.; writing—review and editing, E.K.W., G.V., H.S., and O.G.; visualization, E.K.W.; supervision, O.G.; project administration, O.G.; funding acquisition, O.G. All authors have read and agreed to the published version of the manuscript.

Funding: This research and the APC were funded by the National Science Foundation, grant number MCB-1453529 to O.G.

Conflicts of Interest: The authors declare no conflicts of interest. The funders had no role in the design of the study; in the collection, analyses, or interpretation of data; in the writing of the manuscript, or in the decision to publish the results.

References

1. Djerbal, L.; Lortat-Jacob, H.; Kwok, J. Chondroitin sulfates and their binding molecules in the central nervous system. *Glycoconj. J.* **2017**, *34*, 363–376. [CrossRef] [PubMed]
2. Lander, A.D.; Selleck, S.B. The elusive functions of proteoglycans: In vivo veritas. *J. Cell Biol.* **2000**, *148*, 227–232. [CrossRef] [PubMed]
3. Forsten-Williams, K.; Chu, C.L.; Fannon, M.; Buczek-Thomas, J.A.; Nugent, M.A. Control of growth factor networks by heparan sulfate proteoglycans. *Ann. Biomed. Eng.* **2008**, *36*, 2134–2148. [CrossRef] [PubMed]
4. Bhavanandan, V.P.; Davidson, E.A. Mucopolysaccharides associated with nuclei of cultured mammalian cells. *Proc. Natl. Acad. Sci. USA* **1975**, *72*, 2032–2036. [CrossRef]
5. Fedarko, N.S.; Conrad, H.E. A unique heparan sulfate in the nuclei of hepatocytes: Structural changes with the growth state of the cells. *J. Cell Biol.* **1986**, *102*, 587–599. [CrossRef]
6. Ishihara, M.; Fedarko, N.S.; Conrad, H.E. Transport of heparan sulfate into the nuclei of hepatocytes. *J. Biol. Chem.* **1986**, *261*, 13575–13580.
7. Tumova, S.; Hatch, B.A.; Law, D.J.; Bame, K.J. Basic fibroblast growth factor does not prevent heparan sulphate proteoglycan catabolism in intact cells, but it alters the distribution of the glycosaminoglycan degradation products. *Biochem. J.* **1999**, *337 Pt 3*, 471–481. [CrossRef]
8. Mizumoto, S.; Fongmoon, D.; Sugahara, K. Interaction of chondroitin sulfate and dermatan sulfate from various biological sources with heparin-binding growth factors and cytokines. *Glycoconj. J.* **2012**, *30*, 619–632. [CrossRef]
9. Sugahara, K.; Mikami, T.; Uyama, T.; Mizuguchi, S.; Nomura, K.; Kitagawa, H. Recent advances in the structural biology of chondroitin sulfate and dermatan sulfate. *Curr. Opin. Struct. Biol.* **2003**, *13*, 612–620. [CrossRef]
10. Sugahara, K.; Mikami, T. Chondroitin/dermatan sulfate in the central nervous system. *Curr. Opin. Struct. Biol.* **2007**, *17*, 536–545. [CrossRef]
11. Kawashima, H.; Atarashi, K.; Hirose, M.; Hirose, J.; Yamada, S.; Sugahara, K.; Miyasaka, M. Oversulfated chondroitin/dermatan sulfates containing GlcAβ1/IdoAα1-3GalNAc(4,6-O-disulfate) interact with L- and P-selectin and chemokines. *J. Biol. Chem.* **2002**, *277*, 12921–12930. [CrossRef] [PubMed]
12. Kjellén, L.; Lindahl, U. Specificity of glycosaminoglycan-protein interactions. *Curr. Opin. Struct. Biol.* **2018**, *50*, 101–108. [CrossRef] [PubMed]
13. Kolset, S.O.; Prydz, K.; Pejler, G. Intracellular proteoglycans. *Biochem. J.* **2004**, *379*, 217–227. [CrossRef]
14. Yoneda, A.; Couchman, J.R. Regulation of cytoskeletal organization by syndecan transmembrane proteoglycans. *Matrix Biol.* **2003**, *22*, 25–33. [CrossRef]
15. Stallcup, W.B.; Dahlin, K.; Healy, P. Interaction of the NG2 chondroitin sulfate proteoglycan with type VI collagen. *J. Cell Biol.* **1990**, *111*, 3177–3188. [CrossRef] [PubMed]
16. Van Susante, J.L.C.; Pieper, J.; Buma, P.; van Kuppevelt, T.H.; van Beuningen, H.; van Der Kraan, P.M.; Veerkamp, J.H.; van den Berg, W.B.; Veth, R.P.H. Linkage of chondroitin-sulfate to type I collagen scaffolds stimulates the bioactivity of seeded chondrocytes in vitro. *Biomaterials* **2001**, *22*, 2359–2369. [CrossRef]
17. Streit, A.; Nolte, C.; Rásony, T.; Schachner, M. Interaction of astrochondrin with extracellular matrix components and its involvement in astrocyte process formation and cerebellar granule cell migration. *J. Cell Biol.* **1993**, *120*, 799–814. [CrossRef]

18. Lukas, M.; Susanne, S.; De Santis Rebecca, D.S.; Hajo, H.; Janine, H.; Lena, H.; Klaus, B.; Gernot, M.; Tobias, S. Peptide 19–2.5 inhibits heparan sulfate-triggered inflammation in murine cardiomyocytes stimulated with human sepsis serum. *PLoS ONE* **2015**, *10*, e0127584.
19. Schmidt, E.P.; Overdier, K.H.; Sun, X.; Lin, L.; Liu, X.; Yang, Y.; Ammons, L.A.; Hiller, T.D.; Suflita, M.A.; Yu, Y. Urinary glycosaminoglycans predict outcomes in septic shock and acute respiratory distress syndrome. *Am. J. Respir. Crit. Care Med.* **2016**, *194*, 439–449. [CrossRef]
20. Fawcett, J.W. The extracellular matrix in plasticity and regeneration after CNS injury and neurodegenerative disease. In *Progress in Brain Research*; Elsevier: Amsterdam, The Netherlands, 2015; Volume 218, pp. 213–226.
21. Yang, S.; Hilton, S.; Alves, J.N.; Saksida, L.M.; Bussey, T.; Matthews, R.T.; Kitagawa, H.; Spillantini, M.G.; Kwok, J.C.; Fawcett, J.W. Antibody recognizing 4-sulfated chondroitin sulfate proteoglycans restores memory in tauopathy-induced neurodegeneration. *Neurobiol. Aging* **2017**, *59*, 197–209. [CrossRef]
22. Galtrey, C.M.; Fawcett, J.W. The role of chondroitin sulfate proteoglycans in regeneration and plasticity in the central nervous system. *Brain Res. Rev.* **2007**, *54*, 1–18. [CrossRef] [PubMed]
23. Mizumoto, S.; Kosho, T.; Hatamochi, A.; Honda, T.; Yamaguchi, T.; Okamoto, N.; Miyake, N.; Yamada, S.; Sugahara, K. Defect in dermatan sulfate in urine of patients with Ehlers-Danlos syndrome caused by a CHST14/D4ST1 deficiency. *Clin. Biochem.* **2017**, *50*, 670–677. [CrossRef] [PubMed]
24. Kosho, T. CHST14/D4ST1 deficiency: New form of Ehlers–Danlos syndrome. *Pediatr. Int.* **2016**, *58*, 88–99. [CrossRef] [PubMed]
25. Gallagher, J. Fell-Muir Lecture: Heparan sulphate and the art of cell regulation: A polymer chain conducts the protein orchestra. *Int. J. Exp. Pathol.* **2015**, *96*, 203–231. [CrossRef]
26. Gandhi, N.S.; Mancera, R.L. The Structure of Glycosaminoglycans and their Interactions with Proteins. *Chem. Biol. Drug Des.* **2008**, *72*, 455–482. [CrossRef] [PubMed]
27. Casu, B.; Petitou, M.; Provasoli, M.; Sinay, P. Conformational flexibility: A new concept for explaining binding and biological properties of iduronic acid-containing glycosaminoglycans. *Trends Biochem. Sci.* **1988**, *13*, 221–225. [CrossRef]
28. Mulloy, B.; Forster, M.J.; Jones, C.; Drake, A.F.; Johnson, E.A.; Davies, D.B. The effect of variation of substitution on the solution conformation of heparin: A spectroscopic and molecular modelling study. *Carbohydr. Res.* **1994**, *255*, 1–26. [CrossRef]
29. Zamparo, O.; Comper, W.D. The hydrodynamic frictional coefficient of polysaccharides: The role of the glycosidic linkage. *Carbohydr. Res.* **1991**, *212*, 193–200. [CrossRef]
30. Samsonov, S.A.; Gehrcke, J.-P.; Pisabarro, M.T. Flexibility and explicit solvent in molecular-dynamics-based docking of protein–glycosaminoglycan systems. *J. Chem. Inf. Model.* **2014**, *54*, 582–592. [CrossRef]
31. Shriver, Z.; Raguram, S.; Sasisekharan, R. Glycomics: A pathway to a class of new and improved therapeutics. *Nat. Rev. Drug Discov.* **2004**, *3*, 863–873. [CrossRef]
32. Trottein, F.; Schaffer, L.; Ivanov, S.; Paget, C.; Vendeville, C.; Cazet, A.; Groux-Degroote, S.; Lee, S.; Krzewinski-Recchi, M.-A.; Faveeuw, C.; et al. Glycosyltransferase and sulfotransferase gene expression profiles in human monocytes, dendritic cells and macrophages. *Glycoconj. J.* **2009**, *26*, 1259–1274. [CrossRef] [PubMed]
33. Toyoda, H.; Kinoshita-Toyoda, A.; Selleck, S.B. Structural Analysis of Glycosaminoglycans in Drosophila and Caenorhabditis elegans and Demonstration That tout-velu, a Drosophila Gene Related to EXT Tumor Suppressors, Affects Heparan Sulfate in Vivo. *J. Biol. Chem.* **2000**, *275*, 2269–2275. [CrossRef] [PubMed]
34. Yang, H.O.; Gunay, N.S.; Toida, T.; Kuberan, B.; Yu, G.; Kim, Y.S.; Linhardt, R.J. Preparation and structural determination of dermatan sulfate-derived oligosaccharides. *Glycobiology* **2000**, *10*, 1033–1039. [CrossRef] [PubMed]
35. Zaia, J.; Costello, C.E. Compositional Analysis of Glycosaminoglycans by Electrospray Mass Spectrometry. *Anal. Chem.* **2001**, *73*, 233–239. [CrossRef]
36. Shao, C.; Zhang, F.; Kemp, M.M.; Linhardt, R.J.; Waisman, D.M.; Head, J.F.; Seaton, B.A. Crystallographic analysis of calcium-dependent heparin binding to annexin A2. *J. Biol. Chem.* **2006**, *281*, 31689–31695. [CrossRef]
37. Capila, I.; Hernáiz, M.A.J.; Mo, Y.D.; Mealy, T.R.; Campos, B.; Dedman, J.R.; Linhardt, R.J.; Seaton, B.A. Annexin V–Heparin Oligosaccharide Complex Suggests Heparan Sulfate–Mediated Assembly on Cell Surfaces. *Structure* **2001**, *9*, 57–64. [CrossRef]
38. Dementiev, A.; Petitou, M.; Herbert, J.-M.; Gettins, P.G.W. The ternary complex of antithrombin–anhydrothrombin–heparin reveals the basis of inhibitor specificity. *Nat. Struct. Mol. Biol.* **2004**, *11*, 863–867. [CrossRef]
39. Li, W.; Johnson, D.J.D.; Esmon, C.T.; Huntington, J.A. Structure of the antithrombin–thrombin–heparin ternary complex reveals the antithrombotic mechanism of heparin. *Nat. Struct. Mol. Biol.* **2004**, *11*, 857–862. [CrossRef]

40. Li, Z.; Kienetz, M.; Cherney, M.M.; James, M.N.G.; Brömme, D. The crystal and molecular structures of a cathepsin K: Chondroitin sulfate complex. *J. Mol. Biol.* **2008**, *383*, 78–91. [CrossRef]
41. Imberty, A.; Lortat-Jacob, H.; Pérez, S. Structural view of glycosaminoglycan–protein interactions. *Carbohydr. Res.* **2007**, *342*, 430–439. [CrossRef]
42. Ferro, D.R.; Provasoli, A.; Ragazzi, M.; Torri, G.; Casu, B.; Gatti, G.; Jacquinet, J.C.; Sinay, P.; Petitou, M.; Choay, J. Evidence for conformational equilibrium of the sulfated L-iduronate residue in heparin and in synthetic heparin mono- and oligo-saccharides: NMR and force-field studies. *J. Am. Chem. Soc.* **1986**, *108*, 6773–6778. [CrossRef]
43. Bossennec, V.; Petitou, M.; Perly, B. 1H-n.m.r. investigation of naturally occurring and chemically oversulphated dermatan sulphates. Identification of minor monosaccharide residues. *Biochem. J.* **1990**, *267*, 625–630. [CrossRef] [PubMed]
44. Ferro, D.R.; Provasoli, A.; Ragazzi, M.; Casu, B.; Torri, G.; Bossennec, V.; Perly, B.; Sinaÿ, P.; Petitou, M.; Choay, J. Conformer populations of l-iduronic acid residues in glycosaminoglycan sequences. *Carbohydr. Res.* **1990**, *195*, 157–167. [CrossRef]
45. Yamada, S.; Yamane, Y.; Sakamoto, K.; Tsuda, H.; Sugahara, K. Structural determination of sulfated tetrasaccharides and hexasaccharides containing a rare disaccharide sequence, -3GalNAc(4,6-disulfate)beta1-4IdoAalpha1-, isolated from porcine intestinal dermatan sulfate. *Eur. J. Biochem.* **1998**, *258*, 775–783. [CrossRef]
46. Almond, A.; Sheehan, J.K. Glycosaminoglycan conformation: Do aqueous molecular dynamics simulations agree with x-ray fiber diffraction? *Glycobiology* **2000**, *10*, 329–338. [CrossRef]
47. Silipo, A.; Zhang, Z.; Cañada, F.J.; Molinaro, A.; Linhardt, R.J.; Jiménez-Barbero, J. Conformational analysis of a dermatan sulfate-derived tetrasaccharide by NMR, molecular modeling, and residual dipolar couplings. *ChemBioChem* **2008**, *9*, 240–252. [CrossRef]
48. Sattelle, B.M.; Shakeri, J.; Cliff, M.J.; Almond, A. Proteoglycans and their heterogeneous glycosaminoglycans at the atomic scale. *Biomacromolecules* **2015**, *16*, 951–961. [CrossRef]
49. Gandhi, N.S.; Mancera, R.L. Can current force fields reproduce ring puckering in 2-O-sulfo-α-l-iduronic acid? A molecular dynamics simulation study. *Carbohydr. Res.* **2010**, *345*, 689–695. [CrossRef]
50. Sattelle, B.M.; Hansen, S.U.; Gardiner, J.; Almond, A. Free energy landscapes of iduronic acid and related monosaccharides. *J. Am. Chem. Soc.* **2010**, *132*, 13132–13134. [CrossRef]
51. Balogh, G.; Gyongyosi, T.S.; Timári, I.; Herczeg, M.; Borbás, A.; Fehér, K.; Kövér, K.E. Comparison of carbohydrate force fields using Gaussian Accelerated Molecular Dynamics simulations and development of force field parameters for heparin-analogue pentasaccharides. *J. Chem. Inf. Model.* **2019**, *59*, 4855–4867. [CrossRef]
52. Ng, C.; Nandha Premnath, P.; Guvench, O. Rigidity and Flexibility in the Tetrasaccharide Linker of Proteoglycans from Atomic-Resolution Molecular Simulation. *J. Comput. Chem.* **2017**, *38*, 1438–1446. [CrossRef] [PubMed]
53. Guvench, O. Revealing the Mechanisms of Protein Disorder and N-Glycosylation in CD44-Hyaluronan Binding Using Molecular Simulation. *Front. Immunol.* **2015**, *6*, 305. [CrossRef] [PubMed]
54. Faller, C.E.; Guvench, O. Sulfation and cation effects on the conformational properties of the glycan backbone of chondroitin sulfate disaccharides. *J. Phys. Chem. B* **2015**, *119*, 6063–6073. [CrossRef] [PubMed]
55. Favreau, A.J.; Faller, C.E.; Guvench, O. CD44 receptor unfolding enhances binding by freeing basic amino acids to contact carbohydrate ligand. *Biophys. J.* **2013**, *105*, 1217–1226. [CrossRef] [PubMed]
56. Guvench, O.; Mallajosyula, S.S.; Raman, E.P.; Hatcher, E.; Vanommeslaeghe, K.; Foster, T.J.; Jamison, F.W., 2nd; Mackerell, A.D., Jr. CHARMM additive all-atom force field for carbohydrate derivatives and its utility in polysaccharide and carbohydrate-protein modeling. *J. Chem. Theory Comput.* **2011**, *7*, 3162–3180. [CrossRef] [PubMed]
57. Humphrey, W.; Dalke, A.; Schulten, K. VMD: Visual molecular dynamics. *J. Mol. Graph.* **1996**, *14*, 33–38. [CrossRef]
58. Brannigan, G.; Lin, L.C.; Brown, F.L. Implicit solvent simulation models for biomembranes. *Eur. Biophys. J.* **2006**, *35*, 104–124. [CrossRef]
59. Srinivas, G.; Cheng, X.; Smith, J.C. A Solvent-Free Coarse Grain Model for Crystalline and Amorphous Cellulose Fibrils. *J. Chem. Theory Comput.* **2011**, *7*, 2539–2548. [CrossRef]
60. Ingólfsson, H.I.; Lopez, C.A.; Uusitalo, J.J.; de Jong, D.H.; Gopal, S.M.; Periole, X.; Marrink, S.J. The power of coarse graining in biomolecular simulations. *Wiley Interdiscip. Rev. Comput. Mol. Sci.* **2014**, *4*, 225–248. [CrossRef]
61. Guvench, O.; Greene, S.N.; Kamath, G.; Brady, J.W.; Venable, R.M.; Pastor, R.W.; Mackerell, A.D., Jr. Additive Empirical Force Field for Hexopyranose Monosaccharides. *J. Comput. Chem.* **2008**, *29*, 2543–2564. [CrossRef]

62. Guvench, O.; Hatcher, E.R.; Venable, R.M.; Pastor, R.W.; Mackerell, A.D. CHARMM Additive All-Atom Force Field for Glycosidic Linkages between Hexopyranoses. *J. Chem. Theory Comput.* **2009**, *5*, 2353–2370. [CrossRef] [PubMed]
63. Mallajosyula, S.S.; Guvench, O.; Hatcher, E.; Mackerell, A.D., Jr. CHARMM Additive All-Atom Force Field for Phosphate and Sulfate Linked to Carbohydrates. *J. Chem. Theory Comput.* **2012**, *8*, 759–776. [CrossRef] [PubMed]
64. Singh, A.; Montgomery, D.; Xue, X.; Foley, B.L.; Woods, R.J. GAG Builder: A web-tool for modeling 3D structures of glycosaminoglycans. *Glycobiology* **2019**, *29*, 515–518. [CrossRef] [PubMed]
65. Engelsen, S.B.; Hansen, P.I.; Perez, S. POLYS 2.0: An open source software package for building three-dimensional structures of polysaccharides. *Biopolymers* **2014**, *101*, 733–743. [CrossRef] [PubMed]
66. Kuttel, M.M.; Ståhle, J.; Widmalm, G. CarbBuilder: Software for building molecular models of complex oligo-and polysaccharide structures. *J. Comput. Chem.* **2016**, *37*, 2098–2105. [CrossRef] [PubMed]
67. Clerc, O.; Deniaud, M.; Vallet, S.D.; Naba, A.; Rivet, A.; Perez, S.; Thierry-Mieg, N.; Ricard-Blum, S. MatrixDB: Integration of new data with a focus on glycosaminoglycan interactions. *Nucleic Acids Res.* **2018**, *47*, D376–D381. [CrossRef] [PubMed]
68. Clerc, O.; Mariethoz, J.; Rivet, A.; Lisacek, F.; Pérez, S.; Ricard-Blum, S. A pipeline to translate glycosaminoglycan sequences into 3D models. Application to the exploration of glycosaminoglycan conformational space. *Glycobiology* **2018**, *29*, 36–44. [CrossRef]
69. Brooks, B.R.; Bruccoleri, R.E.; Olafson, B.D.; States, D.J.; Swaminathan, S.; Karplus, M. CHARMM: A Program for Macromolecular Energy, Minimization, and Dynamics Calculations. *J. Comput. Chem.* **1983**, *4*, 187–217. [CrossRef]
70. Brooks, B.R.; Brooks, C.L., III; Mackerell, A.D., Jr.; Nilsson, L.; Petrella, R.J.; Roux, B.; Won, Y.; Archontis, G.; Bartels, C.; Boresch, S.; et al. CHARMM: The Biomolecular Simulation Program. *J. Comput. Chem.* **2009**, *30*, 1545–1614. [CrossRef]
71. MacKerell, A.D., Jr.; Brooks, B.; Brooks, C.L., III; Nilsson, L.; Roux, B.; Won, Y.; Karplus, M. CHARMM: The Energy Function and Its Parameterization with an Overview of the Program. In *Encyclopedia of Computational Chemistry*; von Ragué Schleyer, P., Allinger, N.L., Clark, T., Gasteiger, J., Kollman, P.A., Schaefer, H.F., Schreiner, P.R., Eds.; John Wiley & Sons: Chichester, UK, 1998; Volume 1, pp. 271–277.
72. Jorgensen, W.L.; Chandrasekhar, J.; Madura, J.D.; Impey, R.W.; Klein, M.L. Comparison of simple potential functions for simulating liquid water. *J. Chem. Phys.* **1983**, *79*, 926–935. [CrossRef]
73. Durell, S.R.; Brooks, B.R.; Ben-Naim, A. Solvent-induced forces between two hydrophilic groups. *J. Phys. Chem.* **1994**, *98*, 2198–2202. [CrossRef]
74. Phillips, J.C.; Braun, R.; Wang, W.; Gumbart, J.; Tajkhorshid, E.; Villa, E.; Chipot, C.; Skeel, R.D.; Kale, L.; Schulten, K. Scalable Molecular Dynamics with NAMD. *J. Comput. Chem.* **2005**, *26*, 1781–1802. [CrossRef] [PubMed]
75. Hestenes, M.R.; Stiefel, E. Methods of conjugate gradients for solving linear systems. *J. Res. Nat. Bur. Stand.* **1952**, *49*, 409–436. [CrossRef]
76. Fletcher, R.; Reeves, C.M. Function minimization by conjugate gradients. *Comput. J.* **1964**, *7*, 149–154. [CrossRef]
77. Ryckaert, J.-P.; Ciccotti, G.; Berendsen, H.J. Numerical Integration of the Cartesian Equations of Motion of a System with Constraints: Molecular Dynamics of n-Alkanes. *J. Comput. Phys.* **1977**, *23*, 327–341. [CrossRef]
78. Andersen, H.C. Rattle: A "Velocity" Version of the Shake Algorithm for Molecular Dynamics Calculations. *J. Comput. Phys.* **1983**, *52*, 24–34. [CrossRef]
79. Miyamoto, S.; Kollman, P.A. SETTLE: An analytical version of the SHAKE and RATTLE algorithm for rigid water models. *J. Comput. Chem.* **1992**, *13*, 952–962. [CrossRef]
80. Jones, J.E.; Chapman, S. On the Determination of Molecular Fields. -II. From the Equation of State of a Gas. *Proc. R. Soc. Lond. A-Contain.* **1924**, *106*, 463–477. [CrossRef]
81. Steinbach, P.J.; Brooks, B.R. New spherical-cutoff methods for long-range forces in macromolecular simulation. *J. Comput. Chem.* **1994**, *15*, 667–683. [CrossRef]
82. Allen, M.; Tildesley, D. *Computer Simulation of Liquids*; Clarendon Press: Oxford, UK, 1987.
83. Darden, T.; York, D.; Pedersen, L. Particle mesh Ewald: An N log (N) method for Ewald sums in large systems. *J. Chem. Phys.* **1993**, *98*, 10089–10092. [CrossRef]
84. Vanommeslaeghe, K.; MacKerell, A.D., Jr. CHARMM additive and polarizable force fields for biophysics and computer-aided drug design. *Biochim. Biophys. Acta* **2015**, *1850*, 861–871. [CrossRef] [PubMed]
85. Feller, S.E.; Zhang, Y.; Pastor, R.W.; Brooks, B.R. Constant pressure molecular dynamics simulation: The Langevin piston method. *J. Chem. Phys.* **1995**, *103*, 4613–4621. [CrossRef]

86. Kubo, R.; Toda, M.; Hashitsume, N. Statistical Physics II: Nonequilibrium Statistical Mechanics. In *Springer Series in Solid-State Sciences*, 2nd ed.; Cardona, M., Fulde, P., von Klitzing, K., Queisser, H.-J., Eds.; Springer: Berlin/Heidelberg, Germany; New York, NY, USA, 1991; Volume 31.
87. Eastman, P.; Swails, J.; Chodera, J.D.; McGibbon, R.T.; Zhao, Y.; Beauchamp, K.A.; Wang, L.-P.; Simmonett, A.C.; Harrigan, M.P.; Stern, C.D. OpenMM 7: Rapid development of high performance algorithms for molecular dynamics. *PLoS Comput. Biol.* **2017**, *13*, e1005659. [CrossRef] [PubMed]
88. Friedrichs, M.S.; Eastman, P.; Vaidyanathan, V.; Houston, M.; Legrand, S.; Beberg, A.L.; Ensign, D.L.; Bruns, C.M.; Pande, V.S. Accelerating Molecular Dynamic Simulation on Graphics Processing Units. *J. Comput. Chem.* **2009**, *30*, 864–872. [CrossRef]
89. Eastman, P.; Pande, V. OpenMM: A Hardware-Independent Framework for Molecular Simulations. *Comput. Sci. Eng.* **2010**, *12*, 34–39. [CrossRef] [PubMed]
90. Eastman, P.; Pande, V.S. Constant Constraint Matrix Approximation: A Robust, Parallelizable Constraint Method for Molecular Simulations. *J. Chem. Theory Comput.* **2010**, *6*, 434–437. [CrossRef]
91. Eastman, P.; Friedrichs, M.S.; Chodera, J.D.; Radmer, R.J.; Bruns, C.M.; Ku, J.P.; Beauchamp, K.A.; Lane, T.J.; Wang, L.-P.; Shukla, D. OpenMM 4: A Reusable, Extensible, Hardware Independent Library for High Performance Molecular Simulation. *J. Chem. Theory Comput.* **2012**, *9*, 461–469. [CrossRef]
92. Eastman, P.; Pande, V.S. Efficient Nonbonded Interactions for Molecular Dynamics on a Graphics Processing Unit. *J. Comput. Chem.* **2010**, *31*, 1268–1272. [CrossRef]
93. Kiss, P.T.; Sega, M.; Baranyai, A. Efficient Handling of Gaussian Charge Distributions: An Application to Polarizable Molecular Models. *J. Chem. Theory Comput.* **2014**, *10*, 5513–5519. [CrossRef]
94. Gingrich, T.R.; Wilson, M. On the Ewald summation of Gaussian charges for the simulation of metallic surfaces. *Chem. Phys. Lett.* **2010**, *500*, 178–183. [CrossRef]
95. Verlet, L. Computer "Experiments" on Classical Fluids. I. Thermodynamical Properties of Lennard-Jones Molecules. *Phys. Rev.* **1967**, *159*, 98–103. [CrossRef]
96. Sattelle, B.M.; Shakeri, J.; Roberts, I.S.; Almond, A. A 3D-structural model of unsulfated chondroitin from high-field NMR: 4-sulfation has little effect on backbone conformation. *Carbohydr. Res.* **2010**, *345*, 291–302. [CrossRef] [PubMed]
97. Gatti, G.; Casu, B.; Torri, G.; Vercellotti, J.R. Resolution-enhanced 1H-nmr spectra of dermatan sulfate and chondroitin sulfates: Conformation of the uronic acid residues. *Carbohydr. Res.* **1979**, *68*, C3–C7. [CrossRef]
98. Alibay, I.; Bryce, R.A. Ring puckering landscapes of glycosaminoglycan-related monosaccharides from molecular dynamics simulations. *J. Chem. Inf. Model.* **2019**, *59*, 4729–4741. [CrossRef]
99. Nyerges, B.; Kovacs, A. Density functional study of the conformational space of $^4C_1$ D-glucuronic acid. *J. Phys. Chem. A* **2005**, *109*, 892–897. [CrossRef]
100. Scott, J.E. Supramolecular organization of extracellular matrix glycosaminoglycans, in vitro and in the tissues. *FASEB J.* **1992**, *6*, 2639–2645. [CrossRef]
101. Van Boeckel, C.; Van Aelst, S.; Wagenaars, G.; Mellema, J.R.; Paulsen, H.; Peters, T.; Pollex, A.; Sinnwell, V. Conformational analysis of synthetic heparin-like oligosaccharides containing α-L-idopyranosyluronic acid. *Recl. Trav. Chim. Pays-Bas* **1987**, *106*, 19–29. [CrossRef]
102. Agrawal, P.K.; Jacquinet, J.-C.; Krishna, N.R. NMR and molecular modeling studies on two glycopeptides from the carbohydrate-protein linkage region of connective tissue proteoglycans. *Glycobiology* **1999**, *9*, 669–677. [CrossRef]
103. Choe, B.Y.; Ekborg, G.C.; Roden, L.; Harvey, S.C.; Krishna, N.R. High-resolution NMR and molecular modeling studies on complex carbohydrates: Characterization of O-β-D-Gal-(1→3)-O-β-D-Gal-(1→4)-O-β-D-Xyl-(1→0)-L-Ser, a carbohydrate-protein linkage region fragment from connective tissue proteoglycans. *J. Am. Chem. Soc.* **1991**, *113*, 3743–3749. [CrossRef]
104. Krishna, N.; Choe, B.; Prabhakaran, M.; Ekborg, G.C.; Rodén, L.; Harvey, S. Nuclear magnetic resonance and molecular modeling studies on O-β-D-galactopyranosyl-(1→4)-O-β-D-xylopyranosyl-(1→0)-L-serine, a carbohydrate-protein linkage region fragment from connective tissue proteoglycans. *J. Biol. Chem.* **1990**, *265*, 18256–18262.

 © 2020 by the authors. Licensee MDPI, Basel, Switzerland. This article is an open access article distributed under the terms and conditions of the Creative Commons Attribution (CC BY) license (http://creativecommons.org/licenses/by/4.0/).

Article

GAG-DB, the New Interface of the Three-Dimensional Landscape of Glycosaminoglycans

Serge Pérez [1,*], François Bonnardel [1,2,3], Frédérique Lisacek [2,3,4], Anne Imberty [1], Sylvie Ricard Blum [5] and Olga Makshakova [6]

1. University Grenoble Alpes, CNRS, CERMAV, 38000 Grenoble, France; francois.bonnardel@cermav.cnrs.fr (F.B.); anne.imberty@cermav.cnrs.fr (A.I.)
2. SIB Swiss Institute of Bioinformatics, CH-1227 Geneva, Switzerland; frederique.lisacek@sib.swiss
3. Computer Science Department, University of Geneva, CH-1227 Geneva, Switzerland
4. Section of Biology, University of Geneva, CH-1205 Geneva, Switzerland
5. Institut de Chimie et Biochimie Moléculaires et Supramoléculaires, UMR 5246 CNRS—Université Lyon 1, 69622 Villeurbanne CEDEX, France; sylvie.ricard-blum@univ-lyon1.fr
6. Kazan Institute of Biochemistry and Biophysics, FRC Kazan Scientific Center of RAS, 420111 Kazan, Russia; olga.makshakova@kibb.knc.ru
* Correspondence: spsergeperez@gmail.com

Received: 20 November 2020; Accepted: 9 December 2020; Published: 11 December 2020

Abstract: Glycosaminoglycans (GAGs) are complex linear polysaccharides. GAG-DB is a curated database that classifies the three-dimensional features of the six mammalian GAGs (chondroitin sulfate, dermatan sulfate, heparin, heparan sulfate, hyaluronan, and keratan sulfate) and their oligosaccharides complexed with proteins. The entries are structures of GAG and GAG-protein complexes determined by X-ray single-crystal diffraction methods, X-ray fiber diffractometry, solution NMR spectroscopy, and scattering data often associated with molecular modeling. We designed the database architecture and the navigation tools to query the database with the Protein Data Bank (PDB), UniProtKB, and GlyTouCan (universal glycan repository) identifiers. Special attention was devoted to the description of the bound glycan ligands using simple graphical representation and numerical format for cross-referencing to other databases in glycoscience and functional data. GAG-DB provides detailed information on GAGs, their bound protein ligands, and features their interactions using several open access applications. Binding covers interactions between monosaccharides and protein monosaccharide units and the evaluation of quaternary structure. GAG-DB is freely available.

Keywords: glycosaminoglycans; three-dimensional structure; database; polysaccharide conformation; protein-carbohydrate interactions

1. Introduction

Proteoglycans (PGs) constitute a diverse family of proteins that occur in the extracellular matrix (ECM) and pericellular matrix (PCM) and on the surface of mammalian cells. They consist of a core protein and one or more covalently attached glycosaminoglycan (GAG) chains. PGs play critical roles in numerous biological processes, which are mediated by both their protein part and their GAG chains [1,2].

GAGs refer to six major polysaccharides in mammals: chondroitin sulfate (CS) [3], dermatan sulfate (DS), heparin (HP), heparan sulfate (HS) [4,5], hyaluronan (HA) [6], and keratan sulfate [7,8]. Their molecular mass ranges from a few kDa to several million Da for hyaluronan. Despite significant compositional differences, GAGs also share common features. They are linear polysaccharides made of disaccharide repeats. The disaccharides are composed of uronic acid and an hexosamine, alternatively linked through 1-4 and 1-3 glycosidic bonds (Figure 1), except for keratan sulfate, which involves galactose (Gal*p*) and N-acetylglucosamine (Glc*p*NAc) [7]. In contrast to the five other GAGs, hyaluronan is not sulfated and does not bind covalently to proteins to form proteoglycans. Variations in the pattern of GAG sulfation at various positions, create an impressive structural diversity. Two hundred and two unique disaccharides of mammalian GAGs have been identified so far, including 48 theoretical disaccharides in HS [9].

In addition to their contribution to the physicochemical properties of PGs, GAGs play an essential role in the organization and assembly of the extracellular matrix. They also regulate numerous biological processes by interacting with proteins in the extracellular milieu and at the cell surface. The six mammalian GAGs were shown to interact with 827 proteins in the recently published GAG interactome [10].

Many of these GAG interactions have been investigated and characterized in health and disease. According to [10], they take place in various locations (intracellular, cell surface, secreted, and blood proteins) and the protein partners range from individual growth factors (e.g., fibroblast growth factor-2) to large multidomain extracellular proteins such as collagens I and V, and fibronectin with different affinity and half-life [11,12]. These proteins are involved in a variety of biological processes such as extracellular matrix assembly, cell signaling, development, and angiogenesis [10,13,14]. Besides, glycosaminoglycans play a role in host-pathogen interactions by binding to bacterial, viral, and parasite proteins [15–20]. The significance of the understanding and mastering the molecular features underlying the interaction of GAGs to proteins was magistrally demonstrated by the development of the antithrombotic drugs as reviewed in [21].

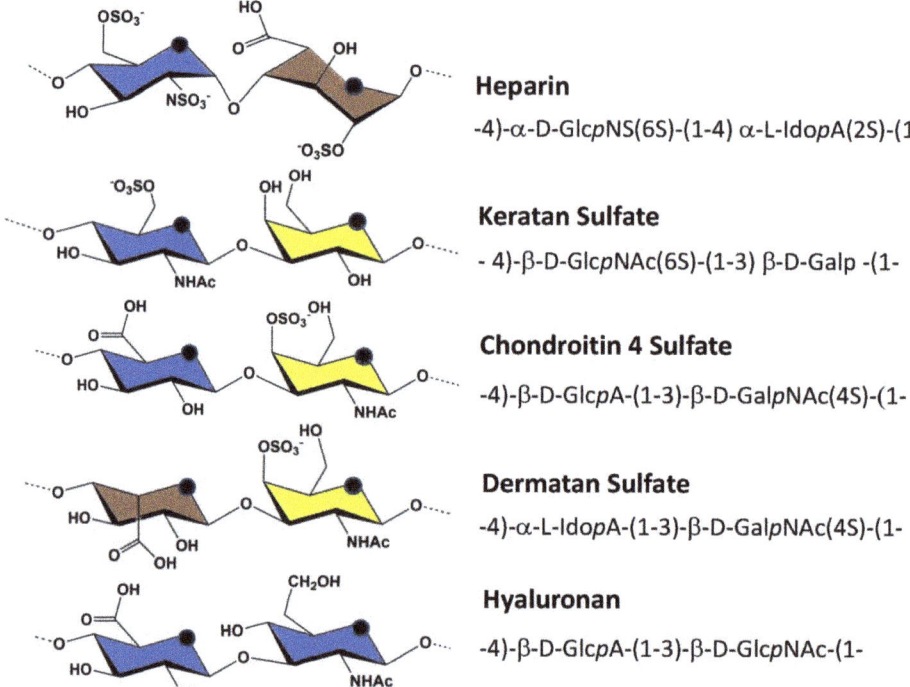

Figure 1. Main repeating units of glycosaminoglycans. The color-coding of the constituting monosaccharide complies with SNFG nomenclature [22]. The abbreviations are as follows: Glc*p* for glucose, Ido*p* for idose), Gal*p* for galactose, N for amine, S for sulfate, A for acid, and NAc for N-acetyl.

The length, sequence, substitution pattern, charge, and shape of GAGs control both their physicochemical properties and their biological functions. Understanding the functions of GAGs first requires methods to accurately assess their molecular weight, their composition and their sequences. This is made possible through ongoing progress in mass spectrometry, and heparan sulfate has been sequenced by liquid chromatography-tandem mass spectrometry (LC-MS/MS) [23–27]. Furthermore, the structural and conformational complexity of GAGs challenges the characterization of their three-dimensional features using either experimental or theoretical methods. In a sense, GAGs concentrate most on the difficulties faced in structural glycoscience. They combine the challenges associated with both glycans and polyelectrolytes. Several experimental techniques have been used to solve GAG structures, including fiber X-ray crystallography, nuclear magnetic resonance (NMR) [28,29], electron microscopy, small-angle X-ray scattering (SAXS) [30], and neutron scattering (elastic incoherent neutron scattering EINS [31], and small-angle neutron scattering SANS [32]. Still, no single technique can cope with such complexity, but, computational methods offer valuable tools to integrate partial information collected experimentally. These, in turn, are useful to validate and improve simulation strategies. However, these approaches remain limited due to the intrinsic properties of GAGs. Like any other complex glycans, they are highly flexible, create many solvent-mediated interactions and have a polyanionic character. Nevertheless, progress in this field is underway, as detailed in [33] that investigates structures from monosaccharides to polysaccharides.

GAG-protein complex structures available in the PDB have been compiled by Samsonov and coworkers [34]. They concluded that this dataset does not represent the diversity of natural GAG sequences. It implies that computational approaches will be critical in understanding GAG structural biology and their mechanisms of interaction with their protein partners [35–37]. Significant progress has been made to investigate GAG structures, isolated and complexed with proteins, both at all-atom and coarse-grained levels [33,38–41]. However, appropriate tools for data mining of GAG-protein interactions are still missing [12,14].

MatrixDB (http://matrixdb.univ-lyon1.fr/) is a biological database focused on molecular interactions between extracellular proteins and polysaccharides [42]. It offers the first step to investigate the molecular mechanisms of GAGs-protein interactions. In this resource, building and displaying the three-dimensional structural models of GAGs was rationalized through an effort to standardize the format of GAGs sequences and group GAG disaccharides into a limited number of families [9]. However, the relative spatial orientations of key GAG chemical groups interacting with (potential) "hot spots" on the proteins was not characterized. The conformational features displayed by the long-chain GAGs polysaccharides were not considered either. To move forward, we collected further evidence of experimental GAG and GAG-protein interaction data, from databases and in the relevant literature.

Experimental details of protein or protein complex three-dimensional structures are comprehensively recorded in the Protein Data Base [43] While being an essential repository, the glycan-related data stored in the PDB is not easily accessible to non-glycoscientists. This difficulty was identified in the glycoscience community and gave rise to several initiatives. Tools were designed to correct inconsistencies in the data [44–46]. Data was organized in publicly available databases, cross-referenced, and interoperable with the glycomic, and other omic, databases to ease data access and analysis, such as Glyco3D [47], UniLectin3D [48], and MatrixDB [42] for GAG-extracellular protein complexes. We now report the development of GAG-DB, a database containing three-dimensional data on GAGs and GAGs-protein complexes retrieved from the PDB. It includes protein sequences and standard nomenclature of GAG composition, sequence, and topology. It provides a family-based classification of GAGs, cross-referenced with glyco-databases, with links to UniProtKB via accession numbers [49]. The 3D visualization of contacts between GAGs and their protein ligands is implemented via the protein-ligand interaction profiler (PLIP) [50] and the nature of the structure that GAG polysaccharides can adopt, either in the solid-state or in solution is also reported. Finally, characterized quaternary structures of the complexes improve understanding if and how GAGs participate in long-range, multivalent, binding with the potential synergy when several chains are involved in interactions.

2. The GAG-DB Database Construction and Utilization

2.1. Database Construction

GAG-DB is available at https://www.gagdb.glycopedia.eu. The database is populated with information extracted from the PDB [51]. It includes the three-dimensional structural information on GAG and GAG oligosaccharides in interaction with proteins. We propose a classification based on the nature of GAGs, e.g., hyaluronan, heparin/heparan sulfate, chondroitin sulfate/dermatan sulfate, and keratan sulfate. GAG mimetics are included, as long as they appear in the PDB. The content of GAG-DB is focused on three-dimensional data, with an appropriate curation of the nomenclature, and extended related information. The entries are structures of GAGs and GAG protein complexes obtained by a wide range of methods.

To avoid any confusion; we note that under the name GAG database, a resource to gather genomic annotation cross-references has been developed and published in 2013 (The GAG database: a new resource to gather genomic annotation cross-references, T Obadia, O Sallou, M Ouedraogo, G Guernec, F Lecerf and published (Gene. 2013;5;527(2):503-9., DOI:10.1016/j.gene.2013.06.063. Epub 16 2013 July). Available annotation data includes all transcripts and their identifiers, functional description of genes, chromosomal localisation, gene symbols, gene homologs for model species (human, chicken, mouse), and several identifiers to link those genes to external databases (UniProt, HGNC).

The GAG-DB database contains 15 entries of long-chain GAGs established from fiber X-ray diffraction. A value of 3.0 Å is assigned to the structural models that have been proposed from X-ray fiber diffraction, and to 0 for those established by solution NMR or X-ray scattering (the structures are not filtered). It also contains 125 manually curated entries extracted from PDBe [52,53] (September 2020 release). These three-dimensional structures have been experimentally determined with methods involving either X-ray single-crystal diffraction, or X-ray fiber diffraction and solution NMR, in conjunction with molecular modeling. The number of GAG-protein complexes amounts to 105. The value of the resolution index indicates the accuracy of the experimental conditions, high values (e.g., 4 Å) indicate a poor resolution and low values (e.g., 1.5 Å) a good resolution. The median resolution for X-ray crystallographic data in the Protein Data Bank is 2.05 Å. Proteins of the database can be grossly separated into enzymes and skeletal proteins. Interestingly, the size distribution of oligosaccharides complexed with proteins varies from 34 disaccharides to only one polysaccharide with a degree of polymerization (DP) of 10 (DP 3 (1), DP 4 (18), DP 5 (13), DP 6 (15), DP 7 (7), DP 8 (8), and DP 9 (1). More than 80% of the GAGs involved in the complexes are heparin and hyaluronic acid oligosaccharides. However, these figures tend to reflect the interest of a community in investigating those GAGs more obviously involved in biological and biomedical applications.

Our collection is far from covering the molecular diversity of GAGs. This lack of data echoes the limitations of carbohydrate synthesis that fails to provide sufficiently long sequences needed to properly investigate the molecular features driving interactions with proteins. Nonetheless, progress is in sight, as recently described in [54,55].

The representation of GAGs sequences complies with recommended nomenclatures and formats, the IUPAC condensed being the reference (http://www.sbcs.qmul.ac.uk/iupac/2carb/38.html) [56]. Each sequence is also encoded in a machine-readable GlycoCT format [57,58], and depicted in SNFG (Symbol Nomenclature for Glycan) [22], following the description provided in [9].

At present, information associated with each entity of the database is added manually. This allows for proper curation and annotation, at the expense of a time lag between the date of deposition and the date of release in the database. Technically, the database was developed with PHP version 7, Bootstrap version 3 and MySQL database version 7. The interface is compatible with all devices and browsers. The pages are dynamically generated to match user-selected search criteria in the query window. Interactive graphics are developed in JavaScripts on D3JS libraries version 3. A tutorial is available on the first page.

2.2. Description of the Search Interface

The database can be searched and explored with an advanced search tool handling a range of criteria.

Figure 2 shows the different fields that can be searched. Possible inputs are:

- The name of the polysaccharide **gag_name**, or its protein ligand, **macromolecule_name**.
- Cross-entries with external databases, such as **pdb**, **UniProt**, and repository **GlytouCan**.
- The biological role, such as **function, process,** or **cellular compartment** (compliant with Gene Ontology terms).
- The origin such as **organism**.
- The experimental condition(s) used to solve the structure: **method and resolution**.
- Characteristics of the GAG such as nature, (is_gag differentiates GAG and mimetics) and size (**gag_max, gag-length,** and **gag-mass**).
- Codes used for ligands in the Protein DataBase, **pdb_ligand** (nomenclature of the Chemical Component Dictionary. www.ebi.ac.uk/pdbe-srv/pdbechem/), or as encoded in **LINUCS** (Linear Notation for Unique description of Carbohydrate Sequences [58]), which provides access to **WURCS** (Web3 Unique Representation of Carbohydrate Structures [59]).

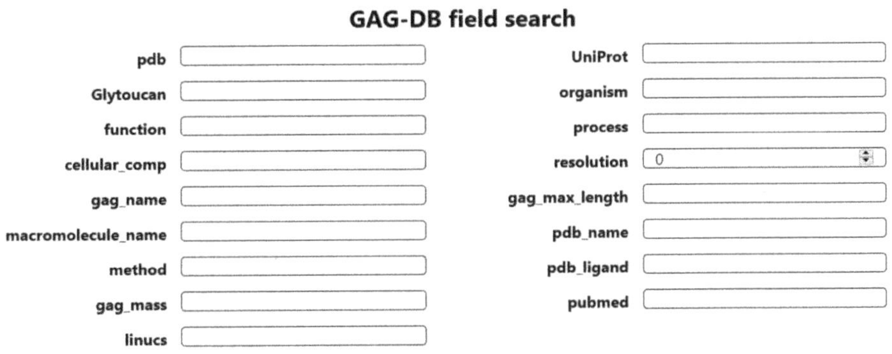

GAG-DB field search

pdb | UniProt
Glytoucan | organism
function | process
cellular_comp | resolution 0
gag_name | gag_max_length
macromolecule_name | pdb_name
method | pdb_ligand
gag_mass | pubmed
linucs

Figure 2. Multiple criteria of the advanced search in the GAG DB database.

2.3. Curated Information for Each GAG Entry

For each entry, a detailed page is available, with 3D visualization, interactions, conformations, nomenclature, and links to external databases (Figure 3). The PDB code assigned to each entry is used to list alternative structures and to display additional information. Each structure is related to a protein with a UniProt accession number [49]. Each oligosaccharide is given a GlyTouCan identifier [60]. The 3D structures of the protein and the interacting GAG are visualized directly and interactively with LiteMol [61] and NGL Viewers [62]. High-resolution images of both the protein-GAG complex and the GAG are available for download. The atomic coordinates of the GAG, isolated from its interaction with the proteins, can also be downloaded for further use.

GAG-DB cross-references to several other databases that rely on a variety of strategies for visualizing the interaction between the GAG ligand and its protein environment (Figure 4). Several applications are available through the four different PDB sites, RSCG ORG, PDBe [51,53], PDBj [63], and PDB SUM [64].

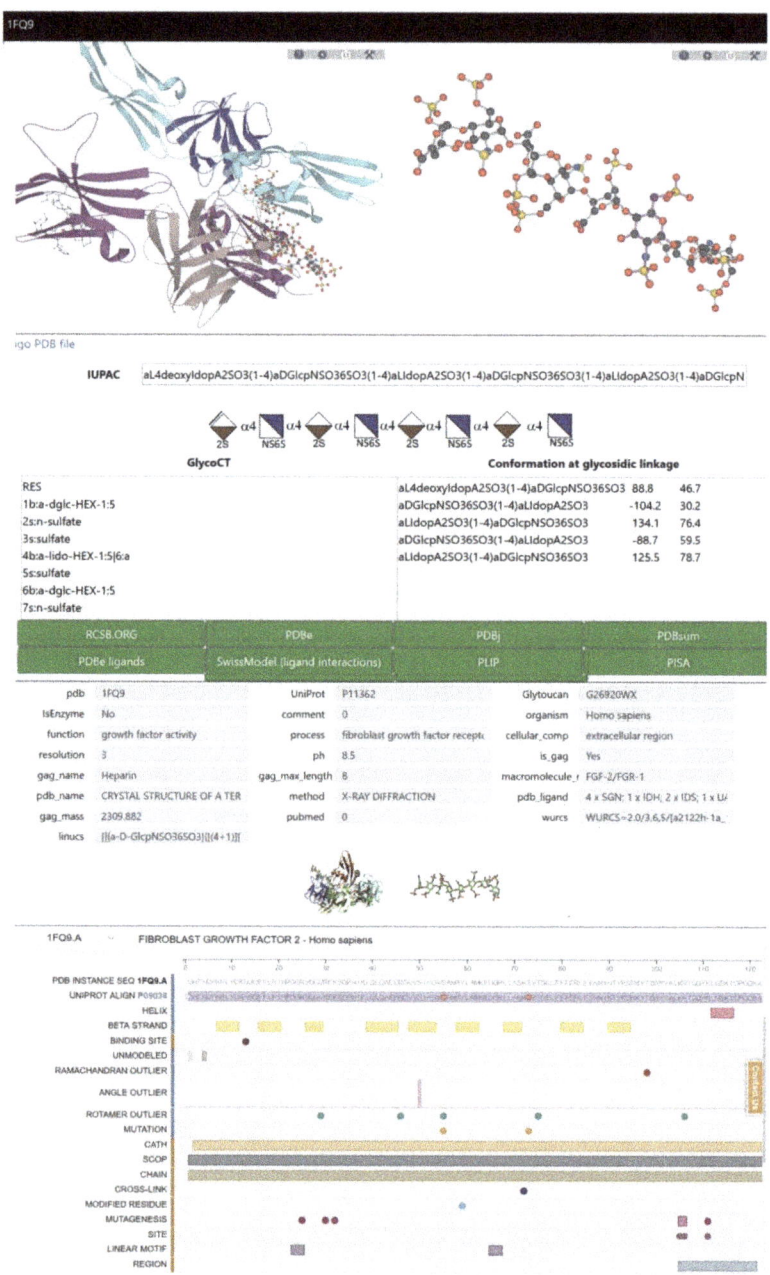

Figure 3. Full results from the search on GAG oligosaccharide present in the database using 1FQ9 in the PDB field. More information can be obtained by clicking on the green bars.

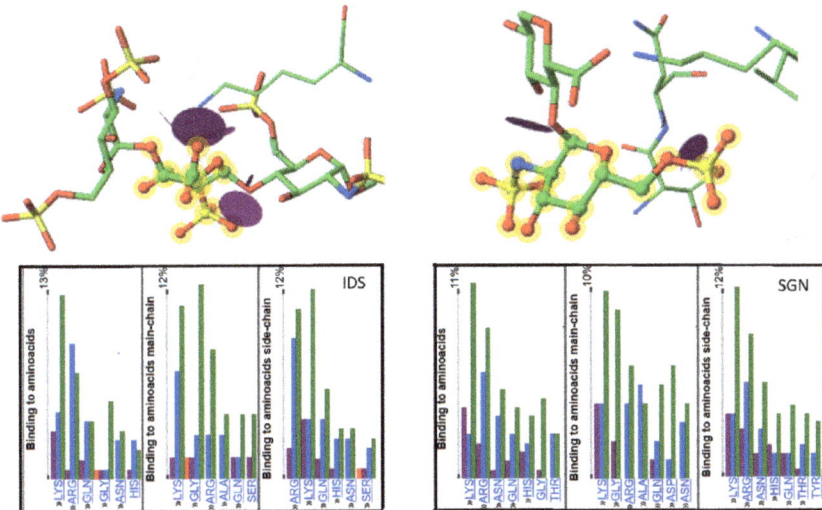

Figure 4. Example of graphical representation of GAG binding site obtained from the PDBe ligand interface. Distribution of the amino acids involved in the binding to GAGs oligosaccharides in cocrystallized complexes. Illustration of two cases showing the hydrogen bonding and electrostatic interaction involved in the interaction. https://www.ebi.ac.uk/pdbe-site/pdbemotif/?tab=ligandbindingstats&ligandCode3letter1=IDS; https://www.ebi.ac.uk/pdbe-site/pdbemotif/?tab=ligandbindingstats&ligandCode3letter1=SGN.

Additional information on the interactions formed between the GAG and the protein can also be obtained using the protein-ligand interaction profiler (PLIP) server [50]. The NGL viewer [62] adapted to SwissModel [65] displays the interactions identified by the PLIP application that calculates and displays atomic level interactions (hydrogen bonds, hydrophobic, water bridge, etc.) occurring between GAGs and proteins. The specific features of the glycans interacting with the surrounding amino acid residues and possible metal ions are shown in 3D. The SwissModel application [65] provides direct access to the PDBsum deployed by the EMBL-EBI [64], CATH [66], and PLIP [50].

A cross-link to the PISA application [67,68] enables the exploration of quaternary structure formation and stability. The potential contribution of GAGs to the formation of quaternary macromolecular complexes requires the evaluation of energetic stability. The structural information relates to the interfaces between the macromolecular entities, the individual monomers, and the resulting assemblies, from which complex stability can be assessed or predicted. Supplementary Figures S1 and S2 provide examples of the interaction features offered by several visualization applications.

3. Utilization of GAG-DB for Analysis of GAGs Structure and Conformation

3.1. Monosaccharides

Repeated disaccharide units of glycosamine and uronic acids with a non-uniform distribution of sulfated and acetylated groups along the chain constitute the main structural features of sulfated GAGs. Despite the high diversity of potential structures, only 28 unique monosaccharide structures occur in GAGs. Three of them correspond to 4,5 unsaturated uronic acids resulting from the eliminative cleavage of GAGs oligo- or polysaccharides containing (1->4)-linked D-glucuronate or L-iduronate residues and (1->4)-alpha-linked 2-sulfoamino-2-deoxy-6-sulfo-D-glucose residues to give oligosaccharides with terminal 4-deoxy-alpha-D-gluco-4-enuronosyl groups at their non-reducing ends.

The cartoon representation of monosaccharides was extended [42] in compliance with the SNFG representation of glycans [22] to link this description with the GlycoCT [58] and condensed IUPAC [56] codes of the monosaccharides.

While these nomenclatures have become widely popular in the field of glycoscience, they are not used to identify and describe monosaccharides in the PDB, which has its carbohydrate nomenclature in its ligand dictionary [69]. Therefore, we established the cross-references between some of these nomenclatures (Figure 5).

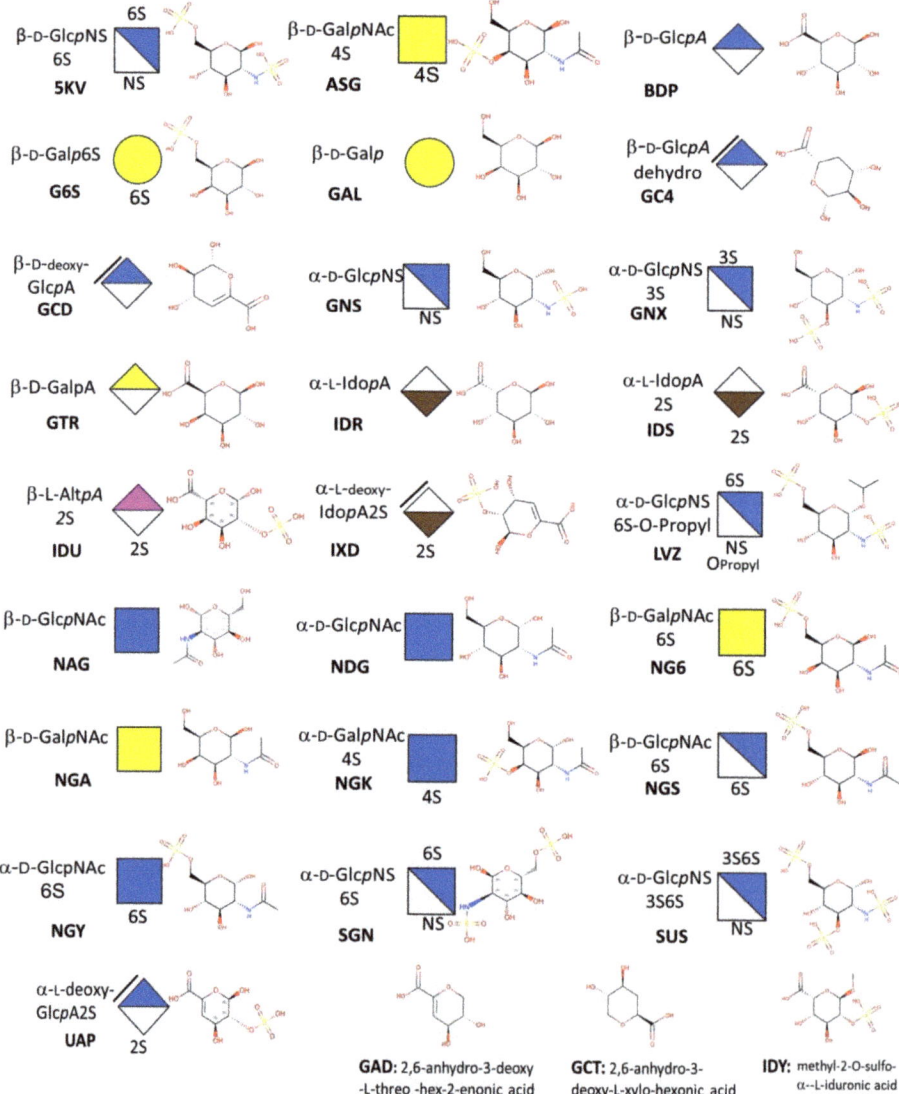

Figure 5. Cross-references between the common representations and nomenclature of monosaccharides: condensed IUAPC nomenclature, symbol nomenclature for glycan; PDB Chemical Component Nomenclature, 2-dimensional structure.

Except for L-idopyranosides, and the 4,5 unsaturated uronic acids, all monosaccharides exist as hexopyranosides. The predominant conformation being $^4C_1$. As for L-idopyranosides, the following $^1C_4$, $^4C_1$, and $^2S_0$ conformations may be found. Figure 6 depicts the 3-dimensional representations of these unusual conformations, along with the corresponding SNFG extensions.

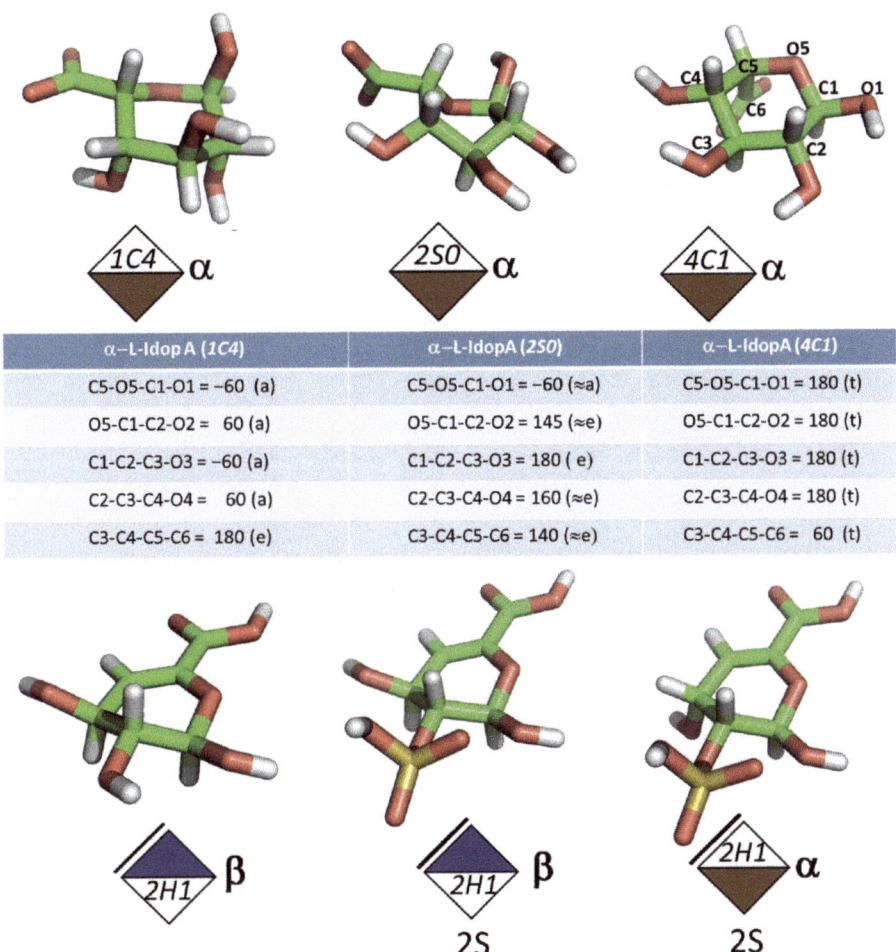

Figure 6. 3-D structures, descriptors, and schematic representations of L-Idopyranosides, the following $^1C_4$, $^4C_1$, and $^2S_0$; and 4,5 unsaturated uronic acids (drawn with pyMol (Schrödinger)).

3.2. Disaccharides

The PDB dataset consisting of 105 proteins-GAG complexes contains 270 disaccharides [9]. Table 1 displays the major disaccharides as extracted from GAG-DB.

Table 1. Major disaccharides found in the GAG-protein complexes.

Major Disaccharides Found in the GAG-Protein Complexes Extracted from the PDB	Number
α-D-GlcpNS (1-4) β-D-GlcpA	21
α-D-GlcpNS(6S) (1-4) α-L-IdopA(2S) [1C4]	42
α-D-GlcpNS(6S) (1-4) α-L-IdopA(2S) [2S0]	15
α-L-IdopA(2S) [1C4] (1-4) α-D-GlcpNS(6S)	40
α-L-IdopA(2S) [2S0] (1-4) α-D-GlcpNS(6S)	16
β-D-GlcpA (1-3) β-D-GlcpNAc	13
β-D-GlcpA (1-4) α-D-GlcpNS	15
β-D-GlcpNAc (1-4) β-D-GlcpA	12

Such a rich set of experimental data provides useful information to validate and improve computational strategies to build GAG models. The determination of the conformational preferences of GAG disaccharides can be assessed by computing potential energy surfaces as a function of their glycosidic torsion angles Φ and Ψ as implemented in the CAT application [70]. As an example, Figure 7 displays two such potential energy surfaces (alternatives are not shown). In all cases, the experimentally observed Φ and Ψ are plotted on the corresponding potential energy surfaces. While being somehow scattered, they are all located on the lowest energy basins.

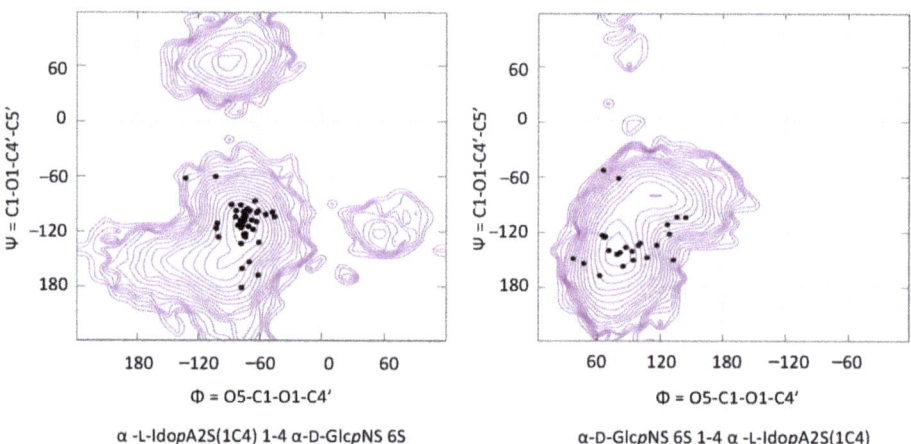

Figure 7. Φ and Ψ angles measured in the 3D structures of cocrystallized GAG protein complexes, reported on conformational maps.

Similar features are observed for all disaccharides (or disaccharide units) irrespective of the presence and the positions of sulfate groups on the monosaccharides. The agreement between the repertoire of the experimentally determined conformations and those predicted by computational methods provided the basis to develop a pipeline to translate glycosaminoglycans sequences into 3D models (http://glycan-builder.cermav.cnrs.fr/gag/) [47].

3.3. GAG Structures in the Solid-State

The solid-state features of chondroitin sulfate, dermatan sulfate, hyaluronan, and keratan sulfate have been established by X-ray fiber diffraction [71] (Table 2) and are available in GAG-DB. They encompass several allomorphs that occur in different experimental conditions, including the nature of the counterions (Na^+, Ca^{++}, and K^+). More structural features such as the polarity of the polysaccharide chains, their interactions with the counterions and packing features can be deduced.

Table 2. Characterization of the helix symmetry of GAGs polysaccharides in the solid-state.

Glycosaminoglycans	Structure of the Main Repeating Disaccharides	Helix Symmetry	Ref.
Hyaluronan	-4)-β-D-Glc$p$A-(1-3)-β-D-Glc$p$NAc-(1-	$2_1, 3_2, 4_3$	[72–74]
Chondroitin-4-sulfate	-4)-β-D-Glc$p$A-(1-3)-β-D-Gal$p$NAc(4S)-(1-	$2_1, 3_2$	[75–77]
Chondroitin-6-sulfate	-4)-β-D-Glc$p$A-(1-3)-β-D-Gal$p$NAc(6S)-(1-	$2_1, 3_2, 8_3$	
Dermatan sulfate	-4)-α-L-Ido$p$A-(1-3)-β-D-Gal$p$NAc(4S)-(1-	$2_1, 3_2, 8_3$	[78]
Heparin	-4)-α-L-Ido$p$A(2S)-(1-4)-α-D-Glc$p$NS(6S)-(1	(Na$^+$) 2_1	
Heparan sulfate	-4)-β-D-Glc$p$A-(1-4)-α-D-Glc$p$NAc-(1-	2_1	[79]
Keratan sulfate	-3)-β-D-Gal$p$-(1-4)-β-D-Glc$p$NAc(6S)-(1-	2_1	[80]

The organization of all these polysaccharide chains in the form of helices seems recurrent. Two parameters, n and h, characterize helical structures, where n is the number of repeat units (disaccharide unit) per turn of the helix and h is the projection of one repeat unit on the helical axis. The sign attributed to n indicates the chirality of the helix. The positive value of n corresponds to the right-handed helix and a negative value to a left-handed helix. Such helical descriptors provide a simple way to classify the secondary structures and their potential allomorphs.

As with the disaccharide segments of GAGs, the values of the Φ and Ψ torsional angles found in all the conformations of GAGs fall in the low energy regions of the corresponding potential energy surfaces. It is therefore relevant to question whether secondary structures other than those derived from crystallographic characterization do occur. The sets of (Φ, Ψ) values corresponding to the low energy conformations can be propagated regularly, to generate structures, which can be further optimized to form integral helices. When applied to hyaluronan structures, the analysis indicates that this polysaccharide display a wide range of energetically stable helices (Figure 8). They span the left-handed 4-fold symmetry to the right-handed five-fold symmetry with a rise per disaccharide between 9.51 and 10.13 Å [81].

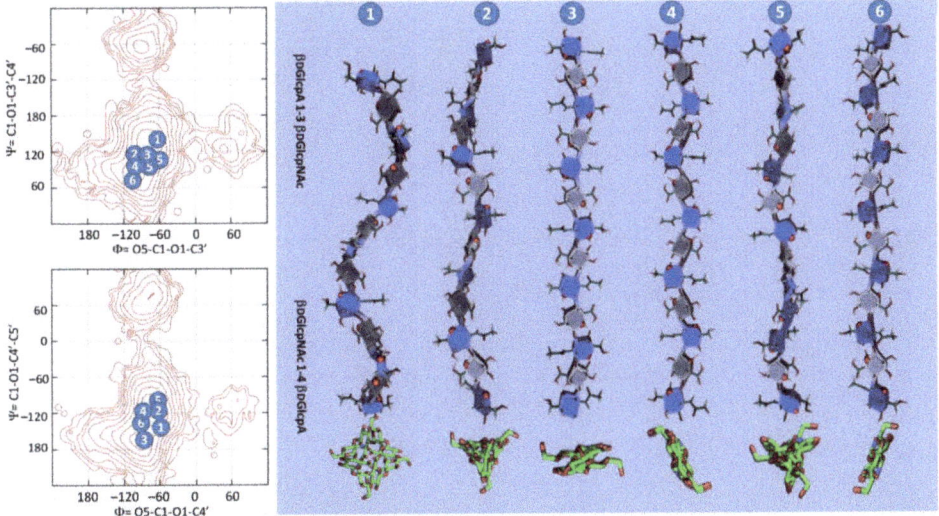

Figure 8. Stable regular helical conformation of single-stranded hyaluronic acid projected parallel and orthogonal to their axes (drawn with SweetUnityMol [82]). The Φ and Ψ conformations are shown on the corresponding potential energy surfaces.

The results indicate that small variations in the glycosidic torsion angles might have a significant influence on the symmetry and pitch of the resulting helices without any noticeable energetic cost. This illustrates the capacity of hyaluronic acid to display different sites available for interactions with proteins and would occur, at no cost in energy, without altering the directionality of the polysaccharide chain.

3.4. GAG Structures in Solution

The database contains the structure of heparin as established by NMR in solution (PDB entry 1HPN, 1XT3) and analogue (2ERM). Other structures have been reported for the solution structures of four different heparin oligosaccharides, determined by a combination of analytical ultracentrifugation, synchrotron X-ray solution scattering that gave the radii of gyration and maximum length extension [30,83] (PDB code 3IRI, 3IRJ, 3IRK, and 3IRL). Constrained molecular modeling of randomized heparin conformers resulted in 9–15 best-fit structures for each degree of polymerization (dp) DP18, DP24, DP30, and DP36 that indicated flexibility and the presence of short linear segments in mildly bent structures. All the conformations of the experimental conformations are somewhat scattered. They are all located in the lowest energy region of the corresponding Φ and Ψ maps (see Figure 9). The idopyranose residues experienced some changes, either $^1C_4$ or $^2S_0$, without any influence on the Φ and Ψ maps. This establishes a model of heparin in solution as a semi-rigid object.

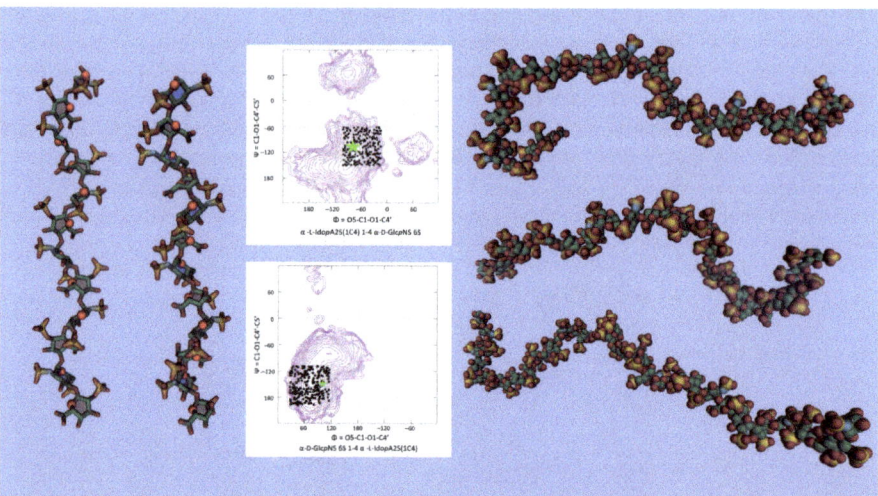

Figure 9. 3D representation of heparin in a helical conformation (left panel) and in disordered conformation. The distribution of the Φ and Ψ angles are reported on the two corresponding potential energy surfaces (drawn with SweetUnityMol [82]).

Such a computational protocol was used to model the disordered features of hyaluronic acid [81] and chondroitin sulfate [84]. As with heparin, the semi-rigid behavior and the stiffness of these GAGs polysaccharides could be established.

4. Conclusions

The aim of the article was to integrate three-dimensional data of GAGs, GAGs oligosaccharides as complexed with proteins. The sources of data are multiple: X-ray fiber diffraction, solution NMR, small angle X-ray scattering for GAGs, and X-ray biomolecular crystallography for protein-GAGs and protein-GAG mimetics complexes. A series of descriptors were selected to guide the search. They include cross-references to PDB, UniProtKB, MatrixDB, and GlyTouCan. GAG-DB opens the possibility of deciphering the full potential of GAGs as bioactive fragments or a structurally important multivalent scaffold for interaction synergy at assembling proteins within quaternary structures. The inspection of the many features of the database supports the reporting of robust facts/knowledge and the determination of what remains to be investigated or discovered. The amount and the quality of the 3D structures of GAG-protein complexes are amenable to comparison between the observed and the calculated 3D descriptors. Such a rich set of experimental information provides a solid basis for validating and improving computational strategies. We could confirm previously described features such as the lack of counterion effect in the interaction between GAGs; the definition of the preferred amino acids bringing the electrostatic neutrality of the interaction; and the lack of influence of sulfate groups on the glycosidic torsion angles. All the observed conformations fell within the low energy basins, thereby comforting the suitability of the computational protocol to model GAGs conformation in a disordered state. An emerging picture is the description of these polysaccharide chains as propagating linearly in a preferred direction, with extended fragments separated by kinks. The semi-rigid character of the chains involves microarchitectural domains. They contain preformed conformation for optimal binding to protein targets. The separation of such domains, at a long enough distance, offers the possibility of multivalent binding to create further spatial arrangements that can induce the formation of functional assemblies of proteins.

Supplementary Materials: The following are available online at http://www.mdpi.com/2218-273X/10/12/1660/s1, Figure S1: Detailed of the analysis of GAG binding in crystal structure of a ternary FGF1-FGFR2-Heparin complex (PDB 2GD4) available from different souces; PDBe; Swiss-Model; PLIP, LIGPOLT. Figure S2: Quaternary organisation in the crystal Structure of the Antithrombin-S195A Factor Xa-Pentasaccharide Complex (PDB 1EO0) computed and displayed by PISA.

Author Contributions: Methodology, S.P., O.M., and S.R.B.; software, F.B. and S.P.; data curation, S.P. and S.R.B.; writing—original draft preparation, S.P., O.M., and S.R.B; writing—review and editing, all; funding acquisition, A.I., S.P., S.R.B., and F.L. All authors have read and agreed to the published version of the manuscript.

Funding: This research profited from Campus Rhodanien Co-Funds (http://campusrhodanien. unige-cofunds.ch. Support from ANR PIA Glyco@Alps [ANR-15-IDEX-02]; Alliance), and Labex Arcane/CBH-EUR-GS [ANR-17-EURE-0003] is acknowledged. Innogly COST Action covered open access charge. O.M thanks for partial financial support from the government assignment for FRC Kazan Scientific Center of Russian Academy of Science.

Conflicts of Interest: The authors declare no conflict of interest.

References

1. Iozzo, R.V.; Schaefer, L. Proteoglycan form and function: A comprehensive nomenclature of proteoglycans. *Matrix Biol.* **2015**, *42*, 11–55. [CrossRef] [PubMed]
2. Karamanos, N.K.; Piperigkou, Z.; Theocharis, A.D.; Watanabe, H.; Franchi, M.; Baud, S.; Brezillon, S.; Gotte, M.; Passi, A.; Vigetti, D.; et al. Proteoglycan chemical diversity drives multifunctional cell regulation and therapeutics. *Chem. Rev.* **2018**, *118*, 9152–9232. [CrossRef] [PubMed]
3. Mikami, T.; Kitagawa, H. Biosynthesis and function of chondroitin sulfate. *Biochim. Biophys. Acta* **2013**, *1830*, 4719–4733. [CrossRef] [PubMed]
4. Gallagher, J. Fell-Muir Lecture: Heparan sulphate and the art of cell regulation: A polymer chain conducts the protein orchestra. *Int. J. Exp. Pathol.* **2015**, *96*, 203–231. [CrossRef] [PubMed]
5. Li, J.P.; Kusche-Gullberg, M. Heparan sulfate: Biosynthesis, structure, and function. *Int. Rev. Cell. Mol. Biol.* **2016**, *325*, 215–273. [CrossRef] [PubMed]
6. Garantziotis, S.; Savani, R.C. Hyaluronan biology: A complex balancing act of structure, function, location and context. *Matrix Biol.* **2019**, *78–79*, 1–10. [CrossRef] [PubMed]
7. Caterson, B.; Melrose, J. Keratan sulfate, a complex glycosaminoglycan with unique functional capability. *Glycobiology* **2018**, *28*, 182–206. [CrossRef]
8. Pomin, V.H. Keratan sulfate: An up-to-date review. *Int. J. Biol. Macromol.* **2015**, *72*, 282–289. [CrossRef] [PubMed]
9. Clerc, O.; Mariethoz, J.; Rivet, A.; Lisacek, F.; Perez, S.; Ricard-Blum, S. A pipeline to translate glycosaminoglycan sequences into 3D models. Application to the exploration of glycosaminoglycan conformational space. *Glycobiology* **2019**, *29*, 36–44. [CrossRef] [PubMed]
10. Vallet, S.D.; Clerc, O.; Ricard-Blum, S. Glycosaminoglycan-protein Interactions: The first draft of the glycosaminoglycan interactome. *J. Histochem. Cytochem.* **2020**, in press. [CrossRef] [PubMed]

11. Peysselon, F.; Ricard-Blum, S. Heparin-protein interactions: From affinity and kinetics to biological roles. Application to an interaction network regulating angiogenesis. *Matrix Biol.* **2014**, *35*, 73–81. [CrossRef] [PubMed]
12. Ricard-Blum, S. Glycosaminoglycans: Major biological players. *Glycoconj. J.* **2017**, *34*, 275–276. [CrossRef] [PubMed]
13. Ori, A.; Wilkinson, M.C.; Fernig, D.G. A systems biology approach for the investigation of the heparin/heparan sulfate interactome. *J. Biol. Chem.* **2011**, *286*, 19892–19904. [CrossRef] [PubMed]
14. Ricard-Blum, S.; Lisacek, F. Glycosaminoglycanomics: Where we are. *Glycoconj. J.* **2017**, *34*, 339–349. [CrossRef] [PubMed]
15. Aquino, R.S.; Park, P.W. Glycosaminoglycans and infection. *Front. Biosci.* **2016**, *21*, 1260–1277. [CrossRef]
16. Burns, J.M.; Lewis, G.K.; DeVico, A.L. Soluble complexes of regulated upon activation, normal T cells expressed and secreted (RANTES) and glycosaminoglycans suppress HIV-1 infection but do not induce Ca(2+) signaling. *Proc. Natl. Acad. Sci. USA* **1999**, *96*, 14499–14504. [CrossRef] [PubMed]
17. Fatoux-Ardore, M.; Peysselon, F.; Weiss, A.; Bastien, P.; Pratlong, F.; Ricard-Blum, S. Large-scale investigation of Leishmania interaction networks with host extracellular matrix by surface plasmon resonance imaging. *Infect. Immun.* **2014**, *82*, 594–606. [CrossRef]
18. Hsiao, F.S.; Sutandy, F.R.; Syu, G.D.; Chen, Y.W.; Lin, J.M.; Chen, C.S. Systematic protein interactome analysis of glycosaminoglycans revealed YcbS as a novel bacterial virulence factor. *Sci. Rep.* **2016**, *6*, 28425. [CrossRef]
19. Jinno, A.; Park, P.W. Role of glycosaminoglycans in infectious disease. *Methods Mol. Biol.* **2015**, *1229*, 567–585. [CrossRef]
20. Merida-de-Barros, D.A.; Chaves, S.P.; Belmiro, C.L.R.; Wanderley, J.L.M. Leishmaniasis and glycosaminoglycans: A future therapeutic strategy? *Parasit. Vect.* **2018**, *11*, 536. [CrossRef]
21. Casu, B.; Naggi, A.; Torri, G. Re-visiting the structure of heparin. *Carbohydr. Res.* **2015**, *403*, 60–68. [CrossRef] [PubMed]
22. Varki, A.; Cummings, R.D.; Aebi, M.; Packer, N.H.; Seeberger, P.H.; Esko, J.D.; Stanley, P.; Hart, G.; Darvill, A.; Kinoshita, T.; et al. Symbol nomenclature for graphical representations of glycans. *Glycobiology* **2015**, *25*, 1323–1324. [CrossRef] [PubMed]
23. Silva, J.C.; Carvalho, M.S.; Han, X.; Xia, K.; Mikael, P.E.; Cabral, J.M.S.; Ferreira, F.C.; Linhardt, R.J. Compositional and structural analysis of glycosaminoglycans in cell-derived extracellular matrices. *Glycoconj. J.* **2019**, *36*, 141–154. [CrossRef] [PubMed]
24. Volpi, N.; Linhardt, R.J. High-performance liquid chromatography-mass spectrometry for mapping and sequencing glycosaminoglycan-derived oligosaccharides. *Nat. Protoc.* **2010**, *5*, 993–1004. [CrossRef] [PubMed]
25. Wu, J.; Wei, J.; Chopra, P.; Boons, G.J.; Lin, C.; Zaia, J. Sequencing heparan sulfate using HILIC LC-NETD-MS/MS. *Anal. Chem.* **2019**, *91*, 11738–11746. [CrossRef] [PubMed]
26. Yu, Y.; Duan, J.; Leach, F.E., III; Toida, T.; Higashi, K.; Zhang, H.; Zhang, F.; Amster, I.J.; Linhardt, R.J. Sequencing the Dermatan Sulfate Chain of Decorin. *J. Am. Chem. Soc.* **2017**, *139*, 16986–16995. [CrossRef]
27. Zaia, J. Glycosaminoglycan glycomics using mass spectrometry. *Mol. Cell Proteom.* **2013**, *12*, 885–892. [CrossRef]
28. Langeslay, D.J.; Beni, S.; Larive, C.K. Detection of the 1H and 15N NMR resonances of sulfamate groups in aqueous solution: A new tool for heparin and heparan sulfate characterization. *Anal. Chem.* **2011**, *83*, 8006–8010. [CrossRef]
29. Pomin, V.H. (1)H and (15)N NMR analyses on heparin, heparan sulfates and related monosaccharides concerning the chemical exchange regime of the N-sulfo-glucosamine sulfamate proton. *Pharmaceuticals* **2016**, *9*, 58. [CrossRef]
30. Khan, S.; Fung, K.W.; Rodriguez, E.; Patel, R.; Gor, J.; Mulloy, B.; Perkins, S.J. The solution structure of heparan sulfate differs from that of heparin: Implications for function. *J. Biol. Chem.* **2013**, *288*, 27737–27751. [CrossRef]
31. Jasnin, M. Use of neutrons reveals the dynamics of cell surface glycosaminoglycans. *Methods Mol. Biol.* **2012**, *836*, 161–169. [CrossRef] [PubMed]
32. Rubinson, K.A.; Chen, Y.; Cress, B.F.; Zhang, F.; Linhardt, R.J. Heparin's solution structure determined by small-angle neutron scattering. *Biopolymers* **2016**, *105*, 905–913. [CrossRef] [PubMed]

33. Scherbinina, S.I.; Toukach, P.V. Three-dimensional structures of carbohydrates and where to find them. *Intern. J. Mol. Sci.* **2020**, *21*, 7702. [CrossRef] [PubMed]
34. Samsonov, S.A.; Pisabarro, M.T. Computational analysis of interactions in structurally available protein-glycosaminoglycan complexes. *Glycobiology* **2016**, *26*, 850–861. [CrossRef] [PubMed]
35. Sankaranarayanan, N.V.; Nagarajan, B.; Desai, U.R. So you think computational approaches to understanding glycosaminoglycan-protein interactions are too dry and too rigid? Think again! *Curr. Opin. Struct. Biol.* **2018**, *50*, 91–100. [CrossRef] [PubMed]
36. Sattelle, B.M.; Almond, A. Microsecond kinetics in model single- and double-stranded amylose polymers. *Phys. Chem. Chem. Phys.* **2014**, *16*, 8119–8126. [CrossRef] [PubMed]
37. Sattelle, B.M.; Shakeri, J.; Cliff, M.J.; Almond, A. Proteoglycans and their heterogeneous glycosaminoglycans at the atomic scale. *Biomacromolecules* **2015**, *16*, 951–961. [CrossRef]
38. Almond, A. Multiscale modeling of glycosaminoglycan structure and dynamics: Current methods and challenges. *Curr. Opin. Struct. Biol.* **2018**, *50*, 58–64. [CrossRef]
39. Kolesnikov, A.L.; Budkov, Y.A.; Nogovitsyn, E.A. Coarse-grained model of glycosaminoglycans in aqueous salt solutions. A field-theoretical approach. *J. Phys. Chem. B* **2014**, *118*, 13037–13049. [CrossRef]
40. Samsonov, S.A.; Bichmann, L.; Pisabarro, M.T. Coarse-grained model of glycosaminoglycans. *J. Chem. Inf. Model.* **2015**, *55*, 114–124. [CrossRef]
41. Whitmore, E.K.; Martin, D.; Guvench, O. Constructing 3-dimensional atomic-resolution models of nonsulfated glycosaminoglycans with arbitrary lengths using conformations from molecular dynamics. *Intern. J. Mol. Sci.* **2020**, *21*. [CrossRef] [PubMed]
42. Clerc, O.; Deniaud, M.; Vallet, S.D.; Naba, A.; Rivet, A.; Perez, S.; Thierry-Mieg, N.; Ricard-Blum, S. MatrixDB: Integration of new data with a focus on glycosaminoglycan interactions. *Nucleic Acids Res.* **2019**, *47*, D376–D381. [CrossRef] [PubMed]
43. Berman, H.M.; Westbrook, J.; Feng, Z.; Gilliland, G.; Bhat, T.N.; Weissig, H.; Shindyalov, I.N.; Bourne, P.E. The Protein Data Bank. *Nucleic Acids Res.* **2000**, *28*, 235–242. [CrossRef] [PubMed]
44. Bagdonas, H.; Ungar, D.; Agirre, J. Leveraging glycomics data in glycoprotein 3D structure validation with Privateer. *Beilstein J. Org. Chem.* **2020**, *16*, 2523–2533. [CrossRef] [PubMed]
45. Lutteke, T.; von der Lieth, C.W. pdb-care (PDB carbohydrate residue check): A program to support annotation of complex carbohydrate structures in PDB files. *BMC Bioinform.* **2004**, *5*, 69. [CrossRef] [PubMed]
46. Sehnal, D.; Svobodova Varekova, R.; Pravda, L.; Ionescu, C.M.; Geidl, S.; Horsky, V.; Jaiswal, D.; Wimmerova, M.; Koca, J. ValidatorDB: Database of up-to-date validation results for ligands and non-standard residues from the Protein Data Bank. *Nucleic Acids Res.* **2015**, *43*, D369–D375. [CrossRef] [PubMed]
47. Perez, S.; Sarkar, A.; Rivet, A.; Breton, C.; Imberty, A. Glyco3D: A portal for structural glycosciences. *Methods Mol. Biol.* **2015**, *1273*, 241–258. [CrossRef]
48. Bonnardel, F.; Mariethoz, J.; Salentin, S.; Robin, X.; Schroeder, M.; Perez, S.; Lisacek, F.; Imberty, A. UniLectin3D, a database of carbohydrate binding proteins with curated information on 3D structures and interacting ligands. *Nucleic Acids Res.* **2019**, *47*, D1236–D1244. [CrossRef]
49. UniProt, C. UniProt: A worldwide hub of protein knowledge. *Nucleic Acids Res.* **2019**, *47*, D506–D515. [CrossRef]
50. Salentin, S.; Schreiber, S.; Haupt, V.J.; Adasme, M.F.; Schroeder, M. PLIP: Fully automated protein-ligand interaction profiler. *Nucleic Acids Res.* **2015**, *43*, W443–W447. [CrossRef]
51. Mir, S.; Alhroub, Y.; Anyango, S.; Armstrong, D.R.; Berrisford, J.M.; Clark, A.R.; Conroy, M.J.; Dana, J.M.; Deshpande, M.; Gupta, D.; et al. PDBe: Towards reusable data delivery infrastructure at protein data bank in Europe. *Nucleic Acids Res.* **2018**, *46*, D486–D492. [CrossRef] [PubMed]
52. Gutmanas, A.; Alhroub, Y.; Battle, G.M.; Berrisford, J.M.; Bochet, E.; Conroy, M.J.; Dana, J.M.; Fernandez Montecelo, M.A.; van Ginkel, G.; Gore, S.P.; et al. PDBe: Protein Data Bank in Europe. *Nucleic Acids Res.* **2014**, *42*, 285–291. [CrossRef] [PubMed]
53. PDBe-KB-Consortium. PDBe-KB: A community-driven resource for structural and functional annotations. *Nucleic Acids Res.* **2020**, *48*, D344–D353. [CrossRef] [PubMed]
54. Pardo-Vargas, A.; Delbianco, M.; Seeberger, P.H. Automated glycan assembly as an enabling technology. *Curr. Opin. Chem. Biol.* **2018**, *46*, 48–55. [CrossRef] [PubMed]

55. Pomin, V.H.; Wang, X. Synthetic oligosaccharide libraries and microarray technology: A powerful combination for the success of current glycosaminoglycan interactomics. *ChemMedChem* **2018**, *13*, 648–661. [CrossRef] [PubMed]
56. McNaught, A.D. Nomenclature of carbohydrates (recommendations 1996). *Adv. Carbohydr. Chem. Biochem.* **1997**, *52*, 43–177. [CrossRef] [PubMed]
57. Herget, S.; Ranzinger, R.; Maass, K.; Lieth, C.W. GlycoCT-a unifying sequence format for carbohydrates. *Carbohydr. Res.* **2008**, *343*, 2162–2171. [CrossRef]
58. Lutteke, T.; Bohne-Lang, A.; Loss, A.; Goetz, T.; Frank, M.; von der Lieth, C.W. GLYCOSCIENCES.de: An Internet portal to support glycomics and glycobiology research. *Glycobiology* **2006**, *16*, 71R–81R. [CrossRef]
59. Tanaka, K.; Aoki-Kinoshita, K.F.; Kotera, M.; Sawaki, H.; Tsuchiya, S.; Fujita, N.; Shikanai, T.; Kato, M.; Kawano, S.; Yamada, I.; et al. WURCS: The Web3 unique representation of carbohydrate structures. *J. Chem. Inf. Model.* **2014**, *54*, 1558–1566. [CrossRef]
60. Tiemeyer, M.; Aoki, K.; Paulson, J.; Cummings, R.D.; York, W.S.; Karlsson, N.G.; Lisacek, F.; Packer, N.H.; Campbell, M.P.; Aoki, N.P.; et al. GlyTouCan: An accessible glycan structure repository. *Glycobiology* **2017**, *27*, 915–919. [CrossRef]
61. Sehnal, D.; Deshpande, M.; Varekova, R.S.; Mir, S.; Berka, K.; Midlik, A.; Pravda, L.; Velankar, S.; Koca, J. LiteMol suite: Interactive web-based visualization of large-scale macromolecular structure data. *Nat. Methods* **2017**, *14*, 1121–1122. [CrossRef] [PubMed]
62. Rose, A.S.; Bradley, A.R.; Valasatava, Y.; Duarte, J.M.; Prlic, A.; Rose, P.W. NGL viewer: Web-based molecular graphics for large complexes. *Bioinformatics* **2018**, *34*, 3755–3758. [CrossRef] [PubMed]
63. Kinjo, A.R.; Bekker, G.J.; Wako, H.; Endo, S.; Tsuchiya, Y.; Sato, H.; Nishi, H.; Kinoshita, K.; Suzuki, H.; Kawabata, T.; et al. New tools and functions in data-out activities at Protein Data Bank Japan (PDBj). *Protein Sci.* **2018**, *27*, 95–102. [CrossRef] [PubMed]
64. Laskowski, R.A.; Jablonska, J.; Pravda, L.; Varekova, R.S.; Thornton, J.M. PDBsum: Structural summaries of PDB entries. *Protein Sci.* **2018**, *27*, 129–134. [CrossRef] [PubMed]
65. Waterhouse, A.; Bertoni, M.; Bienert, S.; Studer, G.; Tauriello, G.; Gumienny, R.; Heer, F.T.; de Beer, T.A.P.; Rempfer, C.; Bordoli, L.; et al. SWISS-MODEL: Homology modelling of protein structures and complexes. *Nucleic Acids Res.* **2018**, *46*, W296–W303. [CrossRef] [PubMed]
66. Sillitoe, I.; Dawson, N.; Lewis, T.E.; Das, S.; Lees, J.G.; Ashford, P.; Tolulope, A.; Scholes, H.M.; Senatorov, I.; Bujan, A.; et al. CATH: Expanding the horizons of structure-based functional annotations for genome sequences. *Nucleic Acids Res.* **2019**, *47*, D280–D284. [CrossRef] [PubMed]
67. Krissinel, E. Crystal contacts as nature's docking solutions. *J. Comput. Chem.* **2010**, *31*, 133–143. [CrossRef]
68. Krissinel, E.; Henrick, K. Inference of macromolecular assemblies from crystalline state. *J. Mol. Biol.* **2007**, *372*, 774–797. [CrossRef]
69. Dimitropoulos, D.; Ionides, J.; Henrick, K. Using MSDchem to search the PDB ligand dictionary. *Curr. Protoc. Bioinform.* **2006**. [CrossRef]
70. Frank, M.; Lutteke, T.; von der Lieth, C.W. GlycoMapsDB: A database of the accessible conformational space of glycosidic linkages. *Nucleic Acids Res.* **2007**, *35*, 287–290. [CrossRef]
71. Smith, P.J.C.; Arnott, S. LALS, a linked-atom least-squares reciprocal-space refinement system incorporating stereochemical restraints to supplement sparse diffraction data. *Acta Crystallogr. Sect. A Found. Crystallogr.* **1978**, *34*, 3–11. [CrossRef]
72. Guss, J.M.; Hukins, D.W.; Smith, P.J.; Winter, W.T.; Arnott, S. Hyaluronic acid: Molecular conformations and interactions in two sodium salts. *J. Mol. Biol.* **1975**, *95*, 359–384. [CrossRef]
73. Mitra, A.K.; Arnott, S.; Sheehan, J.K. Hyaluronic acid: Molecular conformation and interactions in the tetragonal form of the potassium salt containing extended chains. *J. Mol. Biol.* **1983**, *169*, 813–827. [CrossRef]
74. Winter, W.T.; Arnott, S. Hyaluronic acid: The role of divalent cations in conformation and packing. *J. Mol. Biol.* **1977**, *117*, 761–784. [CrossRef]
75. Cael, J.J.; Winter, W.T.; Arnott, S. Calcium chondroitin 4-sulfate: Molecular conformation and organization of polysaccharide chains in a proteoglycan. *J. Mol. Biol.* **1978**, *125*, 21–42. [CrossRef]
76. Millane, R.P.; Mitra, A.K.; Arnott, S. Chondroitin 4-sulfate: Comparison of the structures of the potassium and sodium salts. *J. Mol. Biol.* **1983**, *169*, 903–920. [CrossRef]

77. Winter, W.T.; Arnott, S.; Isaac, D.H.; Atkins, E.D. Chondroitin 4-sulfate: The structure of a sulfated glycosaminoglycan. *J. Mol. Biol.* **1978**, *125*, 1–19. [CrossRef]
78. Mitra, A.K.; Arnott, S.; Atkins, E.D.; Isaac, D.H. Dermatan sulfate: Molecular conformations and interactions in the condensed state. *J. Mol. Biol.* **1983**, *169*, 873–901. [CrossRef]
79. Lee, S.C.; Guan, H.H.; Wang, C.H.; Huang, W.N.; Tjong, S.C.; Chen, C.J.; Wu, W.G. Structural basis of citrate-dependent and heparan sulfate-mediated cell surface retention of cobra cardiotoxin A3. *J. Biol. Chem.* **2005**, *280*, 9567–9577. [CrossRef]
80. Arnott, S.; Gus, J.M.; Hukins, D.W.; Dea, I.C.; Rees, D.A. Conformation of keratan sulphate. *J. Mol. Biol.* **1974**, *88*, 175–184. [CrossRef]
81. Haxaire, K.; Braccini, I.; Milas, M.; Rinaudo, M.; Perez, S. Conformational behavior of hyaluronan in relation to its physical properties as probed by molecular modeling. *Glycobiology* **2000**, *10*, 587–594. [CrossRef] [PubMed]
82. Perez, S.; Tubiana, T.; Imberty, A.; Baaden, M. Three-dimensional representations of complex carbohydrates and polysaccharides—SweetUnityMol: A video game-based computer graphic software. *Glycobiology* **2015**, *25*, 483–491. [CrossRef] [PubMed]
83. Khan, S.; Gor, J.; Mulloy, B.; Perkins, S.J. Semi-rigid solution structures of heparin by constrained X-ray scattering modelling: New insight into heparin-protein complexes. *J. Mol. Biol.* **2010**, *395*, 504–521. [CrossRef] [PubMed]
84. Rodriguez-Carvajal, M.A.; Imberty, A.; Perez, S. Conformational behavior of chondroitin and chondroitin sulfate in relation to their physical properties as inferred by molecular modeling. *Biopolymers* **2003**, *69*, 15–28. [CrossRef] [PubMed]

Publisher's Note: MDPI stays neutral with regard to jurisdictional claims in published maps and institutional affiliations.

© 2020 by the authors. Licensee MDPI, Basel, Switzerland. This article is an open access article distributed under the terms and conditions of the Creative Commons Attribution (CC BY) license (http://creativecommons.org/licenses/by/4.0/).

Article

Glycosaminoglycans and Contrast Agents: The Role of Hyaluronic Acid as MRI Contrast Enhancer

Alfonso Maria Ponsiglione [1,2], **Maria Russo** [2,†], **and Enza Torino** [2,3,*]

1. Department of Electrical Engineering and Information Technology (DIETI), University of Naples "Federico II", Via Claudio 21, 80125 Naples, Italy; alfonsomaria.ponsiglione@unina.it
2. Department of Chemical, Materials and Production Engineering, University of Naples Federico II, Piazzale V. Tecchio 80, 80125 Naples, Italy; maria.russo@espci.psl.eu
3. Interdisciplinary Research Center on Biomaterials, CRIB, Piazzale V. Tecchio 80, 80125 Naples, Italy
* Correspondence: enza.torino@unina.it; Tel.: +39-328-955-8158
† Present address: Microfluidique, MEMS et Nanostructures, Institut Pierre-Gilles de Gennes, CNRS UMR 8231, ESPCI Paris and Paris Sciences et Lettres (PSL) Research University, 75005 Paris, France.

Received: 26 September 2020; Accepted: 26 November 2020; Published: 28 November 2020

Abstract: A comprehensive understanding of the behaviour of Glycosaminoglycans (GAGs) combined with imaging or therapeutic agents can be a key factor for the rational design of drug delivery and diagnostic systems. In this work, physical and thermodynamic phenomena arising from the complex interplay between GAGs and contrast agents for Magnetic Resonance Imaging (MRI) have been explored. Being an excellent candidate for drug delivery and diagnostic systems, Hyaluronic acid (HA) (0.1 to 0.7%w/v) has been chosen as a GAG model, and Gd-DTPA (0.01 to 0.2 mM) as a relevant MRI contrast agent. HA samples crosslinked with divinyl sulfone (DVS) have also been investigated. Water Diffusion and Isothermal Titration Calorimetry studies demonstrated that the interaction between HA and Gd-DTPA can form hydrogen bonds and coordinate water molecules, which plays a leading role in determining both the polymer conformation and the relaxometric properties of the contrast agent. This interaction can be modulated by changing the GAG/contrast agent molar ratio and by acting on the organization of the polymer network. The fine control over the combination of GAGs and imaging agents could represent an enormous advantage in formulating novel multifunctional diagnostic probes paving the way for precision nanomedicine tools.

Keywords: hyaluronic acid; glycosaminoglycan; hydrogel; MRI; hydrodenticity; precision medicine

1. Introduction

Glycosaminoglycans (GAGs) have always attracted the interest of many research groups because of their versatile properties, making them desirable resources for the design of multifunctional materials in biomedicine [1–3]. Compared to other classes of materials, like amino acid sequences, which have been coded and possess well-known properties and characteristics, GAGs represent a still unexplored group of materials, not specifically ascribable to any of the already known chemical and biophysical patterns [1]. This stimulates the investigation of their nature and behavior in biological environments, e.g., nanoscale interactions with proteins, lipids, and other GAGs, in order to fully understand and control their potential in the precision nanomedicine field as drug delivery systems and image contrast enhancers. As naturally derived biomaterials from affordable sources, GAGs represent an abundant, biodegradable, biocompatible class of materials for the synthesis of the new generation of nanomedicines, overcoming some of the toxicity- and stability-related issues of synthetic materials [4]. Physico-chemical properties of GAGs, such as monomer length, reactive groups, molecular weight, and charge, proved to be key features to design engineered nanostructures for biomedical applications [1,4]. In addition, they provide a large polymeric backbone for chemical modification where small molecules, drugs, proteins, or diagnostic agents can be easily conjugated onto the NPs' surface or, alternatively, physically encapsulated into the NPs' core or shell [5,6], thereby improving their targeting efficiency. The easy decoration of NPs, for example with polyethylene glycol (PEG), can prolong their in-vivo circulation time, which increases the possibility of accumulation of the delivered drug in the site of interest [7–9]. Additional advantages, compared to other metallic or silica NPs [10–12], lie in their high tunability thanks to the recent advancements in nanotechnology and material processing techniques, from batch synthesis to high-pressure homogenization and microfluidics. The latter, in particular, proved to be a scalable, low-cost, and high-throughput technique for controlling sizes, shapes, porosity, structure, and functional properties of polymer NPs [13–20]. Among the GAGs, hyaluronic acid (HA) (alone or coupled with other GAGs) proved to be an ideal candidate for designing nanostructured probes for drug delivery and imaging [4]. HA, also called hyaluronan, is an anionic highly hydrophilic GAG ubiquitously presents in tissues and fluids and composed of a repeating disaccharide of d-glucuronic acid and N-acetyl-d-glucosamine. It is present in the extracellular matrix and plays key roles in modulating cellular functions [4]. Moreover, HA can intrinsically target CD44 receptors, which are overexpressed in various tumor cells, thus serving as a targeting moiety for cancer therapy [21–23]. Very recent works demonstrated its potential as hydrogel nanosystem for neural tissue regeneration [5], theranostic agent in breast cancer and atherosclerosis [24–26], engineered nanostructure for multimodal imaging of B-cell lymphoma [27,28], and contrast enhancer in Magnetic Resonance Imaging (MRI) [29,30]. Concerning the use of HA as a contrast enhancer, studies [31,32] highlighted that MRI signal depends on the GAGs' concentration in human tissues, especially in articular cartilage, whose synovial fluid is made up of 98% HA [33] (ranging from 0.25 to 0.4%w/v in healthy adults [34,35]). They showed that the administration of a paramagnetic contrast agent, a metal chelate like Gadolinium diethylene triamine pentaacetic acid (Gd-DTPA), can be used to visualize relative GAG distribution in-vivo since the negative charge of the contrast agent will distribute itself within articular cartilage in a spatially inverse relationship to the concentration of the negatively charged GAG molecules [31,32]. In a more recent study [36], crosslinked HA-based hydrogels at different HA concentrations (ranging from 17%w/v to 30%w/v) have been used as model tissues to investigate the relaxation enhancement of an MRI contrast agent interacting the hydrogel structure at increasing magnetic fields. Such studies are focused on the characteristic correlation times of the metal chelate within the hydrogel but do not take into account the thermodynamic phenomena underlying the HA-contrast agent interaction, which are crucial to understand the mixing process and control the complexation of the two compounds. Furthermore, no tissue models at low HA concentrations (below 1%w/v), which correspond to the physiological range of HA concentrations in human tissues [37–39], have been yet developed nor adopted. The investigation of the relaxation enhancement mechanisms in the presence of a biopolymer network can be fundamental for the rational design of novel nanostructured MRI contrast agents with enhanced properties [40] in the field of drug delivery and precision medicine [2]. Recently, in our previous work, HA-based nanostructures have been investigated [41] and the impact of the structural properties of the hydrogel matrix on the relaxometric properties of an MRI contrast agent has been explained introducing the concept of hydrodenticity, i.e., the complex equilibrium established by the elastic stretches of polymer chains, water osmotic pressure, and hydration degree of the contrast agent, able to boost the relaxometric properties of the contrast agent itself. In other previous works [42,43], we demonstrated how the HA hydrogel structural parameters can impact the relaxivity of MRI contrast agents and then we translated the acquired know-how into a microfluidic flow focusing approach to design and produce functional Gd-loaded nanohydrogels with tunable relaxivity for MRI and multimodal imaging applications [41,44–46].

Herein, based on our previous findings, we investigate from both a physical and thermodynamic perspective the interactions between HA, chosen as a GAG model, and Gd-DTPA, as a linear ionic MRI Gd-based contrast agent, able to boost the relaxometric properties of the metal-chelate. We highlight the importance of understanding and controlling their complex interplay and show how to take advantage of their combination to develop nanosystems with precisely tailored composition. In the foreseeable future, this knowledge can contribute to the innovation of traditional drugs and imaging agents.

2. Materials and Methods

2.1. Materials

Divinyl sulfone (DVS, 118.15 Da), Diethylenetriaminepentaacetic acid gadolinium (III) dihydrogen salt hydrate (Gd-DTPA, 547.57 Da) and Sodium hydroxide pellets (NaOH) are purchased from Sigma Aldrich (Milan, Italy). Sodium Hyaluronate, with an average molecular weight of 42 kDa is supplied Bohus Biotech (Strömstad, Sweden) as dry powder and used without purification. Milli-Q water is systematically used for sample preparation, purification, and analysis.

2.2. Sample Preparation

Non-crosslinked HA samples, from 0.1 to 0.7%w/v, are prepared by dispersing polymer powder in Milli-Q water and then mechanically mixed using a magnetic stirrer (Fisher Scientific Italia, Milan, Italy), 500 rpm at Room Temperature (RT) for 2 h. Crosslinked HA samples are prepared by adding 0.2 M NaOH to the above-described solutions in order to achieve the desired pH for the crosslinking reaction and samples are mechanically stirred for 2 h (RT, 500 rpm). DVS is then added, with a DVS/HA weight ratio ranging from 2 to 11, to chemically crosslink the polymer network. The crosslinking reaction is performed at RT for 24 h in order to obtain a homogeneous gel. The biocompatibility of HA–DVS hydrogels is already confirmed in the literature [47]. Crosslinked and non-crosslinked Gd-DTPA loaded samples are prepared by adding Gd-DTPA at a concentration ranging from 0.01 and 0.2 mM (0.01—0.02—0.03—0.04—0.05—0.06—0.08—0.1—0.13—0.15—0.18—0.2 mM).

2.3. Time-Domain Relaxometry at 20 MHz and 60 MHz

Bruker Minispec (Bruker, Billerica, MA, USA) mq20 and mq60 bench-top relaxometer operating at 20 MHz (magnetic field strength: 0.47 T) and 60 MHz (magnetic field strength: 1.41 T), respectively, are used to measure longitudinal relaxation times (T1). 1 mL and 300 µL of the prepared samples are used for the measurements at 20 and 60 MHz, respectively. Samples are placed into the NMR probe for about 15 min for thermal equilibration. T1 values are determined by both saturation (SR) and inversion recovery (IR) pulse sequences. The relaxation recovery curves are fitted using a multi-exponential model. Relaxivity, r1, is calculated from the slope of the regression line of the relaxation rate, R1 = 1/T1, versus HA concentration with a least-squares method, as showed in the following Equation (1):

$$R1_{HA} = R1_{water} + r1*[HA], \qquad (1)$$

where $R1_{HA}$ is the relaxation rate of the HA sample expressed in s^{-1}, $R1_{water}$ is the relaxation rate of free water expressed in s^{-1}, and [HA] is the polymer concentration expressed in %w/v.

2.4. Measurement of Water Self-Diffusion Coefficient at 20 MHz

Diffusion measurements of water molecules are carried out on a Bruker Minispec (Bruker, Billerica, MA, USA) mq 20 bench-top relaxometer using a pulsed-field gradient spin echo (PFG-SE) sequence [48]. As previously described [42], the water self-diffusion coefficient, D, is calculated by linear regression of the echo attenuation versus the tunable parameter of the PFG-SE sequence, k, as showed in the following Equation (2):

$$k = (\gamma g \delta)^{*}(\Delta - \delta/3) \qquad (2)$$

where γ is the proton's gyromagnetic ratio (equal to 42.58 MHz T^{-1}), δ is the length of the two gradients (set equal to 0.5 ms), g is the strength of the two gradients (varied between 0.5 and 2 T m^{-1}), Δ is the delay between the two gradients (set equal to 7.5 ms).

2.5. Diffusion-Ordered NMR Spectroscopy (DOSY) at 600 MHz

As described in our previous work [44], Diffusion-ordered NMR Spectroscopy (DOSY) measurements are carried out on a Varian Agilent NMR spectrometer (Agilent Technologies, Santa Clara, CA, USA) operating at 600 MHz. Gradient strengths (Gz) are varied from 500 to 32,500 G/cm. The gradient pulse duration (δ) is kept constant to 2 ms while the diffusion delay (Δ) is increased from 7 to 1000 ms. After Fourier transformation and baseline correction, DOSY spectra are processed and analysed using Varian software VNMRJ (Agilent Technologies, Santa Clara, CA, USA) in order to obtain the values of water self-diffusion coefficient, which is then plotted as a function of Δ.

2.6. Isothermal Titration Calorimetry

Titration experiments are performed by using a Nano ITC Low Volume calorimeter from TA Instruments (New Castle, DE, USA) in accordance with our previously adopted protocol [44]. The sample cell (700 µL) and the syringe (50 µL) are filled with aqueous solutions of HA (from 0.1 to 0.7%w/v) and Gd-DTPA (1.5 mM) respectively. Injection volumes and intervals are fixed at 1 µL and 500 s, respectively. Measurements are performed at 25 °C with a stirring rate of 200 rpm. Analysis and modeling of the raw data is carried out using the NanoAnalyze (TA instruments, New Castle, DE, USA). The function adopted to analyze the ITC data is the sum of two models: independent sites model plus a constant used for the blank (i.e., Gd-DTPA in water). The first point is excluded from the analysis. Statistics of the thermodynamic parameters are calculated on 1000 trials with a confidence level equal to 95%.

3. Results and Discussion

Longitudinal relaxation times, T1, measured both at 20 MHz and 60 MHz of HA solutions at increasing DVS/HA mass ratio, is reported in Table 1.

Table 1. Longitudinal relaxation time (T1) at different DVS/HA mass ratio.

HA (%w/v)	DVS/HA (g/g)	T1 (ms) [1] Mean ± std	T1 (ms) [2] Mean ± std
0.25	0	3410 ± 30	3740 ± 60
0.25	2.35	3270 ± 20	3560 ± 70
0.25	4.70	2980 ± 20	3530 ± 50
0.25	7.06	3260 ± 30	3410 ± 50
0.25	9.42	3110 ± 20	3520 ± 50
0.25	11.77	3170 ± 20	3540 ± 50

[1] measured at 20 MHz. [2] measured at 60 MHz.

The experiments show a measurable decrease in T1 at increasing DVS/HA mass ratio, which is more evident at low frequency (20 MHz) when T1 is lower. Indeed, in the range of magnetic field between 0.3 T and 3 T, which is the range of preclinical and clinical MRI applications, T1 increases with the field strength. Therefore, if we read Table 1 horizontally, we will notice the appreciable increase in T1 due to the increase in the magnetic field. This phenomenon occurs because the Larmor frequency scales with field strength and, with increasing Larmor frequencies, the fraction of protons able to interact at the higher Larmor frequency decreases, resulting in longer T1 values.

Then, water self-diffusion coefficient was measured through a Stejskal-Tanner plot (Figure 1a) for each sample reported in Table 1. Diffusion values, D, are also plotted against the DVS/HA mass ratio (HA fixed at 0.25%w/v) in Figure 1b, where the diffusion is measured 8 h and 24 h after the addition of the crosslinker (see also Table S1 of the Supplementary Material).

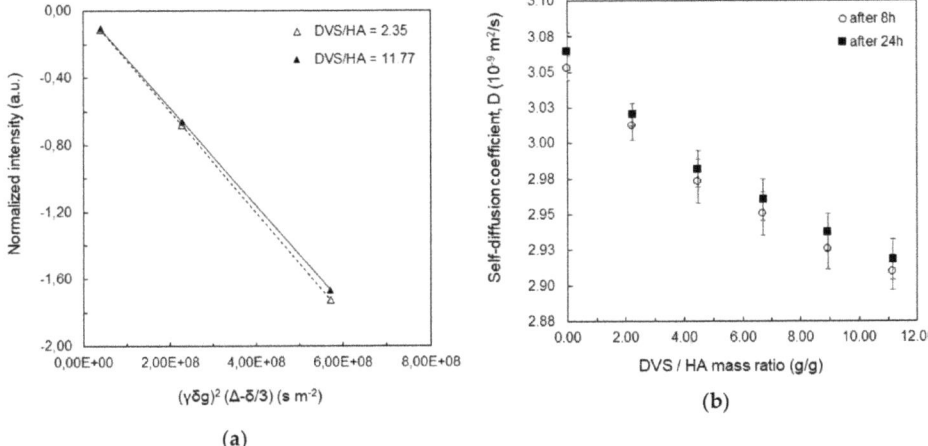

Figure 1. Diffusion measurements at 20 MHz: (**a**) Stejskal-Tanner plot to calculate water self-diffusion coefficient of 0.25%*w/v* HA crosslinked with DVS at 2.35 (empty triangles) and 11.77 (filled triangles) DVS/HA mass ratio after 8 h from the addition of the crosslinker; (**b**) Self-diffusion coefficient as a function of DVS/HA mass ratio measured after 8 h (empty circles) and 24 h (filled squares) from the addition of the crosslinker.

In Figure 1a, water self-diffusion coefficient is determined by the slope of the straight line as already reported elsewhere [49–51]. A reduction in the water mobility with increasing DVS/HA mass ratio can be observed in Figure 1a by looking at the increasing slope of the two regression lines of the Stejskal-Tanner plot. This is far more evident in Figure 1b where an inverse relationship can be observed between the water self-diffusion coefficient and DVS/HA mass ratio, as it also results from studies of solvent molecules within polymer matrices or in confined environments [11,52]. A time-dependent effect of the crosslinking reaction on the mobility of water molecules can be also observed in Figure 1b by comparing D values at 8 h and 24 h. Indeed, at 8 h, the crosslinking reaction is completed [53,54] and the swelling process is ongoing with polymer chains slowly hydrating and relaxing, thus the rate of water diffusion in the polymer networks is still slow while the hydrogel matrix is hydrating [55], binding, and entrapping water molecules. On the other hand, at 24 h, the swelling process is in the later stage, all the sulfonyl-bis-ethyl bridges between the hydroxyl groups of the HA are formed, polymer chains are well-relaxed and the swelling equilibrium is almost reached, as observed in previous studies on swelling time of crosslinked HA [55], i.e., a balance between bound water and bulk water is achieved, thus the contribution from the free diffusing water molecules is higher and the average self-diffusion coefficient assumes slightly higher values than those measured after 8 h [49]. However, this difference is not significantly appreciable especially with growing DVS/HA, since the higher crosslinking density limits polymer chains movement, thus lowering the water uptake and shortening the time to reach the swelling equilibrium [56,57].

T1 changes with water self-diffusion coefficient are evaluated afterwards at 20 MHz and 60 MHz, as shown in Figure 2a,b respectively.

Both Figure 2a,b show how T1 increases with increasing water self-diffusion coefficient. This is due to the higher mobility of the water slowing down the time taken by protons to re-align to the external magnetic field after the stimulation with controlled radiofrequency pulses of the SR sequence. Higher T1 values in Figure 2b compared to Figure 1a are due to the increase of T1 with the applied magnetic field, as from previous considerations about Table 1. It is worth noting how Figures 1 and 2 show the opportunity to obtain a relaxation enhancement by simply increasing the crosslinking degree of the sample, which is responsible for the reduction in the water mobility that thereby shortens the T1.

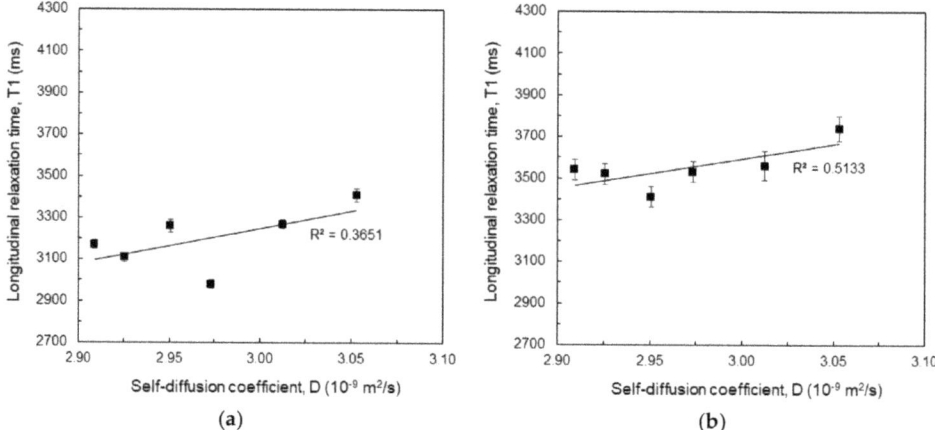

Figure 2. Longitudinal relaxation time (T1) as a function of the water self-diffusion coefficient measured for DVS/HA mass ratio solutions without Gd-DTPA at: (**a**) 20 MHz; (**b**) 60 MHz. Linear regression lines with values of the determination coefficients are displayed.

The further step of our experimental campaign consisted in measuring the relaxivity, r1, as defined from Equation (1), for crosslinked and non-crosslinked samples with addition of Gd-DTPA. Here, DVS/HA is kept equal to 8 and three different HA concentrations are tested: 0.3%w/v, 0.5%w/v, and 0.7%w/v. Results of measurements carried out with SR and IR sequences are plotted in Figure 3a,b respectively (see also Table S1 of the Supplementary Material for the measured longitudinal relaxation times). The r1 values of the samples are normalized against the longitudinal relaxivity of free Gd-DTPA in water (rGd). As a reference, relaxivity of crosslinked and non-crosslinked samples without Gd-DTPA are reported in the Table S3 of the Supplementary Material.

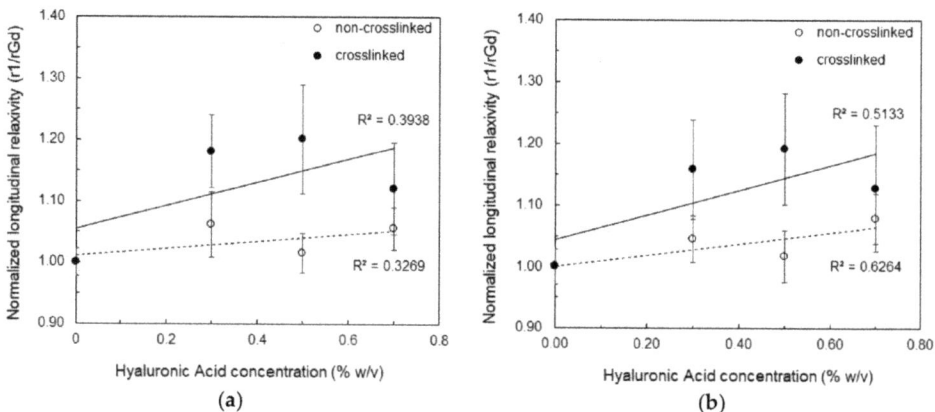

Figure 3. Longitudinal relaxivity (r1) for crosslinked (filled circles) and non-crosslinked (empty circles) samples with addition of Gd-DTPA, normalized against the relaxivity of free Gd-DTPA (rGd), as a function of the HA concentration measured at 60 MHz using: (**a**) Saturation Recovery sequence; (**b**) Inversion Recovery sequence. Linear regression lines with values of the determination coefficients are displayed.

Both SR and IR sequences confirm that r1 increases with the polymer concentration and the values are slightly higher (up to 1.2 folds) than the relaxivity of free Gd-DTPA in water. The higher accuracy of the IR sequence explains the higher determination coefficient (R^2) of the linear regression lines displayed in the both graphs. These values are two and three orders of magnitude higher than the relaxivity of the polymer without Gd-DTPA (see Tables S2 and S3 of the Supplementary Material).

This behavior is explained by the reorientation and residence times of the water molecules interacting with HA. At increasing HA concentration, indeed, the collisions of water molecules and neighboring polymer chains increase the microviscosities of the environment [58] and the percentage of the water molecules with the longest correlation time increases with respect to those with shortest, with a consequent boost in the relaxivity [36]. Hence, the water molecules reorient more slowly, and a stronger influence of the bound water molecules on the water relaxation is expected [58].

Moreover, the presence of Gd-DTPA has a significant impact on the relaxivity by further decreasing the water self-diffusion coefficient within the polymer network. This influence of Gd-DTPA is studied at increasing diffusion delays and observation times through NMR DOSY experiments, as displayed in Figure 4.

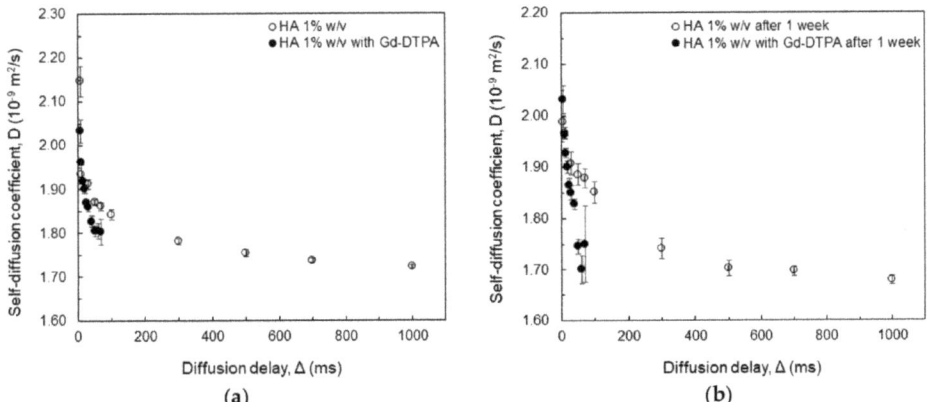

Figure 4. NMR-DOSY measurements of water self-diffusion coefficient at 600 MHz as a function of the diffusion delay in: (**a**) 1%*w/v* HA with and without Gd-DTPA at 18 mM; (**b**) measurements repeated after one week.

Both Figure 4a,b show that the co-existence of polymer and Gd-DTPA affects the water mobility more than the polymer alone, causing an additional reduction in the value of the water self-diffusion coefficient. Compared to our previous study [44], where the mobility of water molecules in presence of HA and Gd-DTPA was investigated at low contrast agent concentrations (below 30 µM), here a relatively high Gd-DTPA concentration (18 mM) is used in order to amplify the impact of the contrast agent on the water self-diffusion coefficient. However, due to the interference of Gadolinium with NMR measurements [59,60], the highest diffusion delay in the case of Gd-DTPA samples was 70 ms, since values above this threshold present a very low signal-to-noise ratio impairing the reliability of the taken measurements. Therefore, up to Δ = 70 ms, which is enough to describe the movement of water molecules within the polymer meshes in the micrometer range (a Δ range from 1 ms to 70 ms corresponds to diffusion distances from 0.5 µm to 40 µm) [61–66], the Gd-DTPA causes a further decrease in the solvent mobility. This is also confirmed after one week (Figure 4b), where a more evident drop in the water self-diffusion occurs in presence of Gd-DTPA at longer diffusion delay (50 ms < Δ < 70 ms), i.e., nearby the polymer chains.

The impact of Gd-DTPA was carried out also from a thermodynamic perspective by investigating the mixing between Gd-DTPA and HA through ITC. The modeling of the collected ITC data is showed in Figure 5 and Table 2.

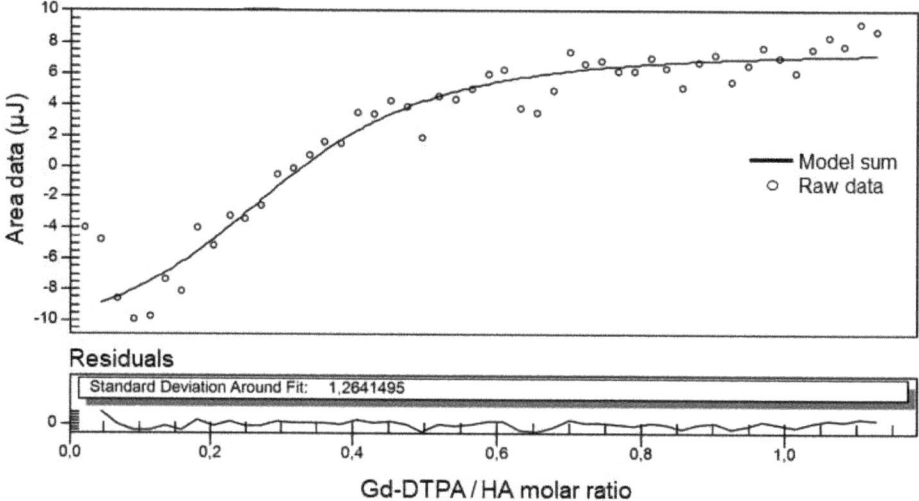

Figure 5. Fitting of ITC data for Gd-DTPA titrated into 0.4%w/v HA. ITC peak area data (empty circles) are plotted as a function of the Gd-DTPA/HA molar ratio. The model curve (solid line) is calculated as the sum of two models: independent sites model plus a constant used for the blank (i.e., Gd-DTPA in water). Residuals of the model and the standard deviation around fit are displayed in the bottom graph.

Table 2. Thermodynamic parameters from the modeling of ITC measurements conducted at 25 °C.

HA (%w/v)	Blank [1] Mean ± std [2]	n [1] Mean ± std [2]	K_d (*10^{-6} M) [1] Mean ± std [2]	ΔH (kJ/mol) [1] Mean ± std [2]	$T \cdot \Delta S$ (kJ/mol) [1]	ΔG (kJ/mol) [3]
0.1	−25.43 ± 24.36	9.99 ± 3.77	26.30 ± 722.3	24.04 ± 49.60	50.15	-26.11
0.2	5.59 ± 4.02	0.71 ± 0.37	7.25 ± 0.58	−10.12 ± 53.83	19.21	29.33
0.3	7.76 ± 1.21	0.30 ± 0.11	3.62 ± 0.12	−10.56 ± 20.03	20.49	31.05
0.4	7.37 ± 0.96	0.33 ± 0.038	3.71 ± 3.84	−11.54 ± 2.84	19.44	30.98
0.5	−2.18 ± 1.16	0.18 ± 0.040	1.60 ± 7.98	−10.06 ± 9.43	23.01	33.07
0.6	9.97 ± 22.97	0.39 ± 0.53	0.24 ± 0.013	−23.84 ± 64.48	2.48	26.31
0.7	1.56 ± 1.71	0.20 ± 0.031	1.96 ± 5.16	−11.79 ± 5.11	20.78	32.57

[1] Values obtained by fitting ITC data with the independent sites model (constant is used for the blank, i.e., Gd-DTPA in water). [2] Statistics are calculated on 1000 trials with a confidence level equal to 95%. [3] $\Delta G = \Delta H - T \cdot \Delta S$.

A representative binding isotherm for the titration of the HA with Gd-DTPA is shown for a single HA concentration (equal to 0.4%w/v). The binding curve shows a slow increase in the enthalpy of binding for the injections before 0.5 mol of Gd-DTPA per mole of HA. After this, a change in the signal is observed for the following injections with the curve reaching a constant value after saturation of the binding sites in the polymer chain, similarly to what is showed in previous works on the synthesis of metal-chelating polymers [67,68]. Similar ITC curves are obtained for the other tested concentrations (from 0.2 to 0.7%w/v) with the exception of HA = 0.1%w/v, which is showed in the Supplementary Material (Figure S1).

Since multiple binding sites are usually present on GAGs [69], thermodynamic parameters are determined by using an independent sites model, i.e., assuming multiple independent sites, and a constant used to model the blank (Gd-DTPA in water). The best fit of the ITC curve gives the following parameters: constant for the blank; reaction stoichiometry or number of binding sites (n); dissociation constant (Kd); enthalpy gain (ΔH); entropy gain (ΔS).

Table 2 includes the thermodynamic parameters (mean and standard deviation) calculated at increasing HA concentrations.

For HA concentrations above 0.1%w/v, the interaction process is exothermic ($\Delta H < 0$); the binding affinity between the HA and Gd-DTPA, expressed by $1/K_d$, is relatively weak and ranges from 0.1×10^6

to 4×10^6 M^{-1}; the reaction stoichiometry n ranges from 0.2 to 0.7 and decreases with the polymer concentration. The fitting parameters allows the calculation of the Gibbs free energy ($\Delta G = \Delta H - T \cdot \Delta S$, being T the temperature), showing the spontaneous nature of the interaction ($\Delta G > 0$), which is mainly driven by the entropy ($|\Delta H| < |T \cdot \Delta S|$).

The favorable enthalpy conditions ($\Delta H < 0$) suggest the formation of new complexes also encouraged by the conformational changes of the HA in presence of Gd-DTPA, which increases the entropy of the ternary system ($\Delta S > 0$). Despite the interaction process being both entropically and enthalpically favorable, nor the enthalpy nor the entropy gains are significantly influenced by the increase in polymer concentration. Conversely, the reaction stoichiometry shows an inverse relationship with the investigated HA concentrations. This can be attributed to the entanglement and conformational changes of the polymer, which are highly dependent on concentration. Indeed, at HA concentrations higher than 0.1%*w/v*, HA chains form a continuous three-dimensional network [37,39] with chains interacting with each other and forming stretches of double helices that makes the network more rigid and increases the fraction of water bound to HA chains and confined within the polymer matrix with respect to the free water not interacting with the polymer [70]. Therefore, the higher the polymer concentration is, the more HA-HA entangles and HA-water hydrogen bonds are formed, thus reducing the number of available sites, n, for the interaction with Gd-DTPA. As already observed in charged hydrophilic GAG [71], the interplay between intra- and inter-molecular solvent hydrogen bonding, along with the entanglement mechanism, plays a major role during interaction processes, and is also responsible for new arrangements of the polymer chains in solution [72].

However, the hydrogen bond network developed in solution around a polysaccharide depends not only on the water layers organization but also on the presence of other solute species capable of hydrogen bonds formation [72], like Gd-DTPA in this study. It is also known that a drug binding to a GAG is expected to cause a decrease in the internal degrees of freedom of the GAG, thus affecting its possible conformational changes [73]. Therefore, in such a continuous reorganization of the HA chains in water, the presence of Gd-DTPA provides an additional contribution confirmed by the large positive entropy changes, which arises from the conformational freedoms of both HA and Gd-DTPA upon mixing, as also measured in other studies on Gd complexation [74].

The entropic gain due to conformational changes predominates but is not the only phenomenon governing the process, since a smaller but still significant enthalpic contribution is measured and suggests the presence of weak interactions between HA and Gd-DTPA, ascribable to non-covalent binding, namely hydrogen bond, hydrophobic, electrostatic, and van der Waals interactions [74,75]. In accordance with Flory's mean field theory [75], such non-covalent interactions are crucial in determining the swelling equilibrium of the polymer network. Therefore, changes in the Gibbs free energy of the system can be interpreted as function of the polymer–solvent mixing, the elastic deformation of the polymer matrix, and the osmotic pressure due to the gradients of solute concentrations.

Among these interactions, the nature of the compounds in solution, both hydrophilic and negatively charged, brings our attention mainly to hydrogen bonding and electrostatic forces.

As also demonstrated elsewhere [76], hydrogen bonding phenomena is a fundamental factor determining the thermodynamics of polymers in aqueous solution and, as previously mentioned, the high hydrophilicity of HA enables the formation of inter- and intra-molecular hydrogen bonds [77]. At the beginning of the titration (Figure 5), the exchanged heat increases since the fraction of HA–water hydrogen bonds increases. Starting from a Gd-DTPA/HA ratio of 0.5, the heat decreases because other interactions take place (intra-molecular HA hydrogen bonding and HA conformational changes) giving opposite enthalpy contributions and bringing the curve to its plateau, when all HA binding sites are saturated and only water–water hydrogen bonding occur. This behavior agrees with observation reported in other studies [78], demonstrating that the enthalpy of mixing of polyelectrolyte complexes decreases at increasing salt concentration and polyelectrolyte complexation is essentially entropy driven. Furthermore, the capability of forming intra-molecular hydrogen bonds impact on the exchange and diffusion of water molecules in the inner and outer coordination sphere of the Gd-DTPA, both factors being responsible for the relaxation enhancement of the metal chelate [36,79].

As far as the electrostatic forces, it is known that the hundreds of negative charges fixed to each polymer chain are responsible for electrostatic interactions with surrounding molecules [39]. These electrostatic interactions play an important role in the mixing process, giving a large positive contribution to the entropy of the system [37,73]. In our system, we can hypothesize that intra-molecular interactions and conformational changes are driven by the presence of Gd-DTPA. Indeed, since HA and Gd-DTPA are both negatively charged, the addition of the contrast agent in solution causes electrostatic repulsion, forcing the HA chains to rearrange in order to reach a new energetically favorable configuration. This agrees with studies on the Gd-DTPA distribution in cartilage [32], showing how the negative fixed charge density of GAGs forces the contrast agent to accumulate more into areas with less GAG concentration. Moreover, as observed in other studies on polyelectrolytes interaction [80], at high enough solute concentrations, a repulsion effect can also be caused by hydration forces. These forces promote the local structuring of several layers of water molecules around the polymer due to electrostatic and hydrogen-bonding interactions. When Gd-DTPA approaches closely to the polymer, a collective disruption of these structured water layers would cost a fair amount of energy, thus producing repulsive forces.

A further non-negligible effect that needs to be taken into account is the osmotic pressure. As it occurs for solutes moving inside and outside the polymer network [75], the presence of Gd-DTPA not only induces rearrangement of the polymer in solution but also generates an osmotic pressure due to clustering of HA chains. Like other GAGs [1], indeed, HA conformational changes create regions of high anionic charge leading to high osmotic pressure, which promotes the taking up of unbound water molecule from the environment and drives the swelling of the polymer matrix [75]. The water uptake is confirmed by the large entropic contributions due to the large number of possible configurations upon swelling [58]. These changes in the osmotic pressure impact the hydration of the contrast agent and contribute to the attainment of that complex equilibrium, called hydrodenticity [41], able to boost the relaxometric properties of the Gd-DTPA, whose enhancement is promoted by the formation of the Gado-mesh, as extensively defined in previous publications [41,44].

4. Conclusions

In this work, diffusion (Figures 1, 2 and 4), thermodynamic (Figure 5 and Table 2), and relaxation properties (Table 1 and Figure 3) of HA and Gd-DTPA mixtures have been presented and discussed.

Following the previous studies on the binding between drugs and GAGs [73] and similarly to what is shown about polyelectrolyte-protein interactions [81], our results suggest that the interaction between HA and Gd-DTPA is mainly mediated by the role played by the water and determined by two factors: (i) non-covalent processes (hydrogen bonding and electrostatic forces); (ii) conformational changes of the polymer. While the former is endothermic and characterized by negative enthalpy gain, the latter is exothermic and brings a positive entropy gain. Since they occur simultaneously, the overall interaction can be described as the combination of the two above-mentioned factors with one predominating on the others during the mixing. Indeed, both HA and Gd-DTPA have the capacity of forming hydrogen bonds and coordinate water molecules, which not only produces conformational changes but also affect the relaxometric properties of the contrast agent.

In conclusion, our results show a representative picture on GAGs interaction with MRI contrast agents and contribute to build a useful framework for the interpretation of their behavior in solution and for the understanding of the fundamental phenomena underlying the MRI relaxation enhancement. Moreover, we also expect that our results can be extended to other liner Gd-based contrast agents since they present analogous chemistry and relaxation mechanisms. Further potential applications extended also to macrocyclic Gd-based contrast agents could be explored in future works. This knowledge could provide insights into the fields of nanomedicine and precision medicine, where the proper choice and combination of GAGs with imaging or therapeutic agents is the key factor for the formulation of effective targeted drug delivery systems.

Supplementary Materials: The following are available online at http://www.mdpi.com/2218-273X/10/12/1612/s1, Figure S1: Fitting of ITC data for Gd-DTPA titrated into 0.1%*w/v* HA, Table S1: Values of water self-diffusion coefficient at different DVS concentrations measured at 20 MHz after 8 h and 24 h from DVS addition, Table S2: Longitudinal relaxation times of the crosslinked and non-crosslinked samples without Gd-DTPA measured with Saturation and Inversion Recovery sequences, Table S3: Relaxivity of the crosslinked and non-crosslinked samples without Gd-DTPA measured with Saturation and Inversion Recovery sequences.

Author Contributions: Conceptualization, A.M.P. and E.T.; Methodology, A.M.P. and E.T.; Validation, A.M.P. and E.T.; Formal Analysis, A.M.P. and M.R.; Investigation, A.M.P. and M.R.; Resources, E.T.; Data Curation, A.M.P.; Writing-Original Draft Preparation, A.M.P.; Writing-Review & Editing, A.M.P., M.R. and E.T.; Visualization, A.M.P.; Supervision, E.T. All authors have read and agreed to the published version of the manuscript.

Funding: This work has been supported by the project "NANOPARTICLES BASED ON A THERANOSTIC APPROACH FOR THE DELIVERY OF A MICRORNA SET TO BE USED FOR BREAST AND THYROID CANCERS RESISTANT TO DRUGS" funded by the MIUR Progetti di Ricerca di Rilevante Interesse Nazionale (PRIN) Bando 2017—grant 2017 MHJJ55.

Conflicts of Interest: The authors declare no conflict of interest.

Acronyms

DVS	Divinyl Sulfone
GAG	Glycosaminoglycan
HA	Hyaluronic Acid
IR	Inversion Recovery
ITC	Isothermal Titration Calorimetry
MRI	Magnetic Resonance Imaging
NMR	Nuclear Magnetic Resonance
NMR-DOSY	Diffusion-ordered NMR Spectroscopy
NPs	Nanoparticles
PFG-SE	Pulsed-Field Gradient Spin Echo
RT	Room Temperature
SR	Saturation Recovery

References

1. Scott, R.A.; Panitch, A. Glycosaminoglycans in biomedicine. *Wiley Interdiscip. Rev. Nanomed. Nanobiotechnol.* **2013**, *5*, 388–398. [CrossRef] [PubMed]
2. Liao, J.; Huang, H. Review on Magnetic Natural Polymer Constructed Hydrogels as Vehicles for Drug Delivery. *Biomacromolecules* **2020**, *21*, 2574–2594. [CrossRef] [PubMed]
3. Albertsson, A.-C.; Percec, S. Future of Biomacromolecules at a Crossroads of Polymer Science and Biology. *Biomacromolecules* **2020**, *21*, 1–6. [CrossRef]
4. Swierczewska, M.; Han, H.S.; Kim, K.; Park, J.H.; Lee, S. Polysaccharide-based Nanoparticles for Theranostic Nanomedicine. *Adv. Drug Deliv. Rev.* **2016**, *99*, 70–84. [CrossRef]
5. Jian, W.-H.; Wang, H.-C.; Kuan, C.-H.; Chen, M.-H.; Wu, H.-C.; Sun, J.-S.; Wang, T.-W. Glycosaminoglycan-based hybrid hydrogel encapsulated with polyelectrolyte complex nanoparticles for endogenous stem cell regulation in central nervous system regeneration. *Biomaterials* **2018**, *174*, 17–30. [CrossRef]
6. Oommen, O.P.; Duehrkop, C.; Nilsson, B.; Hilborn, J.; Varghese, O.P. Multifunctional Hyaluronic Acid and Chondroitin Sulfate Nanoparticles: Impact of Glycosaminoglycan Presentation on Receptor Mediated Cellular Uptake and Immune Activation. *ACS Appl. Mater. Interfaces* **2016**, *8*, 20614–20624. [CrossRef]
7. Zhong, L.; Liu, Y.; Xu, L.; Li, Q.; Zhao, D.; Li, Z.; Zhang, H.; Zhang, H.; Kan, Q.; Sun, J.; et al. Exploring the relationship of hyaluronic acid molecular weight and active targeting efficiency for designing hyaluronic acid-modified nanoparticles. *Asian J. Pharm. Sci.* **2019**, *14*, 521–530. [CrossRef]
8. Huang, G.; Huang, H. Application of hyaluronic acid as carriers in drug delivery. *Drug Deliv.* **2018**, *25*, 766–772. [CrossRef]

9. Kim, K.; Choi, H.; Choi, E.S.; Park, M.-H.; Ryu, J.-H. Hyaluronic Acid-Coated Nanomedicine for Targeted Cancer Therapy. *Pharmaceutics* **2019**, *11*. [CrossRef]
10. Patil-Sen, Y.; Torino, E.; Sarno, F.D.; Ponsiglione, A.M.; Chhabria, V.N.; Ahmed, W.; Mercer, T. Biocompatible superparamagnetic core-shell nanoparticles for potential use in hyperthermia-enabled drug release and as an enhanced contrast agent. *Nanotechnology* **2020**. [CrossRef]
11. Ori, G.; Villemot, F.; Viau, L.; Vioux, A.; Coasne, B. Ionic liquid confined in silica nanopores: molecular dynamics in the isobaric–isothermal ensemble. *Mol. Phys.* **2014**, *112*, 1350–1361. [CrossRef]
12. Debroye, E.; Parac-Vogt, T.N. Towards polymetallic lanthanide complexes as dual contrast agents for magnetic resonance and optical imaging. *Chem. Soc. Rev.* **2014**, *43*, 8178–8192. [CrossRef]
13. Russo, M.; Bevilacqua, P.; Netti, P.A.; Torino, E. A Microfluidic Platform to design crosslinked Hyaluronic Acid Nanoparticles (cHANPs) for enhanced MRI. *Sci. Rep.* **2016**, *6*, 37906. [CrossRef]
14. Torino, E.; Russo, M.; Ponsiglione, A.M. Chapter 6—Lab-on-a-chip preparation routes for organic nanomaterials for drug delivery. In *Microfluidics for Pharmaceutical Applications*; Santos, H.A., Liu, D., Zhang, H., Eds.; Micro and Nano Technologies; Elsevier: Amsterdam, The Netherlands, 2019; pp. 137–153. ISBN 978-0-12-812659-2.
15. Capretto, L.; Carugo, D.; Mazzitelli, S.; Nastruzzi, C.; Zhang, X. Microfluidic and lab-on-a-chip preparation routes for organic nanoparticles and vesicular systems for nanomedicine applications. *Adv. Drug Deliv. Rev.* **2013**, *65*, 1496–1532. [CrossRef]
16. Capretto, L.; Cheng, W.; Carugo, D.; Katsamenis, O.L.; Hill, M.; Zhang, X. Mechanism of co-nanoprecipitation of organic actives and block copolymers in a microfluidic environment. *Nanotechnology* **2012**, *23*, 375602. [CrossRef]
17. Bally, F.; Garg, D.K.; Serra, C.A.; Hoarau, Y.; Anton, N.; Brochon, C.; Parida, D.; Vandamme, T.; Hadziioannou, G. Improved size-tunable preparation of polymeric nanoparticles by microfluidic nanoprecipitation. *Polymer* **2012**, *53*, 5045–5051. [CrossRef]
18. Bazban-Shotorbani, S.; Dashtimoghadam, E.; Karkhaneh, A.; Hasani-Sadrabadi, M.M.; Jacob, K.I. Microfluidic Directed Synthesis of Alginate Nanogels with Tunable Pore Size for Efficient Protein Delivery. *Langmuir* **2016**, *32*, 4996–5003. [CrossRef]
19. Dashtimoghadam, E.; Mirzadeh, H.; Taromi, F.A.; Nyström, B. Microfluidic self-assembly of polymeric nanoparticles with tunable compactness for controlled drug delivery. *Polymer* **2013**, *54*, 4972–4979. [CrossRef]
20. Maimouni, I.; Cejas, C.M.; Cossy, J.; Tabeling, P.; Russo, M. Microfluidics Mediated Production of Foams for Biomedical Applications. *Micromachines* **2020**, *11*, 83. [CrossRef]
21. Yoon, H.Y.; Koo, H.; Choi, K.Y.; Lee, S.J.; Kim, K.; Kwon, I.C.; Leary, J.F.; Park, K.; Yuk, S.H.; Park, J.H.; et al. Tumor-targeting hyaluronic acid nanoparticles for photodynamic imaging and therapy. *Biomaterials* **2012**, *33*, 3980–3989. [CrossRef]
22. Thomas, R.G.; Moon, M.; Lee, S.; Jeong, Y.Y. Paclitaxel loaded hyaluronic acid nanoparticles for targeted cancer therapy: In vitro and in vivo analysis. *Int. J. Biol. Macromol.* **2015**, *72*, 510–518. [CrossRef] [PubMed]
23. Cai, Z.; Zhang, H.; Wei, Y.; Cong, F. Hyaluronan-Inorganic Nanohybrid Materials for Biomedical Applications. *Biomacromolecules* **2017**, *18*, 1677–1696. [CrossRef] [PubMed]
24. Cai, H.; Huang, X.; Xu, Z. Development of Novel Nano Hyaluronic Acid Carrier for Diagnosis and Therapy of Atherosclerosis. *J. Clust. Sci.* **2020**. [CrossRef]
25. De Sarno, F.; Ponsiglione, A.M.; Torino, E. Emerging use of nanoparticles in diagnosis of atherosclerosis disease: A review. *AIP Conf. Proc.* **2018**, *1990*, 020021. [CrossRef]
26. El-Dakdouki, M.H.; Zhu, D.C.; El-Boubbou, K.; Kamat, M.; Chen, J.; Li, W.; Huang, X. Development of Multifunctional Hyaluronan-Coated Nanoparticles for Imaging and Drug Delivery to Cancer Cells. *Biomacromolecules* **2012**, *13*, 1144–1151. [CrossRef]
27. Torino, E.; Auletta, L.; Vecchione, D.; Orlandella, F.M.; Salvatore, G.; Iaccino, E.; Fiorenza, D.; Grimaldi, A.M.; Sandomenico, A.; Albanese, S.; et al. Multimodal imaging for a theranostic approach in a murine model of B-cell lymphoma with engineered nanoparticles. *Nanomedicine Nanotechnol. Biol. Med.* **2018**, *14*, 483–491. [CrossRef]
28. Vecchione, D.; Aiello, M.; Cavaliere, C.; Nicolai, E.; Netti, P.A.; Torino, E. Hybrid core shell nanoparticles entrapping Gd-DTPA and 18F-FDG for simultaneous PET/MRI acquisitions. *Nanomedicine* **2017**, *12*, 2223–2231. [CrossRef]

29. Russo, M.; Bevilacqua, P.; Netti, P.A.; Torino, E. Commentary on "A Microfluidic Platform to Design Crosslinked Hyaluronic Acid Nanoparticles (cHANPs) for Enhanced MRI.". *Mol. Imaging* **2017**, *16*. [CrossRef]
30. Russo, M.; Grimaldi, A.M.; Bevilacqua, P.; Tammaro, O.; Netti, P.A.; Torino, E. PEGylated crosslinked hyaluronic acid nanoparticles designed through a microfluidic platform for nanomedicine. *Nanomedicine* **2017**, *12*, 2211–2222. [CrossRef]
31. Zheng, S.; Xia, Y. The impact of the relaxivity definition on the quantitative measurement of glycosaminoglycans in cartilage by the MRI dGEMRIC method. *Magn. Reson. Med.* **2010**, *63*, 25–32. [CrossRef]
32. Bashir, A.; Gray, M.L.; Hartke, J.; Burstein, D. Nondestructive imaging of human cartilage glycosaminoglycan concentration by MRI. *Magn. Reson. Med.* **1999**, *41*, 857–865. [CrossRef]
33. Fakhari, A.; Berkland, C. Applications and Emerging Trends of Hyaluronic Acid in Tissue Engineering, as a Dermal Filler, and in Osteoarthritis Treatment. *Acta Biomater.* **2013**, *9*, 7081–7092. [CrossRef]
34. Gupta, R.C.; Lall, R.; Srivastava, A.; Sinha, A. Hyaluronic Acid: Molecular Mechanisms and Therapeutic Trajectory. *Front. Vet. Sci.* **2019**, *6*, 192. [CrossRef] [PubMed]
35. Bergamini, G.; Presutti, L.; Molteni, G. *Injection Laryngoplasty*; Springer International Publishing: Cham, Switzerland, 2015; ISBN 978-3-319-20143-6.
36. Fragai, M.; Ravera, E.; Tedoldi, F.; Luchinat, C.; Parigi, G. Relaxivity of Gd-Based MRI Contrast Agents in Crosslinked Hyaluronic Acid as a Model for Tissues. *ChemPhysChem* **2019**, *20*, 2204–2209. [CrossRef] [PubMed]
37. Park, H.; Park, K.; Shalaby, W.S.W. *Biodegradable Hydrogels for Drug Delivery*; CRC Press (Taylor and Francis Group): Boca Raton, FL, USA, 2011; ISBN 978-1-4398-9296-1.
38. Fallacara, A.; Baldini, E.; Manfredini, S.; Vertuani, S. Hyaluronic Acid in the Third Millennium. *Polymers* **2018**, *10*. [CrossRef] [PubMed]
39. Becker, L.C.; Bergfeld, W.F.; Belsito, D.V.; Klaassen, C.D.; Marks, J.G.; Shank, R.C.; Slaga, T.J.; Snyder, P.W.; Cosmetic Ingredient Review Expert Panel; Andersen, F.A. Final report of the safety assessment of hyaluronic acid, potassium hyaluronate, and sodium hyaluronate. *Int. J. Toxicol.* **2009**, *28*, 5–67. [CrossRef] [PubMed]
40. Caro, C.; García-Martín, M.L.; Pernia Leal, M. Manganese-Based Nanogels as pH Switches for Magnetic Resonance Imaging. *Biomacromolecules* **2017**, *18*, 1617–1623. [CrossRef]
41. Russo, M.; Ponsiglione, A.M.; Forte, E.; Netti, P.A.; Torino, E. Hydrodenticity to enhance relaxivity of gadolinium-DTPA within crosslinked hyaluronic acid nanoparticles. *Nanomedicine* **2017**, *12*, 2199–2210. [CrossRef]
42. Ponsiglione, A.M.; Russo, M.; Netti, P.A.; Torino, E. Impact of biopolymer matrices on relaxometric properties of contrast agents. *Interface Focus* **2016**, *6*. [CrossRef]
43. De Sarno, F.; Ponsiglione, A.M.; Grimaldi, A.M.; Netti, P.A.; Torino, E. Effect of crosslinking agent to design nanostructured hyaluronic acid-based hydrogels with improved relaxometric properties. *Carbohydr. Polym.* **2019**, *222*, 114991. [CrossRef]
44. De Sarno, F.; Ponsiglione, A.M.; Russo, M.; Grimaldi, A.M.; Forte, E.; Netti, P.A.; Torino, E. Water-Mediated Nanostructures for Enhanced MRI: Impact of Water Dynamics on Relaxometric Properties of Gd-DTPA. *Theranostics* **2019**, *9*, 1809–1824. [CrossRef] [PubMed]
45. Vecchione, D.; Grimaldi, A.M.; Forte, E.; Bevilacqua, P.; Netti, P.A.; Torino, E. Hybrid Core-Shell (HyCoS) Nanoparticles produced by Complex Coacervation for Multimodal Applications. *Sci. Rep.* **2017**, *7*, 45121. [CrossRef] [PubMed]
46. Tammaro, O.; Costagliola di Polidoro, A.; Romano, E.; Netti, P.A.; Torino, E. A Microfluidic Platform to design Multimodal PEG—crosslinked Hyaluronic Acid Nanoparticles (PEG-cHANPs) for diagnostic applications. *Sci. Rep.* **2020**, *10*, 1–11. [CrossRef]
47. Oh, E.J.; Kang, S.-W.; Kim, B.-S.; Jiang, G.; Cho, I.H.; Hahn, S.K. Control of the molecular degradation of hyaluronic acid hydrogels for tissue augmentation. *J. Biomed. Mater. Res. A* **2008**, *86*, 685–693. [CrossRef] [PubMed]
48. Stejskal, E.O.; Tanner, J.E. Spin Diffusion Measurements: Spin Echoes in the Presence of a Time-Dependent Field Gradient. *J. Chem. Phys.* **1965**, *42*, 288–292. [CrossRef]
49. Horstmann, M.; Urbani, M.; Veeman, W.S. Self-Diffusion of Water in Block Copoly(ether–ester) Polymers: An NMR Study. *Macromolecules* **2003**, *36*, 6797–6806. [CrossRef]

50. Lucas, L.H.; Larive, C.K. Measuring ligand-protein binding using NMR diffusion experiments. *Concepts Magn. Reson. Part A* **2004**, *20A*, 24–41. [CrossRef]
51. Puibasset, J.; Porion, P.; Grosman, A.; Rolley, E. Structure and Permeability of Porous Silicon Investigated by Self-Diffusion NMR Measurements of Ethanol and Heptane. *Oil Gas Sci. Technol.—Rev. D'IFP Energ. Nouv.* **2016**, *71*, 54. [CrossRef]
52. Ori, G.; Massobrio, C.; Pradel, A.; Ribes, M.; Coasne, B. Structure and Dynamics of Ionic Liquids Confined in Amorphous Porous Chalcogenides. *Langmuir* **2015**, *31*, 6742–6751. [CrossRef]
53. Maiz-Fernández, S.; Pérez-Álvarez, L.; Ruiz-Rubio, L.; Pérez González, R.; Sáez-Martínez, V.; Ruiz Pérez, J.; Vilas-Vilela, J.L. Synthesis and Characterization of Covalently Crosslinked pH-Responsive Hyaluronic Acid Nanogels: Effect of Synthesis Parameters. *Polymers* **2019**, *11*. [CrossRef]
54. Shimojo, A.A.M.; Pires, A.M.B.; Lichy, R.; Santana, M.H.A.; Shimojo, A.A.M.; Pires, A.M.B.; Lichy, R.; Santana, M.H.A. The Performance of Crosslinking with Divinyl Sulfone as Controlled by the Interplay Between the Chemical Modification and Conformation of Hyaluronic Acid. *J. Braz. Chem. Soc.* **2015**, *26*, 506–512. [CrossRef]
55. Collins, M.N.; Birkinshaw, C. Physical properties of crosslinked hyaluronic acid hydrogels. *J. Mater. Sci. Mater. Med.* **2008**, *19*, 3335–3343. [CrossRef] [PubMed]
56. Singh, T.R.R.; Laverty, G.; Donnelly, R.; Laverty, G.; Donnelly, R. *Hydrogels: Design, Synthesis and Application in Drug Delivery and Regenerative Medicine*; CRC Press (Taylor and Francis Group): Boca Raton, FL, USA, 2018; ISBN 978-1-315-15222-6.
57. Omidian, H.; Park, K. Introduction to Hydrogels. In *Biomedical Applications of Hydrogels Handbook*; Ottenbrite, R.M., Park, K., Okano, T., Eds.; Springer: New York, NY, USA, 2010; pp. 1–6. ISBN 978-1-4419-5919-5.
58. Lüsse, S.; Arnold, K. Water Binding of PolysaccharidesNMR and ESR Studies. *Macromolecules* **1998**, *31*, 6891–6897. [CrossRef]
59. Strain, S.M.; Fesik, S.W.; Armitage, I.M. Structure and metal-binding properties of lipopolysaccharides from heptoseless mutants of Escherichia coli studied by 13C and 31P nuclear magnetic resonance. *J. Biol. Chem.* **1983**, *258*, 13466–13477. [PubMed]
60. Prudêncio, M.; Rohovec, J.; Peters, J.A.; Tocheva, E.; Boulanger, M.J.; Murphy, M.E.P.; Hupkes, H.-J.; Kosters, W.; Impagliazzo, A.; Ubbink, M. A caged lanthanide complex as a paramagnetic shift agent for protein NMR. *Chem. Weinh. Bergstr. Ger.* **2004**, *10*, 3252–3260. [CrossRef]
61. Heatley, F. 18—Dynamics of Chains in Solutions by NMR Spectroscopy. In *Comprehensive Polymer Science and Supplements*; Allen, G., Bevington, J.C., Eds.; Pergamon: Amsterdam, The Netherlands, 1989; pp. 377–396. ISBN 978-0-08-096701-1.
62. Nordlund, T.M. *Quantitative Understanding of Biosystems: An Introduction to Biophysics*; CRC Press (Taylor and Francis Group): Boca Raton, FL, USA, 2011; ISBN 978-1-4200-8973-8.
63. Chen, F.J. *Progress in Brain Mapping Research*; Nova Publishers Inc.: Hauppauge, NY, USA, 2006; ISBN 978-1-59454-580-1.
64. Knauss, R.; Schiller, J.; Fleischer, G.; Kärger, J.; Arnold, K. Self-diffusion of water in cartilage and cartilage components as studied by pulsed field gradient NMR. *Magn. Reson. Med.* **1999**, *41*, 285–292. [CrossRef]
65. Burstein, D.; Gray, M.L.; Hartman, A.L.; Gipe, R.; Foy, B.D. Diffusion of small solutes in cartilage as measured by nuclear magnetic resonance (NMR) spectroscopy and imaging. *J. Orthop. Res. Off. Publ. Orthop. Res. Soc.* **1993**, *11*, 465–478. [CrossRef]
66. Lüsse, S.; Arnold, K. The Interaction of Poly(ethylene glycol) with Water Studied by 1H and 2H NMR Relaxation Time Measurements. *Macromolecules* **1996**, *29*, 4251–4257. [CrossRef]
67. Majonis, D.; Herrera, I.; Ornatsky, O.; Schulze, M.; Lou, X.; Soleimani, M.; Nitz, M.; Winnik, M.A. Synthesis of a functional metal-chelating polymer and steps toward quantitative mass cytometry bioassays. *Anal. Chem.* **2010**, *82*, 8961–8969. [CrossRef]
68. Gouin, S.; Winnik, F.M. Quantitative assays of the amount of diethylenetriaminepentaacetic acid conjugated to water-soluble polymers using isothermal titration calorimetry and colorimetry. *Bioconjug. Chem.* **2001**, *12*, 372–377. [CrossRef]
69. Dutta, A.K.; Rösgen, J.; Rajarathnam, K. Using isothermal titration calorimetry to determine thermodynamic parameters of protein-glycosaminoglycan interactions. *Methods Mol. Biol. Clifton NJ* **2015**, *1229*, 315–324. [CrossRef]

70. Kim, S.J.; Lee, C.K.; Lee, Y.M.; Kim, I.Y.; Kim, S.I. Electrical/pH-sensitive swelling behavior of polyelectrolyte hydrogels prepared with hyaluronic acid–poly(vinyl alcohol) interpenetrating polymer networks. *React. Funct. Polym.* **2003**, *55*, 291–298. [CrossRef]
71. Termühlen, F.; Kuckling, D.; Schönhoff, M. Isothermal Titration Calorimetry to Probe the Coil-to-Globule Transition of Thermoresponsive Polymers. *J. Phys. Chem. B* **2017**, *121*, 8611–8618. [CrossRef]
72. Ivanov, D.; Neamtu, A. Molecular dynamics evaluation of hyaluronan interactions with dimethylsilanediol in aqueous solution. *Rev. Roum. Chim.* **2013**, *58*, 229–238.
73. Santos, H.A.; Manzanares, J.A.; Murtomäki, L.; Kontturi, K. Thermodynamic analysis of binding between drugs and glycosaminoglycans by isothermal titration calorimetry and fluorescence spectroscopy. *Eur. J. Pharm. Sci.* **2007**, *32*, 105–114. [CrossRef]
74. Othman, M.; Bouchemal, K.; Couvreur, P.; Gref, R. Microcalorimetric investigation on the formation of supramolecular nanoassemblies of associative polymers loaded with gadolinium chelate derivatives. *Int. J. Pharm.* **2009**, *379*, 218–225. [CrossRef] [PubMed]
75. Li, H. Multi-Effect-Coupling Thermal-Stimulus (MECtherm) Model for Temperature-Sensitive Hydrogel. In *Smart Hydrogel Modelling*; Li, H., Ed.; Springer: Berlin, Germany, 2009; pp. 219–293. ISBN 978-3-642-02368-2.
76. Ruggiero, F.; Netti, P.A.; Torino, E. Experimental Investigation and Thermodynamic Assessment of Phase Equilibria in the PLLA/Dioxane/Water Ternary System for Applications in the Biomedical Field. *Langmuir* **2015**, *31*, 13003–13010. [CrossRef] [PubMed]
77. Hargittai, I.; Hargittai, M. Molecular structure of hyaluronan: an introduction. *Struct. Chem.* **2008**, *19*, 697–717. [CrossRef]
78. Schlenoff, J.B.; Rmaile, A.H.; Bucur, C.B. Hydration Contributions to Association in Polyelectrolyte Multilayers and Complexes: Visualizing Hydrophobicity. *J. Am. Chem. Soc.* **2008**, *130*, 13589–13597. [CrossRef]
79. Boros, E.; Srinivas, R.; Kim, H.-K.; Raitsimring, A.M.; Astashkin, A.V.; Poluektov, O.G.; Niklas, J.; Horning, A.D.; Tidor, B.; Caravan, P. Intramolecular Hydrogen Bonding Restricts Gd–Aqua-Ligand Dynamics. *Angew. Chem.* **2017**, *129*, 5695–5698. [CrossRef]
80. Peitzsch, R.M.; Reed, W.F. High osmotic stress behavior of hyaluronate and heparin. *Biopolymers* **1992**, *32*, 219–238. [CrossRef] [PubMed]
81. Wang, X.; Zheng, K.; Si, Y.; Guo, X.; Xu, Y. Protein–Polyelectrolyte Interaction: Thermodynamic Analysis Based on the Titration Method †. *Polymers* **2019**, *11*, 82. [CrossRef] [PubMed]

Publisher's Note: MDPI stays neutral with regard to jurisdictional claims in published maps and institutional affiliations.

© 2020 by the authors. Licensee MDPI, Basel, Switzerland. This article is an open access article distributed under the terms and conditions of the Creative Commons Attribution (CC BY) license (http://creativecommons.org/licenses/by/4.0/).

Review

Aggrecan, the Primary Weight-Bearing Cartilage Proteoglycan, Has Context-Dependent, Cell-Directive Properties in Embryonic Development and Neurogenesis: Aggrecan Glycan Side Chain Modifications Convey Interactive Biodiversity

Anthony J Hayes [1] and James Melrose [2,3,4,*]

1 Bioimaging Research Hub, Cardiff School of Biosciences, Cardiff University, Cardiff CF10 3AX, UK; HayesAJ@cardiff.ac.uk
2 Raymond Purves Laboratory, Institute of Bone and Joint Research, Kolling Institute of Medical Research, Northern Sydney Local Health District, Royal North Shore Hospital, St. Leonards, NSW 2065, Australia
3 Graduate School of Biomedical Engineering, University of New South Wales, Sydney, NSW 2052, Australia
4 Sydney Medical School, Northern, the University of Sydney, Faculty of Medicine and Health at Royal North Shore Hospital, St. Leonards, NSW 2065, Australia
* Correspondence: james.melrose@sydney.edu.au

Received: 21 July 2020; Accepted: 23 August 2020; Published: 27 August 2020

Abstract: This review examines aggrecan's roles in developmental embryonic tissues, in tissues undergoing morphogenetic transition and in mature weight-bearing tissues. Aggrecan is a remarkably versatile and capable proteoglycan (PG) with diverse tissue context-dependent functional attributes beyond its established role as a weight-bearing PG. The aggrecan core protein provides a template which can be variably decorated with a number of glycosaminoglycan (GAG) side chains including keratan sulphate (KS), human natural killer trisaccharide (HNK-1) and chondroitin sulphate (CS). These convey unique tissue-specific functional properties in water imbibition, space-filling, matrix stabilisation or embryonic cellular regulation. Aggrecan also interacts with morphogens and growth factors directing tissue morphogenesis, remodelling and metaplasia. HNK-1 aggrecan glycoforms direct neural crest cell migration in embryonic development and is neuroprotective in perineuronal nets in the brain. The ability of the aggrecan core protein to assemble CS and KS chains at high density equips cartilage aggrecan with its well-known water-imbibing and weight-bearing properties. The importance of specific arrangements of GAG chains on aggrecan in all its forms is also a primary morphogenetic functional determinant providing aggrecan with unique tissue context dependent regulatory properties. The versatility displayed by aggrecan in biodiverse contexts is a function of its GAG side chains.

Keywords: aggrecan; tissue morphogenesis; HNK-1 trisaccharide; glycosaminoglycan; cellular regulation; extracellular matrix

1. Introduction

A vast amount has been written over the last five decades on aggrecan's structure (Figure 1) and function in weight-bearing connective tissues such as hyaline cartilage and intervertebral disc (IVD) in health and disease [1–8]. However, aggrecan also has important roles in tensional connective tissues (e.g., meniscus, tendon and ligaments) [9,10] as well as in non-cartilaginous tissues such as the heart and nervous system [11–22]. While aggrecan has important roles in tissue development and function, surprisingly little has been published on its interactive and cell-directive properties in tissue morphogenesis. This review aims to rectify this deficiency but cannot be meaningfully undertaken without first covering aggrecan's functional attributes in weight-bearing tissues that contribute to matrix stabilisation. This diversity in aggrecan's functional properties is due to modifications in its glycosaminoglycan (GAG) side chains which equip it with unique ligand interactivity in specific developmental contexts.

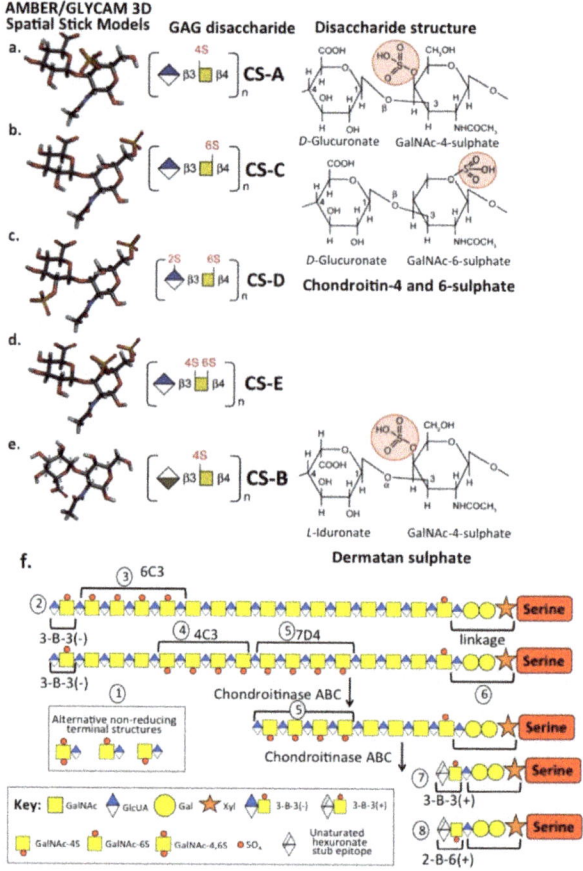

Figure 1. Amber/GLYCAM 3D stick structures of chondroitin sulphate isomers depicting their 3D conformations, disaccharide compositions and Haworth projection disaccharide structures showing sulphation positions (**a–e**). Schematic depiction of the structural organisation of the chondroitin sulphate glycosaminoglycan side chains of aggrecan depicting specific structural features of areas of the chain detected by monoclonal antibodies, putative sulphation patterns, linkage region structure to aggrecan core protein and non-reducing terminal structures (**f**). These regions on the CS side chain are numbered 1–8. Key: (1) Non-reducing terminal groups present on some cartilage aggrecan CS chains; (2) 3-B-3(−) CS sulphation motif is also present as a non-reducing terminal component on some chains; (3) putative region on CS chain identified by MAb 6C3; (4) putative region on CS chain identified by MAb 4C3; (5) putative region on CS chain identified by MAb 7D4; (6) CS linkage attachment region to Serine residues of the aggrecan core protein; (7) 3-B-3(+) CS sulphation stub epitope generated by exhaustive digestion of the CS chain by chondroitinase ABC and recognised by MAb 3-B-3; and (8) 2-B-6(+) CS sulphation stub epitope generated by exhaustive digestion of the CS chain by chondroitinase ABC and recognised by MAb 2-B-6. Note: Regions 3–5 of the CS chains containing the 6-C-3, 4-C-3 and

7-D-4 reactivity are susceptible to chondroitinase ABC digestion; thus, in graded partial digestions, the 6-C-3 and 4-C-3 reactivity can be selectively removed leaving the 7-D-4 reactive region intact. However, this is also susceptible to chondroitinase ABC, and exhaustive digestion conditions eventually lead to generation of the unsaturated 3-B-3(+) and 2-B-6(+) stub epitopes attached to the linkage region, as shown in this diagram. In (f), the structures shown hypothetical many features such as the sulphation positions on GAGs are variable; the depictions shown are thus generalisations based on literature data.

2. CS Sulphation on Aggrecan Is an Important Functional Determinant

Studies have shown that two out of every seven non-reducing termini of normal [23] and chondrosarcoma [24] aggrecan CS chains contain 4, 6-disulphated GalNAc. Non-reducing terminal GalNAc4S or GalNAc4,6S can be linked to either a 4-sulphated or a 6-sulphated disaccharide. In a further study, CS from juvenile and adolescent growth plate cartilage was shown to contain non-reducing terminal GalNAc4S, whereas in adult cartilages approximately half of the non-reducing termini were disulphated GalNAc4,6S [25], representing an increase in aggrecan sulphation with tissue maturation. It is these sulphate groups which provide aggrecan's interactive properties with a number of ligands; a high density of sulphate and carboxylate groups in aggrecan confer its remarkable ability to imbibe water and to provide tissue hydration that allows some tissues to withstand compressive loading (Figure 1a,f).

While clear functional roles for HS-PGs in cell signalling transduction pathways are well established, roles for CS-PGs in such processes have often been given lower importance; nevertheless, CS can also modulate cell-signalling pathways involving hedgehog proteins, wingless-related proteins and fibroblast growth factors [26–35]. Indeed, the co-distribution of these components with aggrecan in growth plate cartilages and localisation of particular CS sulphation motifs with chondroprogenitor cell populations associated with diarthrodial joint development (Figure 2a,f) alludes to multiple interactive possibilities [36]. Studies with brachymorphic mice, nanomelic chick, dyschondroplastic chicken and Cmd mutant mice clearly show the importance of aggrecan in growth plate cartilage development and skeletogenesis. Furthermore, individuals suffering from Kashin–Beck disease, an endemic osteochondropathy that occurs in certain parts of China, is characterised by small stature and deformities of the limbs and digits, distorted growth plates, chondrocyte apoptosis and low levels of aggrecan [37–41]. The correct sulphation of the CS chains of aggrecan is essential to generate functional determinants capable of interacting with growth factors and morphogens [42–45]. Approximately one in three of aggrecan's CS chains have a non-reducing terminal chondroitin-4, 6 disulphated residue (CS-E) in articular cartilage [46]. Highly sulphated CS-E binds the HS binding growth factors midkine and pleiotrophin [47–49].

Approximately two in seven CS chains are terminated in 4, 6 disulphated GalNAc, which varies with the age and cartilage type; four in seven of CS chains are terminated by 4-sulphated GalNAc; and one in seven CS chains are terminated in a GlcUA linked to 4-sulphated GalNAc. Non-reducing terminal 4,6-disulphated GalNAc residues are 60-fold more abundant than 4,6-disulphated GalNAc in interior regions of the CS chain [24].

CS chains terminated in 4-sulphated GalNAc predominate in aggrecan from foetal to 15-year-old knee cartilage, whereas, in 22–72-year-olds, 50% of the CS chains were terminated in 4,6-disulphated GalNAc. GlcUA-4-sulphated GalNAc disaccharides terminated 7% of CS chains in foetal to 15-year-old cartilage but fell to 3% in adults, whereas GlcUA-6-sulphated GalNAc represented 9% of the CS chains in foetal to 72-year-old cartilage. This disaccharide is recognised by MAb 3-B-3 (−) [46].

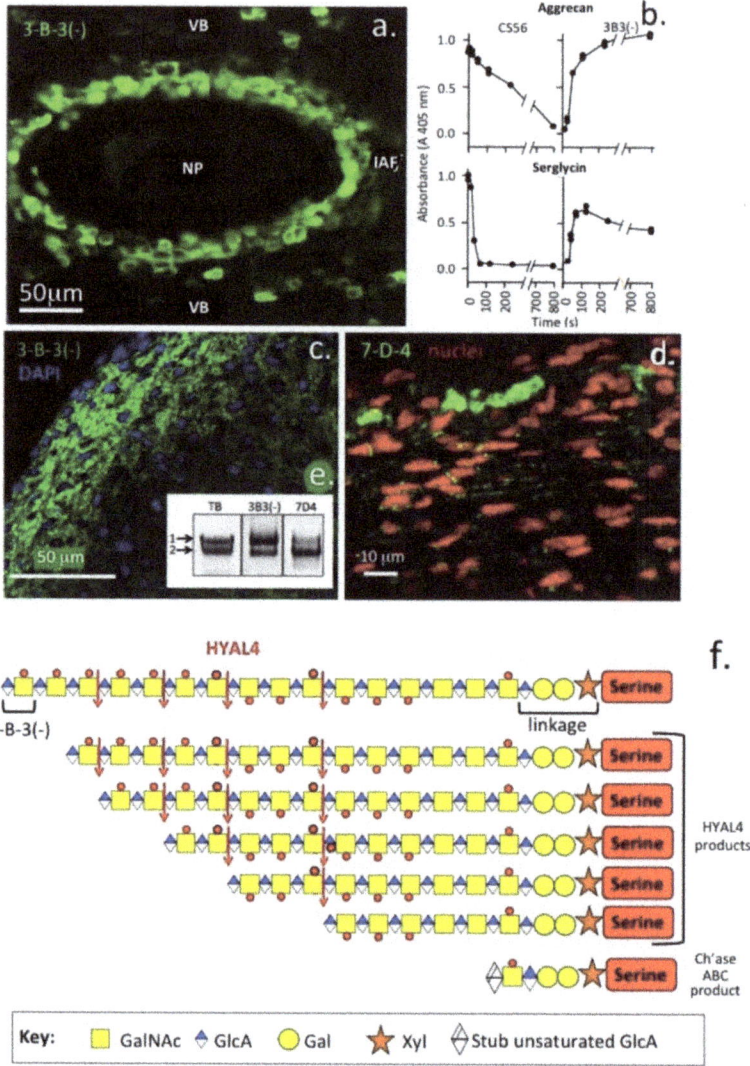

Figure 2. Immunofluorescent localisation of the 3-B-3(−) CS sulphation motifs on aggrecan associated with rudiment cartilage of a rat intervertebral disc (**a**) and demonstration of the generation of the 3-B-3(−) epitope by digestion of aggrecan and serglycin as model proteoglycans with hyaluronidase-4 (HYAL4) (**b**). Immunolocalisation of the 3-B-3(−) and 7-D-4 CS sulphation motifs in developmental human foetal knee joint cartilage (14 weeks gestational age) (**c,d**). The inset of (**e**) shows foetal aggrecan samples separated by native composite agarose polyacrylamide gel electrophoresis and blotted to nitrocellulose for detection of the 3-B-3(−) and 7-D-4 proteoglycan populations. Two aggrecan populations are discernible. The 3-B-3 (−) CS sulphation epitope has a widespread distribution in the developing rudiment cartilage, whereas the 7-D-4 epitope has a more discrete distribution pattern in small stem cell niches in the cartilage surface. A schematic depicting a typical CS chain and digestion products generated by endoglycolytic cleavage by HYAL4 generating the 3-B-3(−) non-reducing terminal on the cleaved CS chain (**f**). Exhaustive digestion of CS by chondroitinase (Ch'ase) ABC also depolymerises the CS chain but generates a 3-B-3(+) stub epitope attached to the CS linkage attachment to aggrecan core protein. Inset image (**e**) modified from [50]. (**a,c,d**) Images supplied courtesy of Prof B. Caterson, University of Cardiff, UK. As already shown in this manuscript approximately ~1–2 in every seven non-

reducing termini of CS chains in cartilage are terminated in the 3-B-3(−) epitope and these vary with age and cartilage type. The 3-B-3(−) epitope is a marker of tissue morphogenesis [36,51,52]. Stem cells are surrounded in proteoglycans decorated with this CS motif [8–10]. This motif is also released into synovial fluid in degenerative conditions such as OA [53–56]. Recently, Farrugia et al. [57] showed that mast cells synthesised HYAL4, a CS hydrolase that could generate the 3-B-3(−) motif in the CS chains of aggrecan and Serglycin in vitro.

The distribution of 4- and 6-sulphated CS epitopes is variable along a CS chain in aggrecan and is influenced by the maturational status of the cartilage or the extent to which the cartilage was sampled from a high or low weight-bearing cartilage region [3]. Certain trends have been observed in the sulphation patterns of CS in aggrecan chains. C-4-S is more predominant in aggrecan from foetal and young articular cartilage and occupies a central region in the CS chain, whereas non-sulphated chondroitin is more predominant towards the linkage region. C-6-S has a predominant distribution towards the non-reducing terminus and is more abundant in mature cartilage to the detriment of C-4-S sulphation [51].

Graded partial digestions of CS chains with chondroitinase ABC or ACII reveals regions along the CS chain where MAbs 6C3, 4C3 and 7D4 are most immunoreactive [51]. MAb 6C3 reacts optimally with regions of CS chains towards the non-reducing terminus where C-6-S predominates, and this reactivity is removed during early stages of chondroitinase digestion. Further digestion removes MAb 4C3 reactivity and continued digestion then removes reactivity to MAb 7-D-4. While the specific epitopes identified by MAb 4C3 and 7D4 are yet to be identified, reactivity of these antibodies in a range of tissues undergoing morphogenetic transition during development displays subtly different immunolocalisation patterns and are of functional significance [52,58–65].

MAb 3-B-3 identifies a non-reducing terminal disaccharide in CS consisting of GlcUA-GalNAc-6-sulphate, which is termed a 3-B-3(−) epitope to distinguish it from the 3-B-3(+) stub epitope disaccharide attached to the linkage region that is generated by exhaustive end-point digestion of CS chains by chondroitinase ABC [64]. As noted above, this non-reducing terminal 3-B-3(−) epitope occurs in approximately two in every seven CS chains; disulphated C-4,6-S and C-6-S GalNAc also occur as components in this non-reducing terminal disaccharide in CS chains [24,46].

3. HNK-1 Aggrecan Regulates Neural Crest Cell Migration during Embryonic Development

Neural crest stem cells (NCSCs) are a transient multipotent migratory embryonic neuroepithelial cell population present in vertebrate embryos [66]. The neural crest (NC) gives rise to the neural tube and notochord, neurons and glia of the peripheral nervous system/central nervous system (PNS/CNS), melanocytes, cartilaginous and bony tissues of the craniofacial skeleton as well as cephalic neuroendocrine organs and some cardiac tissues, including large vessels, valve leaflets and heart tendons. NCSCs express Sox 10 and HNK-1 and transition to a mesenchymal NC tissue during early embryonic development [67]. HNK-1 is a highly interactive functional module participating in homophilic and heterophilic interactions [68] in a number of neural PGs and cell adhesive proteins [69]. HNK-1 is also widely expressed on a number of myelin-associated glycoproteins such as L1, myelin-associated glycoprotein (MAG), TAG-1 (transient axonal glycoprotein) [70] and P0 as well as sulpho-glucuronyl glycolipids (SGGLs, SGGL-1 and SGGL-2) which have important roles to play in the remyelination of damaged axons [69]. P0 and MAG are integral transmembrane glycoprotein components of peripheral nerve myelin (Figure 3a,d). TAG-1, a GPI anchored 135-kDa glycoprotein expressed transiently on the surface of subsets of neurons in the developing mammalian nervous system, has neurite outgrowth promoting activity [70]. HNK-1 also mediates neural cell attachment to laminin in ECM structures [71,72]. During early embryonic development, HNK-1 decorates aggrecan in the notochord, and this form of aggrecan has roles in the directional control of NCSCs in the development of the neural tube, notochord, neural networks and associated tissues (Figure 3e).

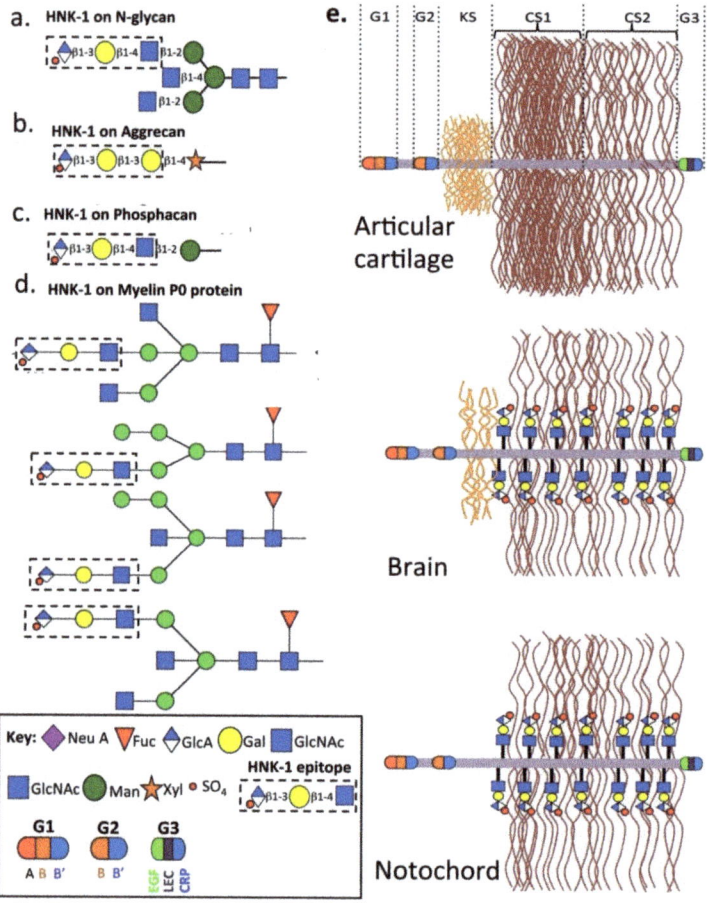

Figure 3. Structure of human natural killer-1 epitope (HNK-1) present on: N-glycans (**a**); notochordal aggrecan (**b**); brain phosphacan (**c**); and myelin Po glycoprotein in nervous tissues (**d**). Schematic depictions of representative aggrecan structures in articular cartilage, brain perineuronal nets and notochord in embryonic developmental tissues showing their variable relative KS contents and the presence of HNK-1 carbohydrate substitution in brain and notochordal aggrecan (**e**).

4. Variation in the CS Chain Fine Structure with Development and Pathology in Health and Disease

Several years ago [58,64,73], MAbs 3-B-3(−) and 7-D-4 were shown to identify chondrocyte-clusters in pathological (osteoarthritic) canine and human articular cartilage. At that time, which pre-dated knowledge of stem/progenitor cell niches in tissues, these cell-clusters were considered a classical feature of the onset of late stage degenerative joint disease and were interpreted to indicate a failed, late-stage, response to replace PGs in a matrix extensively degraded by matrix proteases. An alternative explanation of this cellular phenomenon however has now emerged. It is now believed that these "chondrocyte clusters" arise from adult stem/progenitor cell niches [74–77]. The 3-B-3(−), 4-C-3 and 7-D-4 CS sulphation motifs also occur in foetal development and are markers of anabolic processes in transitional tissues (Figure 2a,d). An important feature of the stem/progenitor cell niche is the sulphation of the PG GAG side chains (Figure 1a,f). Variable expression of GAG sulphotransferases and glycosyl transferases in stem/progenitor cell niches (Figure 2a,c,d) supports such an interpretation [78–80].

Cell clusters have also been shown to express Notch 1 and CD166, biomarkers that are synonymous with the stem cell niche [74,81].

5. Effects of Modulation of CS Sulphation on Gene Expression and Cartilage Development

In chondrocyte cultures, p-nitrophenyl xyloside (PNPX) acts as a competitive acceptor of CS/DS substitution on PG core proteins [82]. PNPX treatment reduces SOX-9, aggrecan and collagen type II gene expression, levels of collagen type II protein synthesis and PG sulphation. It also leads to delayed expression of native CS/DS sulphation motif epitopes and delayed chondrogenic differentiation of bovine MSCs accompanied with reduced tissue development. While the precise role of native CS sulphation motifs identified by MAb 3-B-3(−), 4-C-3, 7-D-4 and 6-C-3 in transitional tissues are not known, they appear to be of importance in the initial stages of chondrogenesis and their distribution patterns indicate they have roles in morphogenetic signalling through the capture and cellular presentation of soluble bioactive molecules (growth factors, morphogens, etc.) of importance in tissue development and morphogenesis [51,58–60,83] (Figure 2a,c,d).

6. Aggrecans Roles in Articular Cartilage, Fibrocartilages, Heart and Neural Tissues

Aggrecan is a large KS and CS substituted lectican PG family member with important space filling and water imbibing properties. In weight-bearing articular cartilages aggrecan forms macro-aggregate structures through interaction of its N-terminal G1 domain with hyaluronan and link protein [2,5,6]. Aggrecan–HA aggregates have important water-imbibing properties that entrap water in tissues in a dynamic manner. These properties allow the aggrecan-rich tissues to resist compression and equips articulating tissues in synovial joints with their weight-bearing properties. Cartilage is also self-lubricating through moisture expelled at the cartilage surface when the joint is loaded arising from aggrecan associated water molecules. This is a dynamic process with moisture returning to the cartilage when load is reduced or removed from the joint. Aggrecan is widely distributed in the articular hyaline cartilages of diarthrodial joints, but also occurs in the elastic and fibrocartilages of rib, nasal and tracheal cartilages, larynx, outer ear and the epiglottis [84–87]. Aggrecan is also important in foetal heart development and is a functional ECM component, which contributes to the resilience of the endocardium, myocardium, epicardium and valve leaflets of mature heart tissue [17,88]. Aggrecan is also found in the CNS and PNS in perineuronal net (PNNs) structures. These are aggrecan–HA–tenascin C aggregate structures which localise around neurons during development, and are specialised forms of neural extracellular matrix (ECM), which have neuroprotective roles and control synaptic plasticity [20,21,89]. Several studies show that, similar to notochordal aggrecan, brain aggrecan does not contain KS; however since most of these studies were conducted in mice and murine cartilage aggrecan does not contain KS, the significance of this statement needs to be carefully evaluated [90–92]. Further studies on bovine, ovine and human aggrecan have shown that, while KS is present on brain aggrecan, its content is significantly reduced compared to cartilage aggrecan [90,92–94] (Figure 3e). The hydrodynamic size of brain aggrecan is smaller both due to this absence of KS chains and replacement of CS chains with the HNK-1 trisaccharide. Embryonic chick cartilage aggrecan contains KS however notochordal aggrecan does not. HNK-1 aggrecan is also found in early embryonic cartilage rudiments but it disappears with tissue maturation.

The notochord is a full-length embryonic midline structure found in the Chordata [95]. In vertebrates, the notochord is critical for development and defines the major axis of the embryo [96]. The notochord is a source of developmental signals that regulate the patterning of tissues surrounding the notochord [97]. Hedgehog proteins (Shh, Ihh and Dhh) secreted by the notochord are central regulators of embryonic development [98] and control the patterning of tissues and proliferation of cell populations which form a wide variety of organs including the brain, heart and kidneys. Aggrecan interacts with the hedgehog morphogens and has key roles in the regulation of cellular proliferation and tissue development by embryonic NCCs in these tissues. Morphogens orchestrate the actions of progenitor cell populations through the regulation of cellular behaviours including migration, proliferation and matrix deposition into the axial embryonic tissues and in the patterning of the surrounding connective tissues.

7. Co-Ordination of Weight-Bearing and Tension-Bearing Properties in Tissues

Aggrecan equips tissues with an ability to withstand compressive loads and provides mechanical support to elastic and collagenous fibre networks within tissues. These supporting fibre networks provide mainly tensile strength within tissues and are weak in compression. The hydrodynamic space-filling properties conferred by aggrecan therefore allow these tissues to function optimally to resist tensional and shear stresses as well as providing elastic and compressive resilience. Elastic fibrillar structures control reversible tissue deformation providing elasticity to otherwise largely inextensible collagen rich tissues such as cartilage [99–102]. Historically, the major emphasis of many aggrecan studies were aimed at understanding how aggrecan conveyed functional properties to the weight-bearing articular tissues of diarthrodial joints. The importance of the high fixed charge density of the aggrecan GAG side chains became apparent as an important contributor to the osmotically driven hydration of cartilage which equipped it with the ability to withstand compressive loads [103,104]. However, a few careful studies on aggrecan GAG composition and structure during development, maturation and degeneration also provided important functional information on the GAG side chains of aggrecan. These studies established the importance of GAG sulphation as a functional determinant required not only for aggrecans role in weight bearing but also equip aggrecan with cell directive properties and an ability to interact with morphogens, growth factors and cytokines of importance in tissue development [27,61,105,106].

8. Modifications to Aggrecan Side Chain Structure Modifies Its Functional Properties in Tissues

In adult articular cartilage, aggrecan contains ~100 CS and ~25–30 KS chains, which collectively represent ~90% of the mass of this PG [4]. CS is the predominant GAG in aggrecan and is localised on the C-terminal half of the core protein in so-called CS1 and CS2 domains (Figure 3). KS is also present in a KS rich region between the N-terminal globular domains and the CS rich region. These are O-linked through Serine residues to the aggrecan core-protein and have been classified as KS-II chains [2,107]. Complete sequencing of the murine core protein [108,109] shows that it does not contain the consensus sequences for attachment of KS as found in human aggrecan core protein (E-(E,K)-P-F-P-S or E-E-P-(S,F)-P-S) [8,110,111]. Humans and bovine aggrecans contain a 4–23 hexapeptide repeat segment where KS is attached, while rats and other rodents lack this region [110,111]. Rodent aggrecan is truncated in the KS rich region thus does not contain a KS rich region such as that found in human or bovine aggrecan. Rodent aggrecan does however contain small N- and O- linked KS chains in the G1, G2 and interglobular domain (IGD); IGD KS chains have been proposed to potentiate aggrecanolysis by ADAMTS4 and ADAMTS5 [112]. The lack of a KS rich region in mouse aggrecan does not appear to be detrimental to its normal properties in mouse articular cartilage.

While much still needs to be learnt of the specific roles played by KS in aggrecan, much has already been uncovered about the interactive properties of this GAG in a number of physiological processes in the last decade. Corneal KS-I is interactive with a number of cell stimulatory molecules [113] such as insulin-like growth factor binding protein-2 (IGFBP2) [114], SHH, FGF1 and FGF2 [115–118]. A proteomics and microarray screen of 8268 proteins and secondary screen of 85 extracellular nerve growth factor epitopes using surface plasmon resonance, micro-array and microsequencing has shown that KS-I interacted with 217 proteins including 75 kinases, membrane and secreted proteins, cytoskeletal proteins, nerve regulatory proteins and nerve receptor proteins [113]. In comparison, chondroitin-4-sulphate interacted with 24 proteins including 10 kinases and 2 cell surface proteins in the same microarrays. Confirmation of these interactions by surface plasmon resonance allowed binding constants to be calculated and the validity of these putative interactions to be determined. Of 85 ECM nerve-related epitopes, KS-I bound 40 proteins, including Slit, two Robos, nine Eph receptors, eight Ephrins, eight Semaphorins and two nerve growth factor receptors. It has yet to be ascertained however if the KS-II chains of aggrecan have similar interactive properties as KS-I.

Antibodies which detect low sulphation KS motifs have now been developed (reviewed in [119]) and have demonstrated KS in a number of tissues previously thought to be KS deficient after labelling with mAbs such as 5-D-4, which is specific for highly sulphated KS epitopes [120]. Roles are emerging for low sulphation KS-epitopes in electro-sensory processes [69,115,121]. Neural tissues are the second richest source of KS in the human body after the cornea [69,115].

While aggrecan has important interactions with growth factors and morphogens which direct chondrocyte proliferation and differentiation in cartilage development and maturational processes essential in endochondral ossification and skeletogenesis, it also has important functional roles to play in weight-bearing and in the stabilisation of the cartilage ECM. Aggrecan, as its name indicates, forms massive mega Dalton aggregate ternary complexes via interaction of its N-terminal HA binding G1 domain with hyaluronan (HA) stabilised by cartilage link protein which shares homology with the G1 domain and also has HA binding properties [2,4,107]. The G3 domain of aggrecan also interacts with tenascin-C via its fibronectin type III repeats, which have lectin binding activity, and these interact with the C-type lectin motifs on the aggrecan G3 domain [19,122–124]. Tenascin-C, R, Fibulin-1 and fibulin-2 also bind to the cartilage aggrecan G3 domain through interactions with its C-type lectin and EGF domains of G3 [15]. The C-type lectin of the aggrecan G3 domain also interacts with cells and activates the Complement system [124]. Complement is a defence system against foreign pathogens and aids in the removal of dying cells, immune-complexes, misfolded proteins and invading microbes [125]. Excessive complement activation can exacerbate autoimmune disorders and pathological inflammatory conditions such as rheumatoid arthritis (RA) [126]. Complexes of matrilin-1 and -3 and biglycan or decorin also connect collagen VI microfibrils to collagen II and aggrecan [127], forming a link between the PG and fibrillar collagenous networks in cartilage and IVD [1,3,5]. Cartilage oligomeric protein (COMP and TSP-5) also binds to aggrecan, providing an extended co-operative network in cartilage [128], which helps to distribute loading stresses throughout this tissue avoiding the point loading which can be damaging to ECM components [3]. This extended collagen–aggrecan network also provides a mechanosensory biosensor system extending far from the cell through the interstitial and inter-territorial matrix, which allows the chondrocyte to perceive and respond to perturbations not only in its local mechanical microenvironment but also to more remote cartilage regions to regulate tissue homeostasis and optimal tissue functional properties [1,3].

The essential role of aggrecan to cartilage function is well illustrated in a naturally occurring Cmd (cartilage matrix deficient) mutant mouse [129], which has a single 7 bp deletion in exon 5 of the aggrecan gene which encodes the B loop of the G1 domain of aggrecan [130]. Homozygote (cmd/cmd) mice display dwarf-like features, spinal deformity, chondrodysplasia, abnormal collagen fibrillogenesis, a cleft palate [130], deafness [131] and die shortly after birth due to respiratory failure [132]. The articular cartilage of the Cmd$^-$/Cmd$^-$ mouse displays tightly packed chondrocytes surrounded by little matrix; growth plate cartilage contains chondrocytes arranged in disorganised columns of diminished length in severely diminished proliferative, pre-hypertrophic zones consistent with the reduced proportions of these mice [132] (Figure 4a). Cultured nanomelic chick chondrocytes synthesise a truncated aggrecan core protein precursor [133] due to a premature stop codon, and this is not translocated to the Golgi apparatus for processing, which leads to an absence of aggrecan in nanomelic cartilage, chondrodysplasia, disrupted organisation of the hyaline and growth plate cartilages, severely diminishing skeletal stature [134–137] (Figure 4b).

While the role of the KS chains in the G1 and G2 domains of aggrecan is largely unknown, some G1 KS chains have been found to sterically obscure an N-terminal T cell attachment site in aggrecan and have a protective effect over autoinflammatory conditions arising from fragmentation of aggrecan (Figure 5a,b). Further T cell interactive sites in the G3 domain of aggrecan have also been identified which may contribute to auto-inflammatory arthritic conditions [138–140]. These G1 KS chains suppress a T cell mediated response initiated by free G1 when it is used as an arthritogen in models of inflammatory arthritis [138–142]. KS chains in the IGD also potentiate aggrecanase activity in this region of the core protein [143]. A few KS chains are also interspersed within the CS rich region. KS-II chains in aggrecan from weight-bearing tissues such as articular cartilage and IVD contain 1-3 fucose and 2-6 *N*-acetyl-neuraminic acid residues [119]; however, these modifications in KS are absent in aggrecan from non-weight-bearing nasal and tracheal cartilage [144]. The significance of these KS modifications and why they only occur in aggrecan from weight-bearing tissues is unknown; antibody 3D12/H7 identifies these KS chains embedded in the CS rich region [145] but they do not share immunological identify with KS chains in the KS rich region. This KS epitope contains three sulphate groups and two fucose residues on GlcNAc residues in a branched fucosylated sialo-KS structure of unknown function.

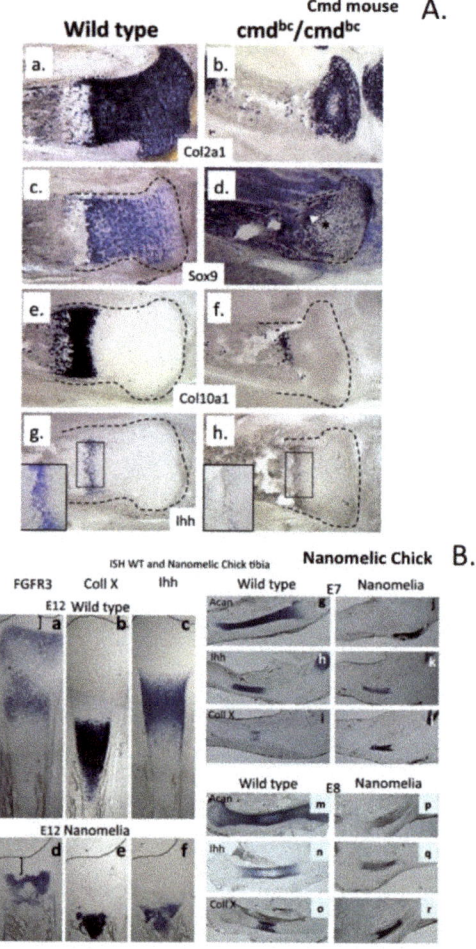

Figure 4. Demonstration of the modulation of growth plate cartilage morphogenesis by aggrecan in wild type (WT) (**A**) (**a–h**) and nanomelic E7-E12 chick tibia (**B**) (**a–r**). The ISH images presented demonstrate the expression of: FGFR3 (**a**); type X collagen (**b**); and Indian Hedgehog (IHH) (**c**) in WT (**a–c**); and nanomelic growth plate (**d–f**) in E12 (**a–f**); E7 (**g–l**); and E8 chick tibia (**m–r**). Images modified from [134] with permission using open access.

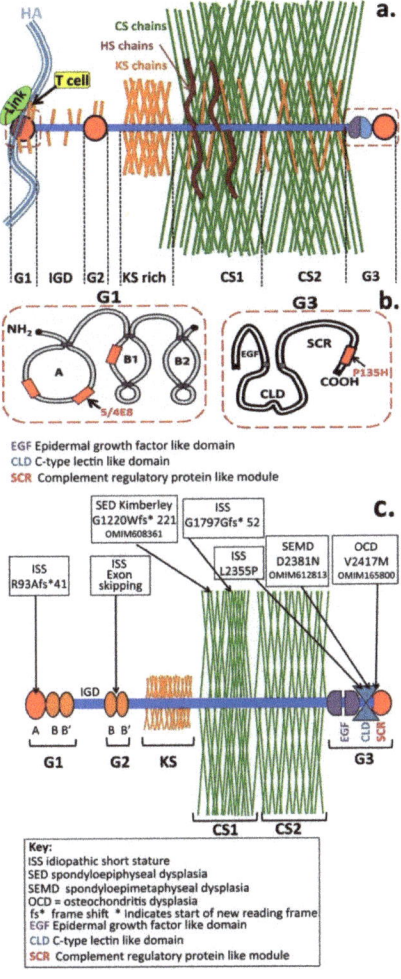

Figure 5. Structural organisation of aggrecan depicting the CS, KS and HS GAG chain distributions (**a**) and T cell receptor epitopes on the G1 and G3 globular domains (**b**). The aggrecanopathies showing regions of aggrecan affected by these mutations and the diseases that result (**c**).

Several mutations in the aggrecan gene have been documented, which affect variable regions in the aggrecan core protein leading to a number of conditions collectively termed the aggrecanopathies (Figure 5c) [146,147]. The aggrecanopathies are a spectrum of non-lethal skeletal dysplasias including spondyloepimetaphyseal (SEMD) and spondyloepiphyseal dysplasia (SED), osteochondritis dissecans (OCD) and a number of accelerated bone maturation disorders that result in short stature and idiopathic short stature (ISS) [146,148]. Skeletal abnormalities are also prominent features of animal models which display deficient levels of cartilage aggrecan such as the Cmd mouse [129,130] or nanomelic chick [137,149]. Brachymorphism [150] also results in reduced PAPS levels, and the aggrecan synthesised is deficiently sulphated and functionally impaired, resulting in abnormal skull development and short squat skeletal frames [151–153]. Manipulation of the diastrophic dysplasia sulphate-transporter gene (*DDST*) also results in the synthesis of aggrecan with deficient sulphation levels and a variety of skeletal abnormalities such as short stature and joint dysplasia in diastrophic dysplasia [151], micromelia in atelosteogenesis Type II [152] and short skeletal proportions due to aberrant trunk and extremity development in achondrogenesis Type II. Heterozygous ACAN mutations result in a phenotypic spectrum of skeletal abnormalities including short stature, accelerated bone maturation, early growth cessation, poor responsiveness to human growth hormone, brachydactyly, early-onset OA and susceptibility to the development of degenerative disc disease due to dysfunctional articular cartilage and IVD tissues [147,154–156]. Osteochondritis dissecans (OCD) is a disabling condition characterised by abnormal deposition of aggrecan in cartilage and the appearance of cracks in the cartilage and subchondral bone. This condition effects juveniles and adults but its aetiology is unknown. Trauma has been suggested as a predisposing factor in juveniles and recent genomic wide studies have identified a cluster of genes associated with this condition suggesting that it may also have a genetic basis [157–159]. Several skeletal dysplasias have been shown to be due to a constitutively activated mutation in a transient receptor potential vanilloid 4 (TRPV4) cation channel protein [160–164]. This results in abnormal cation mediated cell signalling by chondrocytes and altered regulation by BMP2 and TGFβ1 activity [164].

Aggrecan is required for correct growth plate cytoarchitecture and differentiation, endochondral ossification and skeletogenesis [165]. The CS side chains of aggrecan make important contributions to this process and their sulphation status is an important functional determinant [60]. Six CS N-acetylgalactosaminyltransferases (t1–t6) have been described. Initial stages of CS sulphation is undertaken by t1 and t2; t1 and t2 double knockout mice display shortened growth plates, distorted hypertrophic growth plate regions, reduced growth plate chondrocyte proliferation, type X expression, dwarfism, disruption in the postnatal formation of the secondary ossification centres and chondrodysplasia; this is lethal postnatally [166] (Figure 8). Aggrecan aggregates are also formed in the CNS and PNS stabilised by interaction with tenascin-C and tenascin-R, HA and a brain specific link protein variant Bral-1 (HAPLN2) to form perineuronal nets (PNNs), which are structures assembled around neurons (Figure 7) that scavenge oxygen free radicals in neural tissues thus preventing oxidative stress [19,20,93,167–169]. Brain tissue is fatty acid rich and prone to oxidative damage, which produces reactive species that detrimentally affect mitochondrial activity in neurons. Brain tissue is metabolically demanding and requires optimal mitochondrial activity to ensure energy production to power neuronal signalling.

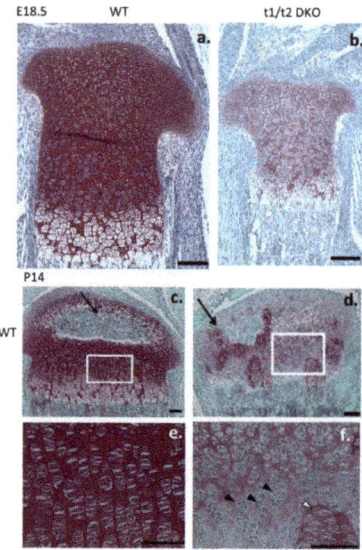

Figure 6. Chondroitin sulphate N-acetylgalactosaminyltransferase-1 and -2 (T1/2) knockout induces

dwarfism in mice and altered cartilage structural organisation of the femoral condyle, its ossification centre and growth plate in wild type mice (**a,c,e**) and T1/2 knockout mice (**b,d,f**). The boxed area in (**c,d**) is depicted at higher magnification in (**e,f**). Safarin O-Fast green stain depicting aggrecan GAG distribution. Arrows depict normal ossification centre in (**c**) and abnormal structural organisation in T1/2 knockout mice in (**d,f**). Figure modified from [166]. Figure reproduced under the terms of the Creative Commons Attribution Licence Copyright: 2017 Shimbo et al. [166].

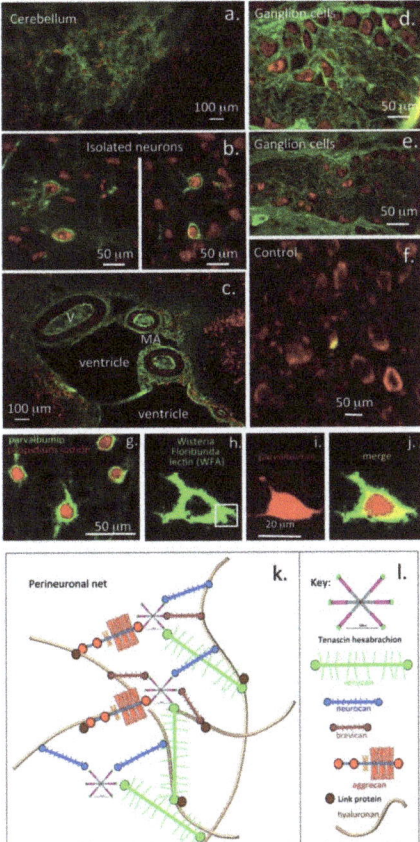

Figure 7. Visualisation of: perineuronal net structures (**a,b**); vascular features (**c**); and ganglion cells (**d,e**) in cerebellum and dorsal root ganglia using MAb 1B5 in confocal images. Immunolocalisation of CS Isomer 1B5 in paraformaldehyde fixed 20-μm cryo-sections of 24-month-old Wistar rat brain and lumbar dorsal root ganglia. Confocal z-stacked images of IB5 CS stub epitope generated by chondroitinase ABC digestion using Alexa 488 secondary antibody for detection and propidium iodide nuclear counterstain, mounted under coverslips using Vectasheld mountant. Images courtesy of Prof B. Caterson, University of Cardiff. Copyright Caterson, Hayes 2012 (**a–f**). Immunolocalisation of perineuronal nets surrounding isolated neurons in murine visual cortex using: antibody to parvalbumin (**g,i**); Wisteria floribunda lectin (**h**); and in a merged image (**j**). A schematic model of the perineuronal net structure in the boxed area in (**k**) showing its constituent lectican proteoglycans (aggrecan, versican, neurocan and brevican) interacting with tenascin hexabrachion and hyaluronan to form an aggregate structure stabilised by link protein. A key is provided to explain items in (**k,l**). Figure modified from [170] under Open Access under the auspices of a Creative Commons Attribution License.

9. Aggrecan–GAG Interactions Are of Importance in Heart Development

During early embryonic development, ectodermal NC cells migrate to form the neural tube and notochord under the direction of HNK-1 substituted aggrecan. This HNK-1 substituted form localises to the peri-notochordal space where its repulsive cell interactive properties guide the NC cells to form the notochord in a co-ordinated pattern-dependant manner (Figure 9). Equally impressive is the direction of assembly of tissue structures through HNK-1 aggrecan with NC cells migrating outwards along specific guidance tracts to form the cardiac neural crest region and the cardiac septae, outflow tracts and aortic arches [171] (Figure 9). Development of the heart valves and cardiac muscle with electroconductive properties from the endocardial cushions is also regulated by distinct spatiotemporal distributions of aggrecan and versican [16,172]. Heart tissues have remarkable mechanical properties of elasticity, compressibility, stiffness, strength and durability achieved through careful guidance of cell-mediated ECM assembly of collagen fibril and versican- and aggrecan-rich tissue regions to provide these tissues with highly specialised functional properties. Cardiac tissues are electroconductive and the charge transfer properties of GAG side chains of heart PGs may contribute to tissue properties in a similar manner to how electrosensory properties are conveyed to neural tissues. Similar developmental processes are also evident in the formation of cartilage, tendon and bone using the same ECM components but in a different manner to effect specialised tissue function [11]. The development of co-ordinated electroconductive cardiomyocyte networks with synchronised pulsatile properties is a particularly impressive achievement [173,174]. The properties of the heart valves and heart strings are equally important to heart function and these have material properties more similar to cartilage and tendon. It is not surprising therefore that aggrecan and transcription factors such as Sox 2, Sox 9 and growth factors/morphogens such as FGFs/BMPs play such prominent roles in the development of cardiac tissues [11].

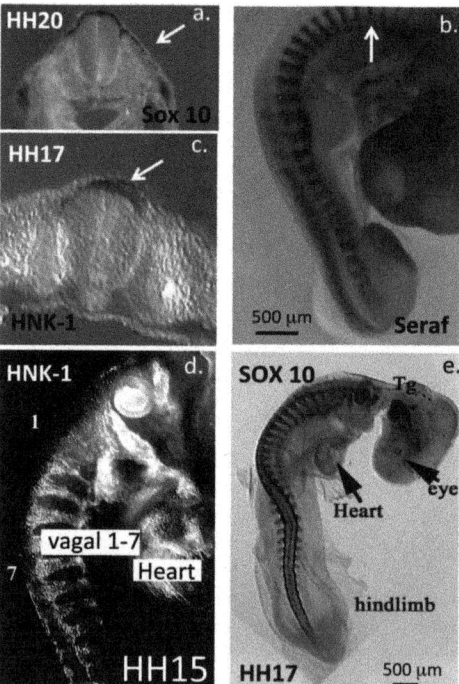

Figure 8. Demonstration of: Sox 10 (**a**,**e**); Seraf (**b**); and HNK-1 epitope (**c**,**d**) expression in migratory

neural crest cells (**a,c**) and in whole mount chick embryos (**b,d,e**). (**a,c,d**) In-situ hybridisation images. (**b,e**) Immunolocalisations with specific antibodies. Seraf (Schwann cell-specific EGF-like repeat autocrine factor) is a unique protein expressed by avian embryo Schwann cell precursor cells [175]. Images reproduced from [176] under the auspices of attribution-non-commercial-no derivatives 4.0 international licence (CC BY-NC-ND 4.0).

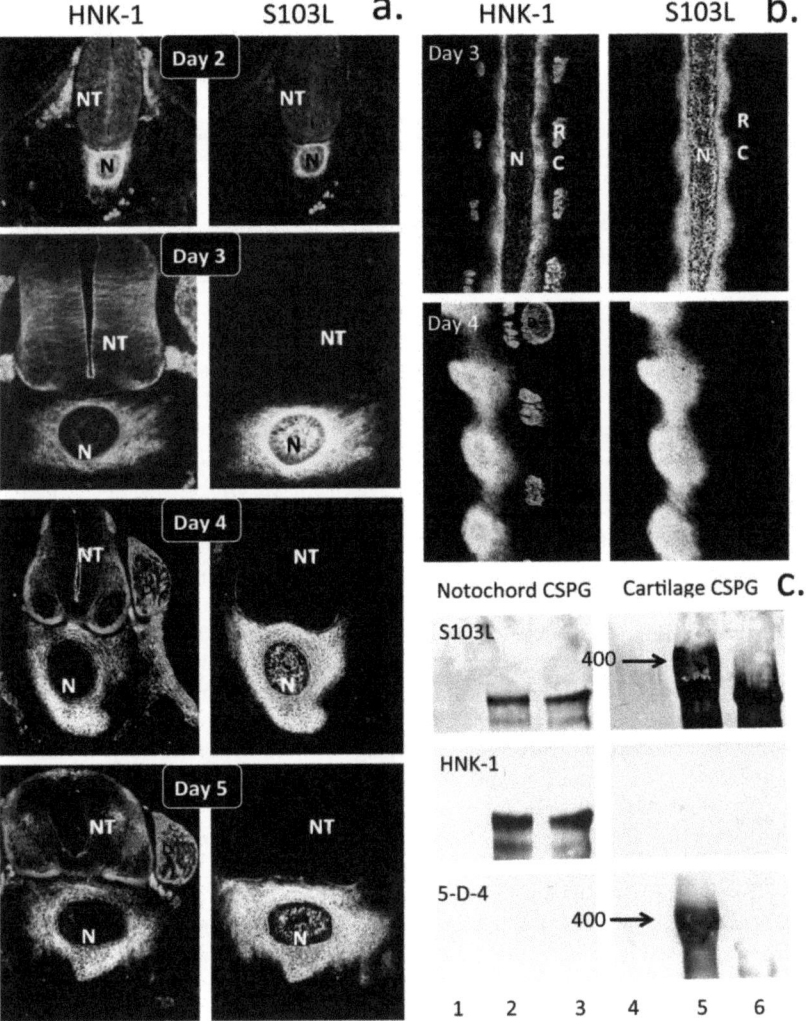

Figure 9. Fluorescent immunolocalisation of HNK-1 and aggrecan S103L epitope in 2–5-day-old chick trunk sections associated with the neural tube (NT) and notochord (N) development (**a,b**) and in Western blots (**c**) of purified chick notochordal and cartilage aggrecan. Keratan sulphate was also immunolocalised on blots using MAb 5-D-4. Notochordal aggrecan was S103L and HNK-1 positive but did not contain KS and was of a smaller molecular weight; the 400-kDa cartilage aggrecan species was not detected. Cartilage aggrecan did not stain with the HNK-1 antibody. The S103L antibody identifies the sequence ^{585}XXX Glu Ileu Ser Gly Phe Leu Ser Gly Asp Arg^{615} in the CS attachment domain of aggrecan. Images reproduced from [177–179].

The calcineurin/nuclear factor of activated T cells (NFatc1), which regulates osteoclast differentiation [180], is also required for valve formation [181,182]. Myocyte-specific enhancer factor 2C (Mef2c), a master transcription factor which regulates hypertrophy and osteogenic differentiation of chondrocytes [183], is also essential for normal cardiovascular development, and loss of function mutations in Mef2c contribute to congenital heart defects [184]. Moreover, activated BMP signalling has been shown to increase expression of cartilage and bone-type collagens, and increased expression of the osteogenic marker Runt-related transcription factor 2 (Runx2)/core-binding factor subunit alpha-1 (CBF α-1) is observed in adult aortic valve disease [185]. Thus, there is considerable overlap in cartilage synonymous transcriptional factors in the integrated development of functional heart tissues. The cell directive properties of HNK-1 aggrecan not only makes a particularly important contribution to the sculpting of embryonic cardiac tissues but it is also a functional component of these tissues.

10. Aggrecan and Cellular Regulation

The form of aggrecan present in the notochord does not contain KS but contains the HNK-1 trisaccharide recognition motif [92] (Figure 3b,e). S103L reactive aggrecan is prominent in the peri-notochordal space and its inhibitory properties on NC cells instructively guides their migration during formation of the neural tube and notochord [90,91] (Figure 9). The absence of KS on aggrecan is not without precedent. Rodent aggrecan has a truncated core protein and also does not contain a KS rich region; however, this is not detrimental to its weight-bearing properties in cartilage or the turnover of aggrecan by MMPs in these tissues. Rodent aggrecan contains small O- and N-linked KS chains in the G1, IGD and G2 domain described in human aggrecan with roles in the potentiation of ADAMTS-4 and -5 activity. Some studies have also reported the existence of populations of aggrecan devoid of KS in brain tissues based on an absence of reactivity with 5D4 anti-KS antibody, although their detailed characterisations have yet to be provided. Aggrecan expressed by embryonic glial cells in the brain is an astrocyte developmental regulator [186]. Chick aggrecan nanomelia mutants display marked increases in the expression of astrocyte differentiation genes in the absence of extracellular aggrecan indicating that aggrecan regulates astrocyte differentiation and controls glial cell maturation during brain development [186]. Heavily sulphated CS chains on aggrecan and other PGs can bind the midkine family members (midkine and pleiotrophin), and some FGF family members (FGF-1, -2, -16 and -18) provide clues as to how they influence cellular processes [48,187]. Appican in brain tissue [188], a CS-PG synthesised by astrocytes but not by neurons [189], contains embedded CS-E motifs [190] within their CS side chains which interact with heparin-binding neuroregulatory factors [187]. Expression of Appican by astrocytes induces morphological changes in C6 glioma cells and promotes adhesion of neural cells to the ECM [191].

In articular cartilage, the CS chains of aggrecan have major roles in the attraction of water into this tissue which forms the basis of its hydrodynamic viscoelastic properties as a weight-bearing tissue. However the non-reducing terminal regions of the CS chains of aggrecan from articular and growth plate cartilages also contain 4, 6-disulphated CS, and these are likely binding candidates for morphogenetic proteins which show a similar distribution to aggrecan in these tissues [192]. BMP-2, FGFR-3 and IHH co-distribute in growth plate cartilage with the pre-hypertrophic cells [193]. BMP-2, BMP-4 and dual BMP-2/4 knockout mice have severely disturbed shortened growth plate organisation due to decreased chondrocyte proliferation and increased apoptosis [194–196]. Type X collagen expression is also severely down regulated as is MMP-13 expression in BMP-2/4 KO mice.

11. Role of IHH in Chondrogenesis

Indian hedgehog (Ihh) [197], a member of the hedgehog protein family along with sonic hedgehog (Shh) [198], regulates chondrocyte differentiation, proliferation and maturation in articular cartilage development [199] and during endochondral ossification through interactions with parathyroid hormone-related peptide (PTHrP) [200] and BMP mediated cell signalling [201]. Ihh has multiple functions during skeletogenesis [202–204]. Mice lacking the Ihh gene exhibit severe skeletal abnormalities, including markedly reduced chondrocyte proliferation and abnormal maturation and an absence of mature osteoblasts, which has detrimental effects on bone development [205]. Ihh and its receptor, smoothened (smo), are expressed in chondrocytes and osteoblasts thus Ihh may have a direct effect on osteoblasts, or its effects may be mediated indirectly through chondrocytes during the process of endochondral ossification.

IHH colocalises with aggrecan in the growth plate (Figure 4A,B; plate A g; B c,f). Aggrecan regulates the expression of growth factors and signalling molecules during cartilage development and is essential for proper chondrocyte organisation, morphology and survival during formation of the axial skeleton. The sulphated GAGs of the CS and KS side chains of aggrecan provide water imbibing properties creating a large hydrophilic molecule important for the hydration of cartilage and the provision of its hydrodynamic weight-bearing properties but also bind growth factors and morphogens crucial to chondrocyte maturation and function [27,206]. Thus, aggrecan should not be considered merely as a space-filling ECM component that provides hydration and weight-bearing properties to tissues but also a cell directive tissue organiser that is capable of modulating the activity of growth factors and morphogenetic proteins, thus mediating tissue development. Indeed, aggrecan knock-out mutants display a range of severe ECM defects, which supports this proposal [134,165].

12. HNK-1 Carbohydrate Epitope as a Recognition Motif

The human natural killer-1 (HNK-1) carbohydrate motif is a unique sulphated trisaccharide of the structure SO_4-3GlcAβ1-3Galβ1-4GlcNAc, which is developmentally and spatially expressed in a cell-type specific manner within the CNS (Figure 3a–d) [207]. HNK-1 is also a well-known CNS glycoprotein epitope with essential roles to play in neural plasticity, higher brain function, synaptic plasticity, spatial learning and memory. However, it is not limited to the CNS and also displays specific localisations in other strategic locations elsewhere in the human body including in the kidney [208], the heart [209,210], retina [211] and as a component of the PNNs identified in the auditory system [212]. As a neural cell marker, HNK-1 plays crucial roles in cell migration and cellular attachment during embryonic nerve development and formation of the notochord from the neural tube (Figure 9). Carbohydrate–protein interactions between HNK-1 reactive sulpho-glucuronyl-glycolipids and PG lectin domains mediate neuronal cell adhesion and neurite outgrowth [213–215]. Some laminin isoforms also bind specifically to sulphated glycolipids [216] and are important in cell adhesion [215], particularly in nerve development. Several neural cell-adhesion molecules contain the HNK-1 epitope including neural cell adhesion molecule (NCAM), myelin associated glycoprotein (MAG), myelin basic protein (MBP), neural-glial adhesion molecule (Ng- CAM, L1), contactin, P0, Tenascin-C, Tenascin-R. Sub-populations of the enzymes *N*-acetylcholinesterase and 5'-nucleotidease also contain the NHK-1 epitope, which is important in their localisation in synaptic vesicles and membranes. Acetyl cholinesterase and 5'-nucleotidase are GPI anchored ecto-enzymes of high catalytic efficiency. Acetylcholinesterase cleaves the neurotransmitter acetylcholine in the neuromuscular junction (NMJ) and this allows muscles to return to a relaxed state following contraction. Acetylcholinesterase cleaves in excess of 5000 molecules of acetylcholine/s per molecule of enzyme. Without such an efficient enzymatic system and co-ordinated expression of neurotransmitters in the NMJ, muscles would be tensed and relaxed in an uncoordinated manner and movements would be jerky and irregular as evident in neuromuscular disorders such as Parkinson's disease or the spastic paralysis evident in Schwartz–Jampel Syndrome. It is not surprising therefore that the expression of NHK-1 is under strict spatial and temporal regulation on migrating neural crest cells, cerebellum, myelinating Schwann cells and motor neurons but not on sensory neurons. A form of HNK-1 substituted aggrecan is synthesised in the notochord and in early foetal rudiment cartilage [90,92] (Figure 9a,b). HNK-1 has been immunolocalised to the electro-receptors of the shark and electric organs of the electric eel (*Electrophorus electricus*), electric catfish (*Malapterurus electricus*) and electric ray (*Torpedo marmorata*). HNK-1 has also been mapped to electroconductive tissue during human foetal heart development and is expressed by cultured cardiomyocytes leading to the development of smart electroconductive polymers for applications in regenerative medicine [217]. HNK-1 sulphotransferase (HNK-1ST) catalyses the transfer of sulphate to position 3 of terminal glucuronic acid in protein and lipid linked oligosaccharides carried by many neural recognition molecules [218,219]. These facilitate cellular interactions during CNS development and in synaptic plasticity. HNK-1 ST acts in combination with two other glucuronyl transferases (GlcAT-P and GlcAT-S) to form a heteromeric complex in the biosynthesis of the HNK-1 trisaccharide epitope [220,221]. HNK-1 ST suppresses the glycosylation of α-dystroglycan in sub-populations of melanoma cells in a number of tissues where neither GlcAT-P nor GlcAT-S is expressed, and this reduces the ligand binding capability of α-dystroglycan establishing a tumour suppressor role for HNK-1 ST in melanoma. The HNK-1 epitopes of acetylcholinesterase and 5′-nucleotidase have roles in cell–cell and cell–matrix communication independent of their enzymatic activities.

HNK-1 (SO_4-3GlcAβ1-3Galβ1-4GlcNAc) is expressed on N-linked and O-mannose linked glycans in the nervous system. Several proteoglycans bear the HNK-1 epitope including phosphacan and aggrecan (Figure 3b,c). NHK-1 sulphotransferase can utilise the xylose-galactose-galactose-glucuronic acid linkage tetrasaccharide as acceptor to attach the 3-O sulphate group to the non-reducing glucuronic acid residue [222] but in so doing inhibits further CS chain elongation; thus, aggrecan substituted with HNK-1 has a lower density of CS chains and has a reduced hydrodynamic size compared to cartilage aggrecan [220,222]. Phosphacan occurs as a soluble PG and as a variant protein tyrosine phosphatase which contains KS and CS side chains in addition to HNK-1 carbohydrate. The HNK-1 motif in phosphacan is O-mannose linked through an Asn on the core protein. Notochordal and early rudiment cartilage cells synthesise a form of aggrecan substituted with the HNK-1 epitope, but this disappears in later stages of skeletal development (Section 9 and Figure 9).

The HNK-1 substituted Tenascin-R and -C splice variant multimeric ECM glycoproteins contain multiple FNIII and EGF repeats and a fibrinogen domain which are interactive with the C type lectin domains of the lectican CS PG family in brain [223,224]. Tenascin-R is a major component of the PNNs which surround neurons in the brain, spinal cord and in specific areas of the hippocampus [225]. Perineuronal nets consist of the lectican CS-PGs assembled into extracellular networks through interaction with HA and link proteins cross-linked by Tenascin-R and are linked to neurons through their C-terminal domains, endowing them with neuroprotective properties [226] (Figure 7a–e,k).

13. The Therapeutic Potential of Aggrecan and Its GAG Side Chain Components

13.1. Analysis of Cartilage Aggrecan and Its GAG Side Chains

As already shown in this review, aggrecan is a large specialised protein which provides weight-bearing or space-filling properties to cartilaginous tissues through its large solvation volume and ability to imbibe water. Cartilage aggrecan has a core protein of ~250 kDa and contains ~100 CS and 25–30 KS side chains, which collectively represent ~90% of its mass. As also shown in this review, aggrecan forms also exist in specialised tissue niches and in developmental contexts which do not contain KS or have some CS chains replaced by the HNK-1 trisaccharide which result in changes in aggrecan's interactive properties. Murine aggrecan contains a truncated core protein and is devoid of a KS rich region; however, this does not impede its functional properties in murine cartilage or the normal turn-over of this proteoglycan. Thus, the functional importance of KS in human aggrecan is unknown at present and the need for two GAG types in aggrecan is a question which has yet to be answered. While corneal KS-I has interactive properties with a range of neuron associated proteins [113,119] such as SHH, FGF-1 and FGF-2 [117], it is not known if the KS-II chains of aggrecan share this property. KS-II differs from KS-I in its capping modifications with L-fucose and N-acetyl neuraminic acid [119], which render KS-II resistant to total depolymerisation by keratanase-I and II and endo-β-D-galactosidase apparently through steric constraints which prevent access of these KS depolymerising enzymes to KS substrate, KS-I is totally depolymerised under the same digestion conditions thus significant differences exist between KS-I and KS-II [227]. This could also sterically impede potential interactions of KS-II with other ligands. Furthermore, cartilage aggrecan also contains a few KS chains interspersed within its CS-1 and CS-2 rich regions. An antibody to these KS chains, (MAb 3D12/H7) identifies trisulphated fucosylated and poly-*N*-acetyllactosamine modifications in the KS linkage regions in these KS chains to aggrecan core protein [145]. These 3D12/H7 positive KS chains do not share immunological identity with the KS-II chains of the KS rich region of aggrecan however their functional properties still have to be ascertained.

To understand the properties of the aggrecan side chain GAGs and how these may contribute to the properties of aggrecan, methods have been developed to isolate aggrecan and its GAG side chains from cartilaginous tissues.

13.2. Aggrecan Isolation Procedures

To isolate aggrecan from cartilage for analysis, it must be dissociated from its ternary complex formation with HA and link protein. This is achieved by using chaotropic agents such as guanidinium hydrochloride (GuHCl), which disrupts the water structure of the tissue, opens up the dense collagenous structure allowing dissociation of the aggrecan–HA–link protein complexes and release of aggrecan monomer which diffuses out of the tissue and is recovered in the extraction solution [228]. Cartilage is initially diced into small pieces to reduce the diffusive pathways out of the tissue for effective extraction; broad spectrum protease inhibitors covering all four mechanistic classes of proteases are included in the extraction solution to protect the aggrecan from proteolysis. Homogenisation procedures, which are commonly used in the extraction of proteins from other soft connective tissues, cannot be used for the isolation of aggrecan in an intact form since their high shear forces fragment the aggrecan. The cartilage extract can then be subjected to anion exchange chromatography on support matrices derivatised with anionic ligands such as diethylaminoethyl (DEAE) or sulphopropyl [229], dissociative size exclusion chromatography in 4-M GuHCl containing buffers using open pore gel chromatographic media such as Sephacryl or Sepharose CL2B [230], or density gradient equilibrium isopycnic ultracentrifugation in high concentrations of CsCl [231]. This latter procedure relies on aggrecan's high buoyant density in CsCl gradients of $\geq$1.55 g/mL to isolate aggrecan; while extracted proteins typically have buoyant densities of 1.3–1.35 g/mL, HA has a buoyant density of 1.4–1.45 g/mL. Density gradient ultracentrifugation can be conducted in the presence of 4M GuHCl to ensure aggrecan is isolated free of other interactive components also present in the cartilage extract [231,232].

13.3. Analysis of Aggrecan's GAG Side Chains

The CS chains of aggrecan are primarily of interest since these make a major contribution to aggrecans physicochemical and biological properties in tissues, whereas the function of the more minor KS chains are currently not known and thus are of lesser interest [116,233–235]. Several qualitative and quantitative methods have been developed for the discriminative measurement of intact GAG chains, including dye specific, thin layer chromatography (TLC), capillary electrophoresis [236–238], high-performance liquid chromatography (HPLC), various mass-spectrophotometric formats including liquid chromatography–tandem mass spectrometry (LC-MS/MS) [239–241], gas chromatography, enzyme linked immunosorption analysis (ELISA) using a wide array of anti-GAG antibodies, GAG microarrays and automated high-throughput mass spectrometric methods. Electrophoretic methods to separate intact GAG chains and GAG disaccharides generated by GAG depolymerising enzymes by capillary electrophoresis and conventional slab gel formats use media such as highly purified agaroses [242], polyacrylamide or mixtures of these as separation media. GAGs can be discriminated enzymatically either by eliminative cleavage with lyases (EC 4.2.2.-) or by hydrolytic cleavage with hydrolases (EC 3.2.1.-). These enzymes can be used in combination with chromatographic or electrophoretic separation methodologies to identify GAG species [243–247]. Following electrophoresis, electroblotting of the separated GAGs can be employed to nylon or nitrocellulose support membranes treated with cationic detergents. A large array of specific anti-GAG antibodies can be used for identification of the blotted GAGs. In gel, detection of separated

GAG disaccharides and oligosaccharide species can also be carried out using fluorophore assisted carbohydrate electrophoresis (FACE) [248]. Capillary electrophoresis, has high resolving power and sensitivity in the analysis of GAG composition, disaccharide sulphation patterns and sequence analysis [236–238].

14. Evaluation of the Aggrecan Content and Distribution in Pathological Cartilage Using Imaging Techniques

This review shows that aggrecan is an important functional component of articular cartilage and is depleted in OA cartilage. Several non-invasive cartilage imaging procedures have been developed that allow the assessment of the spatiotemporal distribution of aggrecan during OA disease progression. MRI of articular cartilage (AC) has been applied to assess osteoarthritic changes occurring in cartilage with the progression of OA. Traditional MRI evaluates AC morphology and measures cartilage thickness over time [249]. More advanced MRI techniques can now be used to assess AC matrix composition non-invasively to detect early articular changes. T2-mapping and T1ρ sequences estimate the relaxation times of water inside AC and have found application in clinical protocols to assess cartilage changes in OA [250]. Diffusion-weighted and diffusion tensor imaging can also be used to assess ECM changes in AC since the movement of water in cartilage is affected by ECM composition and structure. Specific imaging techniques that evaluate cartilage GAGs, such as delayed gadolinium enhanced MRI [251] or Chemical Exchange Saturation Transfer [252,253] and sodium imaging [254,255] or PET (positron emission tomography)-sodium imaging [256] have also shown utility in the non-invasive detection of AC damage [257].

15. Tissue Therapeutic Interventions Involving CS

With the recent publication of the first draft of the GAG Interactome, the GAG–protein interactive properties previously investigated [116,233] have now been extensively catalogued [235]. There are two major areas of therapeutic application involving CS: (i) use of CS as a drug to treat OA cartilage depleted of proteoglycans; and (ii) therapeutic use of CS-depolymerising enzymes to remove the CS side chains of CS-PGs that are laid down in scar tissues, which stabilise spinal cord defects, neural damage in the PNS and neural damage following brain trauma [243,258,259]. While CS-PGs are laid down in neural scar tissues to stabilise the neural defects to prevent further mechanical damage at the defect site, the CS side chains of these PGs inhibit neural outgrowth through the scar tissue, thus functional recovery of the spinal cord or other traumatised neural tissues is prevented [260]. Chondroitinase ABC, ACII and hyaluronidase-4 (HYAL4) [261], which is a CS hydrolase despite its misnaming, have all been evaluated in models of spinal cord injury [262]. Removal of CS from the defect site by these enzymes resulted in recovery of neural functional properties [260,263–265]. Acute trauma to the brain and upper limbs resulting in neural damage have also been treated using chondroitinase ABC [266,267], resulting in neural sprouting through the defect site [268] and functional recovery [269].

Chondroitin sulphate has been of interest as a therapeutic agent for the treatment of OA for at least the last decade. A biochemical study in 2008 showed that CS interfered with progressive degenerative structural changes in joint tissues and thus showed promise in the treatment of OA [270]. A review of several CS preparations subsequently shows variable but generally positive responses in the treatment of OA but emphasises the need for highly purified CS preparations to provide unequivocal results [271]. A further study with pharmaceutical grade CS subsequently confirmed that highly purified CS had beneficial effects in the treatment of OA [272]. Continued assessment of highly purified pharmaceutical grade CS and other CS preparations confirmed safety data on the use of pharmaceutical grade CS for the treatment of OA and set down some guidelines for its use but indicated these were not applicable to lower grade CS preparations [273]. A further study raised doubts on some therapeutic CS preparations primarily focussing on the beneficial functions of CS-based therapeutic supplements and potential harmful effects of some fucosylated CS preparations which may contaminate these in a similar manner to the oversulphated CS species which have previously been identified, as contaminants in some heparin preparations [274]. A current study on highly purified therapeutic commercially available CS (Condrosulf®, IBSA, Biochimique, Lugano, Switzerland) and a literature review on its clinical efficacy confirmed the reduced pain and improvements in joint function afforded by this preparation of CS to OA patients [275]. Condrosulf® is a cost-effective and safe treatment for OA, efficacious after 30 days of administration and has beneficial properties for at least the months after the drug is discontinued. Full safety reports analyses confirmed the safety profile for CS. It has almost no side effects and shows better gastrointestinal tolerance compared to conventional non-steroidal anti-inflammatory drugs used to treat OA [275]. The beneficial properties of CS may explain the resurgence in the use of PGs, recombinant PG sub-domains, GAG and PG mimetic, and the development of neo-PGs [276] for therapeutic repair procedures. CS can also be used to stimulate stem cells and promote the attainment of defined pluripotent stem cell lineages [277]. GAGs also have generalised properties which are useful in tissue repair [278] and have been incorporated into a number of bioscaffolds to promote stem cell regulation and to develop potential new tissue repair strategies (reviewed in [279]).

16. Conclusions

N-terminal interactions of aggrecan with HA forms macro-aggregate structures that are physically entrapped within the type II collagen networks of cartilaginous tissues. These along with the C-terminal G3 mediated interactions of aggrecan EGF-like, Complement-like and C-type lectin domains with Type VI collagen lattices forms an extended tissue-wide cooperative network. This network provides a bio-sensory platform that not only dissipates loading in cartilaginous tissues but also facilitates an appropriate metabolic response by chondrocytes within the tissue to the loads they perceive to assemble an optimal functional and protective matrix, thus preventing cellular damage through point loading. Matrilin-3, SLRPs, COMP and fibulin also stabilise G3 interactions in this network. Collectively, these N- and C-terminal interactions tether aggrecan at both ends in tissues where its dynamic water imbibing properties with HA convey weight-bearing properties to tissues. In cartilaginous tissues, incorporation of aggrecan into an extended mechano-transductive sensory network allows the resident chondrocytes to perceive and respond to biomechanical changes to maintain biosynthetic responses which ensure tissue homeostasis and optimal tissue properties. The interactive properties of the G1 domain of aggrecan with HA undergoes a maturational phase where initially it does not interact with HA for 24 h which allows the newly secreted aggrecan to diffuse away from the cell into the interstitial matrix, as confirmed through pulse-chase radiosulphate labelling experiments. Immunohistochemistry also demonstrates a high density of aggrecan in the pericellular matrix around chondrocytes where network formation transmits regulatory cues to the chondrocyte to effect tissue homeostasis and aid in the stabilisation of cartilage. The variable functional attributes of aggrecan in particular tissue contexts is due to the diverse structure of its attached glycan side chains, allowing it to act as a space-filling molecule with an ability to entrap water in weight-bearing tissues such as articular cartilage and IVD but also as an interactive molecule with morphogens, growth factors and cells in growth plate cartilage and embryonic tissues. This functional diversity arises through substitution and post-translational modifications of aggrecan's attached GAG chains, for example post-translational modification of CS can occur such as disulphation on some non-reducing termini or variation in the disaccharide composition along the CS chain. These disulphated terminal groups regulate collagen fibrillogenesis in growth plate cartilage and are interactive with morphogenetic proteins such as IHH and SHH. These morphogenetic proteins direct maturational changes in the growth plates by regulating spatial and temporal chondrocyte differentiation eventually leading to growth plate closure and mineralisation at the cartilage–bone interface as part of the endochondral ossification process to extend the axial skeleton. The molecular composition of aggrecan thus varies in specific cellular and developmental contexts. Aggrecan in early embryonic development of the neural tube and notochord does not contain KS chains and some of its CS chains are replaced by the HNK-1 trisaccharide motif. This reduces aggrecans charge density and its solvation volume but conveys homophilic and heterophilic HNK-1 mediated properties and interactions with NC cells and ECM glycoproteins which direct NC cell migration, development of the neural tube and notochord and migration of precursor cells involved in the development of the neural network, heart and brain stem. Aggrecan also forms specialised ECM structures such as PNNs in the CNS/PNS, which are neuroprotective and important for synaptic plasticity and cognitive learning. The form of aggrecan in these structures contains KS, CS and the HNK-1 glycan motif; however, the density of attached CS and KS side chains in brain aggrecan is less than in cartilage aggrecan. While aggrecan has essential cell directive properties in embryonic heart formation, it also has important supportive roles to play in mature heart tissues. The heart has a complex structure, aggrecan is an important functional component of a number of its tissues providing mechanical strength and resilience for the demanding continuous cycles of compression and relaxation which occur throughout an animals life-time. Aggrecan also stabilises the attachment points of valve leaflets facilitating the co-ordinated flow of blood between the ventricles and strengthens the heart tendon chordae tendineae which attach the papillary muscles of the internal heart wall to the atrioventricular valve. These are important internal stabilising structural components of the heart. This review shows that aggrecan has a diverse range of functional attributes and is of major importance not only in embryonic skeletal development but also in mature tissues where it maintains homeostasis and functionality. Due recognition of aggrecans attached GAG chains is important and explains its diverse tissue context driven properties. A greater understanding of the glyco-code and its cell directive properties may one day provide important insights as to how specific tissue repair and regenerative strategies may be directed more effectively in repair biology.

Author Contributions: Conceptualisation, J.M. and A.J.H.; Methodology, A.J.H.; Validation, J.M. and A.J.H.; Formal Analysis, A.J.H.; Writing—Original Draft Preparation, J.M.; Writing—Review and Editing, J.M. and A.J.H; and Project Administration, J.M. All authors have read and agreed to the published version of the manuscript.

Funding: This study was funded by Melrose Personal Fund, Sydney, Australia.

Conflicts of Interest: The authors declare no conflict of interest and have no financial disclosures to make.

References

1. Feng, H.; Danfelter, M.; Strömqvist, B.; Heinegård, D. Extracellular Matrix in Disc Degeneration. *J. Bone Jt. Surg. Am. Vol.* **2006**, *88*, 25. [CrossRef]
2. Hardingham, T.E.; Fosang, A.J.; Dudhia, J. The structure, function and turnover of aggrecan, the large aggregating proteoglycan from cartilage. *Eur. J. Clin. Chem. Clin. Biochem.* **1994**, *32*, 249–257. [PubMed]
3. Heinegard, D. Fell-Muir Lecture: Proteoglycans and More—From Molecules to Biology. *Int. J. Exp. Pathol.* **2009**, *90*, 575–586. [CrossRef] [PubMed]
4. Kiani, C.; Chen, L.; Wu, Y.J.; Yee, A.J.; Yang, B.B. Structure and function of aggrecan. *Cell Res.* **2002**, *12*, 19–32. [CrossRef] [PubMed]
5. Roughley, P.J.; Lee, E.R. Cartilage proteoglycans: Structure and potential functions. *Microsc. Res. Tech.* **1994**, *28*, 385–397. [CrossRef]
6. Roughley, P.; Mort, J.S. The role of aggrecan in normal and osteoarthritic cartilage. *J. Exp. Orthop.* **2014**, *1*, 1–11. [CrossRef]
7. Sivan, S.S.; Wachtel, E.; Roughley, P. Structure, function, aging and turnover of aggrecan in the intervertebral disc. *BBA Gen. Subj.* **2014**, *1840*, 3181–3189. [CrossRef]
8. Watanabe, H.; Yamada, Y.; Kimata, K. Roles of aggrecan, a large chondroitin sulfate proteoglycan, in cartilage structure and function. *J. Biochem.* **1998**, *124*, 687–693. [CrossRef]
9. Halper, J. Proteoglycans and Diseases of Soft Tissues. *Pharm. Biotechnol.* **2013**, *802*, 49–58. [CrossRef]
10. Sarbacher, C.A.; Halper, J. Connective Tissue and Age-Related Diseases. *Sub-Cell. Biochem.* **2019**, *91*, 281–310. [CrossRef]
11. Lincoln, J.; Lange, A.W.; Yutzey, K.E. Hearts and bones: Shared regulatory mechanisms in heart valve, cartilage, tendon, and bone development. *Dev. Biol.* **2006**, *294*, 292–302. [CrossRef] [PubMed]
12. Pearson, C.S.; Solano, A.G.; Tilve, S.M.; Mencio, C.P.; Martin, K.R.; Geller, H.M. Spatiotemporal distribution of chondroitin sulfate proteoglycans after optic nerve injury in rodents. *Exp. Eye Res.* **2020**, *190*, 107859. [CrossRef] [PubMed]

13. Rambeau, P.; Faure, E.; Faucherre, A.; Theron, A.; Avierinos, J.-F.; Jopling, C.; Zaffran, S. Reduced aggrecan expression affects cardiac outflow tract development in zebrafish and is associated with bicuspid aortic valve disease in humans. *Int. J. Cardiol.* **2017**, *249*, 340–343. [CrossRef] [PubMed]
14. Schulz, A.; Brendler, J.; Blaschuk, O.; Landgraf, K.; Krueger, M.; Ricken, A.M. Non-pathological Chondrogenic Features of Valve Interstitial Cells in Normal Adult Zebrafish. *J. Histochem. Cytochem.* **2019**, *67*, 361–373. [CrossRef] [PubMed]
15. Yasmin; Al Maskari, R.; McEniery, C.M.; Cleary, S.E.; Li, Y.; Siew, K.; Figg, N.L.; Khir, A.W.; Cockcroft, J.R.; Wilkinson, I.B.; et al. The matrix proteins aggrecan and fibulin-1 play a key role in determining aortic stiffness. *Sci. Rep.* **2018**, *8*, 8550. [CrossRef] [PubMed]
16. Zanin, M.K.; Bundy, J.; Ernst, H.; Wessels, A.; Conway, S.J.; Hoffman, S. Distinct spatial and temporal distributions of aggrecan and versican in the embryonic chick heart. *Anat. Rec.* **1999**, *256*, 366–380. [CrossRef]
17. Fomovsky, G.M.; Thomopoulos, S.; Holmes, J.W. Contribution of extracellular matrix to the mechanical properties of the heart. *J. Mol. Cell. Cardiol.* **2010**, *48*, 490–496. [CrossRef]
18. Miyata, S.; Nadanaka, S.; Igarashi, M.; Kitagawa, H. Structural Variation of Chondroitin Sulfate Chains Contributes to the Molecular Heterogeneity of Perineuronal Nets. *Front. Integr. Neurosci.* **2018**, *12*, 3. [CrossRef]
19. Morawski, M.; Dityatev, A.; Hartlage-Rübsamen, M.; Blosa, M.; Holzer, M.; Flach, K.; Pavlica, S.; Dityateva, G.; Grosche, J.; Brückner, G.; et al. Tenascin-R promotes assembly of the extracellular matrix of perineuronal nets via clustering of aggrecan. *Philos. Trans. R. Soc. B Biol. Sci.* **2014**, *369*, 20140046. [CrossRef]
20. Reichelt, A.C.; Hare, D.J.; Bussey, T.J.; Saksida, L.M. Perineuronal Nets: Plasticity, Protection, and Therapeutic Potential. *Trends Neurosci.* **2019**, *42*, 458–470. [CrossRef]
21. Rowlands, D.; Lensjø, K.K.; Dinh, T.; Yang, S.; Andrews, M.R.; Hafting, T.; Fyhn, M.; Fawcett, J.W.; Dick, G. Aggrecan Directs Extracellular Matrix-Mediated Neuronal Plasticity. *J. Neurosci.* **2018**, *38*, 10102–10113. [CrossRef] [PubMed]
22. Hering, T.M.; Beller, J.A.; Calulot, C.M.; Snow, D.M. Contributions of Chondroitin Sulfate, Keratan Sulfate and N-linked Oligosaccharides to Inhibition of Neurite Outgrowth by Aggrecan. *Biology* **2020**, *9*, 29. [CrossRef] [PubMed]
23. Hardingham, T.E.; Fosang, A.J.; Hey, N.J.; Hazell, P.K.; Kee, W.J.; Ewins, R.J.; Hardingham, T. The sulphation pattern in chondroitin sulphate chains investigated by chondroitinase ABC and ACII digestion and reactivity with monoclonal antibodies. *Carbohydr. Res.* **1994**, *255*, 241–254. [CrossRef]
24. Midura, R.J.; Calabrò, A.; Yanagishita, M.; Hascall, V.C. Nonreducing End Structures of Chondroitin Sulfate Chains on Aggrecan Isolated from Swarm Rat Chondrosarcoma Cultures. *J. Biol. Chem.* **1995**, *270*, 8009–8015. [CrossRef]
25. A West, L.; Roughley, P.; Nelson, F.R.; Plaas, A.H. Sulphation heterogeneity in the trisaccharide (GalNAcSbeta1,4GlcAbeta1,3GalNAcS) isolated from the non-reducing terminal of human aggrecan chondroitin sulphate. *Biochem. J.* **1999**, *342*, 223–229. [CrossRef]
26. Benito-Arenas, R.; Doncel-Perez, E.; Fernández-Gutiérrez, M.; Garrido, L.; García-Junceda, E.; Revuelta, J.; Bastida, A.; Fernández-Mayoralas, A. A holistic approach to unravelling chondroitin sulfation: Correlations between surface charge, structure and binding to growth factors. *Carbohydr. Polym.* **2018**, *202*, 211–218. [CrossRef]
27. Cortes, M.; Baria, A.T.; Schwartz, N.B. Sulfation of chondroitin sulfate proteoglycans is necessary for proper Indian hedgehog signaling in the developing growth plate. *Development* **2009**, *136*, 1697–1706. [CrossRef]
28. Miyachi, K.; Wakao, M.; Suda, Y. Syntheses of chondroitin sulfate tetrasaccharide structures containing 4,6-disulfate patterns and analysis of their interaction with glycosaminoglycan-binding protein. *Bioorganic Med. Chem. Lett.* **2015**, *25*, 1552–1555. [CrossRef]
29. Mizumoto, S.; Yamada, S.; Sugahara, K. Molecular interactions between chondroitin–dermatan sulfate and growth factors/receptors/matrix proteins. *Curr. Opin. Struct. Biol.* **2015**, *34*, 35–42. [CrossRef]
30. Nandini, C.D.; Sugahara, K. Role of the Sulfation Pattern of Chondroitin Sulfate in its Biological Activities and in the Binding of Growth Factors. *Adv. Pharmacol.* **2006**, *53*, 253–279. [CrossRef]
31. Wakao, M.; Obata, R.; Miyachi, K.; Kaitsubata, Y.; Kondo, T.; Sakami, C.; Suda, Y. Synthesis of a chondroitin sulfate disaccharide library and a GAG-binding protein interaction analysis. *Bioorganic Med. Chem. Lett.* **2015**, *25*, 1407–1411. [CrossRef] [PubMed]

32. Whalen, D.M.; Malinauskas, T.; Gilbert, R.J.C.; Siebold, C. Structural insights into proteoglycan-shaped Hedgehog signaling. *Proc. Natl. Acad. Sci. USA* **2013**, *110*, 16420–16425. [CrossRef] [PubMed]
33. Zhang, L. Glycosaminoglycan (GAG) Biosynthesis and GAG-Binding Proteins. *Prog. Mol. Biol. Transl.* **2010**, *93*, 1–17. [CrossRef]
34. Nadanaka, S.; Ishida, M.; Ikegami, M.; Kitagawa, H. Chondroitin 4-O-Sulfotransferase-1 Modulates Wnt-3a Signaling through Control of E Disaccharide Expression of Chondroitin Sulfate. *J. Biol. Chem.* **2008**, *283*, 27333–27343. [CrossRef]
35. Reichsman, F.; Smith, L.; Cumberledge, S. Glycosaminoglycans can modulate extracellular localization of the wingless protein and promote signal transduction. *J. Cell Biol.* **1996**, *135*, 819–827. [CrossRef]
36. Melrose, J.; Isaacs, M.D.; Smith, S.M.; Hughes, C.E.; Little, C.B.; Caterson, B.; Hayes, A.J. Chondroitin sulphate and heparan sulphate sulphation motifs and their proteoglycans are involved in articular cartilage formation during human foetal knee joint development. *Histochem. Cell Biol.* **2012**, *138*, 461–475. [CrossRef]
37. Cao, J.; Li, S.; Shi, Z.; Yue, Y.; Sun, J.; Chen, J.; Fu, Q.; Hughes, C.E.; Caterson, B. Articular cartilage metabolism in patients with Kashin–Beck Disease: An endemic osteoarthropathy in China. *Osteoarthr. Cartil.* **2008**, *16*, 680–688. [CrossRef]
38. Gao, Z.Q.; Guo, X.; Duan, C.; Ma, W.; Xu, P.; Wang, W.; Chen, J.C. Altered aggrecan synthesis and collagen expression profiles in chondrocytes from patients with Kashin-Beck disease and osteoarthritis. *J. Int. Med. Res.* **2012**, *40*, 1325–1334. [CrossRef]
39. Luo, M.; Chen, J.; Li, S.-Y.; Sun, H.; Zhang, Z.; Fu, Q.; Li, J.; Wang, J.; Hughes, C.E.; Caterson, B.; et al. Changes in the metabolism of chondroitin sulfate glycosaminoglycans in articular cartilage from patients with Kashin–Beck disease. *Osteoarthr. Cartil.* **2014**, *22*, 986–995. [CrossRef]
40. Wang, M.; Xue, S.; Fang, Q.; Zhang, M.; He, Y.; Zhang, Y.; Lammi, M.J.; Cao, J.; Chen, J. Expression and localization of the small proteoglycans decorin and biglycan in articular cartilage of Kashin-Beck disease and rats induced by T-2 toxin and selenium deficiency. *Glycoconj. J.* **2019**, *36*, 1–9. [CrossRef]
41. Wu, C.; Lei, R.; Tiainen, M.; Wu, S.-X.; Zhang, Q.; Pei, F.-X.; Guo, X. Disordered glycometabolism involved in pathogenesis of Kashin–Beck disease, an endemic osteoarthritis in China. *Exp. Cell Res.* **2014**, *326*, 240–250. [CrossRef] [PubMed]
42. Asada, M.; Shinomiya, M.; Suzuki, M.; Honda, E.; Sugimoto, R.; Ikekita, M.; Imamura, T. Glycosaminoglycan affinity of the complete fibroblast growth factor family. *BBA Gen. Subj.* **2009**, *1790*, 40–48. [CrossRef] [PubMed]
43. Prinz, R.D.; Willis, C.M.; Van Kuppevelt, T.H.; Kluppel, M. Biphasic Role of Chondroitin Sulfate in Cardiac Differentiation of Embryonic Stem Cells through Inhibition of Wnt/β-Catenin Signaling. *PLoS ONE* **2014**, *9*, e92381. [CrossRef] [PubMed]
44. Sirko, S.; Von Holst, A.; Weber, A.; Wizenmann, A.; Theocharidis, U.; Götz, M.; Faissner, A. Chondroitin Sulfates Are Required for Fibroblast Growth Factor-2-Dependent Proliferation and Maintenance in Neural Stem Cells and for Epidermal Growth Factor-Dependent Migration of Their Progeny. *Stem Cells* **2010**, *28*, 775–787. [CrossRef]
45. Sterner, E.; Meli, L.; Kwon, S.J.; Dordick, J.S.; Linhardt, R.J. FGF–FGFR Signaling Mediated through Glycosaminoglycans in Microtiter Plate and Cell-Based Microarray Platforms. *Biochemistry* **2013**, *52*, 9009–9019. [CrossRef]
46. Plaas, A.; West, L.A.; Wong-Palms, S.; Nelson, F.R. Glycosaminoglycan sulfation in human osteoarthritis. Disease-related alterations at the non-reducing termini of chondroitin and dermatan sulfate. *J. Biol. Chem.* **1998**, *273*, 12642–12649. [CrossRef]
47. De Paz, J.L.; Nieto, P.M. Improvement on binding of chondroitin sulfate derivatives to midkine by increasing hydrophobicity. *Org. Biomol. Chem.* **2016**, *14*, 3506–3509. [CrossRef]
48. Deepa, S.S.; Umehara, Y.; Higashiyama, S.; Itoh, N.; Sugahara, K. Specific Molecular Interactions of Oversulfated Chondroitin Sulfate E with Various Heparin-binding Growth Factors. *J. Biol. Chem.* **2002**, *277*, 43707–43716. [CrossRef]
49. Solera, C.; Macchione, G.; Maza, S.; Kayser, M.M.; Corzana, F.; De Paz, J.L.; Nieto, P.M. Chondroitin Sulfate Tetrasaccharides: Synthesis, Three-Dimensional Structure and Interaction with Midkine. *Chem. Eur. J.* **2016**, *22*, 2356–2369. [CrossRef]

50. Shu, C.C.; Hughes, C.E.; Smith, S.M.; Smith, M.M.; Hayes, A.J.; Caterson, B.; Little, C.B.; Melrose, J. The ovine newborn and human foetal intervertebral disc contain perlecan and aggrecan variably substituted with native 7D4 CS sulphation motif: Spatiotemporal immunolocalisation and co-distribution with Notch-1 in the human foetal disc. *Glycoconj. J.* **2013**, *30*, 717–725. [CrossRef]
51. Caterson, B. Fell-Muir Lecture: Chondroitin sulphate glycosaminoglycans: Fun for some and confusion for others. *Int. J. Exp. Pathol.* **2012**, *93*, 1–10. [CrossRef] [PubMed]
52. Hayes, A.J.; Hughes, C.E.; Ralphs, J.; Caterson, B. Chondroitin sulphate sulphation motif expression in the ontogeny of the intervertebral disc. *Eur. Cell Mater.* **2011**, *21*, 1–14. [CrossRef] [PubMed]
53. Belcher, C.; Yaqub, R.; Fawthrop, F.; Bayliss, M.; Doherty, M. Synovial fluid chondroitin and keratan sulphate epitopes, glycosaminoglycans, and hyaluronan in arthritic and normal knees. *Ann. Rheum. Dis.* **1997**, *56*, 299–307. [CrossRef] [PubMed]
54. Hazell, P.K.; Dent, C.; Fairclough, J.A.; Bayliss, M.T.; Hardingham, T.E. Changes in glycosaminoglycan epitope levels in knee joint fluid following injury. *Arthritis Rheum.* **1995**, *38*, 953–959. [CrossRef]
55. Ratcliffe, A.; Shurety, W.; Caterson, B. The quantitation of a native chondroitin sulfate epitope in synovial fluid lavages and articular cartilage from canine experimental osteoarthritis and disuse atrophy. *Arthritis Rheum.* **1993**, *36*, 543–551. [CrossRef]
56. Ratcliffe, A.; Grelsamer, R.P.; Kiernan, H.; Saed-Nejad, F.; Visco, D. High levels of aggrecan aggregate components are present in synovial fluids from human knee joints with chronic injury or osteoarthrosis. *Acta Orthop. Scand.* **1995**, *66*, 111–115. [CrossRef]
57. Farrugia, B.L.; Mizumoto, S.; Lord, M.S.; O'Grady, R.L.; Kuchel, R.P.; Yamada, S.; Whitelock, J.M. Hyaluronidase-4 is produced by mast cells and can cleave serglycin chondroitin sulfate chains into lower molecular weight forms. *J. Biol. Chem.* **2019**, *294*, 11458–11472. [CrossRef]
58. Caterson, B.; Mahmoodian, F.; Sorrell, J.M.; Hardingham, T.E.; Bayliss, M.T.; Carney, S.L.; Ratcliffe, A.; Muir, H. Modulation of native chondroitin sulphate structure in tissue development and in disease. *J. Cell Sci.* **1990**, *97*, 411–417.
59. Hayes, A.J.; Smith, S.M.; Caterson, B.; Melrose, J. Concise Review: Stem/Progenitor Cell Proteoglycans Decorated with 7-D-4, 4-C-3, and 3-B-3(-) Chondroitin Sulfate Motifs Are Morphogenetic Markers of Tissue Development. *Stem Cells* **2018**, *36*, 1475–1486. [CrossRef]
60. Hayes, A.; Sugahara, K.; Farrugia, B.L.; Whitelock, J.M.; Caterson, B.; Melrose, J. Biodiversity of CS–proteoglycan sulphation motifs: Chemical messenger recognition modules with roles in information transfer, control of cellular behaviour and tissue morphogenesis. *Biochem. J.* **2018**, *475*, 587–620. [CrossRef] [PubMed]
61. Klüppel, M. Maintenance of chondroitin sulfation balance by chondroitin-4-sulfotransferase 1 is required for chondrocyte development and growth factor signaling during cartilage morphogenesis. *Development* **2005**, *132*, 3989–4003. [CrossRef] [PubMed]
62. Mark, M.P.; Baker, J.R.; Kimata, K.; Ruch, J.V. Regulated changes in chondroitin sulfation during embryogenesis: An immunohistochemical approach. *Int. J. Dev. Biol.* **1990**, *34*, 191–204. [PubMed]
63. Melrose, J.; Smith, S.M. The 7D4, 4C3 and 3B3 (-) Chondroitin Sulphation Motifs are expressed at Sites of Cartilage and Bone Morphogenesis during Foetal Human Knee Joint Development. *J. Glycobiol.* **2016**, *5*, 2–9. [CrossRef]
64. Slater, R.R.; Bayliss, M.T.; Lachiewicz, P.F.; Visco, D.M.; Caterson, B. Monoclonal antibodies that detect biochemical markers of arthritis in humans. *Arthritis Rheum.* **1995**, *38*, 655–659. [CrossRef]
65. Sorrell, J.M.; Lintala, A.M.; Mahmoodian, F.; Caterson, B. Epitope-specific changes in chondroitin sulfate/dermatan sulfate proteoglycans as markers in the lymphopoietic and granulopoietic compartments of developing bursae of Fabricius. *J. Immunol.* **1988**, *140*, 4263–4270.
66. Curchoe, C.L.; Maurer, J.; McKeown, S.J.; Cattarossi, G.; Cimadamore, F.; Nilbratt, M.; Snyder, E.Y.; Bronner-Fraser, M.; Terskikh, A.V. Early Acquisition of Neural Crest Competence During hESCs Neuralization. *PLoS ONE* **2010**, *5*, e13890. [CrossRef]
67. Tucker, G.; Delarue, M.; Zada, S.; Boucaut, J.C.; Thiery, J.P. Expression of the HNK-1/NC-1 epitope in early vertebrate neurogenesis. *Cell Tissue Res.* **1988**, *251*, 457–465. [CrossRef]
68. Griffith, L.; Schmitz, B.; Schachner, M. L2/HNK-1 carbohydrate and protein-protein interactions mediate the homophilic binding of the neural adhesion molecule P0. *J. Neurosci. Res.* **1992**, *33*, 639–648. [CrossRef]

69. Melrose, J. Keratan sulfate (KS)-proteoglycans and neuronal regulation in health and disease: The importance of KS-glycodynamics and interactive capability with neuroregulatory ligands. *J. Neurochem.* **2019**, *149*, 170–194. [CrossRef]
70. Furley, A.J.; Morton, S.B.; Manalo, D.; Karagogeos, D.; Dodd, J.; Jessell, T.M. The axonal glycoprotein TAG-1 is an immunoglobulin superfamily member with neurite outgrowth-promoting activity. *Cell* **1990**, *61*, 157–170. [CrossRef]
71. Hall, H.; Liu, L.; Schachner, M.; Schmitz, B. The L2/HNK-1 carbohydrate mediates adhesion of neural cells to laminin. *Eur. J. Neurosci.* **1993**, *5*, 34–42. [CrossRef] [PubMed]
72. Hall, H.; Carbonetto, S.; Schachner, M. L1/HNK-1 carbohydrate- and beta 1 integrin-dependent neural cell adhesion to laminin-1. *J. Neurochem.* **1997**, *68*, 544–553. [CrossRef] [PubMed]
73. Visco, D.M.; Johnstone, B.; Hill, M.A.; Jolly, G.A.; Caterson, B. Immunohistochemical analysis of 3-b-3(−) and 7-d-4 epitope expression in canine osteoarthritis. *Arthritis Rheum.* **1993**, *36*, 1718–1725. [CrossRef] [PubMed]
74. Hiraoka, K.; Grogan, S.; Olee, T.; Lotz, M. Mesenchymal progenitor cells in adult human articular cartilage. *Biorheology* **2006**, *43*, 447–454. [PubMed]
75. Dowthwaite, G.P.; Bishop, J.C.; Redman, S.N.; Khan, I.M.; Rooney, P.; Evans, D.J.R.; Haughton, L.; Bayram-Weston, Z.; Boyer, S.; Thomson, B.; et al. The surface of articular cartilage contains a progenitor cell population. *J. Cell Sci.* **2004**, *117*, 889–897. [CrossRef]
76. Grogan, S.P.; Miyaki, S.; Asahara, H.; D'Lima, D.D.; Lotz, M.K. Mesenchymal progenitor cell markers in human articular cartilage: Normal distribution and changes in osteoarthritis. *Arthritis Res. Ther.* **2009**, *11*, R85. [CrossRef]
77. Hollander, A.P.; Dickinson, S.C.; Kafienah, W. Stem Cells and Cartilage Development: Complexities of a Simple Tissue. *Stem Cells* **2010**, *28*, 1992–1996. [CrossRef]
78. Akatsu, C.; Mizumoto, S.; Kaneiwa, T.; Maccarana, M.; Malmström, A.; Yamada, S.; Sugahara, K. Dermatan sulfate epimerase 2 is the predominant isozyme in the formation of the chondroitin sulfate/dermatan sulfate hybrid structure in postnatal developing mouse brain. *Glycobiology* **2010**, *21*, 565–574. [CrossRef]
79. Akita, K.; Von Holst, A.; Furukawa, Y.; Mikami, T.; Sugahara, K.; Faissner, A. Expression of Multiple Chondroitin/Dermatan Sulfotransferases in the Neurogenic Regions of the Embryonic and Adult Central Nervous System Implies That Complex Chondroitin Sulfates Have a Role in Neural Stem Cell Maintenance. *Stem Cells* **2008**, *26*, 798–809. [CrossRef]
80. Mitsunaga, C.; Mikami, T.; Mizumoto, S.; Fukuda, J.; Sugahara, K. Chondroitin sulfate/dermatan sulfate hybrid chains in the development of cerebellum. Spatiotemporal regulation of the expression of critical disulfated disaccharides by specific sulfotransferases. *J. Biol. Chem.* **2006**, *281*, 18942–18952. [CrossRef]
81. Hayes, A.J.; Tudor, D.; Nowell, M.A.; Caterson, B.; Hughes, C.E. Chondroitin Sulfate Sulfation Motifs as Putative Biomarkers for Isolation of Articular Cartilage Progenitor Cells. *J. Histochem. Cytochem.* **2007**, *56*, 125–138. [CrossRef] [PubMed]
82. Li, S.; Hayes, A.J.; Caterson, B.; Hughes, C.E. The effect of beta-xylosides on the chondrogenic differentiation of mesenchymal stem cells. *Histochem. Cell Biol.* **2012**, *139*, 59–74. [CrossRef] [PubMed]
83. Hayes, A.J.; Hughes, C.E.; Smith, S.M.; Caterson, B.; Little, C.; Melrose, J. The CS Sulfation Motifs 4C3, 7D4, 3B3[−]; and Perlecan Identify Stem Cell Populations and Their Niches, Activated Progenitor Cells and Transitional Areas of Tissue Development in the Fetal Human Elbow. *Stem Cells Dev.* **2016**, *25*, 836–847. [CrossRef] [PubMed]
84. Skandalis, S.S.; Theocharis, A.D.; Vynios, D.H.; a Theocharis, D.; Papageorgakopoulou, N. Proteoglycans in human laryngeal cartilage. Identification of proteoglycan types in successive cartilage extracts with particular reference to aggregating proteoglycans. *Biochimie* **2004**, *86*, 221–229. [CrossRef]
85. Novoselov, V.P.; Savchenko, S.V.; Pyatkova, E.V.; Nadeev, A.P.; Ageeva, T.A.; Chikinev, Y.V.; Polyakevich, A.S. Morphological Characteristics of the Cartilaginous Tissue of Human Auricle in Different Age Periods. *Bull. Exp. Biol. Med.* **2016**, *160*, 840–843. [CrossRef]
86. Theocharis, A.D.; Karamanos, N.K.; Papageorgakopoulou, N.; Tsiganos, C.P.; a Theocharis, D. Isolation and characterization of matrix proteoglycans from human nasal cartilage. Compositional and structural comparison between normal and scoliotic tissues. *BBA Gen. Subj.* **2002**, *1569*, 117–126. [CrossRef]
87. Wilson, C.G.; Nishimuta, J.F.; Levenston, M.E. Chondrocytes and Meniscal Fibrochondrocytes Differentially Process Aggrecan During De Novo Extracellular Matrix Assembly. *Tissue Eng. Part A* **2009**, *15*, 1513–1522. [CrossRef]

88. Lockhart, M.; Wirrig, E.; Phelps, A.; Wessels, A. Extracellular matrix and heart development. *Birth Defects Res. Part A Clin. Mol. Teratol.* **2011**, *91*, 535–550. [CrossRef]
89. Miyata, S.; Kitagawa, H. Formation and remodeling of the brain extracellular matrix in neural plasticity: Roles of chondroitin sulfate and hyaluronan. *BBA Gen. Subj.* **2017**, *1861*, 2420–2434. [CrossRef]
90. Domowicz, M.; Li, H.; Hennig, A.; Henry, J.; Vertel, B.M.; Schwartz, N.B. The Biochemically and Immunologically Distinct CSPG of Notochord Is a Product of the Aggrecan Gene. *Dev. Biol.* **1995**, *171*, 655–664. [CrossRef]
91. Domowicz, M.; Mangoura, D.; Schwartz, N.B. Cell Specific-Chondroitin Sulfate Proteoglycan Expression During CNS Morphogenesis in the Chick Embryo. *Int. J. Dev. Neurosci.* **2000**, *18*, 629–641. [CrossRef]
92. Domowicz, M.; Mueller, M.M.; Novak, T.E.; Schwartz, L.E.; Schwartz, N.B. Developmental expression of the HNK-1 carbohydrate epitope on aggrecan during chondrogenesis. *Dev. Dyn.* **2002**, *226*, 42–50. [CrossRef] [PubMed]
93. Morawski, M.; Bruckner, G.; Arendt, T.; Matthews, R. Aggrecan: Beyond cartilage and into the brain. *Int. J. Biochem. Cell Biol.* **2012**, *44*, 690–693. [CrossRef]
94. Schwartz, N.B.; Domowicz, M.; Krueger, R.C.; Li, H.; Mangoura, D. Brain aggrecan. *Perspect. Dev. Neurobiol.* **1996**, *3*, 291–306. [PubMed]
95. Trapani, V.; Bonaldo, P.; Corallo, D. Role of the ECM in notochord formation, function and disease. *J. Cell Sci.* **2017**, *130*, 3203–3211. [CrossRef]
96. Corallo, D.; Trapani, V.; Bonaldo, P. The notochord: Structure and functions. *Cell. Mol. Life Sci.* **2015**, *72*, 2989–3008. [CrossRef]
97. De Bree, K.; De Bakker, B.S.; Oostra, R.-J. The development of the human notochord. *PLoS ONE* **2018**, *13*, e0205752. [CrossRef]
98. Tada, M. Notochord morphogenesis: A prickly subject for ascidians. *Curr. Biol.* **2005**, *15*, R14–R16. [CrossRef]
99. Yu, J.; Urban, J.P.G. The elastic network of articular cartilage: An immunohistochemical study of elastin fibres and microfibrils. *J. Anat.* **2010**, *216*, 533–541. [CrossRef]
100. He, B.; Wu, J.; Chen, H.H.; Kirk, T.B.; Xu, J. Elastin fibers display a versatile microfibril network in articular cartilage depending on the mechanical microenvironments. *J. Orthop. Res.* **2013**, *31*, 1345–1353. [CrossRef]
101. He, B.; Wu, J.; Chim, S.; Xu, J.; Kirk, T.B. Microstructural analysis of collagen and elastin fibres in the kangaroo articular cartilage reveals a structural divergence depending on its local mechanical environment. *Osteoarthr. Cartil.* **2013**, *21*, 237–245. [CrossRef] [PubMed]
102. He, B.; Wu, J.; Xu, J.; Day, R.E.; Kirk, T.B. Microstructural and Compositional Features of the Fibrous and Hyaline Cartilage on the Medial Tibial Plateau Imply a Unique Role for the Hopping Locomotion of Kangaroo. *PLoS ONE* **2013**, *8*, e74303. [CrossRef] [PubMed]
103. Nguyen, M.K.; Kurtz, I. Physiologic Interrelationships between Gibbs-Donnan Equilibrium, Osmolality of Body Fluid Compartments, and Plasma Water Sodium Concentration. *J. Appl. Physiol.* **2006**. [CrossRef] [PubMed]
104. Nimer, E.; Schneiderman, R.; Maroudas, A. Diffusion and partition of solutes in cartilage under static load. *Biophys. Chem.* **2003**, *106*, 125–146. [CrossRef]
105. Klüppel, M. The Roles of Chondroitin-4-Sulfotransferase-1 in Development and Disease. *Prog. Mol. Biol. Transl. Sci.* **2010**, *93*, 113–132. [CrossRef]
106. Mikami, T.; Kitagawa, H. Biosynthesis and function of chondroitin sulfate. *BBA Gen. Subj.* **2013**, *1830*, 4719–4733. [CrossRef]
107. Hardingham, T.E.; Fosang, A.J. The structure of aggrecan and its turnover in cartilage. *J. Rheumatol. Suppl.* **1995**, *43*, 86–90.
108. Walcz, E.; Deák, F.; Erhardt, P.; Coulter, S.N.; Fülöp, C.; Horvath, P.; Doege, K.J.; Glant, T.T. Complete Coding Sequence, Deduced Primary Structure, Chromosomal Localization, and Structural Analysis of Murine Aggrecan. *Genomics* **1994**, *22*, 364–371. [CrossRef]
109. Watanabe, H.; Gao, L.; Sugiyama, S.; Doege, K.; Kimata, K.; Yamada, Y. Mouse aggrecan, a large cartilage proteoglycan: Protein sequence, gene structure and promoter sequence. *Biochem. J.* **1995**, *308*, 433–440. [CrossRef]
110. Antonsson, P.; Heinegård, D.; Oldberg, A. The keratan sulfate-enriched region of bovine cartilage proteoglycan consists of a consecutively repeated hexapeptide motif. *J. Biol. Chem.* **1989**, *264*, 16170–16173.

111. Doege, K.J.; Sasaki, M.; Kimura, T.; Yamada, Y. Complete coding sequence and deduced primary structure of the human cartilage large aggregating proteoglycan, aggrecan. Human-specific repeats, and additional alternatively spliced forms. *J. Biol. Chem.* **1991**, *266*, 894–902. [PubMed]
112. Fosang, A.; Rogerson, F.; East, C.; Stanton, H. ADAMTS-5: The story so far. *Eur. Cells Mater.* **2008**, *15*, 11–26. [CrossRef] [PubMed]
113. Conrad, A.H.; Zhang, Y.; Tasheva, E.S.; Conrad, G.W. Proteomic analysis of potential keratan sulfate, chondroitin sulfate A, and hyaluronic acid molecular interactions. *Investig. Opthalmol. Vis. Sci.* **2010**, *51*, 4500–4515. [CrossRef]
114. Russo, V.C.; Bach, L.A.; Fosang, A.J.; Baker, N.L.; Werther, G.A. Insulin-like growth factor binding protein-2 binds to cell surface proteoglycans in the rat brain olfactory bulb. *Endocrinology* **1997**, *138*, 4858–4867. [CrossRef]
115. Melrose, J. Functional Consequences of Keratan Sulphate Sulfation in Electrosensory Tissues and in Neuronal Regulation. *Adv. Biosyst.* **2019**, *3*, 1800327. [CrossRef] [PubMed]
116. Gandhi, N.S.; Mancera, R.L. The Structure of Glycosaminoglycans and their Interactions with Proteins. *Chem. Biol. Drug Des.* **2008**, *72*, 455–482. [CrossRef]
117. Weyers, A.; Yang, B.; Solakyildirim, K.; Yee, V.; Li, L.; Zhang, F.; Linhardt, R.J. Isolation of bovine corneal keratan sulfate and its growth factor and morphogen binding. *FEBS J.* **2013**, *280*, 2285–2293. [CrossRef]
118. Zhang, F.; Zheng, L.; Cheng, S.; Peng, Y.; Fu, L.; Zhang, X.; Linhardt, R.J. Comparison of the Interactions of Different Growth Factors and Glycosaminoglycans. *Molecules* **2019**, *24*, 3360. [CrossRef]
119. Caterson, B.; Melrose, J. Keratan sulfate, a complex glycosaminoglycan with unique functional capability. *Glycobiology* **2018**, *28*, 182–206. [CrossRef]
120. Caterson, B.; E Christner, J.; Baker, J.R. Identification of a monoclonal antibody that specifically recognizes corneal and skeletal keratan sulfate. Monoclonal antibodies to cartilage proteoglycan. *J. Biol. Chem.* **1983**, *258*, 8848–8854.
121. Melrose, J. Mucin-like glycopolymer gels in electrosensory tissues generate cues which direct electrolocation in amphibians and neuronal activation in mammals. *Neural Regen. Res.* **2019**, *14*, 1191. [CrossRef]
122. Day, J.M.; Murdoch, A.D.; Hardingham, T.E. The Folded Protein Modules of the C-terminal G3 Domain of Aggrecan Can Each Facilitate the Translocation and Secretion of the Extended Chondroitin Sulfate Attachment Sequence. *J. Biol. Chem.* **1999**, *274*, 38107–38111. [CrossRef] [PubMed]
123. Lundell, A.; Olin, A.I.; Mörgelin, M.; Al-Karadaghi, S.; Aspberg, A.; Logan, D.T. Structural Basis for Interactions between Tenascins and Lectican C-Type Lectin Domains. *Structure* **2004**, *12*, 1495–1506. [CrossRef] [PubMed]
124. Fürst, C.M.; Mörgelin, M.; Vadstrup, K.; Heinegård, D.; Aspberg, A.; Blom, A.M. The C-Type Lectin of the Aggrecan G3 Domain Activates Complement. *PLoS ONE* **2013**, *8*, e61407. [CrossRef]
125. Sjöberg, A.; Trouw, L.A.; Blom, A.M. Complement activation and inhibition: A delicate balance. *Trends Immunol.* **2009**, *30*, 83–90. [CrossRef] [PubMed]
126. Ricklin, D.; Hajishengallis, G.; Yang, K.; Lambris, J.D. Complement: A key system for immune surveillance and homeostasis. *Nat. Immunol.* **2010**, *11*, 785–797. [CrossRef]
127. Wiberg, C.; Klatt, A.R.; Wagener, R.; Paulsson, M.; Bateman, J.; Heinegård, D.; Mörgelin, M. Complexes of Matrilin-1 and Biglycan or Decorin Connect Collagen VI Microfibrils to Both Collagen II and Aggrecan. *J. Biol. Chem.* **2003**, *278*, 37698–37704. [CrossRef]
128. Chen, F.H.; Herndon, M.E.; Patel, N.; Hecht, J.T.; Tuan, R.S.; Lawler, J. Interaction of Cartilage Oligomeric Matrix Protein/Thrombospondin 5 with Aggrecan. *J. Biol. Chem.* **2007**, *282*, 24591–24598. [CrossRef]
129. Krueger, R.C.; Kurima, K.; Schwartz, N.B. Completion of the mouse aggrecan gene structure and identification of the defect in the cmd-Bc mouse as a near complete deletion of the murine aggrecan gene. *Mamm. Genome* **1999**, *10*, 1119–1125. [CrossRef]
130. Watanabe, H.; Kimata, K.; Line, S.; Strong, D.; Gao, L.-Y.; Kozak, C.A.; Yamada, Y. Mouse cartilage matrix deficiency (cmd) caused by a 7 bp deletion in the aggrecan gene. *Nat. Genet.* **1994**, *7*, 154–157. [CrossRef]
131. Yoo, T.J.; Cho, H.; Yamada, Y. Hearing Impairment in Mice with the cmd/cmd (Cartilage Matrix Deficiency) Mutant Gene. *Ann. N. Y. Acad. Sci.* **1991**, *630*, 265–267. [CrossRef] [PubMed]
132. Watanabe, H.; Nakata, K.; Kimata, K.; Nakanishi, I.; Yamada, K.M. Dwarfism and age-associated spinal degeneration of heterozygote cmd mice defective in aggrecan. *Proc. Natl. Acad. Sci. USA* **1997**, *94*, 6943–6947. [CrossRef]

133. Primorac, D.; Stover, M.; Clark, S.; Rowe, D. Molecular basis of nanomelia, a heritable chondrodystrophy of chicken. *Matrix Biol.* **1994**, *14*, 297–305. [CrossRef]
134. Domowicz, M.; Cortes, M.; Henry, J.G.; Schwartz, N.B. Aggrecan modulation of growth plate morphogenesis. *Dev. Biol.* **2009**, *329*, 242–257. [CrossRef] [PubMed]
135. Stirpe, N.S.; Argraves, W.; Goetinck, P.F. Chondrocytes from the cartilage proteoglycan-deficient mutant, nanomelia, synthesize greatly reduced levels of the proteoglycan core protein transcript. *Dev. Biol.* **1987**, *124*, 77–81. [CrossRef]
136. Vertel, B.M.; Grier, B.L.; Li, H.; Schwartz, N.B. The chondrodystrophy, nanomelia: Biosynthesis and processing of the defective aggrecan precursor. *Biochem. J.* **1994**, *301*, 211–216. [CrossRef]
137. Vertel, B.M.; Walters, L.M.; Grier, B.; Maine, N.; Goetinck, P.F. Nanomelic chondrocytes synthesize, but fail to translocate, a truncated aggrecan precursor. *J. Cell Sci.* **1993**, *104*, 104.
138. Glant, T.T.; I Buzás, E.; Finnegan, A.; Negroiu, G.; Cs-Szabó, G.; Mikecz, K. Critical roles of glycosaminoglycan side chains of cartilage proteoglycan (aggrecan) in antigen recognition and presentation. *J. Immunol.* **1998**, *160*, 3812–3819.
139. Glant, T.T.; Fülöp, C.; Cs-Szabo, G.; Buzás, E.I.; Ragasa, D.; Mikecz, K. Mapping of Arthritogenic/Autoimmune Epitopes of Cartilage Aggrecans in Proteoglycan-Induced Arthritis. *Scand. J. Rheumatol.* **1995**, *24*, 43–49. [CrossRef]
140. Glant, T.T.; Radács, M.; Nagyeri, G.; Olasz, K.; László, A.; Boldizsár, F.; Hegyi, A.; Finnegan, A.; Mikecz, K. Proteoglycan-induced arthritis and recombinant human proteoglycan aggrecan G1 domain-induced arthritis in BALB/c mice resembling two subtypes of rheumatoid arthritis. *Arthritis Rheum.* **2011**, *63*, 1312–1321. [CrossRef]
141. De Jong, H.; E Berlo, S.; Hombrink, P.; Otten, H.G.; Van Eden, W.; Lafeber, F.P.; Heurkens, A.H.M.; Bijlsma, J.W.J.; Glant, T.T.; Prakken, B. Cartilage proteoglycan aggrecan epitopes induce proinflammatory autoreactive T-cell responses in rheumatoid arthritis and osteoarthritis. *Ann. Rheum. Dis.* **2009**, *69*, 255–262. [CrossRef] [PubMed]
142. Murad, Y.; Szabó, Z.; Ludanyi, K.; Glant, T.T. Molecular manipulation with the arthritogenic epitopes of the G1 domain of human cartilage proteoglycan aggrecan. *Clin. Exp. Immunol.* **2005**, *142*, 303–311. [CrossRef] [PubMed]
143. Poon, C.J.; Plaas, A.H.; Keene, D.R.; McQuillan, D.J.; Last, K.; Fosang, A.J. N-Linked Keratan Sulfate in the Aggrecan Interglobular Domain Potentiates Aggrecanase Activity. *J. Biol. Chem.* **2005**, *280*, 23615–23621. [CrossRef] [PubMed]
144. A Nieduszynski, I.; Huckerby, T.N.; Dickenson, J.M.; Brown, G.M.; Tai, G.H.; Morris, H.G.; Eady, S. There are two major types of skeletal keratan sulphates. *Biochem. J.* **1990**, *271*, 243–245. [CrossRef] [PubMed]
145. Fischer, D.C.; Haubeck, H.D.; Eich, K.; Kolbe-Busch, S.; Stocker, G.; Stuhlsatz, H.W.; Greiling, H. A novel keratan sulphate domain preferentially expressed on the large aggregating proteoglycan from human articular cartilage is recognized by the monoclonal antibody 3D12/H7. *Biochem. J.* **1996**, *318 Pt 3*, 1051–1056. [CrossRef]
146. Gibson, B.G.; Briggs, M.D. The aggrecanopathies; An evolving phenotypic spectrum of human genetic skeletal diseases. *Orphanet J. Rare Dis.* **2016**, *11*, 86. [CrossRef] [PubMed]
147. Nilsson, O.; Guo, M.H.; Dunbar, N.; Popovic, J.; Flynn, D.; Jacobsen, C.; Lui, J.C.; Hirschhorn, J.N.; Baron, J.; Dauber, A. Short stature, accelerated bone maturation, and early growth cessation due to heterozygous aggrecan mutations. *J. Clin. Endocrinol. Metab.* **2014**, *99*, E1510–E1518. [CrossRef]
148. Schwartz, N.B.; Domowicz, M. Chondrodysplasias due to proteoglycan defects. *Glycobiology* **2002**, *12*, 57R–68R. [CrossRef]
149. Li, H.; Schwartz, N.B.; Vertel, B.M. cDNA cloning of chick cartilage chondroitin sulfate (aggrecan) core protein and identification of a stop codon in the aggrecan gene associated with the chondrodystrophy, nanomelia. *J. Biol. Chem.* **1993**, *268*, 23504–23511.
150. Kurima, K.; Warman, M.L.; Krishnan, S.; Domowicz, M.; Krueger, R.C.; Deyrup, A.; Schwartz, N.B. A member of a family of sulfate-activating enzymes causes murine brachymorphism. *Proc. Natl. Acad. Sci. USA* **1998**, *95*, 8681–8685. [CrossRef]
151. Hästbacka, J.; De La Chapelle, A.; Mahtani, M.M.; Clines, G.; Reeve-Daly, M.P.; Daly, M.; Hamilton, B.A.; Kusumi, K.; Trivedi, B.; Weaver, A.; et al. The diastrophic dysplasia gene encodes a novel sulfate transporter: Positional cloning by fine-structure linkage disequilibrium mapping. *Cell* **1994**, *78*, 1073–1087. [CrossRef]

152. Hästbacka, J.; Superti-Furga, A.; Wilcox, W.R.; Rimoin, D.L.; Cohn, D.H.; Lander, E.S. Atelosteogenesis type II is caused by mutations in the diastrophic dysplasia sulfate-transporter gene (DTDST): Evidence for a phenotypic series involving three chondrodysplasias. *Am. J. Hum. Genet.* **1996**, *58*, 255–262. [PubMed]
153. Superti-Furga, A.; Rossi, A.; Steinmann, B.; Gitzelmann, R. A chondrodysplasia family produced by mutations in the diastrophic dysplasia sulfate transporter gene: Genotype/phenotype correlations. *Am. J. Med. Genet.* **1996**, *63*, 144–147. [CrossRef]
154. Gkourogianni, A.; Andrew, M.; Tyzinski, L.; Crocker, M.; Douglas, J.; Dunbar, N.; Fairchild, J.; Funari, M.F.A.; Heath, K.E.; Jorge, A.A.L.; et al. Clinical characterization of patients with autosomal dominant short stature due to aggrecan mutations. *J. Clin. Endocrinol. Metab.* **2016**, *102*, 460–469. [CrossRef] [PubMed]
155. Sentchordi, L.; Aza-Carmona, M.; Benito-Sanz, S.; Bonis, A.C.B.-; Sánchez-Garre, C.; Prieto-Matos, P.; Ruiz-Ocaña, P.; Lechuga-Sancho, A.M.; Carcavilla-Urquí, A.; Mulero-Collantes, I.; et al. Heterozygous aggrecan variants are associated with short stature and brachydactyly: Description of 16 probands and a review of the literature. *Clin. Endocrinol.* **2018**, *88*, 820–829. [CrossRef] [PubMed]
156. Xu, D.; Sun, C.; Zhou, Z.; Wu, B.-B.; Yang, L.; Chang, Z.; Zhang, M.; Xi, L.; Cheng, R.; Ni, J.; et al. Novel aggrecan variant, p. Gln2364Pro, causes severe familial nonsyndromic adult short stature and poor growth hormone response in Chinese children. *BMC Med. Genet.* **2018**, *19*, 79. [CrossRef] [PubMed]
157. Bates, J.T.; Jacobs, J.C.; Shea, K.G.; Oxford, J. Emerging Genetic Basis of Osteochondritis Dissecans. *Clin. Sports Med.* **2014**, *33*, 199–220. [CrossRef]
158. Stattin, E.-L.; Wiklund, F.; Lindblom, K.; Önnerfjord, P.; Jonsson, B.-A.; Tegner, Y.; Sasaki, T.; Struglics, A.; Lohmander, S.; Dahl, N.; et al. A Missense Mutation in the Aggrecan C-type Lectin Domain Disrupts Extracellular Matrix Interactions and Causes Dominant Familial Osteochondritis Dissecans. *Am. J. Hum. Genet.* **2010**, *86*, 126–137. [CrossRef]
159. Xu, M.; Stattin, E.-L.; Shaw, G.; Heinegård, D.; Sullivan, G.J.; Wilmut, I.; Colman, A.; Önnerfjord, P.; Khabut, A.; Aspberg, A.; et al. Chondrocytes Derived From Mesenchymal Stromal Cells and Induced Pluripotent Cells of Patients With Familial Osteochondritis Dissecans Exhibit an Endoplasmic Reticulum Stress Response and Defective Matrix Assembly. *Stem Cells Transl. Med.* **2016**, *5*, 1171–1181. [CrossRef]
160. Dai, J.; Kim, O.-H.; Cho, T.-J.; Schmidt-Rimpler, M.; Tonoki, H.; Takikawa, K.; Haga, N.; Miyoshi, K.; Kitoh, H.; Yoo, W.-J.; et al. Novel and recurrent TRPV4 mutations and their association with distinct phenotypes within the TRPV4 dysplasia family. *J. Med. Genet.* **2010**, *47*, 704–709. [CrossRef]
161. Hurd, L.; Kirwin, S.M.; Boggs, M.; MacKenzie, W.G.; Bober, M.B.; Funanage, V.L.; Duncan, R.L. A mutation in TRPV4 results in altered chondrocyte calcium signaling in severe metatropic dysplasia. *Am. J. Med. Genet. Part A* **2015**, *167*, 2286–2293. [CrossRef] [PubMed]
162. Kang, S.S.; Shin, S.H.; Auh, C.-K.; Chun, J. Human skeletal dysplasia caused by a constitutive activated transient receptor potential vanilloid 4 (TRPV4) cation channel mutation. *Exp. Mol. Med.* **2012**, *44*, 707–722. [CrossRef] [PubMed]
163. Phan, M.N.; Leddy, H.A.; Votta, B.J.; Kumar, S.; Levy, D.S.; Lipshutz, D.B.; Lee, S.H.; Liedtke, W.; Guilak, F. Functional characterization of TRPV4 as an osmotically sensitive ion channel in porcine articular chondrocytes. *Arthritis Rheum.* **2009**, *60*, 3028–3037. [CrossRef] [PubMed]
164. Saitta, B.; Passarini, J.; Sareen, D.; Ornelas, L.; Sahabian, A.; Argade, S.; Krakow, D.; Cohn, D.H.; Svendsen, C.N.; Rimoin, D.L. Patient-derived skeletal dysplasia induced pluripotent stem cells display abnormal chondrogenic marker expression and regulation by BMP2 and TGFbeta1. *Stem Cells Dev.* **2014**, *23*, 1464–1478. [CrossRef]
165. Lauing, K.; Cortes, M.; Domowicz, M.S.; Henry, J.G.; Baria, A.T.; Schwartz, N.B. Aggrecan is required for growth plate architecture and differentiation. *Dev. Biol.* **2014**, *396*, 224–236. [CrossRef]
166. Shimbo, M.; Suzuki, R.; Fuseya, S.; Sato, T.; Kiyohara, K.; Hagiwara, K.; Okada, R.; Wakui, H.; Tsunakawa, Y.; Watanabe, H.; et al. Postnatal Lethality and Chondrodysplasia in Mice Lacking Both Chondroitin Sulfate N-acetylgalactosaminyltransferase-1 and -2. *PLoS ONE* **2017**, *12*, e0190333. [CrossRef]
167. Bekku, Y.; Saito, M.; Moser, M.; Fuchigami, M.; Maehara, A.; Nakayama, M.; Kusachi, S.; Ninomiya, Y.; Oohashi, T. Bral2 is indispensable for the proper localization of brevican and the structural integrity of the perineuronal net in the brainstem and cerebellum. *J. Comp. Neurol.* **2012**, *520*, 1721–1736. [CrossRef]
168. Ueno, H.; Fujii, K.; Suemitsu, S.; Murakami, S.; Kitamura, N.; Wani, K.; Aoki, S.; Okamoto, M.; Ishihara, T.; Takao, K. Expression of aggrecan components in perineuronal nets in the mouse cerebral cortex. *IBRO Rep.* **2018**, *4*, 22–37. [CrossRef]

169. Yamada, J.; Jinno, S. Molecular heterogeneity of aggrecan-based perineuronal nets around five subclasses of parvalbumin-expressing neurons in the mouse hippocampus. *J. Comp. Neurol.* **2016**, *525*, 1234–1249. [CrossRef]
170. Bozzelli, P.L.; Alaiyed, S.; Kim, E.; Villapol, S.; Conant, K. Proteolytic Remodeling of Perineuronal Nets: Effects on Synaptic Plasticity and Neuronal Population Dynamics. *Neural Plast.* **2018**, *2018*, 1–13. [CrossRef]
171. Creazzo, T.L.; Godt, R.E.; Leatherbury, L.; Conway, S.J.; Kirby, M.L. ROLE OF CARDIAC NEURAL CREST CELLS IN CARDIOVASCULAR DEVELOPMENT. *Annu. Rev. Physiol.* **1998**, *60*, 267–286. [CrossRef] [PubMed]
172. Arciniegas, E.; Neves, C.Y.; Candelle, D.; Parada, D. Differential versican isoforms and aggrecan expression in the chicken embryo aorta. *Anat. Rec. A Discov. Mol. Cell. Evol. Biol.* **2004**, *279*, 592–600. [CrossRef] [PubMed]
173. Foglia, M.; Poss, K.D. Building and re-building the heart by cardiomyocyte proliferation. *Development* **2016**, *143*, 729–740. [CrossRef] [PubMed]
174. Xin, M.; Olson, E.N.; Bassel-Duby, R. Mending broken hearts: Cardiac development as a basis for adult heart regeneration and repair. *Nat. Rev. Mol. Cell. Biol.* **2013**, *14*, 529–541. [CrossRef]
175. Wakamatsu, Y.; Osumi, N.; a Weston, J. Expression of a novel secreted factor, Seraf indicates an early segregation of Schwann cell precursors from neural crest during avian development. *Dev. Biol.* **2004**, *268*, 162–173. [CrossRef]
176. Giovannone, D.; Ortega, B.; Reyes, M.; El-Ghali, N.; Rabadi, M.; Sao, S.; De Bellard, M.E. Chicken trunk neural crest migration visualized with HNK1. *Acta Histochem.* **2015**, *117*, 255–266. [CrossRef]
177. Krueger, R.C.; a Fields, T.; Mensch, J.R.; Schwartz, N.B. Chick cartilage chondroitin sulfate proteoglycan core protein. II. Nucleotide sequence of cDNA clone and localization of the S103L epitope. *J. Biol. Chem.* **1990**, *265*, 12088–12097.
178. Krueger, R.C.; Hennig, A.K.; Schwartz, N.B. Two immunologically and developmentally distinct chondroitin sulfate proteolglycans in embryonic chick brain. *J. Biol. Chem.* **1992**, *267*, 12149–12161.
179. Pettway, Z.; Domowicz, M.; Schwartz, N.B.; Bronner-Fraser, M. Age-Dependent Inhibition of Neural Crest Migration by the Notochord Correlates with Alterations in the S103L Chondroitin Sulfate Proteoglycan. *Exp. Cell Res.* **1996**, *225*, 195–206. [CrossRef]
180. Hirotani, H. The Calcineurin/Nuclear Factor of Activated T Cells Signaling Pathway Regulates Osteoclastogenesis in RAW264.7 Cells. *J. Biol. Chem.* **2004**, *279*, 13984–13992. [CrossRef]
181. De La Pompa, J.L.; Timmerman, L.A.; Takimoto, H.; Yoshida, H.; Elia, A.J.; Samper, E.; Potter, J.; Wakeham, A.; Marengere, L.; Langille, B.L.; et al. Role of the NF-ATc transcription factor in morphogenesis of cardiac valves and septum. *Nature* **1998**, *392*, 182–186. [CrossRef] [PubMed]
182. Ranger, A.M.; Grusby, M.J.; Hodge, M.R.; Gravallese, E.M.; De La Brousse, F.C.; Hoey, T.; Mickanin, C.; Baldwin, H.S.; Glimcher, L.H. The transcription factor NF-ATc is essential for cardiac valve formation. *Nature* **1998**, *392*, 186–190. [CrossRef] [PubMed]
183. Arnold, M.A.; Kim, Y.; Czubryt, M.P.; Phan, D.; McAnally, J.; Qi, X.; Shelton, J.M.; Richardson, J.A.; Bassel-Duby, R.; Olson, E.N. MEF2C Transcription Factor Controls Chondrocyte Hypertrophy and Bone Development. *Dev. Cell* **2007**, *12*, 377–389. [CrossRef] [PubMed]
184. Qiao, X.-H.; Wang, F.; Zhang, X.-L.; Huang, R.-T.; Xue, S.; Wang, J.; Qiu, X.-B.; Liu, X.-Y.; Yang, Y.-Q. MEF2C loss-of-function mutation contributes to congenital heart defects. *Int. J. Med. Sci.* **2017**, *14*, 1143–1153. [CrossRef] [PubMed]
185. Wirrig, E.E.; Hinton, R.B.; Yutzey, K.E. Differential expression of cartilage and bone-related proteins in pediatric and adult diseased aortic valves. *J. Mol. Cell. Cardiol.* **2011**, *50*, 561–569. [CrossRef]
186. Domowicz, M.; Sanders, T.A.; Ragsdale, C.W.; Schwartz, N.B. Aggrecan is expressed by embryonic brain glia and regulates astrocyte development. *Dev. Biol.* **2008**, *315*, 114–124. [CrossRef]
187. Umehara, Y.; Yamada, S.; Nishimura, S.; Shioi, J.; Robakis, N.K.; Sugahara, K. Chondroitin sulfate of appican, the proteoglycan form of amyloid precursor protein, produced by C6 glioma cells interacts with heparin-binding neuroregulatory factors. *FEBS Lett.* **2003**, *557*, 233–238. [CrossRef]
188. Pangalos, M.N.; Shioi, J.; Efthimiopoulos, S.; Wu, A.; Robakis, N.K. Characterization of Appican, the Chondroitin Sulfate Proteoglycan Form of the Alzheimer Amyloid Precursor Protein. *Neurodegeneration* **1996**, *5*, 445–451. [CrossRef]

189. Shioi, J.; Pangalos, M.N.; Ripellino, J.A.; Vassilacopoulou, D.; Mytilineou, C.; Margolis, R.U.; Robakis, N.K. The Alzheimer Amyloid Precursor Proteoglycan (Appican) Is Present in Brain and Is Produced by Astrocytes but Not by Neurons in Primary Neural Cultures. *J. Biol. Chem.* **1995**, *270*, 11839–11844. [CrossRef]
190. Tsuchida, K.; Shioi, J.; Yamada, S.; Boghosian, G.; Wu, A.; Cai, H.; Sugahara, K.; Robakis, N.K. Appican, the Proteoglycan Form of the Amyloid Precursor Protein, Contains Chondroitin Sulfate E in the Repeating Disaccharide Region and 4-O-Sulfated Galactose in the Linkage Region. *J. Biol. Chem.* **2001**, *276*, 37155–37160. [CrossRef]
191. Wu, A.; Pangalos, M.N.; Efthimiopoulos, S.; Shioi, J.; Robakis, N.K. Appican Expression Induces Morphological Changes in C6 Glioma Cells and Promotes Adhesion of Neural Cells to the Extracellular Matrix. *J. Neurosci.* **1997**, *17*, 4987–4993. [CrossRef] [PubMed]
192. Thielen, N.G.; Van Der Kraan, P.M.; Van Caam, A. TGFβ/BMP Signaling Pathway in Cartilage Homeostasis. *Cells* **2019**, *8*, 969. [CrossRef] [PubMed]
193. Salazar, V.S.; Gamer, L.W.; Rosen, V. BMP signalling in skeletal development, disease and repair. *Nat. Rev. Endocrinol.* **2016**, *12*, 203–221. [CrossRef] [PubMed]
194. Wang, R.N.; Green, J.; Wang, Z.; Deng, Y.; Qiao, M.; Peabody, M.; Zhang, Q.; Ye, J.; Yan, Z.; Denduluri, S.; et al. Bone Morphogenetic Protein (BMP) signaling in development and human diseases. *Genes Dis.* **2014**, *1*, 87–105. [CrossRef] [PubMed]
195. Chen, D.; Zhao, M.; Mundy, G.R. Bone morphogenetic proteins. *Growth Factors* **2004**, *22*, 233–241. [CrossRef] [PubMed]
196. Hoffmann, A.; Gross, G. BMP signaling pathways in cartilage and bone formation. *Crit. Rev. Eukaryot. Gene Expr.* **2001**, *11*, 24. [CrossRef]
197. Mak, K.K.; Kronenberg, H.M.; Chuang, P.-T.; Mackem, S.; Yang, Y. Indian hedgehog signals independently of PTHrP to promote chondrocyte hypertrophy. *Development* **2008**, *135*, 1947–1956. [CrossRef]
198. Pathi, S.; Pagan-Westphal, S.; Baker, D.P.; a Garber, E.; Rayhorn, P.; Bumcrot, D.; Tabin, C.J.; Pepinsky, R.B.; Williams, K.P. Comparative biological responses to human Sonic, Indian, and Desert hedgehog. *Mech. Dev.* **2001**, *106*, 107–117. [CrossRef]
199. Vortkamp, A.; Lee, K.; Lanske, B.; Segre, G.V.; Kronenberg, H.M.; Tabin, C.J. Regulation of Rate of Cartilage Differentiation by Indian Hedgehog and PTH-Related Protein. *Science* **1996**, *273*, 613–622. [CrossRef]
200. Kronenberg, H.M. PTHrP and Skeletal Development. *Ann. N. Y. Acad. Sci.* **2006**, *1068*, 1–13. [CrossRef]
201. Grimsrud, C.D.; Romano, P.R.; D'Souza, M.; Puzas, J.E.; Schwarz, E.M.; Reynolds, P.R.; Roiser, R.N.; O'Keefe, R.J. BMP signaling stimulates chondrocyte maturation and the expression of Indian hedgehog. *J. Orthop. Res.* **2001**, *19*, 18–25. [CrossRef]
202. Lai, L.P.; Mitchell, J. Indian hedgehog: Its roles and regulation in endochondral bone development. *J. Cell. Biochem.* **2005**, *96*, 1163–1173. [CrossRef] [PubMed]
203. Chen, X.; Macica, C.M.; Nasiri, A.; Broadus, A.E. Regulation of articular chondrocyte proliferation and differentiation by indian hedgehog and parathyroid hormone-related protein in mice. *Arthritis Rheum.* **2008**, *58*, 3788–3797. [CrossRef] [PubMed]
204. Wei, F.; Zhou, J.; Wei, X.; Zhang, J.; Fleming, B.C.; Terek, R.; Pei, M.; Chen, Q.; Liu, T.; Wei, L. Activation of Indian Hedgehog Promotes Chondrocyte Hypertrophy and Upregulation of MMP-13 in Human Osteoarthritic Cartilage. *Osteoarthr. Cartil.* **2012**, *20*, 755–763. [CrossRef] [PubMed]
205. Ohba, S. Hedgehog Signaling in Endochondral Ossification. *J. Dev. Biol.* **2016**, *4*, 20. [CrossRef] [PubMed]
206. Ruoslahti, E.; Yamaguchi, Y. Proteoglycans as modulators of growth factor activities. *Cell* **1991**, *64*, 867–869. [CrossRef]
207. Yoshihara, Y.; Oka, S.; Watanabe, Y.; Mori, K. Developmentally and spatially regulated expression of HNK-1 carbohydrate antigen on a novel phosphatidylinositol-anchored glycoprotein in rat brain. *J. Cell Biol.* **1991**, *115*, 731–744. [CrossRef]
208. Allory, Y.; Commo, F.; Boccon-Gibod, L.; Sibony, M.; Callard, P.; Ronco, P.; Debiec, H. Sulfated HNK-1 Epitope in Developing and Mature Kidney: A New Marker for Thin Ascending Loop of Henle and Tubular Injury in Acute Tubular Necrosis. *J. Histochem. Cytochem.* **2006**, *54*, 575–584. [CrossRef]
209. A Blom, N.; Groot, A.C.G.-D.; DeRuiter, M.C.; E Poelmann, R.; Mentink, M.M.; Ottenkamp, J. Development of the cardiac conduction tissue in human embryos using HNK-1 antigen expression: Possible relevance for understanding of abnormal atrial automaticity. *Circulation* **1999**, *99*, 800–806. [CrossRef]

210. Nakamura, T.; Ikeda, T.; Shimokawa, I.; Inoue, Y.; Suematsu, T.; Sakai, H.; Iwasaki, K.; Matsuo, T. Distribution of acetylcholinesterase activity in the rat embryonic heart with reference to HNK-1 immunoreactivity in the conduction tissue. *Anat. Embryol. Berl* **1994**, *190*, 367–373. [CrossRef]
211. Kivelä, T. Expression of the HNK-1 carbohydrate epitope in human retina and retinoblastoma. An immunohistochemical study with the anti-Leu-7 monoclonal antibody. *Virchows Arch. A Pathol. Anat. Histopathol.* **1986**, *410*, 139–146. [CrossRef] [PubMed]
212. Sonntag, M.; Blosa, M.; Schmidt, S.; Rübsamen, R.; Morawski, M. Perineuronal nets in the auditory system. *Hear. Res.* **2015**, *329*, 21–32. [CrossRef] [PubMed]
213. Miura, R.; Aspberg, A.; Ethell, I.M.; Hagihara, K.; Schnaar, R.L.; Ruoslahti, E.; Yamaguchi, Y. The Proteoglycan Lectin Domain Binds Sulfated Cell Surface Glycolipids and Promotes Cell Adhesion. *J. Biol. Chem.* **1999**, *274*, 11431–11438. [CrossRef] [PubMed]
214. Miura, R.; Ethell, I.M.; Yamaguchi, Y. Carbohydrate-protein interactions between HNK-1-reactive sulfoglucuronyl glycolipids and the proteoglycan lectin domain mediate neuronal cell adhesion and neurite outgrowth. *J. Neurochem.* **2001**, *76*, 413–424. [CrossRef] [PubMed]
215. Roberts, D.D.; Ginsburg, V. Sulfated glycolipids and cell adhesion. *Arch. Biochem. Biophys.* **1988**, *267*, 405–415. [CrossRef]
216. Roberts, D.D.; Rao, C.N.; Magnani, J.L.; Spitalnik, S.L.; Liotta, L.A.; Ginsburg, V. Laminin binds specifically to sulfated glycolipids. *Proc. Natl. Acad. Sci. USA* **1985**, *82*, 1306–1310. [CrossRef]
217. Balint, R.; Cassidy, N.J.; Cartmell, S. Conductive polymers: Towards a smart biomaterial for tissue engineering. *Acta Biomater.* **2014**, *10*, 2341–2353. [CrossRef]
218. Ong, E.; Suzuki, M.; Bélot, F.; Yeh, J.-C.; Franceschini, I.; Angata, K.; Hindsgaul, O.; Fukuda, M. Biosynthesis of HNK-1 Glycans onO-Linked Oligosaccharides Attached to the Neural Cell Adhesion Molecule (NCAM). *J. Biol. Chem.* **2002**, *277*, 18182–18190. [CrossRef]
219. Ong, E. Structure and Function of HNK-1 Sulfotransferase. IDENTIFICATION OF DONOR AND ACCEPTOR BINDING SITES BY SITE-DIRECTED MUTAGENESIS. *J. Biol. Chem.* **1999**, *274*, 25608–25612. [CrossRef]
220. Nakagawa, N.; Izumikawa, T.; Kitagawa, H.; Oka, S. Sulfation of glucuronic acid in the linkage tetrasaccharide by HNK-1 sulfotransferase is an inhibitory signal for the expression of a chondroitin sulfate chain on thrombomodulin. *Biochem. Biophys. Res. Commun.* **2011**, *415*, 109–113. [CrossRef]
221. Nakagawa, N.; Manya, H.; Toda, T.; Endo, T.; Oka, S. Human Natural Killer-1 Sulfotransferase (HNK-1ST)-induced Sulfate Transfer Regulates Laminin-binding Glycans on α-Dystroglycan. *J. Biol. Chem.* **2012**, *287*, 30823–30832. [CrossRef] [PubMed]
222. Hashiguchi, T.; Mizumoto, S.; Nishimura, Y.; Tamura, J.-I.; Yamada, S.; Sugahara, K. Involvement of Human Natural Killer-1 (HNK-1) Sulfotransferase in the Biosynthesis of the GlcUA(3-O-sulfate)-Gal-Gal-Xyl Tetrasaccharide Found in α-Thrombomodulin from Human Urine. *J. Biol. Chem.* **2011**, *286*, 33003–33011. [CrossRef] [PubMed]
223. Aspberg, A.; Miura, R.; Bourdoulous, S.; Shimonaka, M.; Heinegård, D.; Schachner, M.; Ruoslahti, E.; Yamaguchi, Y. The C-type lectin domains of lecticans, a family of aggregating chondroitin sulfate proteoglycans, bind tenascin-R by protein- protein interactions independent of carbohydrate moiety. *Proc. Natl. Acad. Sci. USA* **1997**, *94*, 10116–10121. [CrossRef] [PubMed]
224. Yagi, H.; Yanagisawa, M.; Suzuki, Y.; Nakatani, Y.; Ariga, T.; Kato, K.; Yu, R.K. HNK-1 Epitope-carrying Tenascin-C Spliced Variant Regulates the Proliferation of Mouse Embryonic Neural Stem Cells. *J. Biol. Chem.* **2010**, *285*, 37293–37301. [CrossRef]
225. Saghatelyan, A.; Gorissen, S.; Albert, M.; Hertlein, B.; Schachner, M.; Dityatev, A. The extracellular matrix molecule tenascin-R and its HNK-1 carbohydrate modulate perisomatic inhibition and long-term potentiation in the CA1 region of the hippocampus. *Eur. J. Neurosci.* **2000**, *12*, 3331–3342. [CrossRef]
226. Suttkus, A.; Rohn, S.; Weigel, S.; Glöckner, P.; Arendt, T.; Morawski, M. Aggrecan, link protein and tenascin-R are essential components of the perineuronal net to protect neurons against iron-induced oxidative stress. *Cell Death Dis.* **2014**, *5*, e1119. [CrossRef]
227. Melrose, J.; Ghosh, P. The quantitative discrimination of corneal type I, but not skeletal type II, keratan sulfate in glycosaminoglycan mixtures by using a combination of dimethylmethylene blue and endo-beta-D-galactosidase digestion. *Anal. Biochem.* **1988**, *170*, 293–300. [CrossRef]
228. Hitchcock, A.; Yates, K.E.; Shortkroff, S.; Costello, C.E.; Zaia, J. Optimized extraction of glycosaminoglycans from normal and osteoarthritic cartilage for glycomics profiling. *Glycobiology* **2007**, *17*, 25–35. [CrossRef]

229. Samiric, T.; Ilic, M.Z.; Handley, C.J. Characterisation of proteoglycans and their catabolic products in tendon and explant cultures of tendon. *Matrix Biol.* **2004**, *23*, 127–140. [CrossRef]
230. Bayliss, M.T. The Organization of Aggrecan in Human Articular Cartilage. EVIDENCE FOR AGE-RELATED CHANGES IN THE RATE OF AGGREGATION OF NEWLY SYNTHESIZED MOLECULES. *J. Biol. Chem.* **2000**, *275*, 6321–6327. [CrossRef]
231. Lyon, M.; Greenwood, J.; Sheehan, J.K.; a Nieduszynski, I. Isolation and characterization of high-buoyant-density proteoglycans from bovine femoral-head cartilage. *Biochem. J.* **1983**, *213*, 355–362. [CrossRef] [PubMed]
232. Vilím, V.; Fosang, A.J. Proteoglycans isolated from dissociative extracts of differently aged human articular cartilage: Characterization of naturally occurring hyaluronan-binding fragments of aggrecan. *Biochem. J.* **1994**, *304*, 887–894. [CrossRef] [PubMed]
233. Hileman, R.E.; Fromm, J.R.; Weiler, J.M.; Linhardt, R.J. Glycosaminoglycan-protein interactions: Definition of consensus sites in glycosaminoglycan binding proteins. *BioEssays* **1998**, *20*, 156–167. [CrossRef]
234. Song, Y.; Zhang, F.; Linhardt, R.J. Analysis of the Glycosaminoglycan Chains of Proteoglycans. *J. Histochem. Cytochem.* **2020**. [CrossRef] [PubMed]
235. Vallet, S.; Clerc, O.; Ricard-Blum, S. Glycosaminoglycan-Protein Interactions: The First Draft of the Glycosaminoglycan Interactome. *J. Histochem. Cytochem.* **2020**. [CrossRef]
236. Karamanos, N.; Hjerpe, A. Strategies for analysis and structure characterization of glycans/proteoglycans by capillary electrophoresis. Their diagnostic and biopharmaceutical importance. *Biomed. Chromatogr.* **1999**, *13*, 507–512. [CrossRef]
237. Lamari, F.N.; Militsopoulou, M.; Mitropoulou, T.N.; Hjerpe, A.; Karamanos, N.K. Analysis of glycosaminoglycan-derived disaccharides in biologic samples by capillary electrophoresis and protocol for sequencing glycosaminoglycans. *Biomed. Chromatogr.* **2002**, *16*, 95–102. [CrossRef]
238. Lu, G.; Crihfield, C.L.; Gattu, S.; Veltri, L.M.; Holland, L.A. Capillary Electrophoresis Separations of Glycans. *Chem. Rev.* **2018**, *118*, 7867–7885. [CrossRef]
239. Gill, V.L.; Aich, U.; Rao, S.; Pohl, C.A.; Zaia, J. Disaccharide Analysis of Glycosaminoglycans Using Hydrophilic Interaction Chromatography and Mass Spectrometry. *Anal. Chem.* **2013**, *85*, 1138–1145. [CrossRef]
240. Staples, G.O.; Zaia, J. Analysis of Glycosaminoglycans Using Mass Spectrometry. *Curr. Proteom.* **2011**, *8*, 325–336. [CrossRef]
241. Zaia, J. Glycosaminoglycan glycomics using mass spectrometry. *Mol. Cell. Proteom.* **2013**, *12*, 885–892. [CrossRef] [PubMed]
242. Volpi, N. Disaccharide Analysis and Molecular Mass Determination to Microgram Level of Single Sulfated Glycosaminoglycan Species in Mixtures Following Agarose-Gel Electrophoresis. *Anal. Biochem.* **1999**, *273*, 229–239. [CrossRef] [PubMed]
243. Ernst, S.; Langer, R.; Cooney, C.L.; Sasisekharan, R. Enzymatic degradation of glycosaminoglycans. *Crit. Rev. Biochem. Mol. Biol.* **1995**, *30*, 387–444. [CrossRef] [PubMed]
244. Hernáiz, M.J.; Linhardt, R.J.; Iozzo, R.V. Degradation of Chondroitin Sulfate and Dermatan Sulfate with Chondroitin Lyases. *Methods Mol. Biol.* **2001**, *171*, 363–371. [CrossRef] [PubMed]
245. Kaneiwa, T.; Mizumoto, S.; Sugahara, K.; Yamada, S. Identification of human hyaluronidase-4 as a novel chondroitin sulfate hydrolase that preferentially cleaves the galactosaminidic linkage in the trisulfated tetrasaccharide sequence. *Glycobiology* **2009**, *20*, 300–309. [CrossRef]
246. Linhardt, R.J.; Avci, F.Y.; Toida, T.; Kim, Y.S.; Cygler, M. CS Lyases: Structure, Activity, and Applications in Analysis and the Treatment of Diseases. *Adv. Pharmacol.* **2006**, *53*, 187–215. [CrossRef]
247. Petit, E.; Delattre, C.; Papy–Garcia, D.; Michaud, P. Chondroitin Sulfate Lyases: Applications in Analysis and Glycobiology. *Adv. Pharmacol* **2006**, *53*, 167–186. [CrossRef]
248. Plaas, A.H.; West, L.; Midura, R.J.; Hascall, V.C.; Iozzo, R.V. Disaccharide Composition of Hyaluronan and Chondroitin/Dermatan Sulfate: Analysis with Fluorophore-Assisted Carbohydrate Electrophoresis. *Methods Mol. Biol.* **2003**, *171*, 117–128. [CrossRef]
249. Komarraju, A.; Goldberg-Stein, S.; Pederson, R.; McCrum, C.; Chhabra, A. Spectrum of common and uncommon causes of knee joint hyaline cartilage degeneration and their key imaging features. *Eur. J. Radiol.* **2020**, *129*, 109097. [CrossRef]

250. Kim, J.; Mamoto, K.; Lartey, R.; Nakamura, K.; Shin, W.; Winalski, C.S.; Obuchowski, N.; Tanaka, M.; Bahroos, E.; Link, T.M.; et al. Multi-vendor multi-site T1ρ and T2 quantification of knee cartilage. *Osteoarthr. Cartil.* **2020**. [CrossRef]
251. Abrar, D.B.; Schleich, C.; Nebelung, S.; Frenken, M.; Ullrich, T.; Radke, K.L.; Antoch, G.; Vordenbäumen, S.; Brinks, R.; Schneider, M.; et al. Proteoglycan loss in the articular cartilage is associated with severity of joint inflammation in psoriatic arthritis—A compositional magnetic resonance imaging study. *Arthritis Res.* **2020**, *22*, 124. [CrossRef] [PubMed]
252. Abrar, D.B.; Schleich, C.; Radke, K.L.; Frenken, M.; Stabinska, J.; Ljimani, A.; Wittsack, H.-J.; Antoch, G.; Bittersohl, B.; Hesper, T.; et al. Detection of early cartilage degeneration in the tibiotalar joint using 3 T gagCEST imaging: A feasibility study. *Magn. Reson. Mater. Physics, Biol. Med.* **2020**, 1–12. [CrossRef] [PubMed]
253. Dou, W.; Lin, C.-Y.E.; Ding, H.; Shen, Y.; Dou, C.; Qian, L.; Wen, B.; Wu, B. Chemical exchange saturation transfer magnetic resonance imaging and its main and potential applications in pre-clinical and clinical studies. *Quant. Imaging Med. Surg.* **2019**, *9*, 1747–1766. [CrossRef] [PubMed]
254. Müller-Lutz, A.; Kamp, B.; Nagel, A.M.; Ljimani, A.; Abrar, D.; Schleich, C.; Wollschläger, L.; Nebelung, S.; Wittsack, H.-J. Sodium MRI of human articular cartilage of the wrist: A feasibility study on a clinical 3T MRI scanner. *Magn. Reson. Mater. Physics Biol. Med.* **2020**. [CrossRef]
255. Zbyn, S.; Schreiner, M.; Juras, V.; Mlynarik, V.; Szomolanyi, P.; Laurent, D.; Scotti, C.; Haber, H.; Deligianni, X.; Bieri, O.; et al. Assessment of Low-Grade Focal Cartilage Lesions in the Knee with Sodium MRI at 7 T. *Investig. Radiol.* **2020**, *55*, 430–437. [CrossRef]
256. Tibrewala, R.; Pedoia, V.; Bucknor, M.; Majumdar, S. Principal Component Analysis of Simultaneous PET-MRI Reveals Patterns of Bone-Cartilage Interactions in Osteoarthritis. *J. Magn. Reson. Imaging* **2020**. [CrossRef]
257. Noguerol, T.M.; Raya, J.G.; E Wessell, D.; Vilanova, J.C.; Rossi, I.; Luna, A. Functional MRI for evaluation of hyaline cartilage extracelullar matrix, a physiopathological-based approach. *Br. J. Radiol.* **2019**, *92*, 20190443. [CrossRef]
258. Muir, E.M.; De Winter, F.; Verhaagen, J.; Fawcett, J. Recent advances in the therapeutic uses of chondroitinase ABC. *Exp. Neurol.* **2019**, *321*, 113032. [CrossRef]
259. Rani, A.; Patel, S.; Goyal, A. Chondroitin Sulfate (CS) Lyases: Structure, Function and Application in Therapeutics. *Curr. Protein Pept. Sci.* **2017**, *19*, 22–33. [CrossRef]
260. Morgenstern, D.; Asher, R.A.; Fawcett, J.W. Chondroitin sulphate proteoglycans in the CNS injury response. *Prog. Brain Res.* **2002**, *137*, 313–332. [CrossRef]
261. Tachi, Y.; Okuda, T.; Kawahara, N.; Kato, N.; Ishigaki, Y.; Matsumoto, T. Expression of Hyaluronidase-4 in a Rat Spinal Cord Hemisection Model. *Asian Spine J.* **2015**, *9*, 7–13. [CrossRef] [PubMed]
262. Bradbury, E.J.; Carter, L.M. Manipulating the glial scar: Chondroitinase ABC as a therapy for spinal cord injury. *Brain Res. Bull.* **2011**, *84*, 306–316. [CrossRef] [PubMed]
263. Barritt, A.W.; Davies, M.; Marchand, F.; Hartley, R.; Grist, J.; Yip, P.; McMahon, S.B.; Bradbury, E.J. Chondroitinase ABC promotes sprouting of intact and injured spinal systems after spinal cord injury. *J. Neurosci.* **2006**, *26*, 10856–10867. [CrossRef] [PubMed]
264. Cafferty, W.B.J.; Bradbury, E.J.; Lidierth, M.; Jones, M.; Duffy, P.J.; Pezet, S.; McMahon, S.B. Chondroitinase ABC-mediated plasticity of spinal sensory function. *J. Neurosci.* **2008**, *28*, 11998–12009. [CrossRef]
265. Bradbury, E.J.; Moon, L.; Popat, R.J.; King, V.R.; Bennett, G.S.; Patel, P.N.; Fawcett, J.W.; McMahon, S.B. Chondroitinase ABC promotes functional recovery after spinal cord injury. *Nature* **2002**, *416*, 636–640. [CrossRef]
266. James, N.D.; Shea, J.; Muir, E.M.; Verhaagen, J.; Schneider, B.; Bradbury, E.J. Chondroitinase gene therapy improves upper limb function following cervical contusion injury. *Exp. Neurol.* **2015**, *271*, 131–135. [CrossRef]
267. Koh, C.H.; Pronin, S.; Hughes, M. Chondroitinase ABC for neurological recovery after acute brain injury: Systematic review and meta-analyses of preclinical studies. *Brain Inj.* **2018**, *32*, 715–729. [CrossRef]
268. Day, P.; Alves, N.; Daniell, E.; Dasgupta, D.; Ogborne, R.; Steeper, A.; Raza, M.; Ellis, C.; Fawcett, J.; Keynes, R.; et al. Targeting chondroitinase ABC to axons enhances the ability of chondroitinase to promote neurite outgrowth and sprouting. *PLoS ONE* **2020**, *15*, e0221851. [CrossRef]
269. Fawcett, J.W. The extracellular matrix in plasticity and regeneration after CNS injury and neurodegenerative disease. *Prog. Brain Res.* **2015**, *218*, 213–226. [CrossRef]

270. Monfort, J.; Pelletier, J.-P.; Garcia-Giralt, N.; Martel-Pelletier, J. Biochemical basis of the effect of chondroitin sulphate on osteoarthritis articular tissues. *Ann. Rheum. Dis.* **2008**, *67*, 735–740. [CrossRef]
271. Martel-Pelletier, J.; Tat, S.K.; Pelletier, J.-P. Effects of chondroitin sulfate in the pathophysiology of the osteoarthritic joint: A narrative review. *Osteoarthr. Cartil.* **2010**, *18*, S7–S11. [CrossRef] [PubMed]
272. Hochberg, M.; Chevalier, X.; Henrotin, Y.; Hunter, D.; Uebelhart, D. Symptom and structure modification in osteoarthritis with pharmaceutical-grade chondroitin sulfate: What's the evidence? *Curr. Med. Res. Opin.* **2013**, *29*, 259–267. [CrossRef] [PubMed]
273. Honvo, G.; Bruyère, O.; Reginster, J.-Y. Update on the role of pharmaceutical-grade chondroitin sulfate in the symptomatic management of knee osteoarthritis. *Aging Clin. Exp. Res.* **2019**, *31*, 1163–1167. [CrossRef] [PubMed]
274. Pomin, V.H.; Vignovich, W.P.; Gonzales, A.V.; Vasconcelos, A.A.; Mulloy, B. Galactosaminoglycans: Medical Applications and Drawbacks. *Molecules* **2019**, *24*, 2803. [CrossRef]
275. Reginster, J.-Y.; Veronese, N. Highly purified chondroitin sulfate: A literature review on clinical efficacy and pharmacoeconomic aspects in osteoarthritis treatment. *Aging Clin. Exp. Res.* **2020**, 1–11. [CrossRef]
276. Hayes, A.J.; Melrose, J. Glycosaminoglycan and Proteoglycan Biotherapeutics in Articular Cartilage Protection and Repair Strategies: Novel Approaches to Visco–supplementation in Orthobiologics. *Adv. Ther.* **2019**, *2*, 1900034. [CrossRef]
277. Farrugia, B.; Hayes, A.J.; Melrose, J. Use of chondroitin sulphate to aid in-vitro stem cell differentiation In Proteoglycans in stem cells: From development to cancer. *Biol. Extracell. Matrix* **2020**, submitted.
278. Melrose, J. Glycosaminoglycans in Wound Healing. *Bone Tissue Regen. Insights* **2016**, *7*, 29–50. [CrossRef]
279. Farrugia, B.L.; Lord, M.S.; Whitelock, J.M.; Melrose, J. Harnessing chondroitin sulphate in composite scaffolds to direct progenitor and stem cell function for tissue repair. *Biomater. Sci.* **2018**, *6*, 947–957. [CrossRef]

 © 2020 by the authors. Licensee MDPI, Basel, Switzerland. This article is an open access article distributed under the terms and conditions of the Creative Commons Attribution (CC BY) license (http://creativecommons.org/licenses/by/4.0/).

Article

Hair Histology and Glycosaminoglycans Distribution Probed by Infrared Spectral Imaging: Focus on Heparan Sulfate Proteoglycan and Glypican-1 during Hair Growth Cycle

Charlie Colin-Pierre [1,2,3,†], Valérie Untereiner [4,†], Ganesh D. Sockalingum [5], Nicolas Berthélémy [3], Louis Danoux [3], Vincent Bardey [3], Solène Mine [3], Christine Jeanmaire [3], Laurent Ramont [1,2,6,‡] and Stéphane Brézillon [1,2,*,‡]

1. Université de Reims Champagne-Ardenne, Laboratoire de Biochimie Médicale et Biologie Moléculaire, 51097 Reims, France; charlie.pierre@univ-reims.fr (C.C.-P.); lramont@chu-reims.fr (L.R.)
2. CNRS UMR 7369, Matrice Extracellulaire et Dynamique Cellulaire-MEDyC, 51097 Reims, France
3. BASF Beauty Care Solutions France SAS, 54425 Pulnoy, France; nicolas.berthelemy@basf.com (N.B.); louis.danoux@basf.com (L.D.); vincent.bardey@basf.com (V.B.); solene.mine@basf.com (S.M.); christine.jeanmaire@basf.com (C.J.)
4. Université de Reims Champagne-Ardenne, PICT, 51097 Reims, France; valerie.untereiner@univ-reims.fr
5. Université de Reims Champagne-Ardenne, BioSpecT EA7506, UFR de Pharmacie, 51097 Reims, France; ganesh.sockalingum@univ-reims.fr
6. CHU de Reims, Service Biochimie-Pharmacologie-Toxicologie, 51097 Reims, France
* Correspondence: stephane.brezillon@univ-reims.fr
† These 2 authors contributed equally to the manuscript.
‡ Co-last authors.

Citation: Colin-Pierre, C.; Untereiner, V.; Sockalingum, G.D.; Berthélémy, N.; Danoux, L.; Bardey, V.; Mine, S.; Jeanmaire, C.; Ramont, L.; Brézillon, S. Hair Histology and Glycosaminoglycans Distribution Probed by Infrared Spectral Imaging: Focus on Heparan Sulfate Proteoglycan and Glypican-1 during Hair Growth Cycle. *Biomolecules* **2021**, *11*, 192. https://doi.org/10.3390/biom11020192

Academic Editors: Dragana Nikitovic and Serge Perez
Received: 1 December 2020
Accepted: 22 January 2021
Published: 30 January 2021

Publisher's Note: MDPI stays neutral with regard to jurisdictional claims in published maps and institutional affiliations.

Copyright: © 2021 by the authors. Licensee MDPI, Basel, Switzerland. This article is an open access article distributed under the terms and conditions of the Creative Commons Attribution (CC BY) license (https://creativecommons.org/licenses/by/4.0/).

Abstract: The expression of glypicans in different hair follicle (HF) compartments and their potential roles during hair shaft growth are still poorly understood. Heparan sulfate proteoglycan (HSPG) distribution in HFs is classically investigated by conventional histology, biochemical analysis, and immunohistochemistry. In this report, a novel approach is proposed to assess hair histology and HSPG distribution changes in HFs at different phases of the hair growth cycle using infrared spectral imaging (IRSI). The distribution of HSPGs in HFs was probed by IRSI using the absorption region relevant to sulfation as a spectral marker. The findings were supported by Western immunoblotting and immunohistochemistry assays focusing on the glypican-1 expression and distribution in HFs. This study demonstrates the capacity of IRSI to identify the different HF tissue structures and to highlight protein, proteoglycan (PG), glycosaminoglycan (GAG), and sulfated GAG distribution in these structures. The comparison between anagen, catagen, and telogen phases shows the qualitative and/or quantitative evolution of GAGs as supported by Western immunoblotting. Thus, IRSI can simultaneously reveal the location of proteins, PGs, GAGs, and sulfated GAGs in HFs in a reagent- and label-free manner. From a dermatological point of view, IRSI shows its potential as a promising technique to study alopecia.

Keywords: hair follicle growth; glycosaminoglycans; infrared spectral imaging; k-means clustering; immunohistochemistry

1. Introduction

A hair follicle (HF) is a real mini-organ that makes hair growth possible. The hair shaft extends under the skin into the HF. Histology of a HF shows that it is organized into two compartments [1]. The first compartment with a dermal origin is composed of connective tissue sheath (CTS) and dermal papillae (DP). The second compartment is of epithelial origin and comprises the hair matrix, the outer (ORS) and inner (IRS) root sheaths, and the hair shaft. The bulb is comprised, at the bottom, of the DP surrounded by the germinative hair matrix, and at the top, of the differentiation zone of the matrix.

In addition, it has appendages, the sebaceous gland, and the arrector pili muscle. The sebaceous gland associated with the HFs protects the hair by sebum secretion, a substance rich in lipids [2].

A HF undergoes cyclic changes over the course of its life [3] allowing the renewal of lost hair (40 to 100 lost hairs per day). One cycle is comprised of three main phases: anagen, catagen, and telogen. The exogen phase is a phase of the hair growth cycle that is controlled separately leading to hair shaft loss [4,5]. The anagen phase is characterized by intense proliferation allowing the generation of new hair shafts [6] and lasts on the average from three to six years. It is divided into six stages during which the morphology of the HF undergoes major remodeling due to the activation of different cell types at the end of the telogen phase [7,8]. As soon as the hair growth is complete, its degeneration begins. The catagen phase corresponds to the cessation of hair shaft growth and a regression in the size of the HF [8]. This phase is characterized by apoptosis of the keratinocytes separating the secondary hair germ (SHG) from the DP. In contrast to the anagen phase, the catagen phase is more rapid, lasting on the average between 15 and 20 days; it is divided into eight stages [8]. The telogen phase is a resting phase during which the hair shaft remains anchored in the hair follicle [6]. At this stage the DP in the dormant state is in contact with the SHG [8]. This phase may last several months until a stimulus causes the HF to return to the anagen phase.

The morphological changes in the HF observed during the hair cycle involve many cell types: keratinocytes, fibroblasts of the DP, endothelial, fat, and immune cells. The presence of these different cell types makes the study of the regulation of the hair cycle complex. It is also known that a fine regulation of growth factors involved in the hair shaft growth is essential for the passage between the different phases of the hair cycle [9].

The mechanisms involved in the regulation of these growth factors are still poorly understood, but it is highly probable that heparan sulfate proteoglycans (HSPGs) are involved because of their capacity to sequester growth factors [10]. HSPGs are composed of linear chains of heparan sulfate glycosaminoglycans (GAGs) covalently attached to a core protein [11]. They are classified according to their localization, either in the cell membrane or secreted in the extracellular matrix (ECM). The secreted HSPGs interact with the macromolecules of the ECM and growth factors and thus play a pivotal role in cell growth, survival, proliferation, adhesion, migration, and differentiation [12]. The cell membrane HSPGs are divided into two major families: syndecans (SDCs), which are transmembrane PGs, and glypicans (GPCs), which are attached to the cell membrane by a glycosylphosphatidylinositol (GPI) anchor. Syndecans interact with the ECM and growth factors and have a role in embryonic development [13]. GPCs are essential for the formation or regeneration of many tissues and organs by regulating many pathways involved in development. For example, they regulate the Hedgehog (Hh) pathway during the embryonic development or long bone formation [14,15], the Wnt pathway during the embryonic development or regeneration of intestinal crypts [16,17], and the bone morphogenetic protein (BMP) pathway involved in osteogenesis [18]. The HF is a true mini-organ regulated by Wnt, BMP, and Hh pathways [19,20]. Therefore, it appears highly likely that HSPGs and particularly GPCs also play a key role in the growth of a new hair shaft.

Previous studies have shown modifications in the distribution of various ECM HSPGs [21] or membrane HSPGs such as syndecan-1 [22] during the hair cycle. This phase-dependent change seems to indicate a role of HSPGs in the regulation of hair shaft growth. Coulson–Thomas et al. showed that complex control of HSPG sulfation is necessary for correct morphogenesis of the hair shaft [23]. However, expression of GPCs in different HF compartments and their potential roles during hair shaft formation are still poorly understood.

HSPG distribution, localization, and quantification in HFs are classically investigated by conventional histology, biochemical analysis, and immunohistochemistry. In this report, a novel approach based on infrared spectral imaging (IRSI) is proposed to assess, in the first part, the HSPG distribution in the HF, and in the second part, to compare HSPG variations in HFs at different phases of hair cycle. Fourier-transform infrared (FTIR) spectroscopy is a vibrational method based on the principle of interaction between mid-IR radiation and matter, which is used to analyze pure samples but also more complex systems such as cells, tissues, or biofluids. It is non-invasive, label- and chemical-free, very sensitive, and requires no specific preparation [24]. The spectral signature of a sample contains vibrations of molecular bonds that are related to its molecular structure and composition. Today, IRSI is a proven technique that is intensively used for cell (spectral cytology) and tissue (spectral histology) characterization [25–28]. In an infrared image, each pixel is associated with an entire IR spectrum and thus both molecular and spatial information can be obtained. Recently, our group has reported studies on vibrational spectroscopic analyses of standard GAGs [29] and of GAGs in cells and tissues [30,31]. These studies permitted identifying specific spectral signatures. As we reported previously, two major spectral ranges were used to characterize HSPG distribution: the spectral window 1800–900 cm^{-1}, also known as the fingerprint region, shown to be the most specific range for GAG studies [24], and the spectral window 1350–1190 cm^{-1}, centered at 1248 cm^{-1}, that is specific for GAG sulfation [29].

Based on this knowledge, we propose in this original study to probe the distribution of HSPGs in HFs at different phases of the hair growth cycle by IRSI using sulfation as a spectral marker. IRSI might constitute a promising technique for early diagnosis and prevention in alopecia. The goal was to study HSPG, GAG, and sulfated GAG distribution and variation in HFs without any staining or labeling. It allows investigating these changes during the hair growth cycle. In parallel, our data were supported by Western immunoblotting and immunohistochemistry assays, more specifically by analyzing the GPC1 expression and distribution in HFs at different phases of the cycle.

2. Materials and Methods

The workflow of the present study is illustrated in Figure 1.

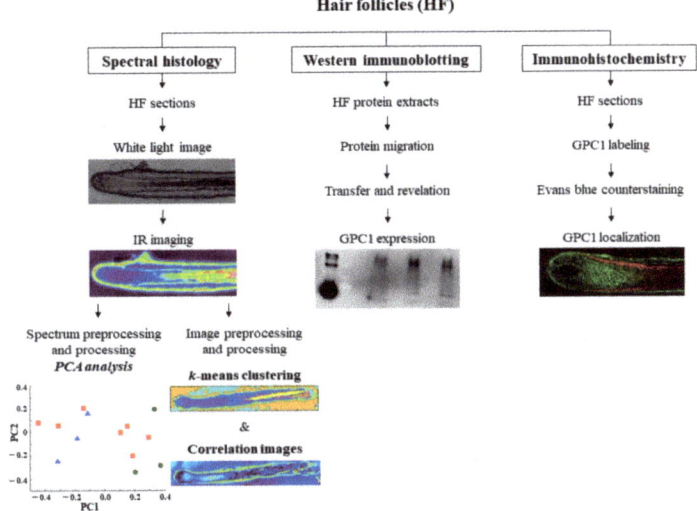

Figure 1. Workflow showing the methodological approach of hair follicle analysis by label-free infrared spectral histology, Western immunoblotting, and immunohistochemistry. Pictures shown here are only illustrated examples. PCA, principal component analysis.

2.1. Ethics Statement

Human scalp biopsy samples were obtained from human donors during surgeries after obtaining informed consent under applicable ethics guidelines and regulations and were provided by Alphenyx (Marseille, France).

2.2. Hair Follicle Sample Isolation and Preparation

2.2.1. Hair Follicle Isolation

Two different methods were used for obtaining human HFs. The first one consists in maintaining the hair follicle in its phase to investigate the three major phases of hair growth cycle (anagen, catagen, and telogen) and the second one permits inducing intermediate phases, in particular, to study intermediate anagen stages of the hair growth cycle. HFs were isolated from the human scalp according to Philpott's method [32].

For all our experiments, four donors were involved with two to three hair samples on the average per donor. For each hair sample, two to three sections were obtained. More precisely, for the first method, 32 anagen A1, 25 anagen A3, 26 catagen C1, 28 catagen C2, 31 telogen T1, and 41 telogen T3 HF sections were analyzed. For the second method, 9 early anagen D0, 18 intermediate anagen D3, and 11 catagen D6 HF sections were analyzed.

In the first approach, the HFs were isolated from the scalp at different hair growth cycle phases (anagen, catagen, and telogen) and maintained in culture in the William's E medium (W4128, Sigma-Aldrich, Saint-Louis, MO, USA) supplemented with 0.5% antibiotics, 2 mM L-glutamine (49420, Sigma-Aldrich), 10 ng/mL hydrocortisone (H-0396, Sigma-Aldrich), 10 µg/mL transferrin (T8158, Sigma-Aldrich), and 10 ng/mL selenite (S-5261, Sigma-Aldrich) for one day (named A1, C1 and T1, respectively) and for three days (named A3, C3, and T3, respectively) before analysis.

In the second approach, the different phases of the hair cycle (early anagen, intermediate anagen, and catagen) were induced in culture. To do so, early anagen HFs were isolated and selected from the scalp; one part was directly analyzed (D0, early anagen HFs), while the other part was maintained in culture for three days (D3, intermediate anagen HFs) and for six days (D6, catagen HFs) in the cell culture medium described above, supplemented with 0.5 µg/mL insulin (91077C, Sigma-Aldrich).

2.2.2. Hair Follicle Preparation for Infrared Analysis

For the different approaches described above, HFs were embedded individually in cryoprotective Tissue-Tek® O.C.T.™ Compound (Sakura, Alphen aan den Rijn, Netherlands) and snap-frozen at −80 °C. Seven micron-thick longitudinal sections of HFs were prepared using a cryostat (Leica Biosystems, Nanterre, France). The sections were then placed on an IR-transparent calcium fluoride (CaF_2) window (Crystran Ltd., Dorset, UK) for IRSI analysis (2 to 4 sections per window). The sections were imaged directly without any staining.

2.2.3. Hair Follicle Preparation for Immunohistochemistry

The same process described for HF preparation for infrared analysis was used to prepare sections. Sections were placed on glass slides and air-dried; then, they were fixed in acetone for 10 min at −20 °C. After three washes in PBS, the sections were placed in a sheep serum solution (Thermo Fisher Scientific, Illkirch-Graffenstaden, France). Primary antibody anti-GPC1 (Proteintech, Rosemont, IL, USA) was incubated overnight at 4 °C. After several washes with PBS, the secondary antibody coupled with Alexa 488 was applied for 45 min at room temperature and in the dark. The Evans blue counterstain was applied after several washes for 5 min at room temperature. After the final washes, the glass slides were mounted under a coverslip using Fluoprep. The observations were performed using a confocal microscope (TCS-SPE, Leica Biosystems).

2.3. IR Spectral Imaging of Hair Follicle Sections

All sections of HFs were imaged in the transmission mode using the Spotlight 400 infrared imaging system (PerkinElmer, Villebon-sur-Yvette, France). The acquisition parameters were as follows: spatial resolution with a projected pixel size of 6.25 µm × 6.25 µm, spectral range from 4000 to 900 cm^{-1}, spectral resolution of 4 cm^{-1}, and at 16 scans/pixel. IRSI acquisition was performed on whole hair follicles. Prior to this, for each section, a background spectrum was measured using 90 scans in a region of the CaF$_2$ window without sample or optimal cutting temperature (OCT) compound and was automatically removed from each pixel spectrum of the image. A total of four spectral images were also acquired on the OCT compound using the same experimental conditions. All acquisitions were performed using the Spectrum Image 1.7.1 software (PerkinElmer, Villebon-sur-Yvette, France). All IR images underwent atmospheric correction using the same software. This step reduces the interferences due to molecules present in the sample environment such as carbon dioxide or water vapor.

2.4. IR Spectral Preprocessing

Pixel spectra were extracted from specific regions of interest of the IR images, smoothed (Savitsky–Golay, 7 points), baseline-corrected (elastic), vector-normalized, and offset-corrected using the OPUS 5.5 software (Bruker Optics, Ettlingen, Germany).

2.5. IR Spectral Processing by Principal Component Analysis

For spectral processing, principal component analysis (PCA), an unsupervised exploratory method, was used. This method allowed spectral data reduction, replacing original and correlated variables with synthetic and uncorrelated variables called principal components (PCs). These PCs contain the information of interest and are ranked from the highest to lowest variance in the dataset. In this study, PCA was performed on mean-centered spectra to remove redundant information and using the spectral ranges 1800–900 or 1350–1190 cm^{-1}.

2.6. IR Image Preprocessing

All images were preprocessed with the extended multiplicative scatter correction (EMSC) algorithm which included correction of the baseline, variations related to the difference in sample thickness, and vector normalization. Indeed, this step allowed removing the outliers and artifacts that can influence spectral image analysis. All spectral images were preprocessed in the 1800 to 900 cm^{-1} spectral region using a developed in-house routine in Matlab (The Mathworks, Natick, MA, USA).

2.7. IR Image Processing by k-Means Algorithm

Spectral image analysis was performed using the *k*-means clustering algorithm, which is one of the most popular unsupervised classification methods [33]. It aims at separating a set of N unlabeled points of d dimensions into k clusters. Thus, it allows grouping pixel spectra into distinct classes (k clusters) and usually the choice of k is driven by a priori knowledge of the structure of the studied dataset. A trial-and-error procedure can also be used by iteratively increasing the number of clusters until obtaining a coherent partition of the studied phenomenon. Each cluster is represented by its barycenter, also called centroid, and the algorithm starts by randomly choosing k initial points as centroids at the beginning of the process. Each pixel spectrum is assigned to the cluster whose centroid is the nearest in terms of spectral similarity, based on a chosen distance metric, which is often the Euclidean distance. These steps are repeated until point assignment is stabilized. The *k*-means algorithm thus converges to a local minimum. Finally, each class is represented by a color and the cluster image is reconstituted as a false-color map. In order to compare different HF section images, a common *k*-means was applied using 10 classes in the 1800–800 cm^{-1} spectral range.

2.8. IR Correlation Images Using Spectra of Standard Compounds

The spectral images of HF sections obtained previously were processed to produce correlation maps corresponding to heparan sulfate (HS), GPC1, and hyaluronic acid (HA) distribution. To do so, solutions of standard GAGs (HS, Celsus Laboratories, Cincinnati, OH, USA; and HA, MP Biomedicals, Illkirch-Graffenstaden, France) and recombinant human GPC1 (R&D Systems, Minneapolis, MN, USA) were deposited on a CaF_2 window, dried, and imaged in the transmission mode using a Spotlight 400 infrared imaging system. For each image, the spectra were averaged to obtain the mean representative spectrum of each standard. Each standard spectrum was then in turn correlated pixel by pixel with HF images corrected for atmospheric contribution using the Spectrum Image 1.7.1 software. The resulting correlation maps show the distribution of the standards with a correlation scale ranging from 0 (dark color) to 1 (white color), corresponding to low and high levels of correlation, respectively.

2.9. Hair Follicle Protein Extract Preparation for Western Immunoblotting

HF proteins were extracted from a pool of five HFs and prepared in a radioimmunoprecipitation assay buffer (RIPA) buffer supplemented with 1% protease inhibitor cocktail added (Sigma-Aldrich) using the FastPrep 24™ (MP Biomedicals) six times at 6.0 m/s for 40 s. The obtained lysates were incubated 20 min on ice with vortex-mixing every 5 min. Cell debris were precipitated by centrifugation at 10,000 g for 10 min at 4 °C. Proteins in the supernatant were collected and assayed using the Bradford (Bio-Rad, Marne-la-Coquette, France) technique [34].

2.10. Western Immunoblotting

Samples were deposited onto polyacrylamide gels as previously described [35]. The GPC1 primary antibody used was 16700-1-AP (Proteintech, Rosemont, IL, USA). The appropriate peroxidase-coupled secondary antibody (1/10,000) was the anti-rabbit NA934V (GE Healthcare Life Sciences, Marlborough, MA, USA).

3. Results and Discussion

The present report describes the evolution of GAGs, sulfated GAGs, and, more specifically, the HS-type GAGs and GPC1 distribution in hair follicles at different phases of the hair growth cycle using spectral imaging. IRSI presents the advantage of being label-free, reagent-free, and rapid. The data obtained by this novel approach were compared to the immunohistochemistry and Western immunoblotting data.

3.1. Discrimination of HF Structures, Distribution of HSPG and GPC1 Using IR Spectral Imaging Analysis

Anagen, catagen, and telogen HFs were individually isolated and maintained in culture for one day (A1, C1, and T1, respectively) or three days (A3, C3, and T3, respectively) before analysis. The white light image in Figure 2A shows a section of a HF and its different structures: germinative matrix (1), differentiation zone (2), IRS (3), ORS (4), and hair shaft (5). OCT is indicated by number 6.

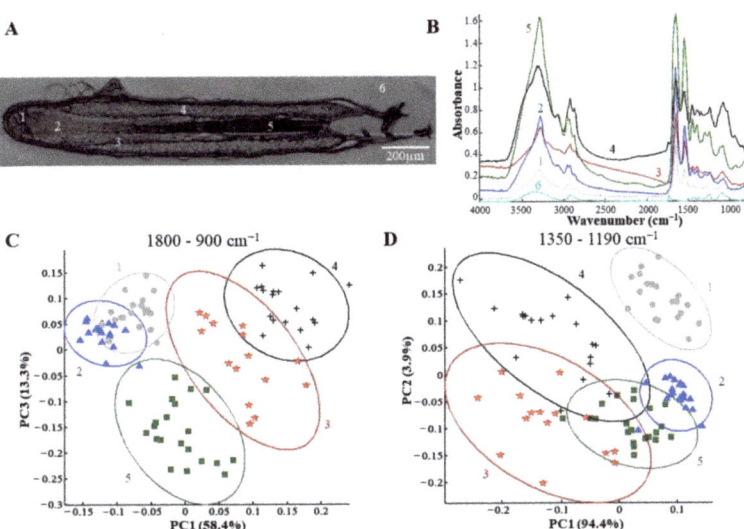

Figure 2. PCA analysis of spectra from specific hair follicle regions. (**A**) White light image of a hair follicle. Five hair follicle structures were analyzed: 1, germinative matrix; 2, differentiation zone of the matrix; 3, IRS; 4, ORS; 5, hair shaft. The number 6 corresponds to OCT. (**B**) Pixel spectrum representative of each structure and OCT. (**C,D**) PCA score plot performed on normalized mean spectra of the above regions in the 1800–900 cm^{-1} (**C**) or 1350–1190 cm^{-1} (**D**) range and carried out on anagen A1/A3, catagen C1/C3, and telogen T1/T3 hair follicles. Ellipses represent the 95% confidence intervals.

3.1.1. Different Structures of the Hair Follicle are Discriminated by IR Spectral Imaging Analysis Based on the Variation of GAGs and Their Sulfation

The whole section was analyzed by IRSI and representative spectra of the HF structures are illustrated in Figure 2B. These spectra represent the means calculated from a small zone from each HF structure (indicated by 1 to 5). For comparison, the mean spectrum of OCT was also represented (6). The analysis of the GAG spectral profiles in the different HF structures was performed by PCA first in the 1800–900 cm^{-1} spectral range (Figure 2C) corresponding to the proteins and PGs and then in the 1350–1190 cm^{-1} spectral range (Figure 2D), focusing on sulfated GAGs. Figure 2C shows the PCA score plot built with the principal components PC1 and PC3 carrying 58.4% and 13.3% of the total explained variance, respectively. This score plot revealed five structural groups distributed in three well-differentiated zones. The first zone was composed of an overlap of the germinative matrix (group 1) and the differentiation zone (group 2). This overlap can be explained by the fact that the matrix cells of the differentiation zone are the result of differentiation of the progenitor cells present in the germinative matrix derived from stem cells of SHG [36–39]. They therefore share similar characteristics, such as proteins and polysaccharides, explaining their similarity. The second zone is constituted by the ORS (group 4), the third—by the hair shaft zones of HFs (group 5). It can be noted that the IRS structure (group 3) lies between the ORS and the hair shaft zones. Interestingly, IRS histologically separates the hair shaft from the ORS, which corroborates spectral data.

Figure 2D shows the PCA score plot built with the principal components PC1 and PC2 carrying 94.4% and 3.9% of the total explained variance, respectively. The PCA performed in the sulfated GAG absorption range (1350–1190 cm^{-1}) succeeded to precisely discriminate the germinative matrix structure (group 1) from the four other groups and, in particular, from the differentiation zone (group 2), in contrast with the overlap observed in Figure 2C. The other structures (differentiation zone, IRS, ORS, and hair shaft) were not clearly separated using sulfated GAG spectral signatures. Interestingly, different groups obtained by this PCA reflect the scheme of differentiation of the multipotent HF cells [36–39]. Indeed, the progenitor hair cells present in the germinative matrix provide, on the one hand, the ORS cells and, on the other hand, the transit-amplifying cells of the differentiation zone that will differentiate into IRS and hair shaft cells.

The PCA results shown in Figure 2C,D allowed discriminating several features of the HFs, the germinative matrix, the differentiation zone, the ORS, the IRS, and the hair shaft, by the contribution of proteins, PGs, GAGs, and sulfated GAGs. The biochemical spectral information is confirmed by the loading vectors shown in the region of 1800–900 cm^{-1} (Figure S1A) and in the region of 1350–1190 cm^{-1} (Figure S1B). The sulfate group vibrations of GAGs are mainly represented by the IR absorption band centered at 1248 cm^{-1}. This peak is assigned an antisymmetric stretching S=O vibrations as we have previously reported [29]. This discrimination is in agreement with the difference in PG composition in the different parts of HFs observed by immunostaining or immunofluorescence [21,22,40]. Moreover, the IRSI technique does not require any staining or chemicals while keeping good discrimination of histological structures.

3.1.2. Focus on GPC1 Distribution in Hair Follicles by Spectral Imaging Analysis and Immunohistochemistry

In order to better understand the contribution of HSPGs in the differences observed by the PCA in the spectral range corresponding to PGs/GAGs and sulfated GAGs, a correlation image was computed with a spectrum of standard HS. This analysis permitted obtaining a representative image of the HS contribution in the section of HFs. As shown in Figure 3, the coefficient of HS contribution was most important in the ORS. The IRS displayed a good correlation with HS. The germinative matrix structure presented a low correlation with HS, while the differentiation zone and the hair shaft did not correlate with the HS. This result correlates well with HS distribution in the HFs obtained by immunohistochemistry reported by Malgouries et al., where several HSPGs were found in the ORS part of HFs [21,40].

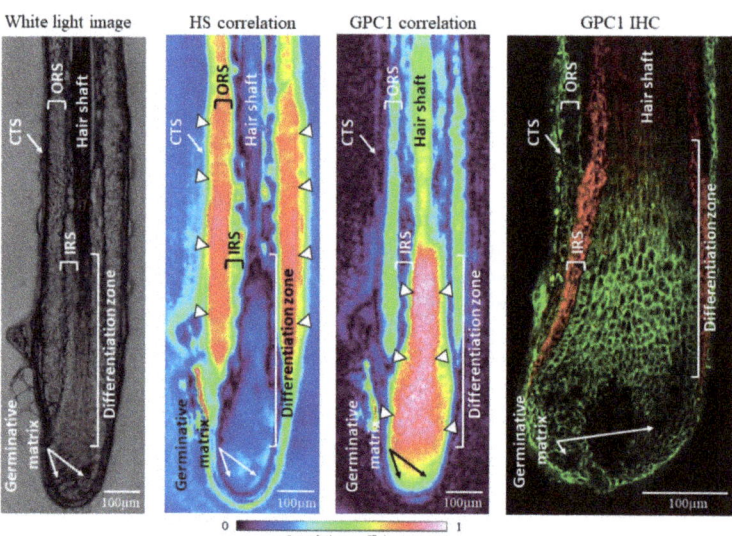

Figure 3. Characterization of hair follicle structures by different imaging approaches. From left to right: white light image, HS- and GPC1-correlated IR images, and immunohistochemical labeling of GPC1 (green) counterstained with the Evans blue dye (red). CTS, connective tissue sheath; IRS, inner root sheath; ORS, outer root sheath. Arrowheads indicate high level of correlation. Scale bar: 100 μm.

A correlation image was also created with a representative spectrum of GPC1. The coefficient of GPC1 contribution is most important in the differentiation zone (Figure 3). The hair shaft, the ORS, and the germinative matrix also presented a good correlation with GPC1, but not the IRS. The GPC1 correlation map is confirmed by the GPC1 green labeling obtained by immunohistochemistry of HFs. These results obtained by the IRSI technique demonstrate its ability to discriminate at the molecular scale in a comparable manner to immunohistochemistry without any labeling.

These results support the PCA analyses carried out on the different HF structures. The use of both HS and GPC1 correlation mapping may partly explain the discrimination potential of the PCA analyses.

3.2. Different Phases of Hair Growth Cycle and Spatial Distribution of HS and GPC1 Determined by IR Spectral Imaging Analysis

3.2.1. k-Means Clustering Identifies the Hair Follicle Structures in the Different Phases Based on the Protein and HSPG Spectral Information

The IR images obtained from HFs in different phases of the hair growth cycle were analyzed by common k-means clustering in the 1800–800 cm^{-1} range (Figure 4). The protein, HSPG, and GAG contribution varied within HFs forming five different clusters (each associated with a pseudo-color) corresponding to different parts of the HF (the CTS, the ORS, the IRS, the bulb (germinative matrix and differentiation zone), and the hair shaft), independently of the phase of the hair growth cycle (Figure 4A–C). The five other clusters (1, 3, 5, 6, and 10) were attributed to the OCT spectral signatures and formed a separate group. The dendrogram constructed on centroid spectra illustrated the correlation between clusters and histological structures (Figure 4D,E).

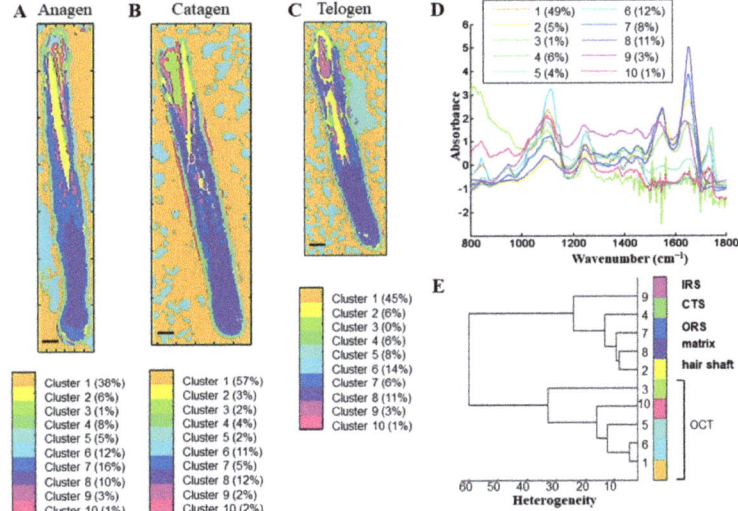

Figure 4. Discrimination of hair follicle structures at different phases of hair growth cycle by k-means clustering. (**A–C**) Representative color-coded k-means clustering images using 10 classes in the 1800–800 cm^{-1} spectral range for hair follicles in anagen A1 (**A**), catagen C1 (**B**), and telogen T1 (**C**) phases. (**D**) Centroid spectra corresponding to each cluster. (**E**) Dendrogram of centroid spectra and assignment of corresponding hair follicle structures. CTS, connective tissue sheath; IRS, inner root sheath; ORS, outer root sheath. Scale bar: 100 μm.

IRSI combined with k-means clustering allows separating the five principal structures of HFs using the protein, HSPG, and GAG spectral signatures and the OCT embedding.

3.2.2. IR Correlation Maps Highlight Spatial Distribution of HS and GPC1 in the Different Phases

In order to investigate the impact of HS and GPC1 in each phase of the hair growth cycle, IR correlation maps were calculated using their representative spectra (Figure 5). The spectra of HS and GPC1 used for the correlation maps are presented in Figure 5A. HS was mostly detected in the ORS in the anagen phase, and in the ORS and IRS in the catagen and telogen phases (see arrowheads in Figure 5B–D). HS correlation coefficient seems to be lower in the ORS in the anagen phase of the HF. GPC1 was majorly distributed in the differentiation zone and less in the hair shaft and in the germinative matrix without any significant differences between the three phases. The presence of GPC1 in the ORS was specifically detected in the catagen phase in agreement with immunohistochemical labeling (see Figure 3). In addition, the spectrum of HA was also used for the correlation map (Figure S2A). HA is detected in the IRS and the hair shaft in the anagen and catagen phases but not in the ORS and the differentiation zone (Figure S2B,C). A faint correlation is observed in the lower part of the differentiation zone in the telogen phase (Figure S2D).

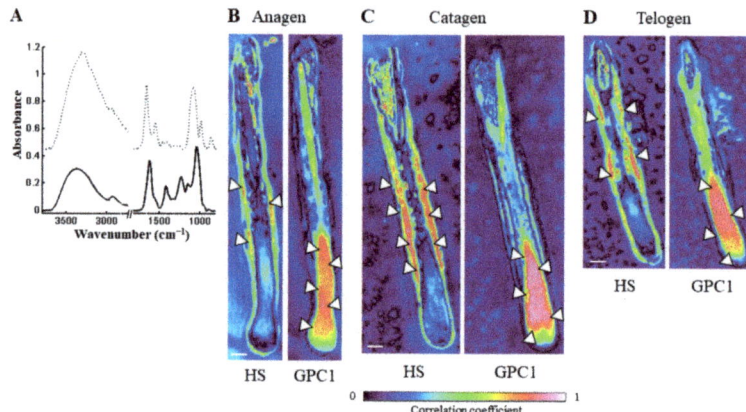

Figure 5. Contribution of the heparan sulfate GAG chain and glypican-1 in a hair follicle. (**A**) Mean spectra of standard HS (continuous line) and GPC1 (dashed line). (**B–D**) IR correlation maps of hair follicles in the anagen A1 (**B**), catagen C1 (**C**), and telogen T1 (**D**) phases using mean spectra from standard heparan sulfate (left) and human recombinant glypican-1 (right). HS, heparan sulfate; GPC1, glypican-1. Arrowheads indicate high level of correlation. Scale bar: 100 µm.

These correlation maps inform on the HS-type GAG and GPC1 distribution within the HF. The results suggest that GPC1, mainly detected in the differentiation zone and hair shaft, does not exhibit HS chains in contrast to HS, which is detected in the ORS and IRS. Based on this GPC1 distribution, at this stage, it is not possible to differentiate the different phases of the hair growth cycle. However, GPC1 correlation in the ORS (Figure 5C) requires further studies to eventually validate GPC1 as a spectral biomarker of the catagen phase. In contrast, the anagen phase, corresponding to a growing phase of the HF, is identified using IRSI as it is associated to a decreased HS correlation coefficient. However, the catagen phase is characterized by an increase in HS correlation. Moreover, HA, the non-sulfated GAG, exhibits a different distribution in the HF as compared to HSPG independently of the phases of the hair growth cycle. Malgouries et al. have previously described the capacity to discriminate the phases of the hair growth cycle by the distribution of several HSPGs using immunofluorescence [21].

The differences observed in the distribution of HS and GPC1 in the three phases may partly account for the different classes obtained by *k*-means clustering.

3.3. Discrimination of Hair Growth Cycle Phases, Distribution of Sulfated GAG and GPC1 Using IR Spectral Imaging Analysis

3.3.1. Different Phases of the Hair Growth Cycle are Discriminated by PCA of IR Spectra Based on the Variation of GAGs and Their Sulfation

Spectra used for this analysis were taken from the ORS (central zone of the HF). This zone was chosen because it is rich in HS (see Figure 3). PCA was performed on the A1, C1/C3, and T1 HF spectra in the spectral range of 1800–900 cm^{-1}, which encompasses the protein and PG absorptions. Figure 6A shows the PCA score plot, built with the principal components PC2 and PC3 carrying 30.3% and 11.5% of the total variance, respectively. PC2 permitted clearly separating the anagen HFs from the telogen HFs. The catagen HFs were dispersed in these two groups, but in a specific manner. Interestingly, the catagen C1 HFs were in the same group as the anagen HFs, while the catagen C3 HFs were in the telogen HF group. The same pattern was observed when the PCA score plot was recalculated using the sulfate absorption (1350–1190 cm^{-1}) spectral range (Figure 6B; PC1 and PC2 carrying 93.2% and 3.0% of the total variance, respectively). The second and third loading vectors shown in the region of 1800–900 cm^{-1} (Figure S3A) and the first and second loading vectors shown in the region of 1350–1190 cm^{-1} (Figure S3B) reveal the biochemical spectral information permitting to separate the different phases of the hair growth cycle.

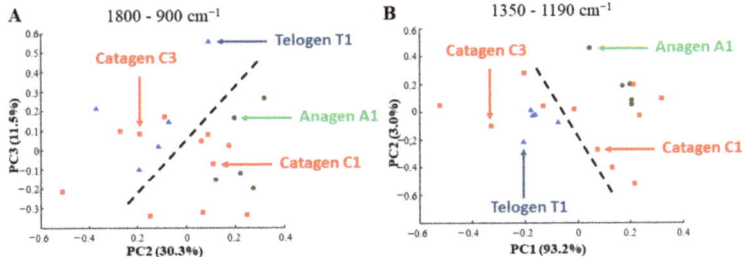

Figure 6. Discrimination of hair follicles at different phases of hair growth cycle by PCA. (**A**,**B**) PCA score plot performed on normalized mean spectra of the ORS region 4 in the 1800–900 cm^{-1} (**A**) or 1350–1190 cm^{-1} (**B**) range and carried out on anagen A1, catagen C1/C3, and telogen T1 hair follicles.

The PCA results, on the one hand, suggest the discriminating power of the IR method capable of differentiating HFs at different phases of the hair growth cycle *via* their spectra and, on the other hand, demonstrate that the protein, PG, GAG, and sulfated GAG content varies during the phases of the hair growth cycle.

3.3.2. GPC1 Expression Corroborated with IR Spectral Maps of Sulfated GAG Distribution, Discriminates Different Phases during the Hair Growth Cycle

In order to support the PCA results obtained in the HFs of different phases of hair growth cycle, a PCA was performed on the HF phases induced by culture. This method allowed obtaining intermediate phases such as early and intermediate anagen.

After isolation, the early anagen HFs at day 0 (D0) were analyzed by spectral imaging and Western immunoblotting. Before analysis, the intermediate anagen (D3) and catagen (D6) HFs were maintained in culture for three and six days, respectively. As previously, spectra used for PCA analysis were extracted from the ORS region.

Figure 7A displays the PCA score plot constructed with the first two principal components PC1 and PC2 carrying 95.8% and 2.7% of the total variance, respectively. The variance carried by PC2 permitted separating the early anagen D0 from the catagen D6 HFs. The intermediate anagen D3 HFs were dispersed in these two groups, but in a specific manner. The observed tendency was that intermediate anagen D3 HFs with the hair shaft seemed to be closer to the early anagen D0 HF group, while those without the hair shaft were closer to catagen D6 HFs. The biochemical spectral information is supported by the loading vectors shown in the region of 1350–1190 cm^{-1} (Figure S3C).

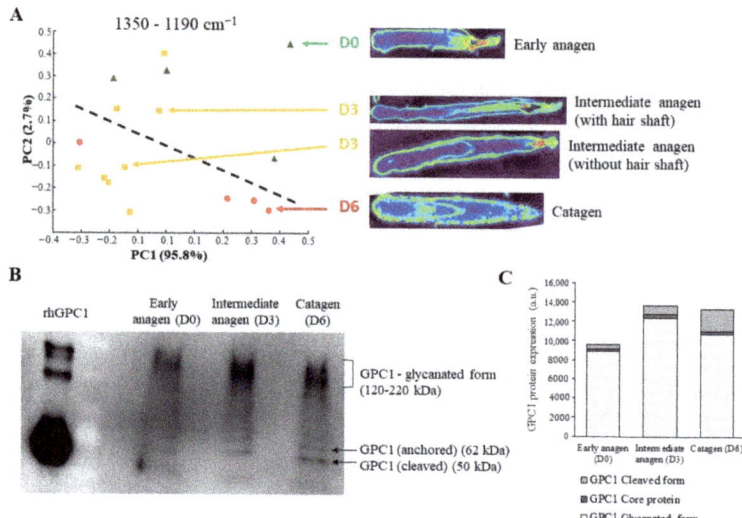

Figure 7. Discrimination of hair follicles by PCA compared to GPC1 expression during the hair growth cycle. (**A**) PCA score plot performed on normalized mean spectra of the ORS region 4 in the 1350–1190 cm^{-1} range and carried out on early anagen D0, intermediate anagen D3, and catagen D6 hair follicles. (**B**) GPC1 protein expression analyzed by Western immunoblotting of early anagen D0, intermediate anagen D3, and catagen D6 hair follicles. (**C**) Quantification of each GPC1 form.

This result confirms the capacity of the PCA analysis to discriminate HFs at different hair growth cycle phases based on the sulfated GAG spectral signature. Moreover, data tend to indicate that the sulfated GAG evolve throughout the hair growth cycle.

In order to confirm the difference in GAG sulfation observed by IR imaging between HFs at different phases of the hair growth cycle, a GPC1 protein expression analysis was performed on HFs in the early anagen D0, intermediate anagen D3, and catagen D6 HFs.

Protein analysis carried out on HFs at different phases of the hair growth cycle shows that the GPC1 forms (cleaved, anchored, and glycanated) changed with the phases. In particular, cleaved and glycanated forms exhibited a noticeable difference observed between D0, D3, and D6 HFs (Figure 7B). Indeed, the glycanated form of GPC1 was more abundant in the intermediate anagen D3 and catagen D6 compared to early anagen D0 HFs. Moreover, a slight decrease was observed in catagen D6 compared to intermediate anagen D3 HFs. An accumulation of the cleaved GPC1 form was observed from D0 to D6 (Figure 7C).

According to the different HF phases, differential glycation of GPC1 could be demonstrated by Western immunoblotting. These observed differences may explain the PCA grouping of HF spectra with regard to the different phases based on sulfated GAG absorption.

Altogether, the IR spectral images processed by different chemometric approaches such as PCA, *k*-means clustering, and correlation mapping, could identify the histological structures and the different phases of the hair growth cycle. In addition, these approaches corroborate immunohistochemical and biochemical analyses of the present report and from the literature.

4. Conclusions

This report demonstrates the capacity of IRSI to identify the different tissue structures of the HFs and to show protein, PG, GAG, and sulfated GAG distribution in these structures. In addition, the comparison between the anagen, catagen, telogen phases of the hair growth cycle shows the qualitative and/or quantitative evolution of GAGs as supported by Western immunoblotting. Thus, IRSI can reveal some information on the location of proteins, PGs, GAGs and sulfated GAGs in HFs in a reagent- and label-free manner. In the long term, from a dermatological point of view, IRSI might constitute a promising technique for the early diagnosis and prevention in alopecia.

Supplementary Materials: The following are available online at https://www.mdpi.com/2218-273X/11/2/192/s1, Figure S1: PCA analysis of spectra from specific hair follicle regions. (A-B) Loading vectors 1 and 3 obtained from Figure 2C (A) and loading vectors 1 and 2 obtained from Figure 2D (B), Figure S2: Contribution of hyaluronic acid in hair follicle. (A) Spectrum of standard HA. (B–D) IR correlation maps of hair follicles in anagen A1 (B), catagen C1 (C) and telogen T1 (D) phases using spectrum from standard hyaluronan. HA, hyaluronan. Arrowheads indicate good level of correlation. Scale bar: 100 µm, Figure S3: PCA analysis of spectra from specific hair follicle regions. (A–C) Loading vectors 2 and 3 obtained from Figure 6A (A), loading vectors 1 and 2 obtained from Figure 6B (B), and loading vectors 1 and 2 obtained from Figure 7A (C).

Author Contributions: Conceptualization, S.B., L.R., and G.D.S.; methodology, C.C.-P., V.U., S.B., L.R., and G.D.S.; software, V.U.; validation, S.B., L.R., G.D.S., N.B., L.D., V.B., S.M. and C.J.; formal Analysis, C.C.-P., V.U., S.B., L.R., and G.D.S.; investigation, C.C.-P., V.U., S.B., L.R., and G.D.S.; resources, N.B.; data curation, C.C.-P., V.U., S.B., L.R., and G.D.S.; writing—original draft preparation, C.C.-P., V.U., S.B., L.R., and G.D.S.; writing—review and editing, C.C.-P., V.U., S.B., L.R., G.D.S., N.B., L.D., V.B., S.M. and C.J.; supervision, S.B., L.R., and G.D.S.; project Administration, S.B., L.R., and G.D.S. All authors have read and agreed to the published version of the manuscript.

Funding: This study was made in collaboration with BASF Beauty Care Solutions. Ms Charlie Colin–Pierre is a BASF/CNRS (Centre national de la recherche scientifique)-funded PhD fellow.

Institutional Review Board Statement: In our study, commercial human scalp samples were ordered and provided by Alphenyx (Marseille, France) after surgery of human donors following informed consent, and applicable ethical guidelines and regulations.

Informed Consent Statement: Informed consent was obtained by Alphenyx from all subjects involved in the study.

Data Availability Statement: Not Applicable.

Acknowledgments: The authors thank the PICT-IBiSA platform of the University of Reims Champagne–Ardenne for instrument facilities. We wish to thank Carine Tedeschi for her technical support in performing hair follicle sections and immunohistochemistry and Nathalie Andres for isolation and culture of hair follicles.

Conflicts of Interest: The authors declare no competing financial interest.

References

1. Bernard, B.A. La vie révélée du follicule de cheveu humain. *Med. Sci. (Paris)* **2006**, *22*, 138–143. [CrossRef] [PubMed]
2. Zouboulis, C.C.; Picardo, M.; Ju, Q.; Kurokawa, I.; Törőcsik, D.; Bíró, T.; Schneider, M.R. Beyond Acne: Current Aspects of Sebaceous Gland Biology and Function. *Rev. Endocr. Metab. Disord.* **2016**, *17*, 319–334. [CrossRef] [PubMed]
3. Sada, A.; Tumbar, T. New Insights into Mechanisms of Stem Cell Daughter Fate Determination in Regenerative Tissues. *Int. Rev. Cell Mol. Biol.* **2013**, *300*, 1–50. [CrossRef] [PubMed]
4. Stenn, K. Exogen Is an Active, Separately Controlled Phase of the Hair Growth Cycle. *J. Am. Acad. Dermatol.* **2005**, *52*, 374–375. [CrossRef] [PubMed]
5. Milner, Y.; Sudnik, J.; Filippi, M.; Kizoulis, M.; Kashgarian, M.; Stenn, K. Exogen, Shedding Phase of the Hair Growth Cycle: Characterization of a Mouse Model. *J. Investig. Dermatol.* **2002**, *119*, 639–644. [CrossRef] [PubMed]
6. Bernard, B.A.; Commo, S.; Gerst, C.; Mahé, F.Y.; Pruche, F. Données récentes sur la biologie du cheveu. *Bull. Esthet. Dermatol. Cosmetol.* **1996**, *4*, 55–64.
7. Chase, H.B.; Rauch, R.; Smith, V.W. Critical Stages of Hair Development and Pigmentation in the Mouse. *Physiol. Zool.* **1951**, *24*, 1–8. [CrossRef]
8. Müller-Röver, S.; Handjiski, B.; van der Veen, C.; Eichmüller, S.; Foitzik, K.; McKay, I.A.; Stenn, K.S.; Paus, R. A Comprehensive Guide for the Accurate Classification of Murine Hair Follicles in Distinct Hair Cycle Stages. *J. Investig. Dermatol.* **2001**, *117*, 3–15. [CrossRef]

9. Botchkarev, V.A.; Paus, R. Molecular Biology of Hair Morphogenesis: Development and Cycling. *J. Exp. Zool. B Mol. Dev. Evol.* **2003**, *298*, 164–180. [CrossRef]
10. Filmus, J.; Capurro, M.; Rast, J. Glypicans. *Genome Biol.* **2008**, *9*, 224. [CrossRef]
11. Yung, S.; Chan, T.M. Glycosaminoglycans and Proteoglycans: Overlooked Entities? *Perit. Dial. Int.* **2007**, *27*, S104–S109. [CrossRef] [PubMed]
12. Schaefer, L.; Iozzo, R.V. Biological Functions of the Small Leucine-Rich Proteoglycans: From Genetics to Signal Transduction. *J. Biol. Chem.* **2008**, *283*, 21305–21309. [CrossRef] [PubMed]
13. Tkachenko, E.; Rhodes, J.M.; Simons, M. Syndecans: New Kids on the Signaling Block. *Circ. Res.* **2005**, *96*, 488–500. [CrossRef] [PubMed]
14. Ayers, K.L.; Gallet, A.; Staccini-Lavenant, L.; Thérond, P.P. The Long-Range Activity of Hedgehog Is Regulated in the Apical Extracellular Space by the Glypican Dally and the Hydrolase Notum. *Dev. Cell* **2010**, *18*, 605–620. [CrossRef]
15. Capurro, M.; Izumikawa, T.; Suarez, P.; Shi, W.; Cydzik, M.; Kaneiwa, T.; Gariepy, J.; Bonafe, L.; Filmus, J. Glypican-6 Promotes the Growth of Developing Long Bones by Stimulating Hedgehog Signaling. *J. Cell Biol.* **2017**, *216*, 2911–2926. [CrossRef]
16. Kreuger, J.; Perez, L.; Giraldez, A.J.; Cohen, S.M. Opposing Activities of Dally-like Glypican at High and Low Levels of Wingless Morphogen Activity. *Dev. Cell* **2004**, *7*, 503–512. [CrossRef]
17. Yamamoto, S.; Nakase, H.; Matsuura, M.; Honzawa, Y.; Matsumura, K.; Uza, N.; Yamaguchi, Y.; Mizoguchi, E.; Chiba, T. Heparan Sulfate on Intestinal Epithelial Cells Plays a Critical Role in Intestinal Crypt Homeostasis via Wnt/β-Catenin Signaling. *Am. J. Physiol. Gastrointest. Liver Physiol.* **2013**, *305*, G241–G249. [CrossRef]
18. Dwivedi, P.P.; Grose, R.H.; Filmus, J.; Hii, C.S.T.; Xian, C.J.; Anderson, P.J.; Powell, B.C. Regulation of Bone Morphogenetic Protein Signalling and Cranial Osteogenesis by Gpc1 and Gpc3. *Bone* **2013**, *55*, 367–376. [CrossRef]
19. Hsu, Y.-C.; Li, L.; Fuchs, E. Transit-Amplifying Cells Orchestrate Stem Cell Activity and Tissue Regeneration. *Cell* **2014**, *157*, 935–949. [CrossRef]
20. Sennett, R.; Rendl, M. Mesenchymal-Epithelial Interactions during Hair Follicle Morphogenesis and Cycling. *Semin. Cell Dev. Biol.* **2012**, *23*, 917–927. [CrossRef]
21. Malgouries, S.; Thibaut, S.; Bernard, B.A. Proteoglycan Expression Patterns in Human Hair Follicle. *Br. J. Dermatol.* **2008**, *158*, 234–242. [CrossRef] [PubMed]
22. Bayer-Garner, I.B.; Sanderson, R.D.; Smoller, B.R. Syndecan-1 Is Strongly Expressed in the Anagen Hair Follicle Outer Root Sheath and in the Dermal Papilla but Expression Diminishes with Involution of the Hair Follicle. *Am. J. Dermatopathol.* **2002**, *24*, 484–489. [CrossRef] [PubMed]
23. Coulson-Thomas, V.J.; Gesteira, T.F.; Esko, J.; Kao, W. Heparan Sulfate Regulates Hair Follicle and Sebaceous Gland Morphogenesis and Homeostasis. *J. Biol. Chem.* **2014**, *289*, 25211–25226. [CrossRef] [PubMed]
24. Mainreck, N.; Brézillon, S.; Sockalingum, G.D.; Maquart, F.-X.; Manfait, M.; Wegrowski, Y. Characterization of Glycosaminoglycans by Tandem Vibrational Microspectroscopy and Multivariate Data Analysis. In *Proteoglycans: Methods and Protocols*; Methods in Molecular Biology; Rédini, F., Ed.; Humana Press: Totowa, NJ, USA, 2012; pp. 117–130. ISBN 978-1-61779-498-8.
25. Draux, F.; Jeannesson, P.; Gobinet, C.; Sule-Suso, J.; Pijanka, J.; Sandt, C.; Dumas, P.; Manfait, M.; Sockalingum, G.D. IR Spectroscopy Reveals Effect of Non-Cytotoxic Doses of Anti-Tumour Drug on Cancer Cells. *Anal. Bioanal. Chem.* **2009**, *395*, 2293–2301. [CrossRef] [PubMed]
26. Townsend, D.; Miljković, M.; Bird, B.; Lenau, K.; Old, O.; Almond, M.; Kendall, C.; Lloyd, G.; Shepherd, N.; Barr, H.; et al. Infrared Micro-Spectroscopy for Cyto-Pathological Classification of Esophageal Cells. *Analyst* **2015**, *140*, 2215–2223. [CrossRef] [PubMed]
27. Mittal, S.; Bhargava, R. A Comparison of Mid-Infrared Spectral Regions on Accuracy of Tissue Classification. *Analyst* **2019**, *144*, 2635–2642. [CrossRef]
28. Kallenbach-Thieltges, A.; Großerüschkamp, F.; Mosig, A.; Diem, M.; Tannapfel, A.; Gerwert, K. Immunohistochemistry, Histopathology and Infrared Spectral Histopathology of Colon Cancer Tissue Sections. *J Biophotonics* **2013**, *6*, 88–100. [CrossRef]
29. Mainreck, N.; Brézillon, S.; Sockalingum, G.D.; Maquart, F.-X.; Manfait, M.; Wegrowski, Y. Rapid Characterization of Glycosaminoglycans Using a Combined Approach by Infrared and Raman Microspectroscopies. *J. Pharm. Sci.* **2011**, *100*, 441–450. [CrossRef]
30. Brézillon, S.; Untereiner, V.; Lovergne, L.; Tadeo, I.; Noguera, R.; Maquart, F.-X.; Wegrowski, Y.; Sockalingum, G.D. Glycosaminoglycan Profiling in Different Cell Types Using Infrared Spectroscopy and Imaging. *Anal. Bioanal. Chem.* **2014**, *406*, 5795–5803. [CrossRef]
31. Mohamed, H.T.; Untereiner, V.; Sockalingum, G.D.; Brézillon, S. Implementation of Infrared and Raman Modalities for Glycosaminoglycan Characterization in Complex Systems. *Glycoconj. J.* **2017**, *34*, 309–323. [CrossRef]
32. Philpott, M.P.; Sanders, D.A.; Kealey, T. Whole Hair Follicle Culture. *Dermatol. Clin.* **1996**, *14*, 595–607. [CrossRef]
33. Macqueen, J. Some Methods for Classification and Analysis of Multivariate Observations. In Proceedings of the Fifth Berkeley Symposium on Mathematical Statistics and Probability, Beckeley, CA, USA; 1965.
34. Bradford, M.M. A Rapid and Sensitive Method for the Quantitation of Microgram Quantities of Protein Utilizing the Principle of Protein-Dye Binding. *Anal. Biochem.* **1976**, *72*, 248–254. [CrossRef]

35. Perrot, G.; Colin-Pierre, C.; Ramont, L.; Proult, I.; Garbar, C.; Bardey, V.; Jeanmaire, C.; Mine, S.; Danoux, L.; Berthélémy, N.; et al. Decreased Expression of GPC1 in Human Skin Keratinocytes and Epidermis during Ageing. *Exp. Gerontol.* **2019**, *126*, 110693. [CrossRef] [PubMed]
36. Inoue, K.; Aoi, N.; Sato, T.; Yamauchi, Y.; Suga, H.; Eto, H.; Kato, H.; Araki, J.; Yoshimura, K. Differential Expression of Stem-Cell-Associated Markers in Human Hair Follicle Epithelial Cells. *Lab. Investig.* **2009**, *89*, 844–856. [CrossRef]
37. Panteleyev, A.A. Functional Anatomy of the Hair Follicle: The Secondary Hair Germ. *Exp. Dermatol.* **2018**, *27*, 701–720. [CrossRef]
38. Purba, T.S.; Haslam, I.S.; Poblet, E.; Jiménez, F.; Gandarillas, A.; Izeta, A.; Paus, R. Human Epithelial Hair Follicle Stem Cells and Their Progeny: Current State of Knowledge, the Widening Gap in Translational Research and Future Challenges. *Bioessays* **2014**, *36*, 513–525. [CrossRef]
39. Kloepper, J.E.; Tiede, S.; Brinckmann, J.; Reinhardt, D.P.; Meyer, W.; Faessler, R.; Paus, R. Immunophenotyping of the Human Bulge Region: The Quest to Define Useful in Situ Markers for Human Epithelial Hair Follicle Stem Cells and Their Niche. *Exp. Dermatol.* **2008**, *17*, 592–609. [CrossRef]
40. Couchman, J.R.; King, J.L.; McCarthy, K.J. Distribution of Two Basement Membrane Proteoglycans through Hair Follicle Development and the Hair Growth Cycle in the Rat. *J. Investig. Dermatol.* **1990**, *94*, 65–70. [CrossRef]

Review

Glycosaminoglycans: Carriers and Targets for Tailored Anti-Cancer Therapy

Aikaterini Berdiaki [1], Monica Neagu [2], Eirini-Maria Giatagana [1], Andrey Kuskov [3], Aristidis M. Tsatsakis [4], George N. Tzanakakis [1,5] and Dragana Nikitovic [1,*]

1. Laboratory of Histology-Embryology, School of Medicine, University of Crete, 71003 Heraklion, Greece; berdiaki@uoc.gr (A.B.); eirini_gt@hotmail.com (E.-M.G.); tzanakak@uoc.gr (G.N.T.)
2. Department of Immunology, Victor Babes National Institute of Pathology, 050096 Bucharest, Romania; neagu.monica@gmail.com
3. Department of Technology of Chemical Pharmaceutical and Cosmetic Substances, D. Mendeleev University of Chemical Technology of Russia, 125047 Moscow, Russia; a_n_kuskov@mail.ru
4. Laboratory of Toxicology, School of Medicine, University of Crete, 71003 Heraklion, Greece; tsatsaka@uoc.gr
5. Laboratory of Anatomy, School of Medicine, University of Crete, 71003 Heraklion, Greece
* Correspondence: nikitovic@uoc.gr

Citation: Berdiaki, A.; Neagu, M.; Giatagana, E.-M.; Kuskov, A.; Tsatsakis, A.M.; Tzanakakis, G.N.; Nikitovic, D. Glycosaminoglycans: Carriers and Targets for Tailored Anti-Cancer Therapy. *Biomolecules* **2021**, *11*, 395. https://doi.org/10.3390/biom11030395

Academic Editor: Vladimir N. Uversky

Received: 27 January 2021
Accepted: 4 March 2021
Published: 8 March 2021

Publisher's Note: MDPI stays neutral with regard to jurisdictional claims in published maps and institutional affiliations.

Copyright: © 2021 by the authors. Licensee MDPI, Basel, Switzerland. This article is an open access article distributed under the terms and conditions of the Creative Commons Attribution (CC BY) license (https://creativecommons.org/licenses/by/4.0/).

Abstract: The tumor microenvironment (TME) is composed of cancerous, non-cancerous, stromal, and immune cells that are surrounded by the components of the extracellular matrix (ECM). Glycosaminoglycans (GAGs), natural biomacromolecules, essential ECM, and cell membrane components are extensively altered in cancer tissues. During disease progression, the GAG fine structure changes in a manner associated with disease evolution. Thus, changes in the GAG sulfation pattern are immediately correlated to malignant transformation. Their molecular weight, distribution, composition, and fine modifications, including sulfation, exhibit distinct alterations during cancer development. GAGs and GAG-based molecules, due to their unique properties, are suggested as promising effectors for anticancer therapy. Considering their participation in tumorigenesis, their utilization in drug development has been the focus of both industry and academic research efforts. These efforts have been developing in two main directions; (i) utilizing GAGs as targets of therapeutic strategies and (ii) employing GAGs specificity and excellent physicochemical properties for targeted delivery of cancer therapeutics. This review will comprehensively discuss recent developments and the broad potential of GAG utilization for cancer therapy.

Keywords: glycosaminoglycans; cancer; cancer therapy; hyaluronan; heparan sulfate; heparin; chondroitin sulfate; drug carriers; nanomaterial; therapy targets

1. Introduction

The tumor microenvironment (TME) is composed of cancerous, non-cancerous, stromal, and immune cells that are surrounded by the components of the extracellular matrix (ECM) [1]. The ECM is a significant component of the TME with a vital role in cancer's pathogenesis [2,3]. It is well established that TME plays an essential role in tumorigenesis. Indeed, tumor growth and metastasis steps, e.g., primary lesion development, intravasation, extravasation, and metastasis to anatomically distant sites, are executed via the discrete interplay of the tumor cells with their microenvironment [4]. Glycosaminoglycans (GAGs), natural biomacromolecules, and essential ECM and cell membrane components are extensively altered in cancer tissues [5]. Indeed, these heteropolysaccharides vital in supporting homeostasis have also been established to participate in inflammatory, fibrotic, and pro-tumorigenic processes [6–9]. Both free GAGs and GAGs bound into the protein cores of proteoglycans- (PG) are crucial mediators of cellular and ECM microenvironments, with the ability to specifically bind and regulate the function of ligands and receptors crucial to cancer genesis [4,10,11].

Structurally, GAGs are linear, long-chained polysaccharides consisting of repeating disaccharide units linked by glycosidic bonds. These building blocks are composed of N-acetylated hexosamine and uronic acid. The type of the disaccharide repeating unit and its modifications, including discrete sulfation patterns, allows the classification of GAGs into specific categories, e.g., chondroitin sulfate (CS)/dermatan sulfate (DS), heparin (Hep)/heparan sulfate (HS), keratan sulfate (KS) and hyaluronan (HA) [12–15]. KS chains contain galactose instead of uronic acid in their disaccharide building blocks [15]. CS/DS, HS/Hep, and KS chains are covalently bound into the protein cores of proteoglycans [6]. On the other hand, the non-sulfated GAG HA is not bound into the proteoglycan core but is secreted to the ECM of almost all tissues [13].

Bound GAGs are initially synthesized on core proteins at the Golgi lumen. Their glucuronic acid—N-acetylglucosamine/N-acetylgalactosamine(GlcA-GlcNAc/GalNAc) or, in the case of KS, galactose-N-acetylglucosamine (Gal-GlcNAc) repeating units are subjected to significant structural modification, including sulfation and in the case of HS/CS epimerization at the Golgi apparatus. Moreover, the desulfationof HS chains is performed at the cell membrane compartment [16]. The fine modifications result in an astonishing number of divergent GAG structures.

The GAG fine modifications define, to no small degree, the specificity of their binding with proteins. Notably, GAGs have been shown to interact with more than 500 proteins [17]. The interactions of GAGs with membrane receptors, ECM proteins, chemokines, and cytokines, as well as enzymes and enzyme inhibitors, are crucial in both development and homeostasis [18,19]. Likewise, GAGs' interactions with the above, both soluble and insoluble ligands, play a vital role in various diseases, including cancer [20]. By modulating numerous signaling pathways, GAGs exert distinct effects on cancer cells' functions, cancer stroma interactions, and cancer-associated inflammation, thus regulating essential processes for tumor progression and metastasis [1,4,6,21].

During disease progression, the GAG fine structure changes in a manner associated with disease evolution. Thus, changes in the GAG sulfation pattern are immediately correlated to malignant transformation [22]. Their molecular weight, distribution, composition, and subtle modifications, including sulfation, exhibit distinct alterations during cancer development [23,24]. Thus, most tumor types exhibit increased CS content with an increase in the 6-O-sulfated and/or unsulfated disaccharide content and a decrease in the 4-O-sulfation level due to changes in relevant enzyme activities [23,24]. Likewise, an aberrant HS sulfation pattern has been correlated to tumorigenesis. It was shown that the N-sulfation of GlcNresidues in specific domains along the HS chain facilitate tumor angioegenesis [25]. The expression of HS 6O-sulphated disaccharide content was shown to be increased during chondrosarcoma [26] and colon carcinoma progression [27].

GAGs and GAG-based molecules, due to their unique properties, are suggested as promising effectors for anticancer therapy [28]. Considering their participation in tumorigenesis, their utilization in drug development has been the focus of both industry and academic research efforts [29]. These efforts have been developing in two main directions; (i) utilizing GAGs as targets of therapeutic strategies and (ii) employing GAGs exquisite specificity and excellent physicochemical properties for targeted delivery of cancer therapeutics.

This review will discuss recent developments and the broad potential of GAG utilization for cancer therapy.

2. Focus on GAGs' Structure and Roles

GAG polymers are assembled through several consecutive steps with different enzymes' involvement at each separate stage. Sulfated GAGs are synthesized by specific enzymes in the cell's Golgi apparatus, whereas HA is synthesized by transmembrane proteins called HA synthases (HASs). While HA is not linked to a protein and is produced from its reducing end, the sulfated GAGs are built up from the non-reducing end and synthesized as side chains attached to a protein core of PGs [5].

In the case of KS, GlcA is replaced by GalN. Henceforth, the growing GAG chain's modifications, e.g., deacetylation/N-sulfation and epimerization of GlcA to IdoA followed by O-sulfation, are performed [30,31]. Therefore, the individualized functionalization of GAGs results in their unique structures. Indeed, distinct sulfation patterns have been identified at the disaccharide unit's functionalization sites, hexosamine, and IdoA components, facilitating great complexity and structural diversity [32,33].

Different variations in the expressions/activities of enzymes involved in GAG synthesis have been described. One example is that the levels of exostoses (multiple)-like 1 (EXTL1) and CS N-acetylgalactosaminyltransferase 1 (CSGalNAcT-1), which participate in the production of HS and CS, respectively, were shown to exhibit an inverse ratio of expression. The inverse expressions identified in the process of B-cell differentiation have been suggested to act as a switch enabling either CS or HS synthesis observed during these cell differentiations [34].

2.1. Heparin and Heparan Sulfate

Both Hep and HS chains are synthesized as a modification of a PG protein core, sharing a biosynthetic scheme but exhibiting some disparities [35,36]. Thus, initially, the sequential addition of four sugar residues by different glycosyltransferases will give rise to the linker tetrasaccharide (for Hep/HSXyl-Gal-Gal-GlcA) connected to the core protein's serine residue as a linker region [37]. Notably, the linkage region also serves as a primer for the initiation of the CS chains biosynthesis. In the case of HS, the members of the EXTL family of glycosyltransferases trigger chain creation by transferring an N-acetylglucosamine (GlcNAc), whereas in the case of CS chains, a β-N-acetylgalactosamine (β-GalNAc) residue is attached to the linkage primer by a CSGalNAc-transferase [37]. Polymerization of HS then takes place by the alternating addition of GlcAβ1,4 and GlcNAcα1,4 residues through the action of designated glycosyltransferases [38]. Modifications, such as N-deacetylation and N-sulfation of glucosamine, and O-sulfations are subsequently performed. The GlcA residues can, on some occasions, be epimerized to iduronic acid (IdoA)[35,36].

The two GAGs differ, as the main HS disaccharide unit comprises a GlcA and N-acetylated GlcN(GlcNAc). In contrast, the main Hep disaccharide consists of sulfated, at the carbon 2 IdoA(IdoA2S), and N-sulfated GlcN also sulfated at C6 (GlcNS6S). Due to the high Hep sulfation level, this GAG is characterized as a biomacromolecule with the highest negative charge density [39]. The functionalization with sulfate is uniformly distributed along the Hep chain, whereas HS chains exhibit alternatively exchanging regions of high sulfation with lower or non-sulfated sequences [40]. Indeed, Sulf-1 and Sulf-2, sulfatase enzymes, are active at the extracellular compartment and trim the 6-O-sulfates partially from HS, but do not affect Hep, which is not located at the cells' membranes [41]. As a result, the Hep chain mainly comprises trisulfated disaccharides (80%) consisting of sulfated IdoA and sulfated GlcN.

The HS chains predominantly consist of disaccharide repeats comprised of GlcA and GlcNAc, with a much lower sulfation level [42]. Notably, the "fully sulfated" HS sequences, denominated as S domains, commonly exhibit the highest binding propensity to Hep/HS-binding proteins [43]. Indeed, the binding between proteins and HS/Hep is most commonly executed by charge–charge interactions between the proteins' basic amino acids and the anionic sulfate and/or carboxylate [18,44]. The interaction between respective binding proteins and HS is likewise affected by the GAG heterogeneity and cationic association [19]. Moreover, posttranslational modifications, such as N-glycosylation, of the HS/Hep binding proteins can regulate ligand and HS/Hep binding as shown for the fibroblast growth factor receptor 1 [45]. Notably, its disaccharide unit's extensive modifications render HS the most complex animal polysaccharide [19].

HS chains are synthesized by almost all mammalian cells in the forms of HSPG and are localized to the cell membrane (e.g., syndecans) and pericellular space/basement membranes (e.g., perlecan) or extracellular matrices. Despite the HS chain's extensive functionalization, its fine structure is notably conserved in a given cell type [46,47]. HS's composition varies both spatially and temporally during development and in a celltype-dependent manner. The involved regulating mechanisms remain poorly elucidated.

Significant changes occur in HS composition during carcinogenesis, and vitally, both tumor growth and tumor-dependent angiogenesis depend on HS growth factor interactions [48].

Hep is synthesized only in connective tissue-type mast cells or basophils [49]. The Hep chain is synthesized during the core protein modification of the PG, seglycin. Seglycine exhibits a small protein core but undergoes extensive glycosylation, resulting in a molecular weight up to 750 kDa [50]. The bound Hep chains' molecular weight varies between 60 KDa and 75 kDa. These Hep chains are cleaved into 5–25 kDa fragments when mast cells and basophils are degranulated [51,52]. Mast cells release Hep by exocytosis upon binding specific antigens to the IgE antibodies attached to their cell-surface receptors [53]. However, Mast cell serglycin can also be decorated by other GAG chains, such as CS and DS [54].

Hep, however, can be uptaken by various cells, including endothelial cells, as the primary site for removing unfractionated Hep from the circulation is the liver sinus endothelial cells [55].

In mammalians, HS/Hep are enzymatically degraded by heparanase, a strict endo-β-glucuronidase [56].

2.2. Chondroitin Sulfate/Dermatan Sulfate

The CS chains consist of disaccharides comprising β(1-4) GlcA and β(1-3) GalNAc. The sulfation pattern of the GlcA and GalNAc determinesthe type of CS. Thus, CS-A is characterized by single sulfation at C4 of the GalNAc, whereas CS-C is determined by single sulfation at C6 of GlcA. Other functionalizations exist, as GalNAc can be sulfated at the carbon 4 and/or 6, whereas GlcA can also be sulfated at the C2 and/or C3 [57,58]. On the other hand, CS-B denominated similarly to DS, consisting of alternating GlcA or IdoA, which can be sulfated at C2, and GalNAc, which can be functionalized by sulfation at C4 or C6 [58]. Both CS and DS exhibit vast differences regarding chain length and MW, with the latter being in the 5–70 kDa range [59]. The prominent heterogeneity of the CS/DS chains is directly correlated to these GAGs' biological roles [60,61].

An example is the altered functionalization of CS/DS in gastric cancer as the sulfation at C4 is downregulated, and sulfation at C6 increased in tumor cells compared to normal gastric cells. Additionally, the chain length of CS/DS and the GAG content of the PGs, decorin, and versican was decreased significantly.

2.3. Keratan Sulfate

The KS chains consist of disaccharides containing β(1-4) GlcNAc and β(1-3) Gal. This specific glycosidic binding results in a GAG chain formation, unique for its lack of a carboxyl group. KS', binding into the protein core of PGs differs compared to HS/CS. Thus, corneal KS denominated as KS-I binds to an Asn in the core proteins through an N-linked complex, branched oligosaccharide. On the other hand, in cartilage, the KS chains denominated as KS-II utilize their N-Acetylgalactosamine (GalNAc) to establish an O-link with the Ser or Thr residues of the protein cores [62]. The type III KS (KS-III), initially identified in the brain tissue, links a mannose to a Ser residue of the protein core [63]. KS chains have a molecular weight ranging from 5–25 kD [64].

KS structure is mostly dependent on the tissue type as corneal KS-I exhibits longer chain length and a lower degree of sulfation than the cartilage KS-II. KS-III is mainly bound to PGs localized to the brain and nervous tissues [65,66]. The expression of KS is also deregulated in cancer. Indeed, it was suggested that KS's aberrant expression could be utilized as a marker of pancreatic cancer progression and metastasis [67] and that highly sulfated KS is produced by malignant astrocytic tumors [68].

2.4. Hyaluronan

Transmembrane enzymes denominated HA synthases (HAS) produce HA chains. The three HAS isoforms, HAS1, HAS2, and HAS3, use cytoplasmic UDP-glucuronic acid and UDP-N-acetylglucosamine as substrates. Their active site is localized intracellularly, whereas the synthesized HA chain extrudes into the ECM [13]. This non-sulfated GAG is composed of repeating units of GlcNAc and GlcA combined by β-1.3 and β-1.4 linkages, with an average mass of 100–2000 kDa [13]. HAS1 and HAS2 synthesize a high molecular weight polymer, whereas HAS3 produces shorter chains ($\sim 2 \times 10^6$ Da vs. $\sim 2 \times 10^5$ Da, respectively) [69]. HA's biological information is translated to the length of its polymers and defines its effects [70]. The UDPsugar precursors and holistic cell metabolism responsible for producing HAS substrates critically regulate HASs activities [71]. HA-mediated effects are executed through various mechanisms that involve the binding of HA to surface receptors such as CD44 and RHAMM [72–74] and the internalization of HA through receptor-mediated endosomal pathways [75].

The human genome contains five active hyaluronidases (Hyals) (Hyal1–Hyal4 and PH-20) and the non-transcribed Hyal pseudogene (HyalP1). Hyal 2 and 3 exhibit degrading activity, exclusively for HA [76]. Some human Hyals exhibit degrees of CS-degrading activity. Thus, PH20 shows high activity for HA and low CS-degrading activity. On the other hand, Hyal1 degrades CS-A more swiftly than HA [77]. Hyal-4 is misnamed, as it shows specificity for CS and no ability to degrade HA [78].

Hyal1 is widely expressed and localized to lysosomes or trafficking vesicles [79]. However, Hyal 1 can also be secreted to the ECM by tumor cells [80]. Hyal1 is upregulated in many human cancers and has been correlated with tumorigenesis [81].

In contrast, Hyal2 is bound onto the cell membrane via a GPI anchor and is usually associated with lipid rafts [82], wherein, in common with CD44 and Hyal1, it promotes HA cellular uptake and endocytic internalization [75].

3. Types of Nanoparticles and Materials Utilized for Targeted Drug Delivery—Focus on GAG-Based Nanoparticles

The development of targeted delivery systems for anticancer drugs in the form of nanoparticles has been prioritized since classical methods, namely chemotherapy, radiation therapy, surgery, and their combination, still do not benefit a significant number of patients [83].

Micro- and nanoencapsulation [84–86], micellar [87,88], and liposomal [89] forms, dendrimers [90], mesoporous particles [91], and nanogels [92] are used most often for targeted drug delivery. A wide range of compounds, both synthetic [93,94] and natural [86,87,89,92], are used as materials, and each of the groups has several advantages and disadvantages (Table 1).

The delivery of nanoparticles to the tumors rests on a series of both specific and nonspecific interactions with cells. The specific interactions are based on functionalizing the surface of nanoparticles with ligands that are specific for the target tumor tissue, including tumor cells, intracellular targets, intratumoral and peritumoral blood vessels, and the ECM. The nonspecific nanoparticles are coated solely with stabilizing agents. Most of the studies suggest that the crossing of the tumor blood vessel barrier by nanoparticles is mostly perpetrated through intercellular gaps. Their restraint to the tumor site is dependent on the pressure produced by inadequate lymphatic drainage, commonly denominated as the enhanced permeability and retention (EPR) process [95]. Recent developments suggest that more than 90% of nanoparticles actively enter solid tumor tissue through endothelial cells, challenging the current rationale for nanomaterial synthesis [96]. Nanoparticles targeting specific tumor-associated antigens exhibit superior delivery and effects [96]. A new stage in developing nanomaterials is utilizing patient-derived macromolecules, as recently shown by Lazarovits et al. [97].

In common with others, GAG-based nanoparticles have to overcome the mononuclear phagocytic system's action, which attenuates their efficiency through sequestration and elimination. Notably, nanoparticles carrying a negative charge are more prone to phagocytosis than positive surface charge carrying nanoparticles. Thus, modulating CS charges with competent functionalization can attenuate their phagocytosis [98]. Renal excretion function is another obstacle as it can severely attenuate nanoparticles' actual delivery efficiency. Indeed, renal excretion function seems to be facilitated by incorporating GAG components even though it does not seem to affect tumor accumulation [99]. Modification of the hydrodynamic diameter to the 5.5 nm–100 nm range minimizes kidney excretion and enhances delivery efficiency [100].

Nanocarriers obtained using biocompatible natural polymers such as GAGs do not exhibit adverse effects on cell viability in cell cultures. They show good biocompatibility in animal experiments [92,101,102]. In addition to biocompatibility and specificity, GAG-based nanocarriers, when their GAG components are specifically modified, exhibit other properties, such as high stability, adjustable particle size, and the ability to respond to external stimuli, such as temperature, light, pH, and ionic strength [103–105], enabling multifunctional utilization [106,107]. GAGs, such as CS and HA, have been utilized as therapeutic agents for various pathologies, including osteoarthritis [108,109], with no significant side effects, suggesting their long-term safety. The broad utilization of HA in dermatological clinical practice has not been associated with side effects [110].

The resulting nano-systems' properties depend on the type and concentration of polymer used for their production and the type and degree of intermolecular interaction or crosslinking. Thus, HA can generate self-assembling micelles with the ability to create amphiphilic nanocarriers. Indeed, HA micelles can effectively deliver hydrophobic drugs to target cancer cells while simultaneously facilitating bioavailability and the half-life of the utilized drugs [111]. Importantly, nanoparticles can be loaded with various types of drugs, both hydrophilic and lipophilic, as well as DNA, RNA, peptides, and proteins [85,88,112,113].

Table 1. Types of nanoparticles and materials utilized for targeted drug delivery.

Nanoparticle System	Material	Nanocarriers Type	Examples of Carried Agents	Reference
Lipids	Phospholipids	Liposomes, solid lipid particles	RGD peptide, apatinib	[89]
Synthetic polymers	Poly(N-isopropylacrylamide, poly-N-vinylpyrrolidone, poly(lactic-co-glycolic acid)	Micelles, nanoparticles,	Doxorubicin, curcumin, indocyanine green	[85,88,93,94,104,112]
Natural polymers	HA, alginate, chitosan, heparosan, carboxymethyl starch, CS, Hep	Microcapsules nanospheres, nanoparticles, nanogel, micelles	Doxorubicin, BSA, tirapazamine, cisplatin	[84–87,92,101,106,111] Section 1
Dendrimer	Polyester, Polyacetal/polyketal	Micelles	Camptothecin, methotrexate	[90]
Silica	Mesoporous silica	Nanoparticles	Doxorubicin, fluorescein isothiocyanate	[91]
Metal	Gold	Nanoparticles, nanorods	Doxorubicin, bleomycin	[105]

3.1. Heparin and Heparan Sulfate for Anticancer Drug Delivery

Hep and low-molecular-weight heparins (LMWHs) are widely used as a clinical anticoagulant due to their ability to bind with and inhibit the serine-threonine antithrombin protease [114]. Hep is also studied and used for applications in other therapeutic areas due to its biocompatibility, for example, wound healing, burn injury treatment, inhibition of inflammation, and metastatic spread of tumor cells [115]. Hep's chemical and physical properties connected with the large surface area of its chain and the presence of reactive functional groups allow efficient binding of different anti-tumor agents. Nanoparticles based on Hep can be applied as efficient anticancer agent carriers with versatile surface chemistry for functionalization and the introduction of biomolecules [116]. Some of the Hep derivatives are used to deliver imaging agents, such as iron oxide nanoparticles, to detect tumor cells in humans [117]. Sodium deoxycholate (DOC)-conjugated Hep derivatives (DOC-heparin) were used to prepare nanoparticles for in vivo tumor targeting and inhibition of angiogenesis based on chemical conjugation and the EPR effect [118]. More substantial anti-tumor effects of the DOC-heparin were achieved in animal studies compared to Hep alone. Obtained results confirmed that the conjugated Hep retained its ability to inhibit binding with the angiogenic factors, inducing a significant decrease in endothelial tubular formation. In a separate study, dendronized Hep–doxorubicin (Dox) conjugates were prepared, exhibiting a combination of Dox and Hep features and characterized as pH-sensitive drug delivery vehicles [119]. The prepared nanoparticles showed potent anti-tumor activity, induced apoptosis, and significant antiangiogenesis effects in the 4T1 breast tumor model. Additionally, dendronized Hep and the derived nanoparticles with the loaded drug demonstrated no significant toxicity to the healthy organs of both tumor-bearing and healthy mice, which was confirmed by histological analysis.

Park et al. first attached low molecular weight Hep to stearylamine to obtain amphiphilic polymer that was used to prepare self-assembled micelle-like nanoparticles, loaded with docetaxel in their hydrophobic core. The obtained preparation was tested in MCF-7 and MDAMD 231 human breast cancer cells. This approach demonstrated that Hep retained about 30% of its anticoagulant activity, increased the half-life time of docetaxel in the novel preparation used, and significantly inhibited tested cells' viability [120]. Park et al. synthesized an amphiphilic biopolymer made of Hep and deoxycholic acid and prepared nanoparticles loaded with Dox. These nanoparticles were tested for cytotoxicity and anti-tumor effects. The investigated system showed high loading efficiency and a substantial anti-tumor effect [121].

Other studies describe the characteristic properties of Hep-based nanoparticles as potential drug delivery systems, not focusing on specific types of cancer [122].

In summary, Hep is capable of forming nanoparticles upon the introduction of amphiphilic or hydrophobic molecules [116,123,124]. It can also interact with proteins, which leads to the formation of complexes with various biological effects [125,126]. Nevertheless, absorption of blood proteins upon the introduction of Hep nanoparticles into the human body needs to be controlled.

3.1.1. Micellar Heparin Nanoparticles

Studies showed that it is necessary to modify the Hep surface of nanoparticles to reduce blood elements' absorption. Moreover, it is possible to introduce additional specific receptors for targeted delivery directly to the tumor [127,128]. In a study on the development of Hep-based micelles, multifunctional self-assembling nanoparticles were created that combine the following properties: the carrier material is non-toxic, and the resulting micelles had high stability and sensitivity to pH. Intravenous injection of the Hep/Dox combined micelles increased Dox blood circulation time and enhanced its accumulation at the animal model's tumor site [129].

Hep nanoparticles can penetrate body barriers. Thus, a study showed that Hep particles 100 nm in size effectively overcame the blood brain barrier (BBB), as evidenced by an increase in the concentration of drugs in the brain target tissue [130]. However, particles with a small size very quickly left the circulation, which indicated the need to select the functionalization of their surface specifically.

3.1.2. Heparin-Coated Metal NanoParticles

Another important direction is the development of targeted delivery systems based on magnetic metal nanoparticles. The main disadvantage of such nanomaterials is that we need to select a proper stabilizer or coating that will contribute to the constant particle size, reduce their toxic effects, increase biocompatibility, and overcome physiological barriers maintaining their high magnetic properties. Hep was found to be a sound basis for these coatings [131]. Another study demonstrated that with Hep's utilization, it is possible to create stable magnetic nanoparticles, based on iron oxide, exhibiting low polydispersity [132]. The introduction of cis-platin to the composition of Hep and iron oxide created Hep-coated metal nanoparticles, which exhibited a cytotoxic effect on cancer cells but lowered toxic side-effects compared to the free drug [133].

In similar studies, magnetic nanoparticles were modified with polyethylene glycol (PEG) and Hep, after which they were functionalized with additional targeting agents. PEGylation enables longer circulation time but can also render metal nanoparticles increased passive targeting via the EPR effect. PEGylated metal nanoparticles were, furthermore, modified with a Hep layer to enable the carrying of the highly cationic CPP-linked protein drug [134]. Further studies demonstrated that the resulting nanoparticles have an increased recirculation time in the blood, retain their high magnetic properties, and overcome the BBB. It was also shown that in a mouse 9L glioma model, particles with a size of more than 50 nm accumulate at high concentrations in the tumor tissues [135].

3.1.3. Heparin Nanogels

Delivery systems based on liposomes, micelles, and magnetic nanoparticles are relatively well-studied systems for which specific rules and dependencies have already been developed, but depot forms based on nanogels represent a new milestone in this field [136]. Most scientific research, in this area, is devoted to creating matrices based on natural polymers, including Hep, chitosan, alginic acid salts, and others.

The majority of the studies were dedicated to delivering genes and proteins. There are also several reports in which Hep nanogels have been developed for the targeted delivery of anticancer drugs [100]. A polymer matrix is typically produced by covalent crosslinking to form strong and stable structures. Due to the polymer's properties, the delivery system can be sensitive to a wide range of external factors, and thus, fine-tuned release of the drug load can be accomplished [124]. Melanoma is characterized by a high metastatic potential of the transformed melanocytes, which also become "invisible" to the immune cells. This "invisibility" is sustained by many mechanisms, one of them being the formation of a platelet cloak. The heterogeneous mixture of GAGs can inhibit this process by blocking P-selectin-mediated intercellular adhesion. LMWHep-coated with Dox and loaded in liposomes (LMWHep-Dox-Lip) was studied in the B16F10 melanoma cell line. This nanomaterial exerted both a cytotoxic effect and inhibited the adhesion between tumor cells and platelets mediated by P-selectin. It was demonstrated in vivo that the pulmonary metastases of melanoma are prevented by LMWHep-Dox-Lip treatment [137].

This type of drug-delivery system can be utilized for combination chemotherapy, where more than two drugs with different properties and mechanisms of action are applied to boost the cancer treatment. Thus, Joung et al.produced Hep-Pluronic (Hep-Pr) nanogel loaded with paclitaxel and DNAase [138]. The nanogel allowed robust intracellular delivery to facilitate these drugs' synergistic effects in a dose-dependent manner and inhibited tumor cells' growth. Notably, the synthesized matrix can bind to high concentrations of both hydrophilic and hydrophobic drugs. Nanogels exhibit some disadvantages due to their high polydispersity, hence the uneven distribution of the active substance in the volume [139].

Some approaches utilize HS for nanoparticle preparation. A recent drug delivery strategy conjugated the chemotherapeutic agent, docetaxel, onto HS. Due to its antimetastatic and T cells infiltration enhancing properties, Aspirin (ASP) was encapsulated into the HS-docetaxel micelle followed by the cationic polyethyleneimine (PEI)-polyethylene glycol (PEG) copolymer binding to HS via electrostatic force. This approach results in an ASP-loaded HS-docetaxel micelle (AHD)/PEI-PEG nanocomplex (PAHD). PAHD exhibits a long half-life in the blood due to the PEG shell. As TME is characterized by weakly acidic pH, the PEI-PEG polymers detach from AHD and increase tumor cells' permeability due to their positive charge. Heparanase, overexpressed by tumor cells, degrades HS, thus delivering the active ASP and docexatel to tumor cells. Indeed, PAHD exhibits targeted toxicity toward tumor cells but not normal cells and is bestowed with superior ability to suppress tumor growth and lung metastasis in 4T1 breast cancer tumor-bearing mice [140].

3.1.4. Summary

Significant progress in utilizing Hep-based nanoparticles as novel venues or in combination with existing therapies, such as chemotherapy or photodynamic therapy [131,138,141], has been achieved. Indeed, many recent studies have proven that Hep-based nano-scaled systems have great potential as drug carriers, the ability for specific delivery to cancer tissues, and excellent biocompatibility [122].

However, even though significant achievements have been obtained in the synthesis of Hep-based nanoparticles, no such nanomaterials have made their way to clinical trials. A hurdle to clinical transition is the anticoagulant properties of Hep, which can lead to bleeding complications. Chemically modifying Hep can attenuate its anticoagulant activities; however, the mechanisms of its anticancer and anticoagulant abilities are not fully understood, and a more profound comprehension of the interplay between structure and activity is needed [142]. Furthermore, one has to respond to difficulties in controlling Hep's quality due to its poly component and holistic pharmacologic characteristics [143]. Indeed, Hep' preparations contaminations have even resulted in patient death [144].

3.2. Chondroitin Sulfate and Dermatan Sulfate-Based Nanoparticles as Drug Delivery Systems

CS exhibits high biocompatibility and specific localization, being bound to PGs in ECMs of tissues such as cartilage, blood vessel walls, skin, and tendons [48]. In line with increasing nanotechnology application, optimally designed nano-scaled carriers on the base of CS have been developed, exhibiting unique properties, such as biocompatibility, low toxicity, and active and passive targeting. Their specific properties and discrete modalities make them promising drug delivery vehicles for cancer therapy [145].

Because CS, like all GAGs, is a specific anionic acid polysaccharide, it couples well with cationic poly-saccharides, including chitosan, which as a natural molecule is likewise characterized with good bioactivity [146]. Thus, a CS–chitosan nanoparticle carrier encapsulating black rice anthocyanins exhibited significant apoptosis-inducing effects in colon cancer cells [147], whereas loading these nanoparticles with curcumin induced a cytotoxic effect in the lung cancer model [148].

Moreover, loaded with camptothecin (CPT) polymeric nanoparticles functionalized with CS exerted targeted colon cancer drug delivery with superior anticancer effects compared to non-targeted nanoparticles [149]. This approach utilized CS's affinity for the CD44 HA receptor overexpressed in various tumors [107].

Notably, CS can lower the adverse side effects of chemotherapeutic drugs as CS-Dox-poly(lactic-co-glycolic acid) (PLGA)conjugated nanoparticles exhibited lower cardiotoxicity and enhanced tumor inhibition compared with free Dox [150]. This development is an important achievement as cardiac toxicity through various mechanisms is a severe drawback of Dox utilization [151,152].

Summary

In summary, CS-derived drug-loaded nanomaterial has been shown to have a reasonable encapsulation efficiency, an appropriate hydrodynamic diameter, manageable surface charge, low toxicity, and improved anticancer properties compared to the free drug [149,153,154].

3.3. Keratan Sulfate in Anticancer Drug Delivery

KS, another perspective GAG for drug delivery, is localized in the ECMs of different tissues, such as cartilages, cornea, and bone [15]. Besides acting as a constitutive molecule of the ECMs, KS also plays a role as a hydrating and signaling agent in cartilage and cornea tissues. KS chains are structurally bound to a protein core, forming PGs. Unlike other GAGs, KS lacks uronic acid and contains galactose in its disaccharide building blocks. Moreover, the unsulfated Gal residue is essential for binding mediated through non-electrostatic interactions [155], such as hydrophobic and/or van der Waals forces [156].

These data suggest that protein binding strategies may need to be chosen based on the GAG class to be incorporated in the drug delivery vehicle [157–159].

Summary

Several reviews describe the structures and functions of KS proteoglycans, but their potential role in drug delivery has not yet been determined [157–159].

3.4. Hyaluronic Acid-Based Nanoparticles for Controlled Drug Release in Cancer

HA is an abundant GAG, deposited to most tissues' ECM [160]. Its properties, biodegradation, biocompatibility, water-solubility, non-toxicity, and non-immunogenicity and its chemical characteristics, enabling modifications with functional groups, define HA as a suitable molecule carrier to deliver low molecular weight drugs [161]. Furthermore, its specific ligation with cell surface receptors such as CD44 and RHAMM [111] enables HA-based nanoparticles to target diseased cells that express these receptors. Indeed, CD44 and RHAMM receptors are overexpressed by many tumor types [162–164]. High production of HA has been determined in many solid tumors, but it is the combination of HA production and Hyal overexpression that facilitate both carcinogenesis and metastasis [165]. In prostate cancer, the increased release of low molecular weight HA (LMWHA) due to Hyal1 overexpression and increased HASs activity results in enhanced autocrine proliferation [166]. The naked mole-rat example, the only mammal resistant to cancer, argues the importance of HA. This rodent produces high amounts of very high molecular weight HA and simultaneously exhibits low Hyals expression, correlated to the minuscule ability to cleave HA [167].

Therefore, the involvement of HA in tumorigenesis processes is of crucial significance. This finding has ignited vibrant research efforts directed at HA metabolism and focusing on the inhibition of HA degradation and on blocking HA-receptors interaction. The HA-degrading enzymes Hyals have been identified as attractive anticancer therapy targets due to their cell surface or extracellular deposition. HA localization enables their inhibition in the ECM [81].

The use of HA-based nanoparticles requires knowledge of HA pharmacokinetics. Thus, it is well established that blood and lymphatic transport system are responsible for HA distribution in the body [168,169]. The utilization of isotopes showed that high molecular weight HA (HMWHA) mainly accumulates in the liver, while LMWHA is secreted in urine [170]. Notably, many studies indicate that the differences in HA-based nanoparticles' targeting efficiency depend on their molecular weight. For example, HMWHA-coated lipid nanoparticles exhibited a stronger binding affinity to the CD44 receptor of murine melanoma cells in vitro than theLMWHA nanoparticles [162].

Different types of HA-based nanoparticles with discrete features have been used as drug carriers (summarized in Figure 1).

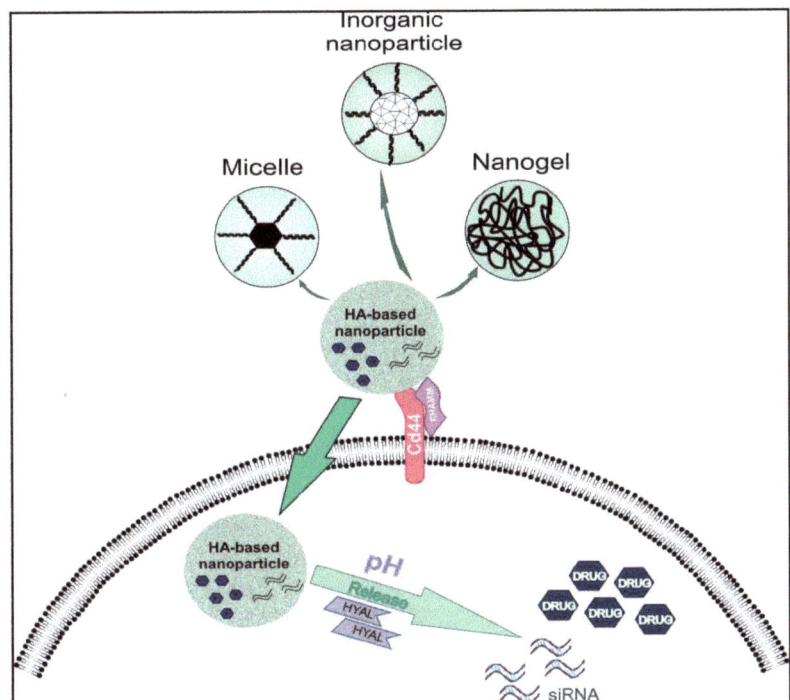

Figure 1. Mechanism of action of HA-based nanoparticles: HA (Hyaluronic acid)-based nanomedicines are used to mediate targeted delivery of therapeutic compounds (DRUGS or siRNA: small interfering RNA) in cancer cells. Nanoparticle targeting is enhanced by HA-specific interaction with CD44 or RHAMM, which are overexpressed in different cancer cell types. These receptors also mediate the internalization of the nanoparticles. After their uptake, each type of nanoparticle is degraded either by enzymatic lysis of HA by hyaluronidase (HYAL) action or by a pH-dependent mechanism.

3.4.1. Hyaluronic Acid-Based Micelles

HA-based micelles were shown to target CD44 positive breast cancer cells with high-affinity in vitro and in vivo [103]. The hydrophilic HA backbone is modified with hydrophobic groups to form an amphiphilic compound, which can self-assemble into a micelle in an aqueous solution and encapsulate or conjugate drugs After reaching cancer cells, drug release is achieved using various mechanisms, such as through pH dependence or enzyme action [171,172].

HA-conjugated hexadecylamine micelles for the docetaxel delivery to MDA-MB-231 breast cancer cells are examples of a study testing HA-conjugated micelle drug formation [173]. Results showed that HA conjugation of micelles enhanced cellular uptake. Moreover, treating mice bearing xenograft MCF-7 human breast cancer tumors with HA-shelled-paclitaxel prodrug micelles resulted in 100% mouse survival and tumor-specific accumulation of the micelles [174].

In pancreatic cancer, the use of HA-engineered nano-micelles loaded with 3,4-difluorobenzylidene curcumin were tested in CD44-positive MiaPaCa-2 and AsPC-1 pancreatic cancer cell lines [175]. The existence of pancreatic cancer stem cells overexpressing CD44 was identified, contributing to a tumor's resistance to chemotherapy [175,176]. Recent studies in vivo continued to prove the success of HA:Sucrose nanoparticles in the delivery of anticancer treatments (such as EF2-Kinase inhibitor) to pancreatic cancer cells, leading to significant inhibition of division and tumor formation [177].

Recognized, HA-based micelles' disadvantages are the limited drug circulation in the blood and the fast uptake by liver endothelial cells [178]. Therefore, the conjugation of HA-micelles with PEG has been tested to improve their blood circulation time. Although PEGylation may affect the micelle interaction with cancer cells HA-receptors [179], the simultaneous use of these two types of micelles showed no significant variances as to their delivery efficiency in vivo [180].

3.4.2. Hyaluronic Acid-Based Nanogels

These types of nanoparticles with physically or chemically crosslinked polymer chains possess pores that can be filled with macromolecules to target cancer cells, as initially demonstrated in vitro [181,182]. HA-based nanogels are used to improve the activity of delivering compounds, enhancing stability, and increasing the biological half-life of HA [178,183]. Indeed, several studies demonstrate the efficiency of HA-based nanogels as drug carriers [184,185]. Moreover, it is possible to link HA with coiled-coil peptide, creating a pH-sensitive nanogel for controled drug release to increase the anti-tumor effect on MCF-7 cells in vitro [186]. Furthermore, it was shown that HA nanogels, fabricated by the methacrylation strategy, are sensitive to enzyme action. The nanogels target cancer cells in a manner dependent on HA receptors expression and are deconstructed by the action of Hyals, releasing their drug load [185]. The introduction of cholesterol to crosslinked nanogels confers hydrophobicity to HA, increasing cell membranes' permeability to HA-based nanogels [187]. Another way to enhance the hydrophobic features of HA nanogels is to acetylate the HA backbone. Indeed, acetylation's degree affects Hela cells' drug loading efficiency and targeting in an in vitro experimental model [188].

3.4.3. Inorganic Hyaluronic Acid-Based Nanoparticles

Another category of HA-based nanoparticles is the metal-organic framework NPs conjugated with HA. These porous materials carry many metal-binding sites that can be used for specific functionalization [106,179]. Their advantages are connected to their high drug loading efficiency due to increased binding surface area [189]. This type of NP is sensitive to different pH conditions, allowing fine-tuning of their release, as demonstrated in a prostate cancer cell model [190].

Combining metal with HA allows the exploitation of specific characteristics of both materials. Thus, AuNPs improve radiotherapy due to gold's ability to adsorb X-rays, demonstrated in animal models [191]. On the other hand, HA allows the restructuring of AuNPs surface, enhancing the ability of the hybrid NPs to conjugate with drugs [192]. Moreover, the combined NPs exhibit superior stability and high-affinity targeting of CD44 positive liver cancer cells in vitro, as shown by Kumar et al. [193]. Furthermore, Dox-HA-super-paramagnetic iron oxide nanoparticles (Dox-HA-SPION) were suggested to enhance the drug efficacy and to minimize off-target effects in MDA-MB-231 human triple-negative breast cancer cells (TNBC) [194]. Liu et al. designed tumor-targeting HA-titanium dioxide (HA-TiO2) nanoparticles loaded with cisplatin (CDDP) with significant anticancer activity in the A2780 ovarian cancer cells [195]. Silica NPs, also classified as inorganic nanomedicines, have many advantages, such as controllable shape and size, low toxicity, and good biocompatibility [196]. Modified with HA silica NPs exhibit increased delivery to HA-receptor expressing cancer cells, as demonstrated in vitro and in vivo [197].

3.4.4. Clinical Trials Implementing Hyaluronic Acid-Based Nanoparticles

Primary studies in cell cultures and animal trials have shown promising results of HA-based nanoparticle efficiency in anticancer therapy. Some of these compounds have already been tested in Phase 2 or Phase 3 clinical trials with positive outcomes regarding efficiency and safety. A phase 2 trial tested HA-irinotecan plus cetuximab in 45 patients with KRAS wild-type metastatic colon cancer to examine the compound's safety and efficacy. However, the results of the study have not yet been announced [198]. Another phase 2 study involving 39 patients with extensive-stage SCLC indicates that HA-irinotecan treatment provides survival benefits for patients bearing CD44 positive tumors [199].

Furthermore, phase 1 and 2 clinical trials utilizing HA-cisplatin nanoconjugates (HA-Pt) in dogs with naturally occurring anal sac carcinoma, oral squamous cell carcinoma, oral melanoma, nasal carcinoma, or digital squamous cell carcinoma have been conducted. The obtained results demonstrated the beneficial effects of HA-Pt drug formulations for the treatment of canine squamous cell carcinomas. Moreover, nephrotoxicity, a serious side-effect of Pt therapy, was not evident in any canine subject. Notably, canine oral SCC's similarity to human HNSCC regarding progression and drug response gives essential information for developing human treatments [200]. Examples of HA-based nanoparticle types tested in different cancer models are shown in Table 2.

Table 2. Types of HA-based nanoparticles tested in different cancer models.

HA-Based NP Types	Composition	Drug/Conjugate	Human Cancer Type	Reference
Micelles	HA-b-dendritic oligoglycerol	paclitaxel	breast	[174]
	HA-copoly(styrene maleic acid)	3,4-difluorobenzylidene curcumin	pancreatic	[175]
Nanogels	Coiled-coil-peptide-cross-linked-HA	GY(EIAALEK)3GC (E3) and GY(KIAALKE)3GC (K3)	breast	[186]
	Acetylated HA with low molecular weight 1,2,3-with degrees of acetylation 0.8, 2.1, 2.6 acetyl groups per unit (2 glucose rings)	Doxorubicin	cervical	[188]
Inorganic	HA super-paramagnetic iron oxide	Doxorubicin	breast	[194]
	HA-titanium dioxide	Cisplatin	ovarian	[195]

4. Targeting GAGs in Cancer—New Prospective

4.1. Targeting Heparan Sulfate/Heparin

HS, expressed by all mammalian cells in homeostasis [31], has been determined to be the most complex GAG [19]. This highly variable GAG is critical in cellular signaling and extensively remodeled during cancer progression. In its natural state, Hep is a heterogeneous mixture composed of polysaccharide chains that exhibit varying lengths and different sulfation patterns. Hep, compared to HS, is more homogeneous and its main function is storage. HS and Hep chains can establish specific interactions with various protein mediators regulating critical cellular signaling [18]. The affinity of HS/Hep chains to proteins such as growth factors seems to be significantly affected by their sulfation status and resulting electrostatic interactions [157–159]. Moreover, analysis by the polyelectrolyte theory demonstrated that the binding of FGF-2 to Hep is primarily accomplished through the more specific nonionic interactions, such as van der Waals packing and hydrogen bonding [201]. Therefore, inherent properties of the GAG chains need to be taken into account when designing novel drug carriers [157–159].

To date, more than 400 HS-binding proteins have been identified, including cytokines, growth factors, chemokines, ECM proteins, as well as enzymes and enzyme inhibitors [18]. Thus, the targeting of HS protein interactions is an essential developing therapy approach.

The strategies that have been examined for cancer-oriented therapy are based on (i) the utilization of GAG mimetics as competitive agents to block HS–protein interactions (ii) the utilization of enzymatic methods to cleave or modify HS to inhibit HS–protein interactions.

The utilization of unfractionated Hep and LMWHs is standard clinical practice for the protection against venous thromboembolism in cancer patients [202]. This clinical practice's implementation has also demonstrated a beneficial effect of Hep on cancer patient survival discrete from its anticoagulant properties [203]. Indeed, Hep has now been recognized as a multifunctional drug [50]. Hep mimetics are commonly described as synthetic or semi-synthetic compounds that are anionic, usually highly sulfated, and structurally defined as distinct GAG analogs [204].

Research efforts focused on the synthesis of Hep derivatives with attenuated polypharmacy traits and anticoagulant activity, exhibiting enhanced potency and specificity while downregulating unwanted side effects, e.g., anticoagulation [204]. This approach has been facilitated by significant development in carbohydrate synthesis, including one-pot multi-step procedures and coupling reactions, enabling the synthesis of complex oligosaccharides [205].

A recently synthesized, multitargeting Hep-based mimetic, necuparanib, was shown to attenuate pancreatic cancer tumor cell growth and invasion in a three-dimensional (3D) culture model. In contrast, in vivo, it facilitated survival and attenuated the metastatic ability of pancreatic cancer cells. Furthermore, the proteomic analysis demonstrated that necuparinib, among others, targeted ECM-originating mediators, well established to affect cancer cell growth and metastasis. Specifically, necuparanib attenuated the expression of metalloproteinase 1 (MMP1) and facilitated the expression of tissue inhibitor of metalloproteinase 3 (TIMP3) in the 3D pancreatic cancer model [206]. Moreover, the levels of TIMP3 in the plasma of patients with metastatic pancreatic cancer who were participating in a phase I/II study treatment with necuparanib plus standard therapy were found to be substantially enhanced [206].

A crucial therapeutic target is cancer-associated angiogenesis. Both fibroblast growth factors (FGFs) and vascular endothelial growth factor (VEGF) can form ternary complexes with HS and their respective cell-membrane receptors, initiating signaling cascades that facilitate angiogenesis [207]. These growth factors are characterized as important cancer therapy targets with Hep mimetics' possible implementation [208,209]. The D-mannose-based sulfated oligosaccharide mixture, PI-88 (Muparfostat) is one such inhibitor. It is developed from the oligosaccharide phosphate fraction obtained from the extracellular phosphomannan, initially derived from the yeast *Pichia (Hansenula) holstii* (NRRL Y-2448) and subsequently extensively sulfated [210,211].

Modified LMWH functionalized by polystyrene (NAC-HCPS) exhibited increased affinity to HS binding growth factors and attenuated anticoagulant properties, decreased endothelial cell growth, and formation of endothelial tubes [212]. Moreover, SST0001 Hep derivatives, characterized by 100% N-acetylated, 25% glycol split Hep SST0001 (100NA-ROH, roneparstat), efficiently reduced FGF2-mediated proliferation of endothelial and lymphoid cells and displayed a limited capacity to release FGF from the ECM. This effect is associated with the N-acetylation of GlcN.SST0001 and was also reported to counteract human sarcoma cell invasion induced by exogenous FGF2 [213]. Interestingly, Hep is actively uptaken by melanoma cells and affects their migration and adhesion [214].

The disadvantages of using Hep derivatives, discussed above, are mostly correlated to the intrinsic Hep anticoagulant properties to initiate severe hemorrhagic effects.

4.2. Enzymatic Modulation of HS–Protein Interactions

Heparanase, the only mammalian enzyme responsible for HS/Hep cleavage, is a strict endo-β-glucuronidase, favoring the fission of a GlcA linked to 6O-sulfated GlcN, which can either be N-sulfated or N-acetylated [56]. However, advances have implicated the potential controlling role of the surrounding saccharide sequences and their sulfation pattern in regulating the extent of substrate degradation [56].This plasticity of substrate specificity enhances the execution of various heparanases' roles [215]. The cleavage of HS chains bound into PGs releases latent growth factors, including FGF2, hepatocyte growth factor (HGF), keratinocyte growth factor (FGF4), and TGF-β, which are sequestered to the matrix and cell surface, but also inherently modulates the protein-GAG interactions and downstream signaling [216]. Indeed, trimming of HS can enhance the binding of growth factors to their respective receptors, as in the case of FGF-2 where the creation of tertiary FGF2-FGFR-HS complex is increased by moderate heparanase activity [217]. Moreover, heparanase was found to reside and accumulate in lysosomes suggesting that it also exhibits intracellular functions [218].

Heparanase strongly affects protein–HS interactions, whereas tumor-associated activated fibroblasts, endothelial cells, and immune cells exhibit increased heparanase activity [219]. The overexpression of heparanase results in vivo in increased tumor metastasis, whereas downregulating heparanase markedly decreases cancer cells' ability to metastasize [220].

Heparanase expression was shown to be upregulated in all cancer types, including sarcomas, carcinomas, and hematological neoplasms [221]. Notably, heparanase activity has been correlated to various human cancers' metastatic potential. Thus, the examination of the Cancer Genome Atlas (TCGA) data on heparanase expression in breast cancer clinical samples showed its upregulation in the majority of specimens. Furthermore, increased heparanase expression was correlated with poor patient survival [222]. Similar results have been obtained for other cancer types, including multiple myeloma [223] and bladder cancer [224]. Moreover, heparanase has been shown to affect cancer angiogenesis [225], invasion, and autophagy [226] and partly through syndecan-1-dependent mechanisms to modulate inflammation-associated tumorigenesis [227].

Heparanase can affect the response to chemotherapy. Thus, anti-myeloma chemotherapeutic agents, including bortezomib (proteasome inhibitor) or melphalan (alkylating agent), were shown to increase the expression and secretion of heparanase in an in vitro myeloma model. The subsequent uptake of soluble heparanase by tumor cells initiated ERK and Akt signaling pathways, stimulated the expression of vascular endothelial growth factor (VEGF), HGF, and MMP-9, and was correlated with an aggressive tumor phenotype [228].

An essential mechanism of heparanase action is promoting exosome secretion, which affects both tumor and host cells' biological behavior and finally drives tumor progression [229]. In a myeloma model, it was shown that chemotherapeutic drugs increase exosome secretion. Notably, chemoexosomes have an increased heparanase load, enhancing cell HS's cleaving activity and initiating ERK signaling and syndecan-1 shedding. These authors suggest that anti-myeloma therapy stimulates the secretion of high heparanase content exosomes, facilitates ECM remodeling, changes tumor and stroma cell behavior, and contributes to chemoresistance [230].

Several therapeutic approaches have been tested to develop efficient inhibitors of heparanase activity. Non-anticoagulant heparin derivatives such as SST0001 or roneparstat significantly downregulated heparanase-dependent cleavage of syndecan-1 HS chains, attenuated HGF, VEGF, and MMP-9 expression resulting in decreased tumor growth and angiogenesisinvivo [231,232]. Preclinical evidence resulted in the first human study (NCT01764880) assessing the safety and tolerability of roneparstat in patients with relapsed or refractory multiple myeloma (MM). Patients treated with Roneparstat exhibited acceptable tolerance at clinically significant doses [233].

PI-88 is an inhibitor of heparanase, in addition to its antagonist of angiogenic growth factors function [234]. Even though it exerted adjuvant properties in hepatocellular carcinoma and melanoma patients [235,236], PI88 has been correlated with bleeding events, and thus, did not progress to clinical practice [237].

A series of PI-88 analogs have been synthesized, exhibiting superior performance. The improved analogs comprise single, characterized oligosaccharides with discrete functionalizations and exhibit more efficient antagonism of angiogenic growth factors and respective receptors binding with HS. These properties are translated into potent inhibition of growth factor-dependent endothelial cell growth and strong downregulation of the endothelial tube formation [234]. PG545 is the outstanding member of the PI88 analogs series exhibiting significant anti-angiogenic, anti-proliferation, and antimetastatic effects through potent heparanase inhibitory and angiogenic growth factor antagonist effects [238]. Moreover, PG545 was shown to exert anti-tumor effects discrete from heparanase inhibition as it induces lymphoma cell apoptosis in a non-heparanase-dependent manner [239]. PG545 (pixatimod) is currently being tested in clinical trials [238]. However, despite promising breakthroughs, the development of heparanase inhibitors with beneficial clinical performance and acceptable adverse effects is still elusive. Therapeutics targeting HS are summarized in Table 3.

Table 3. Therapeutics targeting HS.

Therapy Target	Drug	Cancer Type	Stage	Reference
Antagonists of angiogenic growth factors	necuparanib	Pancreatic cancer	3D model, animal tumor model, Phase I/II clinical trial in combination with standard therapy	[206]
	PI-88 (muparfosfat)	General tumor angiogenesis	In vitro, animal models	[210,211]
	NAC-HCPS	Lung tumor	Animal model	[212]
	Hep SST0001 (roneparstat)	Sarcoma	Animal models Section 2	[213]
Heparanase Inhibitors Section 3	SST0001 (roneparstat)	Multiple myeloma Section 4	Animal model, Clinical trial Section 5	[232,233]
	PI-88 (muparfosfat)	Hepatocellular Carcinoma. melanoma	Clinical trial	[235,236]
	PI-88 analogs (PC545-pixatimod)	Human lymphoma	Animal model, Clinical trial	[237,238]

However, some studies targeting heparanase demonstrated contradictory results. In some model systems, inactive heparanase facilitated adhesion and migration of endothelial cells and induced factors that promote angiogenesis, such as vascular endothelial growth factor [240]. The enzyme has a C-terminus domain involved in the molecule's signaling capacity. The human heparanase variant (T5) lacking enzymatic activity has protumorigenic properties, indicating the enzyme's complex role in cancer pathogenesis [240].

5. GAGs and Immunological Aspects of Cancer Therapy

The involvement of glycobiology in cancer and the anti-tumoral immune response can be analyzed at several levels. GAGs are involved in the immune response; they can constitute new biomarkers and offer possibilities to develop new immune-therapy targets.

The interconnection of the immune system and various aspects of tumorigenesis are described in all types of cancers [241,242]. An array of immune cells, mainly from the myeloid lineage, macrophages, and dendritic cells, modulate tumor neoangiogenesis. HA is an essential component of the TME, and its abnormal deposition has been assessed in different tumor types. As HA is one of the modulators of tumor angiogenesis, it can influence various immune cells' physiopathology within the TME. HA-induced effects depend on both its polymer size and its complexes with other molecules. Under healthy conditions, HASs and Hyals are a tightly regulated molecular network that keeps HA ECM levels within physiological limits. When pathological conditions appear, and HA homeostasis is perturbed, the enzymes that regulate its characteristics aid the pro-tumoral processes in TME and induce resistance to therapy [243].

Inflammation and tumorigenesis are intertwined processes [244], and inflammatory mechanisms are involved in both the tumors' initiation and progression. In the continuous communication with the ECM, GAGs regulate the cell/matrix interface and the immune-related mechanisms [2].

5.1. GAGs Roles in Tumor Immunology

CD44-HA constitutes a molecular tandem that can affect tumor immunology by utilizing complex mechanisms [6]. Macrophages have a different expression of CD44 related to their functions, and their capacity to bind HA is variable. CD44 has the highest expression in M1 polarized macrophages, followed by M0 type of macrophages, whereas M2 type expression is similar to the latter. The higher CD44 expression in M1 induces increased binding of HA. On the other hand, the lower M2 expression favors a better internalization of HA. Therefore, the molecular mechanisms in the CD44-HA tandem exhibit subtleties to predict targeting behavior [245].

The presence of tumor-associated macrophages (TAMs) is correlated with the poor outcome in tumor-bearing organisms because these cells sustain the immune-suppression and enhance pro-tumoral mechanisms, and, last but not least, inhibit the actions of anti-tumoral drugs. Therefore, TAMs are preferred targets in tumor therapies [246]. Several years ago, a nanoparticle was designed comprising poly(lactic acid-co-glycolic acid-grafted HA (HA-g-PLGA) that carried a cytostatic, an active metabolite of irinotecan. At the acidic pH of the TME, HA is exposed, and by linking to the CD44 expressed on tumor cells and TAMs, it delivers the cytostatic intracellularly. Tumor cells continue to recruit TAMs that encounter carriers with the cytostatics; hence, the anti-proliferative effect is propagated [247].

In a recent study, novel carrier molecules were tested. Within the tested compounds, oligomeric HA (oHA) targeted CD44 receptors on TAMs for the delivery of curcumin (Cur) and baicalin (Bai) to overcome tumor resistance. The carrier had good cellular penetration and cytotoxicity upon tumor cells. In in vivo animal models of A549 tumor-bearing nude mice, the significant anti-tumor effect was re-confirmed [248]. HA-based nanoparticles were tested as drug carriers in epithelial ovarian cancers to target TAMs specifically. Thus, HA nanoparticles that encapsulate miR-125b (HA-PEI-miR-125b) targeted TAMs in an experimental mouse model of syngeneic ID8-VEGF ovarian cancer and induced these cells to an immune-activating phenotype [249]. In the 4T1 breast cancer animal model, mesoporous Prussian blue (MPB) nanoparticles and LMWHA (LMWHA-MPB) were tested. This approach demonstrated that LMWHA-MPB penetrates M2 macrophages (pro-tumoral macrophages), which are subsequently diverted toward the M1 phenotype exhibiting anti-tumoral action. Therefore, LMWHA-MPB can induce TAMs pro-tumoral potential and can likewise be used in situ for microenvironmental tumoral regulation [250].

LMWHA per se was demonstrated to have an anti-tumoral effect in colorectal carcinoma. The immune response involves activated dendritic cells (DC). Authors have shown that preconditioning DC from tumors with LMWHA increased their ability to migrate in vitro and enhanced DC in vivo recruitment to regional lymph nodes. In a mouse animal model, tumor lysate-pulsed DC (DC/LMWHA) was administered, and a potent anti-tumor response was obtained. Splenocytes from animals treated with DC/LMWHA displayed higher proliferative capacity, enhanced IFN-γ production, and lower immunosuppressive cytokine levels. Therefore, LMWHA can be considered a new adjuvant candidate for DC-based anticancer vaccines [251].

Using HA's ability in reprogramming pro-tumoral M2 type TAMs to anti-tumor M1 macrophages, other nanoparticles with MnO2 were used to decrease tumor hypoxia chemoresistance in the breast cancer experimental model. Increased tumor oxygenation was obtained in conjunction with hypoxia-inducible factor-1 α (HIF-1α) and VEGF down-regulation. When these nanoparticles were combined with classical cytostatic Dox, tumor growth/proliferation was inhibited [252].

As cancer immunotherapy has recently gained unprecedented momentum, HA's involvement as a drug carrier was tested. TC-1, a polymeric conjugate formed by HA and ovalbumin (OVA) as a foreign antigen, was tested using mouse lung tumor cells. This model showed that OVA257-264 peptide is presented complexed with MHC class I on the cells' surface. With this approach, the foreign antigen could induce an anti-tumor effect by enhancing the immune cells' attack. The mouse model's systemic administration showed that the conjugate is accumulated into tumor tissue and facilitates the cytotoxic T lymphocytes' (CTLs) attack of the tumor cells, thus inhibiting tumor proliferation [253].

OVA-loaded micelle consisting of PEGylated HA was tested for increasing the OVA uptake. The HA-coated micelle targeted CD44 on tumor cells and increased OVA cellular uptake more than 10 times. Loading tumor cells with a foreign antigen, such as OVA, would increase their recognition by CTLs, and thus, enhance destruction. In animal models, tumor growth was significantly inhibited, and the authors point out that in the case of cutaneous melanoma, this can be another approach to enhance immune-therapy [254]. The same principle was implemented in TC-1 mouse cells and lung cancer epithelial cells, using MMP9-responsive conjugates consisting of PEGylated HA and OVA. The complex was taken up through CD44-expressing cancer cells via receptor-mediated endocytosis. In an in vivo animal model, the tumor growth was significantly inhibited, antigen presentation on the tumor cells enhanced, and T cytotoxic anti-tumoral action increased [255]. A complex using HA and OVA on gold nanoparticles (AuNPs) was used to increase antigen uptake, by DC, via receptor-mediated endocytosis. The complex HA-OVA-AuNPs has enhanced near-infrared (NIR) absorption and thermal energy translation, so after engulfment, the cytosolic antigen will be delivered through the photothermally targeted process. Proteasome activity is increased, and the MHC I antigen presentation is enhanced; thus, the CD8+ cytotoxic T-cell response is triggered. This protocol can be fruitfully expanded in the cancer vaccine development area [256].

As some of the tumor cells and primary lymphocytes have low HS expression, other carriers need to be utilized. Thus, proteins complexed with nanosize cholesteryl group-bearing pullulans (cCHP) can be efficiently delivered to myeloma cells and to primary CD4+ T cells by macropinocytosis. When using these new types of nanoparticles to deliver the anti-apoptotic protein Bcl-xL, T cells' functional regulation is achieved. These nanoparticles can bypass the lack of HS expression and deliver anticancer effectors and modulators of immune regulation [257].Figure 2 outlines the main mechanisms GAGs utilize to hinder immune anti-tumoral action.

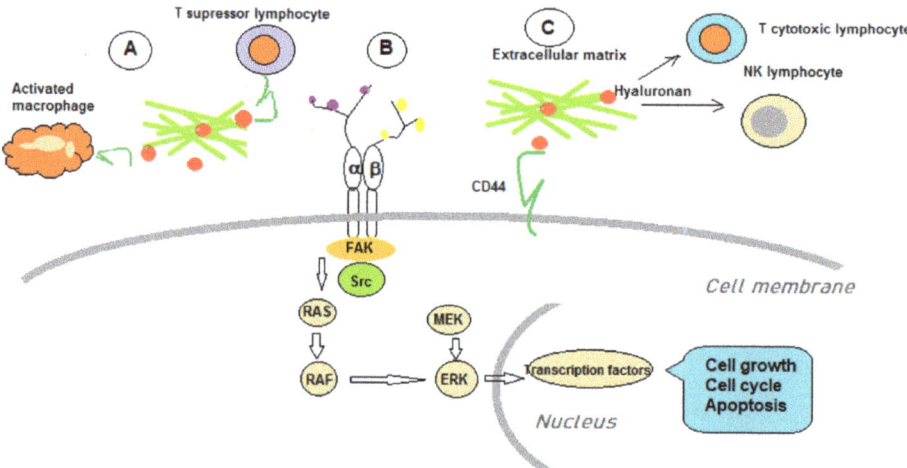

Figure 2. The main mechanisms through which GAGs hinder immune anti-tumoral action. A. HA binds t oCD44-expressing T suppresor cells and to the pool of tumor-associated macrophages contributing to the immuno-suppressive milieu in TME; B. Specific enzymes (e.g., β1,4-N-acetylgalactosaminyltransferase 3 and β1,4-galactosyltransferase 3) induce modification of β1 integrin expressed by tumor cells, triggering intracellular signaling that favor pro-tumorigenic effects in cell growth, cell cycle, and apoptosis. C. HA in the TME binds toCD44 expressed by tumor cells to physically block NK and T cytotoxic lymphocytes' access to tumor cells.

5.2. GAGs as Immunotherapy Targets

TME is complex and consists of immune cells (mainly lymphocytes and myeloid cells), non-immune cells (mainly endothelial cells and fibroblasts), and a complex array of structures, such as ECM, and various molecules that are either secreted or append to the cell membrane [258]. TME sustains molecules that hinder the potential effector function of NK lymphocytes. Transforming growth factor (TGF)-β and members of its superfamily downregulate NK cell cytotoxicity functions, cytokine secretion, metabolism, and proliferation. Likewise, galectins, a family of carbohydrate-binding proteins produced by different sources within the TME, downregulate NK cell functions. Various ECM components and associated enzymes (e.g., MMPs) can hinder NK cells' activation and become future therapy targets [259]. Pancreatic cancer, TME, contains various possible therapy targets, such as HA, focal adhesion kinase (FAK), connective tissue growth factor (CTGF), CD40, and chemokine (C-X-C motif) receptor 4 (CXCR-4), which could be utilized in future clinical applications [260].

Immune checkpoint inhibitor immunotherapies that had achieved broad clinical applicability in recent years [261] face the gaining of resistance. It is supposed that the HA accumulation influences tumor cells' sensitivity to chemotherapy and immunotherapy. A semiquantitative grouping of non-small lung cancer tissue demonstrated that HA deposition predicts the tumor response to pegylated hyaluronidase (PEGPH20) in animal models [262]. Thus, HA degradation facilitates tumor cells' exposure to drugs. Notably, utilization of PEGPH20, in a phase I clinical study demonstrated safety and good tolerability [263]. A phase I clinical trial, combining PEGPH20 with an immunotherapeutic agent, pembrolizumab, is currently ongoing in a cohort of metastatic gastric adenocarcinoma and non-small cell lung carcinoma patients [264], The reasoning behind this approach is the combination of facilitating drug access to tumor cells with the hypothesis that HA may modulate regulatory T cells and antitumor immune responses [265]. Clift et al. have shown that upon degrading HA, the anti-programmed death-ligand 1 (PD-L1) antibody accumulates more intensely in breast cancer tissues in vivo. An increased accumulation of T and NK cells was noticed upon HA degradation. The authors point out that decreasing HA in TME would enhance anti-tumoral immune cell infiltration and increase checkpoint inhibitor therapy efficacy [266].

Heparanase has also been linked to tumor immunology. It was shown that heparanase is implicated in chronic inflammatory bowel conditions and, consequently, in colon carcinoma initiation [222–224]. There is a clear correlation between intestinal heparanase and immune cells, mainly macrophages, which sustain the chronic inflammation and create a pro-tumoral microenvironment. Therapies that can re-equilibrate this enzyme's function and re-establish the physiological crosstalk between immune and epithelial cells would hinder colon cancer development [267]. Leukocyte-derived heparanase is versatile; therefore, subtle changes in the TME can direct the enzyme to either pro-or anti-tumoral action. Thus, in immune cancer therapy, heparanase could be a vital therapy target by either exploiting or inhibiting its activity [268].

Along these lines, heparanase inhibitors were tested in various hematological cancer models. Weissmann et al. showed in 2019 that PG545, a heparanase inhibitor, had a strong effect on human lymphoma. The inhibitor induces tumor cell apoptosis, ER stress response, and increased autophagy. PG545 did not affect naïve splenocytes but induced apoptosis even in lymphoma cells deployed of heparanase activity [239]. Another approach was utilizing heparanase-neutralizing monoclonal antibodies that strongly attenuate lymphoma cell tumor load in mouse bones due to tumor cell growth inhibition and reduced angiogenesis [269].

In chronic lymphocytic leukemia (CLL), stromal cells secrete and present CXCL12, a CXC chemokine ligand, through cellsurface-bound GAGs. By using this mechanism, CLL cells are protected from cytotoxic drugs and sustain the residual disease. The GAG mimetic, NOX-A12, binds and neutralizes CXCL12 and was tested to affect tumor cell migration. NOX-A12 inhibited CLL cell chemotaxis generated through CXCL12. Thus,

NOX-A12 competes with GAGs (e.g., Hep) for CXCL12 binding and sensitizes CLL cells toward chemotherapeutic drugs [270]. An outline of the main immune-therapy targets is summarized in Table 4.

Table 4. Developing GAG-associated immune-therapies.

Target	Therapy	Cancer Type	Stage	Reference
Hyaluronan	PEGylated recombinant hyaluronidase Section 6	Solid tumors	phase I study	[263]
		Non-small lung cancer	Animal model	[262]
		Refractory locally advanced or metastatic gastric adenocarcinoma and Non-small cell lung carcinoma	A phase 1b trial of PEGPH20 with pembrolizumab (NCT02563548)	[264]
Heparanase	Heparanase inhibitors	Colon carcinoma	Animal model	[267,271]
		Human lymphoma	In vitro cellular model	[239]
	Heparanase neutralizing antibody	Human follicular and diffused non-Hodgkin's B-lymphomas	Animal model	[269]

6. GAGs as Potential Cancer Therapy Response Biomarkers

The physical barrier represented by HA in the TME restricts immune therapy efficacy by hindering antibody and immune cell access. It was shown in 50% of HER2(3+) primary breast tumors and almost 50% of EGFR(+) head and neck squamous cell carcinomas that the tumor tissue characterized by high HA expression is associated with immune therapy resistance. The matrix containing high HA deposition hinders NK immune cell access to tumor cells. The depletion of HA by PEGPH20 (pegylated recombinant human PH20 hyaluronidase) propagates NK cells' access to these tumors. In vitro, the same mechanisms enhanced trastuzumab- or cetuximab-dependent antibody-dependent cellular toxicity (ADCC), while the in vivo experiments also demonstrated treatment efficacy. Considering that the tumor HA deposition can be used as a marker for immune therapy resistance, other clinical management protocols can be developed [271].

In colorectal cancer, it was established that glycosylation alters over 80% of human proteins and that aberrant glycosylation is involved in cancer development and progression. Glycan changes (e.g.,carbohydrate antigen CA 19-9 or carcinoembryonic antigen) are already established biomarkers in this cancer. Recent reports have shown that altered glycosylations can be involved in drug resistance mechanisms and indicate new predictive biomarkers [272].

GAGs are utilized as biomarkers in other disease types, including mucopolysaccharidoses (MPSs) [273]. The MPSs present approximately 30% of lysosomal storage diseases and are induced by inefficient GAG breakdown due to active enzyme deficiencies [274]. Without treatment options, patients exhibiting severe MPS forms die within the first two decades of life [273].

7. Conclusions

GAGs are versatile molecules that play multifaceted roles in the human body. They are involved in all biological functions and are acrucial mediator of homeostasis. Alterations in both the expression and GAG fine chemical structure are evident during cancer development and progression. Research efforts directed at the role of GAGs in cancerogenesis are rapidly increasing, and some of the findings have made their way into clinical practice.

The field has been facilitated by essential developments in available technologies, including imaging technologies, mass spectrometry, microarrays, and bioinformatics tools [275–277]. Therefore, we can now deepen our studies of the glycome, leading to an improved understanding of the glycobiology field. Indeed, the recent advancements in the GAG structure/function relationship have allowed a better appreciation of the

GAGs role in tumorigenesis and the utilization of this knowledge for cancer detection, prognosis, and therapy implementation. GAGs are now being employed as biomarkers for disease progression and tumor aggressiveness [278].They are involved in the tumor immune response, can be used by themselves or in the form of hybrid PGs therapeutic targets, and offer targeted drug delivery [1,279]. As drug carriers, GAGs are characterized by high specificity, multi-functionality, and good biocompatibility, the key to the success of new therapies in oncology [279]. Considering that GAGs are critical molecules of the complex cellular and molecular TME network, their multi-factorial utilization could enable personalized therapy implementation. However, some obstacles still need to be overcome as the heterogeneity of native GAG preparations has introduced the need for producing synthetic or semi-synthetic GAG mimetics with improved pharmacokinetic properties, higher selectivity, and attenuated or even abolished adverse side-effects. Future research efforts will enhance GAG implementations in the clinic and hopefully improve therapeutic strategies for some cancer types.

Author Contributions: Conceptualization, D.N.; writing—original draft preparation, D.N., M.N., A.B., E.-M.G. and A.K.; writing—review and editing, D.N., A.M.T. and G.N.T. All authors have read and agreed to the published version of the manuscript.

Funding: D.N. was partially funded by the Research Committee of University of Crete (ELKE), grant number (KA:10028), M.N. was partially funded by UEFISCDI grant number (PN-III-P1-1.2-PCCDI-2017-0341/2018).

Acknowledgments: This article is part of the Innogly Cost action initiative.

Conflicts of Interest: The authors declare no conflict of interest.

References

1. Tzanakakis, G.; Giatagana, E.M.; Kuskov, A.; Berdiaki, A.; Tsatsakis, A.M.; Neagu, M.; Nikitovic, D. Proteoglycans in the Pathogenesis of Hormone-Dependent Cancers: Mediators and Effectors. *Cancers* **2020**, *12*, 2401. [CrossRef]
2. Tzanakakis, G.; Neagu, M.; Tsatsakis, A.; Nikitovic, D. Proteoglycans and Immunobiology of Cancer-Therapeutic Implications. *Front. Immunol.* **2019**, *10*, 875. [CrossRef]
3. Karamanos, N.K.; Piperigkou, Z.; Theocharis, A.D.; Watanabe, H.; Franchi, M.; Baud, S.; Brezillon, S.; Gotte, M.; Passi, A.; Vigetti, D.; et al. Proteoglycan Chemical Diversity Drives Multifunctional Cell Regulation and Therapeutics. *Chem. Rev.* **2018**, *118*, 9152–9232. [CrossRef]
4. Nikitovic, D.; Papoutsidakis, A.; Karamanos, N.K.; Tzanakakis, G.N. Lumican affects tumor cell functions, tumor-ECM interactions, angiogenesis and inflammatory response. *Matrix Biol.* **2014**, *35*, 206–214. [CrossRef] [PubMed]
5. Afratis, N.; Gialeli, C.; Nikitovic, D.; Tsegenidis, T.; Karousou, E.; Theocharis, A.D.; Pavao, M.S.; Tzanakakis, G.N.; Karamanos, N.K. Glycosaminoglycans: Key players in cancer cell biology and treatment. *FEBS J.* **2012**, *279*, 1177–1197. [CrossRef] [PubMed]
6. Nikitovic, D.; Tzardi, M.; Berdiaki, A.; Tsatsakis, A.; Tzanakakis, G.N. Cancer microenvironment and inflammation: Role of hyaluronan. *Front. Immunol.* **2015**, *6*, 169. [CrossRef] [PubMed]
7. Nikitovic, D.; Berdiaki, A.; Spyridaki, I.; Krasanakis, T.; Tsatsakis, A.; Tzanakakis, G.N. Proteoglycans-Biomarkers and Targets in Cancer Therapy. *Front. Endocrinol.* **2018**, *9*, 69. [CrossRef] [PubMed]
8. Karangelis, D.E.; Kanakis, I.; Asimakopoulou, A.P.; Karousou, E.; Passi, A.; Theocharis, A.D.; Triposkiadis, F.; Tsilimingas, N.B.; Karamanos, N.K. Glycosaminoglycans as key molecules in atherosclerosis: The role of versican and hyaluronan. *Curr. Med. Chem.* **2010**, *17*, 4018–4026. [CrossRef] [PubMed]
9. Groux-Degroote, S.; Cavdarli, S.; Uchimura, K.; Allain, F.; Delannoy, P. Glycosylation changes in inflammatory diseases. *Adv. Protein Chem. Struct. Biol.* **2020**, *119*, 111–156. [PubMed]
10. Kouvidi, K.; Berdiaki, A.; Nikitovic, D.; Katonis, P.; Afratis, N.; Hascall, V.C.; Karamanos, N.K.; Tzanakakis, G.N. Role of receptor for hyaluronic acid-mediated motility (RHAMM) in low molecular weight hyaluronan (LMWHA)-mediated fibrosarcoma cell adhesion. *J. Biol. Chem.* **2011**, *286*, 38509–38520. [CrossRef]
11. Schwertfeger, K.L.; Cowman, M.K.; Telmer, P.G.; Turley, E.A.; McCarthy, J.B. Hyaluronan, Inflammation, and Breast Cancer Progression. *Front. Immunol.* **2015**, *6*, 236. [CrossRef] [PubMed]
12. Carlsson, P.; Presto, J.; Spillmann, D.; Lindahl, U.; Kjellen, L. Heparin/heparan sulfate biosynthesis: Processive formation of N-sulfated domains. *J. Biol. Chem.* **2008**, *283*, 20008–20014. [CrossRef]
13. Vigetti, D.; Karousou, E.; Viola, M.; Deleonibus, S.; De Luca, G.; Passi, A. Hyaluronan: Biosynthesis and signaling. *Biochim. Biophys. Acta* **2014**, *1840*, 2452–2459. [CrossRef] [PubMed]
14. Mikami, T.; Kitagawa, H. Biosynthesis and function of chondroitin sulfate. *Biochim. Biophys. Acta* **2013**, *1830*, 4719–4733. [CrossRef]

15. Pomin, V.H. Keratan sulfate: An up-to-date review. *Int. J. Biol. Macromol.* **2015**, *72*, 282–289. [CrossRef]
16. Habuchi, H.; Habuchi, O.; Kimata, K. Sulfation pattern in glycosaminoglycan: Does it have a code? *Glycoconj. J.* **2004**, *21*, 47–52. [CrossRef]
17. Sarkar, A.; Desai, U.R. A Simple Method for Discovering Druggable, Specific Glycosaminoglycan-Protein Systems. Elucidation of Key Principles from Heparin/Heparan Sulfate-Binding Proteins. *PLoS ONE* **2015**, *10*, e0141127. [CrossRef] [PubMed]
18. Xu, D.; Esko, J.D. Demystifying heparan sulfate-protein interactions. *Annu. Rev. Biochem.* **2014**, *83*, 129–157. [CrossRef]
19. Kjellen, L.; Lindahl, U. Specificity of glycosaminoglycan-protein interactions. *Curr. Opin. Struct. Biol.* **2018**, *50*, 101–108. [CrossRef]
20. Morla, S. Glycosaminoglycans and Glycosaminoglycan Mimetics in Cancer and Inflammation. *Int. J. Mol. Sci.* **2019**, *20*, 1963. [CrossRef]
21. Sasisekharan, R.; Shriver, Z.; Venkataraman, G.; Narayanasami, U. Roles of heparan-sulphate glycosaminoglycans in cancer. *Nat. Rev. Cancer* **2002**, *2*, 521–528. [CrossRef] [PubMed]
22. Kreuger, J.; Matsumoto, T.; Vanwildemeersch, M.; Sasaki, T.; Timpl, R.; Claesson-Welsh, L.; Spillmann, D.; Lindahl, U. Role of heparan sulfate domain organization in endostatin inhibition of endothelial cell function. *EMBO J.* **2002**, *21*, 6303–6311. [CrossRef]
23. Lv, H.; Yu, G.; Sun, L.; Zhang, Z.; Zhao, X.; Chai, W. Elevate level of glycosaminoglycans and altered sulfation pattern of chondroitin sulfate are associated with differentiation status and histological type of human primary hepatic carcinoma. *Oncology* **2007**, *72*, 347–356. [CrossRef]
24. Pudelko, A.; Wisowski, G.; Olczyk, K.; Kozma, E.M. The dual role of the glycosaminoglycan chondroitin-6-sulfate in the development, progression and metastasis of cancer. *FEBS J.* **2019**, *286*, 1815–1837. [CrossRef] [PubMed]
25. Fuster, M.M.; Wang, L.; Castagnola, J.; Sikora, L.; Reddi, K.; Lee, P.H.; Radek, K.A.; Schuksz, M.; Bishop, J.R.; Gallo, R.L.; et al. Genetic alteration of endothelial heparan sulfate selectively inhibits tumor angiogenesis. *J. Cell Biol.* **2007**, *177*, 539–549. [CrossRef] [PubMed]
26. Waaijer, C.J.; de Andrea, C.E.; Hamilton, A.; van Oosterwijk, J.G.; Stringer, S.E.; Bovee, J.V. Cartilage tumour progression is characterized by an increased expression of heparan sulphate 6O-sulphation-modifying enzymes. *Virchows Arch.* **2012**, *461*, 475–481. [CrossRef] [PubMed]
27. Hatabe, S.; Kimura, H.; Arao, T.; Kato, H.; Hayashi, H.; Nagai, T.; Matsumoto, K.; De Velasco, M.; Fujita, Y.; Yamanouchi, G.; et al. Overexpression of heparan sulfate 6-O-sulfotransferase-2 in colorectal cancer. *Mol. Clin. Oncol.* **2013**, *1*, 845–850. [CrossRef]
28. Jeney, A.; Timar, J.; Pogany, G.; Paku, S.; Moczar, E.; Mareel, M.; Otvos, L.; Kopper, L.; Lapis, K. Glycosaminoglycans as novel target in antitumor therapy. *Tokai J. Exp. Clin. Med.* **1990**, *15*, 167–177.
29. Kowitsch, A.; Zhou, G.; Groth, T. Medical application of glycosaminoglycans: A review. *J. Tissue Eng. Regen. Med.* **2018**, *12*, e23–e41. [CrossRef] [PubMed]
30. Prydz, K.; Dalen, K.T. Synthesis and sorting of proteoglycans. *J. Cell Sci.* **2000**, *113 Pt 2*, 193–205.
31. Bishop, J.R.; Schuksz, M.; Esko, J.D. Heparan sulphate proteoglycans fine-tune mammalian physiology. *Nature* **2007**, *446*, 1030–1037. [CrossRef]
32. Karamanos, N.K.; Vanky, P.; Tzanakakis, G.N.; Tsegenidis, T.; Hjerpe, A. Ion-pair high-performance liquid chromatography for determining disaccharide composition in heparin and heparan sulphate. *J. Chromatogr. A* **1997**, *765*, 169–179. [CrossRef]
33. Nikitovic, D.; Tsatsakis, A.M.; Karamanos, N.K.; Tzanakakis, G.N. The effects of genistein on the synthesis and distribution of glycosaminoglycans/proteoglycans by two osteosarcoma cell lines depends on tyrosine kinase and the estrogen receptor density. *Anticancer Res.* **2003**, *23*, 459–464. [PubMed]
34. Duchez, S.; Pascal, V.; Cogne, N.; Jayat-Vignoles, C.; Julien, R.; Cogne, M. Glycotranscriptome study reveals an enzymatic switch modulating glycosaminoglycan synthesis during B-cell development and activation. *Eur. J. Immunol.* **2011**, *41*, 3632–3644. [CrossRef] [PubMed]
35. Sugahara, K.; Kitagawa, H. Heparin and heparan sulfate biosynthesis. *IUBMB Life* **2002**, *54*, 163–175. [CrossRef]
36. Feyerabend, T.B.; Li, J.P.; Lindahl, U.; Rodewald, H.R. Heparan sulfate C5-epimerase is essential for heparin biosynthesis in mast cells. *Nat. Chem. Biol.* **2006**, *2*, 195–196. [CrossRef]
37. Lodish, H.; Berk, A.; Zipursky, L.S.; Matsudaira, P.; Baltimore, D.; Darnell, J. *Molecular Cell Biology*, 4th ed.; Freeman, W.H.: New York, NY, USA, 2000.
38. Lindahl, U.; Kusche-Gullberg, M.; Kjellen, L. Regulated diversity of heparan sulfate. *J. Biol. Chem.* **1998**, *273*, 24979–24982. [CrossRef] [PubMed]
39. Casu, B.; Lindahl, U. Structure and biological interactions of heparin and heparan sulfate. *Adv. Carbohydr. Chem. Biochem.* **2001**, *57*, 159–206.
40. Powell, A.K.; Yates, E.A.; Fernig, D.G.; Turnbull, J.E. Interactions of heparin/heparan sulfate with proteins: Appraisal of structural factors and experimental approaches. *Glycobiology* **2004**, *14*, 17R–30R. [CrossRef]
41. Nagamine, S.; Tamba, M.; Ishimine, H.; Araki, K.; Shiomi, K.; Okada, T.; Ohto, T.; Kunita, S.; Takahashi, S.; Wismans, R.G.; et al. Organ-specific sulfation patterns of heparan sulfate generated by extracellular sulfatases Sulf1 and Sulf2 in mice. *J. Biol. Chem.* **2012**, *287*, 9579–9590. [CrossRef]
42. Gallangher, J.T. Heparan sulfate: A heparin in miniature. In *Handbook Experimental Pharmacology*; Lever, R., Mulloy, B., Page, C.P., Eds.; Springer: Berlin/Heidelberg, Germany, 2012; Volume 207, pp. 347–360.
43. Casu, B.; Naggi, A.; Torri, G. Heparin-derived heparan sulfate mimics to modulate heparan sulfate-protein interaction in inflammation and cancer. *Matrix Biol.* **2010**, *29*, 442–452. [CrossRef]

44. Rudd, T.R.; Preston, M.D.; Yates, E.A. The nature of the conserved basic amino acid sequences found among 437 heparin binding proteins determined by network analysis. *Mol. Biosyst.* **2017**, *13*, 852–865. [CrossRef]
45. Duchesne, L.; Tissot, B.; Rudd, T.R.; Dell, A.; Fernig, D.G. N-glycosylation of fibroblast growth factor receptor 1 regulates ligand and heparan sulfate co-receptor binding. *J. Biol. Chem.* **2006**, *281*, 27178–27189. [CrossRef] [PubMed]
46. Multhaupt, H.A.; Couchman, J.R. Heparan sulfate biosynthesis: Methods for investigation of the heparanosome. *J. Histochem. Cytochem.* **2012**, *60*, 908–915. [CrossRef] [PubMed]
47. Nikitovic, D.; Chatzinikolaou, G.; Tsiaoussis, J.; Tsatsakis, A.; Karamanos, N.K.; Tzanakakis, G.N. Insights into targeting colon cancer cell fate at the level of proteoglycans/glycosaminoglycans. *Curr. Med. Chem.* **2012**, *19*, 4247–4258. [CrossRef]
48. Lindahl, U.; Kjellen, L. Pathophysiology of heparan sulphate: Many diseases, few drugs. *J. Intern. Med.* **2013**, *273*, 555–571. [CrossRef]
49. Rabenstein, D.L. Heparin and heparan sulfate: Structure and function. *Nat. Prod. Rep.* **2002**, *19*, 312–331. [CrossRef]
50. Mulloy, B.; Hogwood, J.; Gray, E.; Lever, R.; Page, C.P. Pharmacology of Heparin and Related Drugs. *Pharmacol. Rev.* **2016**, *68*, 76–141. [CrossRef] [PubMed]
51. Mulloy, B.; Lever, R.; Page, C.P. Mast cell glycosaminoglycans. *Glycoconj. J.* **2017**, *34*, 351–361. [CrossRef] [PubMed]
52. Kolset, S.O.; Tveit, H. Serglycin–structure and biology. *Cell Mol. Life Sci.* **2008**, *65*, 1073–1085. [CrossRef] [PubMed]
53. Jacobsson, K.G.; Lindahl, U. Degradation of heparin proteoglycan in cultured mouse mastocytoma cells. *Biochem. J.* **1987**, *246*, 409–415. [CrossRef]
54. Olivera, A.; Beaven, M.A.; Metcalfe, D.D. Mast cells signal their importance in health and disease. *J. Allergy Clin. Immunol.* **2018**, *142*, 381–393. [CrossRef] [PubMed]
55. Oie, C.I.; Olsen, R.; Smedsrod, B.; Hansen, J.B. Liver sinusoidal endothelial cells are the principal site for elimination of unfractionated heparin from the circulation. *Am. J. Physiol. Gastrointest. Liver Physiol.* **2008**, *294*, G520–G528. [CrossRef]
56. Peterson, S.B.; Liu, J. Multi-faceted substrate specificity of heparanase. *Matrix Biol.* **2013**, *32*, 223–227. [CrossRef]
57. Lauder, R.M. Chondroitin sulphate: A complex molecule with potential impacts on a wide range of biological systems. *Complement. Ther. Med.* **2009**, *17*, 56–62. [CrossRef] [PubMed]
58. Lauder, R.M.; Huckerby, T.N.; Nieduszynski, I.A. A fingerprinting method for chondroitin/dermatan sulfate and hyaluronan oligosaccharides. *Glycobiology* **2000**, *10*, 393–401. [CrossRef]
59. Malavaki, C.; Mizumoto, S.; Karamanos, N.; Sugahara, K. Recent advances in the structural study of functional chondroitin sulfate and dermatan sulfate in health and disease. *Connect. Tissue Res.* **2008**, *49*, 133–139. [CrossRef] [PubMed]
60. Sugahara, K.; Mikami, T. Chondroitin/dermatan sulfate in the central nervous system. *Curr. Opin. Struct. Biol.* **2007**, *17*, 536–545. [CrossRef] [PubMed]
61. Bao, X.; Mikami, T.; Yamada, S.; Faissner, A.; Muramatsu, T.; Sugahara, K. Heparin-binding growth factor, pleiotrophin, mediates neuritogenic activity of embryonic pig brain-derived chondroitin sulfate/dermatan sulfate hybrid chains. *J. Biol. Chem.* **2005**, *280*, 9180–9191. [CrossRef]
62. Hopwood, J.J.; Robinson, H.C. The molecular-weight distribution of glycosaminoglycans. *Biochem. J.* **1973**, *135*, 631–637. [CrossRef]
63. Krusius, T.; Finne, J.; Margolis, R.K.; Margolis, R.U. Identification of an O-glycosidic mannose-linked sialylatedtetrasaccharide and keratan sulfate oligosaccharides in the chondroitin sulfate proteoglycan of brain. *J. Biol. Chem.* **1986**, *261*, 8237–8242. [CrossRef]
64. Funderburgh, J.L. Keratan sulfate biosynthesis. *IUBMB Life* **2002**, *54*, 187–194. [CrossRef] [PubMed]
65. Zhang, H.; Muramatsu, T.; Murase, A.; Yuasa, S.; Uchimura, K.; Kadomatsu, K. N-Acetylglucosamine 6-O-sulfotransferase-1 is required for brain keratan sulfate biosynthesis and glial scar formation after brain injury. *Glycobiology* **2006**, *16*, 702–710. [CrossRef]
66. Uchimura, K. Keratan sulfate: Biosynthesis, structures, and biological functions. *Methods Mol. Biol.* **2015**, *1229*, 389–400.
67. Leiphrakpam, P.D.; Patil, P.P.; Remmers, N.; Swanson, B.; Grandgenett, P.M.; Qiu, F.; Yu, F.; Radhakrishnan, P. Role of keratan sulfate expression in human pancreatic cancer malignancy. *Sci. Rep.* **2019**, *9*, 9665. [CrossRef]
68. Kato, Y.; Hayatsu, N.; Kaneko, M.K.; Ogasawara, S.; Hamano, T.; Takahashi, S.; Nishikawa, R.; Matsutani, M.; Mishima, K.; Narimatsu, H. Increased expression of highly sulfated keratan sulfate synthesized in malignant astrocytic tumors. *Biochem. Biophys. Res. Commun.* **2008**, *369*, 1041–1046. [CrossRef]
69. Itano, N.; Sawai, T.; Yoshida, M.; Lenas, P.; Yamada, Y.; Imagawa, M.; Shinomura, T.; Hamaguchi, M.; Yoshida, Y.; Ohnuki, Y.; et al. Three isoforms of mammalian hyaluronan synthases have distinct enzymatic properties. *J. Biol. Chem.* **1999**, *274*, 25085–25092. [CrossRef] [PubMed]
70. Stern, R.; Asari, A.A.; Sugahara, K.N. Hyaluronan fragments: An information-rich system. *Eur. J. Cell Biol.* **2006**, *85*, 699–715. [CrossRef]
71. Vigetti, D.; Viola, M.; Karousou, E.; De Luca, G.; Passi, A. Metabolic control of hyaluronan synthases. *Matrix Biol.* **2014**, *35*, 8–13. [CrossRef] [PubMed]
72. Bourguignon, L.Y.; Zhu, H.; Shao, L.; Chen, Y.W. CD44 interaction with tiam1 promotes Rac1 signaling and hyaluronic acid-mediated breast tumor cell migration. *J. Biol. Chem.* **2000**, *275*, 1829–1838. [CrossRef]
73. Kouvidi, K.; Berdiaki, A.; Tzardi, M.; Karousou, E.; Passi, A.; Nikitovic, D.; Tzanakakis, G.N. Receptor for hyaluronic acid-mediated motility (RHAMM) regulates HT1080 fibrosarcoma cell proliferation via a beta-catenin/c-myc signaling axis. *Biochim. Biophys. Acta* **2016**, *1860*, 814–824. [CrossRef]

74. Kavasi, R.M.; Berdiaki, A.; Spyridaki, I.; Papoutsidakis, A.; Corsini, E.; Tsatsakis, A.; Tzanakakis, G.N.; Nikitovic, D. Contact allergen (PPD and DNCB)-induced keratinocyte sensitization is partly mediated through a low molecular weight hyaluronan (LMWHA)/TLR4/NF-kappaB signaling axis. *Toxicol. Appl. Pharmacol.* **2019**, *377*, 114632. [CrossRef] [PubMed]
75. Harada, H.; Takahashi, M. CD44-dependent intracellular and extracellular catabolism of hyaluronic acid by hyaluronidase-1 and -2. *J. Biol. Chem.* **2007**, *282*, 5597–5607. [CrossRef] [PubMed]
76. Wang, W.; Wang, J.; Li, F. Hyaluronidase and Chondroitinase. *Adv. Exp. Med. Biol.* **2017**, *925*, 75–87. [PubMed]
77. Yamada, S. Role of hyaluronidases in the catabolism of chondroitin sulfate. *Adv. Exp. Med. Biol.* **2015**, *842*, 185–197.
78. Stern, R.; Jedrzejas, M.J. Hyaluronidases: Their genomics, structures, and mechanisms of action. *Chem. Rev.* **2006**, *106*, 818–839. [CrossRef]
79. Puissant, E.; Gilis, F.; Dogne, S.; Flamion, B.; Jadot, M.; Boonen, M. Subcellular trafficking and activity of Hyal-1 and its processed forms in murine macrophages. *Traffic* **2014**, *15*, 500–515. [CrossRef]
80. Tan, J.X.; Wang, X.Y.; Su, X.L.; Li, H.Y.; Shi, Y.; Wang, L.; Ren, G.S. Upregulation of HYAL1 expression in breast cancer promoted tumor cell proliferation, migration, invasion and angiogenesis. *PLoS ONE* **2011**, *6*, e22836. [CrossRef]
81. McAtee, C.O.; Barycki, J.J.; Simpson, M.A. Emerging roles for hyaluronidase in cancer metastasis and therapy. *Adv. Cancer Res.* **2014**, *123*, 1–34.
82. Andre, B.; Duterme, C.; Van Moer, K.; Mertens-Strijthagen, J.; Jadot, M.; Flamion, B. Hyal2 is a glycosylphosphatidylinositol-anchored, lipid raft-associated hyaluronidase. *Biochem. Biophys. Res. Commun.* **2011**, *411*, 175–179. [CrossRef]
83. Karthikeyan, R.; Koushik, O. Nano drug delivery systems to overcome cancer drug resistance-a review. *J. Nanomed. Nanotechnol.* **2016**, *7*, 2.
84. Jiao, Y.; Pang, X.; Zhai, G. Advances in Hyaluronic Acid-Based Drug Delivery Systems. *Curr. Drug Targets* **2016**, *17*, 720–730. [CrossRef] [PubMed]
85. Liu, M.; Song, X.; Wen, Y.; Zhu, J.L.; Li, J. Injectable Thermoresponsive Hydrogel Formed by Alginate-g-Poly(N-isopropylacrylamide) That Releases Doxorubicin-Encapsulated Micelles as a Smart Drug Delivery System. *ACS Appl. Mater. Interfaces* **2017**, *9*, 35673–35682. [CrossRef] [PubMed]
86. Sedyakina, N.; Kuskov, A.; Velonia, K.; Feldman, N.; Lutsenko, S.; Avramenko, G. Modulation of Entrapment Efficiency and In Vitro Release Properties of BSA-Loaded Chitosan Microparticles Cross-Linked with Citric Acid as a Potential Protein-Drug Delivery System. *Materials* **2020**, *13*, 1989. [CrossRef] [PubMed]
87. Cong, Z.; Shi, Y.; Wang, Y.; Niu, J.; Chen, N.; Xue, H. A novel controlled drug delivery system based on alginate hydrogel/chitosan micelle composites. *Int. J. Biol. Macromol.* **2018**, *107 Pt A*, 855–864. [CrossRef]
88. Kuskov, A.N.; Kulikov, P.P.; Goryachaya, A.V.; Shtilman, M.I.; Tzatzarakis, M.N.; Tsatsakis, A.M.; Velonia, K. Self-assembled amphiphilic poly-N-vinylpyrrolidone nanoparticles as carriers for hydrophobic drugs: Stability aspects. *J. Appl. Polym. Sci.* **2018**, *135*, 45637. [CrossRef]
89. Song, Z.; Lin, Y.; Zhang, X.; Feng, C.; Lu, Y.; Gao, Y.; Dong, C. Cyclic RGD peptide-modified liposomal drug delivery system for targeted oral apatinib administration: Enhanced cellular uptake and improved therapeutic effects. *Int. J. Nanomed.* **2017**, *12*, 1941–1958. [CrossRef]
90. Huang, D.; Wu, D. Biodegradable dendrimers for drug delivery. *Mater. Sci. Eng. C Mater. Biol. Appl.* **2018**, *90*, 713–727. [CrossRef]
91. Mehmood, A.; Ghafar, H.; Yaqoob, S.; Gohar, F.; Ahmad, B. Mesoporous Silica Nanoparticles: A Review. *J. Dev. Drugs* **2017**, *6*, 2. [CrossRef]
92. Mohammed, M.A.; Syeda, J.T.M.; Wasan, K.M.; Wasan, E.K. An Overview of Chitosan Nanoparticles and Its Application in Non-Parenteral Drug Delivery. *Pharmaceutics* **2017**, *9*, 53. [CrossRef]
93. Tsatsakis, A.; Stratidakis, A.K.; Goryachaya, A.V.; Tzatzarakis, M.N.; Stivaktakis, P.D.; Docea, A.O.; Berdiaki, A.; Nikitovic, D.; Velonia, K.; Shtilman, M.I.; et al. In vitro blood compatibility and in vitro cytotoxicity of amphiphilic poly-N-vinylpyrrolidone nanoparticles. *Food Chem. Toxicol.* **2019**, *127*, 42–52. [CrossRef]
94. Kuskov, A.N.; Kulikov, P.P.; Shtilman, M.I.; Rakitskii, V.N.; Tsatsakis, A.M. Amphiphilic poly-N-vynilpyrrolidone nanoparticles: Cytotoxicity and acute toxicity study. *Food Chem. Toxicol.* **2016**, *96*, 273–279. [CrossRef]
95. Maeda, H.; Nakamura, H.; Fang, J. The EPR effect for macromolecular drug delivery to solid tumors: Improvement of tumor uptake, lowering of systemic toxicity, and distinct tumor imaging in vivo. *Adv. Drug Deliv. Rev.* **2013**, *65*, 71–79. [CrossRef]
96. Sindhwani, S.; Syed, A.M.; Ngai, J.; Kingston, B.R.; Maiorino, L.; Rothschild, J.; MacMillan, P.; Zhang, Y.; Rajesh, N.U.; Hoang, T.; et al. The entry of nanoparticles into solid tumours. *Nat. Mater.* **2020**, *19*, 566–575. [CrossRef] [PubMed]
97. Lazarovits, J.; Chen, Y.Y.; Song, F.; Ngo, W.; Tavares, A.J.; Zhang, Y.N.; Audet, J.; Tang, B.; Lin, Q.; Tleugabulova, M.C.; et al. Synthesis of Patient-Specific Nanomaterials. *Nano Lett.* **2019**, *19*, 116–123. [CrossRef] [PubMed]
98. Escareno, N.; Topete, A.; Taboada, P.; Daneri-Navarro, A. Rational Surface Engineering of Colloidal Drug Delivery Systems for Biological Applications. *Curr. Top. Med. Chem.* **2018**, *18*, 1224–1241. [CrossRef] [PubMed]
99. Wyss, P.P.; Lamichhane, S.P.; Abed, A.; Vonwil, D.; Kretz, O.; Huber, T.B.; Sarem, M.; Shastri, V.P. Renal clearance of polymeric nanoparticles by mimicry of glycan surface of viruses. *Biomaterials* **2020**, *230*, 119643. [CrossRef] [PubMed]
100. Wu, W.; Yao, W.; Wang, X.; Xie, C.; Zhang, J.; Jiang, X. Bioreducible heparin-based nanogel drug delivery system. *Biomaterials* **2015**, *39*, 260–268. [CrossRef]

101. Rippe, M.; Stefanello, T.F.; Kaplum, V.; Britta, E.A.; Garcia, F.P.; Poirot, R.; Companhoni, M.V.P.; Nakamura, C.V.; Szarpak-Jankowska, A.; Auzely-Velty, R. Heparosan as a potential alternative to hyaluronic acid for the design of biopolymer-based nanovectors for anticancer therapy. *Biomater. Sci.* **2019**, *7*, 2850–2860. [CrossRef]
102. Li, X.M.; Wu, Z.Z.; Zhang, B.; Pan, Y.; Meng, R.; Chen, H.Q. Fabrication of chitosan hydrochloride and carboxymethyl starch complex nanogels as potential delivery vehicles for curcumin. *Food Chem.* **2019**, *293*, 197–203. [CrossRef]
103. Wang, J.; Ma, W.; Guo, Q.; Li, Y.; Hu, Z.; Zhu, Z.; Wang, X.; Zhao, Y.; Chai, X.; Tu, P. The effect of dual-functional hyaluronic acid-vitamin E succinate micelles on targeting delivery of doxorubicin. *Int. J. Nanomed.* **2016**, *11*, 5851–5870. [CrossRef]
104. Niu, C.; Xu, Y.; An, S.; Zhang, M.; Hu, Y.; Wang, L.; Peng, Q. Near-infrared induced phase-shifted ICG/Fe$_3$O$_4$ loaded PLGA nanoparticles for photothermal tumor ablation. *Sci. Rep.* **2017**, *7*, 5490. [CrossRef] [PubMed]
105. Zhang, Y.; Zhan, X.; Xiong, J.; Peng, S.; Huang, W.; Joshi, R.; Cai, Y.; Liu, Y.; Li, R.; Yuan, K.; et al. Temperature-dependent cell death patterns induced by functionalized gold nanoparticle photothermal therapy in melanoma cells. *Sci. Rep.* **2018**, *8*, 8720. [CrossRef] [PubMed]
106. Kim, K.; Choi, H.; Choi, E.S.; Park, M.H.; Ryu, J.H. Hyaluronic Acid-Coated Nanomedicine for Targeted Cancer Therapy. *Pharmaceutics* **2019**, *11*, 301. [CrossRef] [PubMed]
107. Li, M.; Sun, J.; Zhang, W.; Zhao, Y.; Zhang, S. Drug delivery systems based on CD44-targeted glycosaminoglycans for cancer therapy. *Carbohydr. Polym.* **2021**, *251*, 117103. [CrossRef]
108. Bishnoi, M.; Jain, A.; Hurkat, P.; Jain, S.K. Chondroitin sulphate: A focus on osteoarthritis. *Glycoconj. J.* **2016**, *33*, 693–705. [CrossRef]
109. Hulsopple, C. Musculoskeletal Therapies: Musculoskeletal Injection Therapy. *FP Essent.* **2018**, *470*, 21–26. [PubMed]
110. Keen, M.A. Hyaluronic Acid in Dermatology. *Skinmed* **2017**, *15*, 441–448. [PubMed]
111. Choi, K.Y.; Saravanakumar, G.; Park, J.H.; Park, K. Hyaluronic acid-based nanocarriers for intracellular targeting: Interfacial interactions with proteins in cancer. *Colloids Surf. B Biointerfaces* **2012**, *99*, 82–94. [CrossRef]
112. Berdiaki, A.; Perisynaki, E.; Stratidakis, A.; Kulikov, P.P.; Kuskov, A.N.; Stivaktakis, P.; Henrich-Noack, P.; Luss, A.L.; Shtilman, M.M.; Tzanakakis, G.N.; et al. Assessment of Amphiphilic Poly-N-vinylpyrrolidone Nanoparticles' Biocompatibility with Endothelial Cells in Vitro and Delivery of an Anti-Inflammatory Drug. *Mol. Pharm.* **2020**, *17*, 4212–4225. [CrossRef] [PubMed]
113. Hajebi, S.; Rabiee, N.; Bagherzadeh, M.; Ahmadi, S.; Rabiee, M.; Roghani-Mamaqani, H.; Tahriri, M.; Tayebi, L.; Hamblin, M.R. Stimulus-responsive polymeric nanogels as smart drug delivery systems. *Acta Biomater.* **2019**, *92*, 1–18. [CrossRef]
114. Linhardt, R.J.; Claude, S. Hudson Award address in carbohydrate chemistry. Heparin: Structure and activity. *J. Med. Chem.* **2003**, *46*, 2551–2564. [CrossRef]
115. Zhao, F.; Ma, M.L.; Xu, B. Molecular hydrogels of therapeutic agents. *Chem. Soc. Rev.* **2009**, *38*, 883–891. [CrossRef] [PubMed]
116. Nurunnabi, M.; Khatun, Z.; Moon, W.C.; Lee, G.; Lee, Y.K. Heparin based nanoparticles for cancer targeting and noninvasive imaging. *Quant. Imaging Med. Surg.* **2012**, *2*, 219–226.
117. Min, K.A.; Yu, F.; Yang, V.C.; Zhang, X.; Rosania, G.R. Transcellular Transport of Heparin-coated Magnetic Iron Oxide Nanoparticles (Hep-MION) Under the Influence of an Applied Magnetic Field. *Pharmaceutics* **2010**, *2*, 119–135. [CrossRef]
118. Cho, K.J.; Moon, H.T.; Park, G.E.; Jeon, O.C.; Byun, Y.; Lee, Y.K. Preparation of sodium deoxycholate (DOC) conjugated heparin derivatives for inhibition of angiogenesis and cancer cell growth. *Bioconjug. Chem.* **2008**, *19*, 1346–1351. [CrossRef] [PubMed]
119. She, W.; Li, N.; Luo, K.; Guo, C.; Wang, G.; Geng, Y.; Gu, Z. Dendronized heparin-doxorubicin conjugate based nanoparticle as pH-responsive drug delivery system for cancer therapy. *Biomaterials* **2013**, *34*, 2252–2264. [CrossRef]
120. Park, K.; Kim, K.; Kwon, I.C.; Kim, S.K.; Lee, S.; Lee, D.Y.; Byun, Y. Preparation and characterization of self-assembled nanoparticles of heparin-deoxycholic acid conjugates. *Langmuir* **2004**, *20*, 11726–11731. [CrossRef] [PubMed]
121. Park, K.; Lee, G.Y.; Kim, Y.S.; Yu, M.; Park, R.W.; Kim, I.S.; Kim, S.Y.; Byun, Y. Heparin-deoxycholic acid chemical conjugate as an anticancer drug carrier and its antitumor activity. *J. Control. Release* **2006**, *114*, 300–306. [CrossRef] [PubMed]
122. Baier, G.; Winzen, S.; Messerschmidt, C.; Frank, D.; Fichter, M.; Gehring, S.; Mailander, V.; Landfester, K. Heparin-based nanocapsules as potential drug delivery systems. *Macromol. Biosci.* **2015**, *15*, 765–776. [CrossRef] [PubMed]
123. Li, J.P. Glucuronyl C5-epimerase an enzyme converting glucuronic acid to iduronic acid in heparan sulfate/heparin biosynthesis. *Prog. Mol. Biol. Transl. Sci.* **2010**, *93*, 59–78. [PubMed]
124. Yang, X.; Du, H.; Liu, J.; Zhai, G. Advanced nanocarriers based on heparin and its derivatives for cancer management. *Biomacromolecules* **2015**, *16*, 423–436. [CrossRef] [PubMed]
125. Ghiselli, G. Heparin Binding Proteins as Therapeutic Target: An Historical Account and Current Trends. *Medicines* **2019**, *6*, 80. [CrossRef] [PubMed]
126. Weiss, R.J.; Esko, J.D.; Tor, Y. Targeting heparin and heparan sulfate protein interactions. *Org. Biomol. Chem.* **2017**, *15*, 5656–5668. [CrossRef]
127. Ghofrani, M.; Shirmard, L.R.; Dehghankelishadi, P.; Amini, M.; Dorkoosh, F.A. Development of Octreotide-Loaded Chitosan and Heparin Nanoparticles: Evaluation of Surface Modification Effect on Physicochemical Properties and Macrophage Uptake. *J. Pharm. Sci.* **2019**, *108*, 3036–3045. [CrossRef]
128. Duckworth, C.A.; Guimond, S.E.; Sindrewicz, P.; Hughes, A.J.; French, N.S.; Lian, L.Y.; Yates, E.A.; Pritchard, D.M.; Rhodes, J.M.; Turnbull, J.E.; et al. Chemically modified, non-anticoagulant heparin derivatives are potent galectin-3 binding inhibitors and inhibit circulating galectin-3-promoted metastasis. *Oncotarget* **2015**, *6*, 23671–23687. [CrossRef]

129. Mei, L.; Liu, Y.; Zhang, H.; Zhang, Z.; Gao, H.; He, Q. Antitumor and Antimetastasis Activities of Heparin-based Micelle Served As Both Carrier and Drug. *ACS Appl. Mater. Interfaces* **2016**, *8*, 9577–9589. [CrossRef]
130. Wang, J.; Yang, Y.; Zhang, Y.; Huang, M.; Zhou, Z.; Luo, W.; Tang, J.; Wang, J.; Xiao, Q.; Chen, H.; et al. Dual-Targeting Heparin-Based Nanoparticles that Re-Assemble in Blood for Glioma Therapy through Both Anti-Proliferation and Anti-Angiogenesis. *Adv. Funct. Mater.* **2016**, *26*, 7873–7885. [CrossRef]
131. Rodriguez-Torres, M.D.P.; Diaz-Torres, L.A.; Millan-Chiu, B.E.; Garcia-Contreras, R.; Hernandez-Padron, G.; Acosta-Torres, L.S. Antifungal and Cytotoxic Evaluation of Photochemically Synthesized Heparin-Coated Gold and Silver Nanoparticles. *Molecules* **2020**, *25*, 2849. [CrossRef] [PubMed]
132. Groult, H.; Poupard, N.; Herranz, F.; Conforto, E.; Bridiau, N.; Sannier, F.; Bordenave, S.; Piot, J.M.; Ruiz-Cabello, J.; Fruitier-Arnaudin, I.; et al. Family of Bioactive Heparin-Coated Iron Oxide Nanoparticles with Positive Contrast in Magnetic Resonance Imaging for Specific Biomedical Applications. *Biomacromolecules* **2017**, *18*, 3156–3167. [CrossRef]
133. Fazilati, M. Anti-neoplastic applications of heparin coated magnetic nanoparticles against human ovarian cancer. *J. Inorg. Organomet. Polym. Mater.* **2014**, *24*, 551–559. [CrossRef]
134. Zhang, J.; Shin, M.C.; David, A.E.; Zhou, J.; Lee, K.; He, H.; Yang, V.C. Long-circulating heparin-functionalized magnetic nanoparticles for potential application as a protein drug delivery platform. *Mol. Pharm.* **2013**, *10*, 3892–3902. [CrossRef] [PubMed]
135. Zhang, J.; Shin, M.C.; Yang, V.C. Magnetic targeting of novel heparinized iron oxide nanoparticles evaluated in a 9L-glioma mouse model. *Pharm. Res.* **2014**, *31*, 579–592. [CrossRef] [PubMed]
136. Yin, Y.; Hu, B.; Yuan, X.; Cai, L.; Gao, H.; Yang, Q. Nanogel: A Versatile Nano-Delivery System for Biomedical Applications. *Pharmaceutics* **2020**, *12*, 290. [CrossRef]
137. Chen, Y.; Peng, J.; Han, M.; Omar, M.; Hu, D.; Ke, X.; Lu, N. A low-molecular-weight heparin-coated doxorubicin-liposome for the prevention of melanoma metastasis. *J. Drug Target.* **2015**, *23*, 335–346. [CrossRef]
138. Joung, Y.K.; Jang, J.Y.; Choi, J.H.; Han, D.K.; Park, K.D. Heparin-conjugated pluronic nanogels as multi-drug nanocarriers for combination chemotherapy. *Mol. Pharm.* **2013**, *10*, 685–693. [CrossRef]
139. Kandil, R.; Merkel, O.M. Recent Progress of Polymeric Nanogels for Gene Delivery. *Curr. Opin. Colloid Interface Sci.* **2019**, *39*, 11–23. [CrossRef]
140. Liu, Y.; Lang, T.; Zheng, Z.; Cheng, H.; Huang, X.; Wang, G.; Yin, Q.; Li, Y. In Vivo Environment-Adaptive Nanocomplex with Tumor Cell-Specific Cytotoxicity Enhances T Cells Infiltration and Improves Cancer Therapy. *Small* **2019**, *15*, e1902822. [CrossRef]
141. Tran, T.H.; Bae, B.C.; Lee, Y.K.; Na, K.; Huh, K.M. Heparin-folate-retinoic acid bioconjugates for targeted delivery of hydrophobic photosensitizers. *Carbohydr. Polym.* **2013**, *92*, 1615–1624. [CrossRef]
142. Park, I.K.; Tran, T.H.; Oh, I.H.; Kim, Y.J.; Cho, K.J.; Huh, K.M.; Lee, Y.K. Ternary biomolecular nanoparticles for targeting of cancer cells and anti-angiogenesis. *Eur. J. Pharm. Sci.* **2010**, *41*, 148–155. [CrossRef]
143. Liu, H.; Zhang, Z.; Linhardt, R.J. Lessons learned from the contamination of heparin. *Nat. Prod. Rep.* **2009**, *26*, 313–321. [CrossRef]
144. Mousa, S.A.; Petersen, L.J. Anti-cancer properties of low-molecular-weight heparin: Preclinical evidence. *Thromb. Haemost.* **2009**, *102*, 258–267.
145. Khan, A.R.; Yang, X.; Du, X.; Yang, H.; Liu, Y.; Khan, A.Q.; Zhai, G. Chondroitin sulfate derived theranostic and therapeutic nanocarriers for tumor-targeted drug delivery. *Carbohydr. Polym.* **2020**, *233*, 115837. [CrossRef] [PubMed]
146. Wang, X.; Pei, X.; Du, Y.; Li, Y. Quaternized chitosan/rectorite intercalative materials for a gene delivery system. *Nanotechnology* **2008**, *19*, 375102. [CrossRef] [PubMed]
147. Liang, T.; Zhang, Z.; Jing, P. Black rice anthocyanins embedded in self-assembled chitosan/chondroitin sulfate nanoparticles enhance apoptosis in HCT-116 cells. *Food Chem.* **2019**, *301*, 125280. [CrossRef] [PubMed]
148. Jardim, K.V.; Joanitti, G.A.; Azevedo, R.B.; Parize, A.L. Physico-chemical characterization and cytotoxicity evaluation of curcumin loaded in chitosan/chondroitin sulfate nanoparticles. *Mater. Sci. Eng.* **2015**, *56*, 294–304. [CrossRef] [PubMed]
149. Zu, M.; Ma, L.; Zhang, X.; Xie, D.; Kang, Y.; Xiao, B. Chondroitin sulfate-functionalized polymeric nanoparticles for colon cancer-targeted chemotherapy. *Colloids Surf. B Biointerfaces* **2019**, *177*, 399–406. [CrossRef] [PubMed]
150. Liu, P.; Chen, N.; Yan, L.; Gao, F.; Ji, D.; Zhang, S.; Zhang, L.; Li, Y.; Xiao, Y. Preparation, characterisation and in vitro and in vivo evaluation of CD44-targeted chondroitin sulphate-conjugated doxorubicin PLGA nanoparticles. *Carbohydr. Polym.* **2019**, *213*, 17–26. [CrossRef]
151. Nikitovic, D.; Juranek, I.; Wilks, M.F.; Tzardi, M.; Tsatsakis, A.; Tzanakakis, G.N. Anthracycline-dependent cardiotoxicity and extracellular matrix remodeling. *Chest* **2014**, *146*, 1123–1130. [CrossRef]
152. Germanakis, I.; Kalmanti, M.; Parthenakis, F.; Nikitovic, D.; Stiakaki, E.; Patrianakos, A.; Vardas, P.E. Correlation of plasma N-terminal pro-brain natriuretic peptide levels with left ventricle mass in children treated with anthracyclines. *Int. J. Cardiol.* **2006**, *108*, 212–215. [CrossRef]
153. Wang, X.F.; Ren, J.; He, H.Q.; Liang, L.; Xie, X.; Li, Z.X.; Zhao, J.G.; Yu, J.M. Self-assembled nanoparticles of reduction-sensitive poly (lactic-co-glycolic acid)-conjugated chondroitin sulfate A for doxorubicin delivery: Preparation, characterization and evaluation. *Pharm. Dev. Technol.* **2019**, *24*, 794–802. [CrossRef]
154. Liang, S.; Duan, Y.; Zhang, J.; Xing, Z.; Chen, X.; Yang, Y.; Li, Q. Chemically conjugating poly(amidoamine) with chondroitin sulfate to promote CD44-mediated endocytosis for miR-34a delivery. *J. Control. Release* **2015**, *213*, e95–e96. [CrossRef]
155. Miller, T.; Goude, M.C.; McDevitt, T.C.; Temenoff, J.S. Molecular engineering of glycosaminoglycan chemistry for biomolecule delivery. *Acta Biomater.* **2014**, *10*, 1705–1719. [CrossRef]

156. Iwaki, J.; Minamisawa, T.; Tateno, H.; Kominami, J.; Suzuki, K.; Nishi, N.; Nakamura, T.; Hirabayashi, J. Desulfatedgalactosamino-glycans are potential ligands for galectins: Evidence from frontal affinity chromatography. *Biochem. Biophys. Res. Commun.* **2008**, *373*, 206–212. [CrossRef]
157. Yip, G.W.; Smollich, M.; Gotte, M. Therapeutic value of glycosaminoglycans in cancer. *Mol. Cancer Ther.* **2006**, *5*, 2139–2148. [CrossRef] [PubMed]
158. Belting, M. Glycosaminoglycans in cancer treatment. *Thromb. Res.* **2014**, *133* (Suppl. 2), S95–S101. [CrossRef]
159. Caterson, B.; Melrose, J. Keratan sulfate, a complex glycosaminoglycan with unique functional capability. *Glycobiology* **2018**, *28*, 182–206. [CrossRef]
160. Lee, J.Y.; Spicer, A.P. Hyaluronan: A multifunctional, megaDalton, stealth molecule. *Curr. Opin. Cell Biol.* **2000**, *12*, 581–586. [CrossRef]
161. Zhong, L.; Liu, Y.; Xu, L.; Li, Q.; Zhao, D.; Li, Z.; Zhang, H.; Kan, Q.; Sun, J.; He, Z. Exploring the relationship of hyaluronic acid molecular weight and active targeting efficiency for designing hyaluronic acid-modified nanoparticles. *Asian J. Pharm. Sci.* **2019**, *14*, 521–530. [CrossRef]
162. Mizrahy, S.; Goldsmith, M.; Leviatan-Ben-Arye, S.; Kisin-Finfer, E.; Redy, O.; Srinivasan, S.; Shabat, D.; Godin, B.; Peer, D. Tumor targeting profiling of hyaluronan-coated lipid based-nanoparticles. *Nanoscale* **2014**, *6*, 3742–3752. [CrossRef] [PubMed]
163. Landesman-Milo, D.; Goldsmith, M.; Leviatan Ben-Arye, S.; Witenberg, B.; Brown, E.; Leibovitch, S.; Azriel, S.; Tabak, S.; Morad, V.; Peer, D. Hyaluronan grafted lipid-based nanoparticles as RNAi carriers for cancer cells. *Cancer Lett.* **2013**, *334*, 221–227. [CrossRef]
164. Yang, C.; Li, C.; Zhang, P.; Wu, W.; Jiang, X. Radox responsive hyaluronic acid nanogels for treating RHAMM (CD168) over-expressive cancer, both primary and metastatic tumors. *Theranostics* **2017**, *7*, 1719–1734. [CrossRef] [PubMed]
165. Bharadwaj, A.G.; Kovar, J.L.; Loughman, E.; Elowsky, C.; Oakley, G.G.; Simpson, M.A. Spontaneous metastasis of prostate cancer is promoted by excess hyaluronan synthesis and processing. *Am. J. Pathol.* **2009**, *174*, 1027–1036. [CrossRef] [PubMed]
166. Simpson, M.A. Concurrent expression of hyaluronan biosynthetic and processing enzymes promotes growth and vascularization of prostate tumors in mice. *Am. J. Pathol.* **2006**, *169*, 247–257. [CrossRef] [PubMed]
167. Tian, X.; Azpurua, J.; Hine, C.; Vaidya, A.; Myakishev-Rempel, M.; Ablaeva, J.; Mao, Z.; Nevo, E.; Gorbunova, V.; Seluanov, A. High-molecular-mass hyaluronan mediates the cancer resistance of the naked mole rat. *Nature* **2013**, *499*, 346–349. [CrossRef] [PubMed]
168. Balogh, L.; Polyak, A.; Mathe, D.; Kiraly, R.; Thuroczy, J.; Terez, M.; Janoki, G.; Ting, Y.; Bucci, L.R.; Schauss, A.G. Absorption, uptake and tissue affinity of high-molecular-weight hyaluronan after oral administration in rats and dogs. *J. Agric. Food Chem.* **2008**, *56*, 10582–10593. [CrossRef] [PubMed]
169. Laurent, T.C.; Fraser, J.R. Hyaluronan. *FASEB J.* **1992**, *6*, 2397–2404. [CrossRef] [PubMed]
170. Rao, N.V.; Rho, J.G.; Um, W.; Ek, P.K.; Nguyen, V.Q.; Oh, B.H.; Kim, W.; Park, J.H. Hyaluronic Acid Nanoparticles as Nanomedicine for Treatment of Inflammatory Diseases. *Pharmaceutics* **2020**, *12*, 931. [CrossRef] [PubMed]
171. Lee, H.; Lee, K.; Park, T.G. Hyaluronic acid-paclitaxel conjugate micelles: Synthesis, characterization, and antitumor activity. *Bioconjug. Chem.* **2008**, *19*, 1319–1325. [CrossRef]
172. Din, F.U.; Aman, W.; Ullah, I.; Qureshi, O.S.; Mustapha, O.; Shafique, S.; Zeb, A. Effective use of nanocarriers as drug delivery systems for the treatment of selected tumors. *Int. J. Nanomed.* **2017**, *12*, 7291–7309. [CrossRef]
173. Zheng, S.; Jin, S.; Han, J.; Cho, S.; Nguyen, V.D.; Ko, S.Y.; Park, J.O.; Park, S. Preparation of HIFU-triggered tumor-targeted hyaluronic acid micelles for controlled drug release and enhanced cellular uptake. *Colloids Surf. B Biointerfaces* **2016**, *143*, 27–36. [CrossRef] [PubMed]
174. Zhong, Y.; Goltsche, K.; Cheng, L.; Xie, F.; Meng, F.; Deng, C.; Zhong, Z.; Haag, R. Hyaluronic acid-shelled acid-activatable paclitaxel prodrug micelles effectively target and treat CD44-overexpressing human breast tumor xenografts in vivo. *Biomaterials* **2016**, *84*, 250–261. [CrossRef] [PubMed]
175. Kesharwani, P.; Banerjee, S.; Padhye, S.; Sarkar, F.H.; Iyer, A.K. Hyaluronic Acid Engineered Nanomicelles Loaded with 3,4-Difluorobenzylidene Curcumin for Targeted Killing of CD44+ Stem-Like Pancreatic Cancer Cells. *Biomacromolecules* **2015**, *16*, 3042–3053. [CrossRef] [PubMed]
176. Clevers, H. The cancer stem cell: Premises, promises and challenges. *Nat. Med.* **2011**, *17*, 313–319. [CrossRef]
177. ComertOnder, F.; SagbasSuner, S.; Sahiner, N.; Ay, M.; Ozpolat, B. Delivery of Small Molecule EF2 Kinase Inhibitor for Breast and Pancreatic Cancer Cells Using Hyaluronic Acid Based Nanogels. *Pharm. Res.* **2020**, *37*, 63. [CrossRef] [PubMed]
178. Ossipov, D.A. Nanostructured hyaluronic acid-based materials for active delivery to cancer. *Expert Opin. Drug Deliv.* **2010**, *7*, 681–703. [CrossRef] [PubMed]
179. Kim, S.; Moon, M.J.; PoililSurendran, S.; Jeong, Y.Y. Biomedical Applications of Hyaluronic Acid-Based Nanomaterials in Hyperthermic Cancer Therapy. *Pharmaceutics* **2019**, *11*, 306. [CrossRef]
180. Wang, J.; Li, Y.; Wang, L.; Wang, X.; Tu, P. Comparison of hyaluronic acid-based micelles and polyethylene glycol-based micelles on reversal of multidrug resistance and enhanced anticancer efficacy in vitro and in vivo. *Drug Deliv.* **2018**, *25*, 330–340. [CrossRef]
181. Warren, D.S.; Sutherland, S.P.H.; Kao, J.Y.; Weal, G.R.; Mackay, S.M. The Preparation and Simple Analysis of a Clay Nanoparticle Composite Hydrogel. *J. Chem. Educ.* **2017**, *94*, 1772–1779. [CrossRef]
182. Lee, H.; Mok, H.; Lee, S.; Oh, Y.K.; Park, T.G. Target-specific intracellular delivery of siRNA using degradable hyaluronic acid nanogels. *J. Control. Release* **2007**, *119*, 245–252. [CrossRef]

183. Yanqi, Y.; Jicheng, Y.; Zhen, G. Versatile Protein Nanogels Prepared by In Situ Polymerization. *Macromol. Chem. Phys.* **2015**, *217*, 333–343.
184. Chen, Y.Y.; Wu, H.C.; Sun, J.S.; Dong, G.C.; Wang, T.W. Injectable and thermoresponsive self-assembled nanocomposite hydrogel for long-term anticancer drug delivery. *Langmuir* **2013**, *29*, 3721–3729. [CrossRef] [PubMed]
185. Yang, C.; Wang, X.; Yao, X.; Zhang, Y.; Wu, W.; Jiang, X. Hyaluronic acid nanogels with enzyme-sensitive cross-linking group for drug delivery. *J. Control. Release* **2015**, *205*, 206–217. [CrossRef] [PubMed]
186. Ding, L.; Jiang, Y.; Zhang, J.; Klok, H.A.; Zhong, Z. pH-Sensitive Coiled-Coil Peptide-Cross-Linked Hyaluronic Acid Nanogels: Synthesis and Targeted Intracellular Protein Delivery to CD44 Positive Cancer Cells. *Biomacromolecules* **2018**, *19*, 555–562. [CrossRef] [PubMed]
187. Yeagle, P.L. Cholesterol and the cell membrane. *Biochim. Biophys. Acta* **1985**, *822*, 267–287. [CrossRef]
188. Park, W.; Kim, K.S.; Bae, B.C.; Kim, Y.H.; Na, K. Cancer cell specific targeting of nanogels from acetylated hyaluronic acid with low molecular weight. *Eur. J. Pharm. Sci.* **2010**, *40*, 367–375. [CrossRef] [PubMed]
189. Simon-Yarza, T.; Mielcarek, A.; Couvreur, P.; Serre, C. Nanoparticles of Metal-Organic Frameworks: On the Road to In Vivo Efficacy in Biomedicine. *Adv. Mater.* **2018**, *30*, e1707365. [CrossRef] [PubMed]
190. Shu, F.; DaojunLv, A.; Song, X.-L.; Huang, B.; Wang, C.; Yu, Y.; Zhao, S.-C. Fabrication of a hyaluronic acid conjugated metal organic framework for targeted drug delivery and magnetic resonance imaging. *RSC Adv.* **2018**, *8*, 6581–6589. [CrossRef]
191. Hainfeld, J.F.; Dilmanian, F.; Slatkin, D.N.; Smilowitz, H.M. Radiotherapy enhancement with gold nanoparticles. *J. Pharm. Pharmacol.* **2008**, *60*, 977–985. [CrossRef] [PubMed]
192. Manju, S.; Sreenivasan, K. Gold nanoparticles generated and stabilized by water soluble curcumin-polymer conjugate: Blood compatibility evaluation and targeted drug delivery onto cancer cells. *J. Colloid Interface Sci.* **2012**, *368*, 144–151. [CrossRef] [PubMed]
193. Kumar, C.S.; Raja, M.D.; Sundar, D.S.; GoverAntoniraj, M.; Ruckmani, K. Hyaluronic acid co-functionalized gold nanoparticle complex for the targeted delivery of metformin in the treatment of liver cancer (HepG2 cells). *Carbohydr. Polym.* **2015**, *128*, 63–74. [CrossRef] [PubMed]
194. Vyas, D.; Lopez-Hisijos, N.; Gandhi, S.; El-Dakdouki, M.; Basson, M.D.; Walsh, M.F.; Huang, X.; Vyas, A.K.; Chaturvedi, L.S. Doxorubicin-Hyaluronan Conjugated Super-Paramagnetic Iron Oxide Nanoparticles (DOX-HA-SPION) Enhanced Cytoplasmic Uptake of Doxorubicin and Modulated Apoptosis, IL-6 Release and NF-kappaB Activity in Human MDA-MB-231 Breast Cancer Cells. *J. Nanosci. Nanotechnol.* **2015**, *15*, 6413–6422. [CrossRef]
195. Liu, E.; Zhou, Y.; Liu, Z.; Li, J.; Zhang, D.; Chen, J.; Cai, Z. Cisplatin Loaded Hyaluronic Acid Modified TiO_2 Nanoparticles for Neoadjuvant Chemotherapy of Ovarian Cancer. *J. Nanomater.* **2015**, *2015*, e390358. [CrossRef]
196. Liberman, A.; Mendez, N.; Trogler, W.C.; Kummel, A.C. Synthesis and surface functionalization of silica nanoparticles for nanomedicine. *Surf. Sci. Rep.* **2014**, *69*, 132–158. [CrossRef] [PubMed]
197. Nairi, V.; Magnolia, S.; Piludu, M.; Nieddu, M.; Caria, C.A.; Sogos, V.; Vallet-Regi, M.; Monduzzi, M.; Salis, A. Mesoporous silica nanoparticles functionalized with hyaluronic acid. Effect of the biopolymer chain length on cell internalization. *Colloids Surf. B Biointerfaces* **2018**, *168*, 50–59. [CrossRef]
198. Trial of FOLF(HA)Iri With Cetuximab in mCRC (Chime). Available online: https://clinicaltrials.gov/ct2/show/NCT02216487 (accessed on 9 December 2020).
199. Alamgeer, M.; Neil Watkins, D.; Banakh, I.; Kumar, B.; Gough, D.J.; Markman, B.; Ganju, V. A phase IIa study of HA-irinotecan, formulation of hyaluronic acid and irinotecan targeting CD44 in extensive-stage small cell lung cancer. *Investig. New Drugs* **2018**, *36*, 288–298. [CrossRef]
200. Cai, S.; Zhang, T.; Forrest, W.C.; Yang, Q.; Groer, C.; Mohr, E.; Aires, D.J.; Axiak-Bechtel, S.M.; Flesner, B.K.; Henry, C.J.; et al. Phase I-II clinical trial of hyaluronan-cisplatin nanoconjugate in dogs with naturally occurring malignant tumors. *Am. J. Vet. Res.* **2016**, *77*, 1005–1016. [CrossRef]
201. Thompson, L.D.; Pantoliano, M.W.; Springer, B.A. Energetic characterization of the basic fibroblast growth factor-heparin interaction: Identification of the heparin binding domain. *Biochemistry* **1994**, *33*, 3831–3840. [CrossRef] [PubMed]
202. Lyman, G.H.; Bohlke, K.; Khorana, A.A.; Kuderer, N.M.; Lee, A.Y.; Arcelus, J.I.; Balaban, E.P.; Clarke, J.M.; Flowers, C.R.; Francis, C.W.; et al. Venous thromboembolism prophylaxis and treatment in patients with cancer: American society of clinical oncology clinical practice guideline update 2014. *J. Clin. Oncol.* **2015**, *33*, 654–656. [CrossRef]
203. Laubli, H.; Varki, A.; Borsig, L. Antimetastatic Properties of Low Molecular Weight Heparin. *J. Clin. Oncol.* **2016**, *34*, 2560–2561. [CrossRef] [PubMed]
204. Dulaney, S.B.; Huang, X. Strategies in synthesis of heparin/heparan sulfate oligosaccharides: 2000-present. *Adv. Carbohydr. Chem. Biochem.* **2012**, *67*, 95–136. [PubMed]
205. Mohamed, S.; Ferro, V. Synthetic Approaches to L-Iduronic Acid and L-Idose: Key Building Blocks for the Preparation of Glycosaminoglycan Oligosaccharides. *Adv. Carbohydr. Chem. Biochem.* **2015**, *72*, 21–61.
206. MacDonald, A.; Priess, M.; Curran, J.; Guess, J.; Farutin, V.; Oosterom, I.; Chu, C.L.; Cochran, E.; Zhang, L.; Getchell, K.; et al. Multitargeting Heparan Sulfate Mimetic, Targets Tumor and Stromal Compartments in Pancreatic Cancer. *Mol. Cancer Ther.* **2019**, *18*, 245–256. [CrossRef]
207. Gacche, R.N.; Meshram, R.J. Targeting tumor micro-environment for design and development of novel anti-angiogenic agents arresting tumor growth. *Prog. Biophys Mol. Biol.* **2013**, *113*, 333–354. [CrossRef] [PubMed]

208. Rusnati, M.; Presta, M. Fibroblast growth factors/fibroblast growth factor receptors as targets for the development of anti-angiogenesis strategies. *Curr. Pharm. Des.* **2007**, *13*, 2025–2044. [CrossRef]
209. Kessler, T.; Fehrmann, F.; Bieker, R.; Berdel, W.E.; Mesters, R.M. Vascular endothelial growth factor and its receptor as drug targets in hematological malignancies. *Curr. Drug Targets* **2007**, *8*, 257–268. [CrossRef] [PubMed]
210. Ferro, V.; Fewings, K.; Palermo, M.C.; Li, C. Large-scale preparation of the oligosaccharide phosphate fraction of Pichia holstii NRRL Y-2448 phosphomannan for use in the manufacture of PI-88. *Carbohydr. Res.* **2001**, *332*, 183–189. [CrossRef]
211. Ferro, V.; Li, C.; Fewings, K.; Palermo, M.C.; Linhardt, R.J.; Toida, T. Determination of the composition of the oligosaccharide phosphate fraction of Pichia (Hansenula) holstii NRRL Y-2448 phosphomannan by capillary electrophoresis and HPLC. *Carbohydr. Res.* **2002**, *337*, 139–146. [CrossRef]
212. Ono, K.; Ishihara, M.; Ishikawa, K.; Ozeki, Y.; Deguchi, H.; Sato, M.; Hashimoto, H.; Saito, Y.; Yura, H.; Kurita, A.; et al. Periodate-treated, non-anticoagulant heparin-carrying polystyrene (NAC-HCPS) affects angiogenesis and inhibits subcutaneous induced tumour growth and metastasis to the lung. *Br. J. Cancer* **2002**, *86*, 1803–1812. [CrossRef]
213. Cassinelli, G.; Favini, E.; Dal Bo, L.; Tortoreto, M.; De Maglie, M.; Dagrada, G.; Pilotti, S.; Zunino, F.; Zaffaroni, N.; Lanzi, C. Antitumor efficacy of the heparan sulfate mimic roneparstat (SST0001) against sarcoma models involves multi-target inhibition of receptor tyrosine kinases. *Oncotarget* **2016**, *7*, 47848–47863. [CrossRef]
214. Chalkiadaki, G.; Nikitovic, D.; Berdiaki, A.; Katonis, P.; Karamanos, N.K.; Tzanakakis, G.N. Heparin plays a key regulatory role via a p53/FAK-dependent signaling in melanoma cell adhesion and migration. *IUBMB Life* **2011**, *63*, 109–119. [CrossRef] [PubMed]
215. Lindahl, U.; Li, J.-P. Heparanase–Discovery and Targets. In *Heparanase: From Basic Research to Clinical Applications*; Vlodavsky, I., Sanderson, R.D., Ilan, N., Eds.; Springer International Publishing: Cham, Switzerland, 2020; pp. 61–69.
216. Mohan, C.D.; Hari, S.; Preetham, H.D.; Rangappa, S.; Barash, U.; Ilan, N.; Nayak, S.C.; Gupta, V.K.; Basappa; Vlodavsky, I.; et al. Targeting Heparanase in Cancer: Inhibition by Synthetic, Chemically Modified, and Natural Compounds. *iScience* **2019**, *15*, 360–390. [CrossRef]
217. Hammond, E.; Khurana, A.; Shridhar, V.; Dredge, K. The Role of Heparanase and Sulfatases in the Modification of Heparan Sulfate Proteoglycans within the Tumor Microenvironment and Opportunities for Novel Cancer Therapeutics. *Front. Oncol.* **2014**, *4*, 195. [CrossRef]
218. Vlodavsky, I.; Gross-Cohen, M.; Weissmann, M.; Ilan, N.; Sanderson, R.D. Opposing Functions of Heparanase-1 and Heparanase-2 in Cancer Progression. *Trends Biochem. Sci.* **2018**, *43*, 18–31. [CrossRef]
219. Gutter-Kapon, L.; Alishekevitz, D.; Shaked, Y.; Li, J.P.; Aronheim, A.; Ilan, N.; Vlodavsky, I. Heparanase is required for activation and function of macrophages. *Proc. Natl. Acad. Sci. USA* **2016**, *113*, E7808–E7817. [CrossRef]
220. Edovitsky, E.; Elkin, M.; Zcharia, E.; Peretz, T.; Vlodavsky, I. Heparanase gene silencing, tumor invasiveness, angiogenesis, and metastasis. *J. Natl. Cancer Inst.* **2004**, *96*, 1219–1230. [CrossRef] [PubMed]
221. Vlodavsky, I.; Singh, P.; Boyango, I.; Gutter-Kapon, L.; Elkin, M.; Sanderson, R.D.; Ilan, N. Heparanase: From basic research to therapeutic applications in cancer and inflammation. *Drug Resist. Updates* **2016**, *29*, 54–75. [CrossRef]
222. Sun, X.; Zhang, G.; Nian, J.; Yu, M.; Chen, S.; Zhang, Y.; Yang, G.; Yang, L.; Cheng, P.; Yan, C.; et al. Elevated heparanase expression is associated with poor prognosis in breast cancer: A study based on systematic review and TCGA data. *Oncotarget* **2017**, *8*, 43521–43535. [CrossRef] [PubMed]
223. Purushothaman, A.; Sanderson, R.D. Heparanase: A Dynamic Promoter of Myeloma Progression. *Adv. Exp. Med. Biol.* **2020**, *1221*, 331–349. [PubMed]
224. Gohji, K.; Okamoto, M.; Kitazawa, S.; Toyoshima, M.; Dong, J.; Katsuoka, Y.; Nakajima, M. Heparanase protein and gene expression in bladder cancer. *J. Urol.* **2001**, *166*, 1286–1290. [CrossRef]
225. Purushothaman, A.; Uyama, T.; Kobayashi, F.; Yamada, S.; Sugahara, K.; Rapraeger, A.C.; Sanderson, R.D. Heparanase-enhanced shedding of syndecan-1 by myeloma cells promotes endothelial invasion and angiogenesis. *Blood* **2010**, *115*, 2449–2457. [CrossRef] [PubMed]
226. Tatsumi, Y.; Miyake, M.; Shimada, K.; Fujii, T.; Hori, S.; Morizawa, Y.; Nakai, Y.; Anai, S.; Tanaka, N.; Konishi, N.; et al. Inhibition of Heparanase Expression Results in Suppression of Invasion, Migration and Adhesion Abilities of Bladder Cancer Cells. *Int. J. Mol. Sci.* **2020**, *21*, 3789. [CrossRef] [PubMed]
227. Teixeira, F.; Gotte, M. Involvement of Syndecan-1 and Heparanase in Cancer and Inflammation. *Adv. Exp. Med. Biol.* **2020**, *1221*, 97–135.
228. Ramani, V.C.; Vlodavsky, I.; Ng, M.; Zhang, Y.; Barbieri, P.; Noseda, A.; Sanderson, R.D. Chemotherapy induces expression and release of heparanase leading to changes associated with an aggressive tumor phenotype. *Matrix Biol.* **2016**, *55*, 22–34. [CrossRef] [PubMed]
229. Thompson, C.A.; Purushothaman, A.; Ramani, V.C.; Vlodavsky, I.; Sanderson, R.D. Heparanase regulates secretion, composition, and function of tumor cell-derived exosomes. *J. Biol. Chem.* **2013**, *288*, 10093–10099. [CrossRef] [PubMed]
230. Bandari, S.K.; Purushothaman, A.; Ramani, V.C.; Brinkley, G.J.; Chandrashekar, D.S.; Varamballly, S.; Mobley, J.A.; Zhang, Y.; Brown, E.E.; Vlodavsky, I.; et al. Chemotherapy induces secretion of exosomes loaded with heparanase that degrades extracellular matrix and impacts tumor and host cell behavior. *Matrix Biol.* **2018**, *65*, 104–118. [CrossRef]

231. Ritchie, J.P.; Ramani, V.C.; Ren, Y.; Naggi, A.; Torri, G.; Casu, B.; Penco, S.; Pisano, C.; Carminati, P.; Tortoreto, M.; et al. SST 0001, a chemically modified heparin, inhibits myeloma growth and angiogenesis via disruption of the heparanase/syndecan-1 axis. *Clin. Cancer Res.* **2011**, *17*, 1382–1393. [CrossRef]
232. Pala, D.; Rivara, S.; Mor, M.; Milazzo, F.M.; Roscilli, G.; Pavoni, E.; Giannini, G. Kinetic analysis and molecular modeling of the inhibition mechanism of roneparstat (SST0001) on human heparanase. *Glycobiology* **2016**, *26*, 640–654. [CrossRef] [PubMed]
233. Galli, M.; Chatterjee, M.; Grasso, M.; Specchia, G.; Magen, H.; Einsele, H.; Celeghini, I.; Barbieri, P.; Paoletti, D.; Pace, S.; et al. Phase I study of the heparanase inhibitor roneparstat: An innovative approach for ultiple myeloma therapy. *Haematologica* **2018**, *103*, e469–e472. [CrossRef]
234. Ferro, V.; Dredge, K.; Liu, L.; Hammond, E.; Bytheway, I.; Li, C.; Johnstone, K.; Karoli, T.; Davis, K.; Copeman, E.; et al. PI-88 and novel heparan sulfate mimetics inhibit angiogenesis. *Semin. Thromb. Hemost.* **2007**, *33*, 557–568. [CrossRef]
235. Dredge, K.; Hammond, E.; Davis, K.; Li, C.P.; Liu, L.; Johnstone, K.; Handley, P.; Wimmer, N.; Gonda, T.J.; Gautam, A.; et al. The PG500 series: Novel heparan sulfate mimetics as potent angiogenesis and heparanase inhibitors for cancer therapy. *Investig. New Drugs* **2010**, *28*, 276–283. [CrossRef]
236. Lewis, K.D.; Robinson, W.A.; Millward, M.J.; Powell, A.; Price, T.J.; Thomson, D.B.; Walpole, E.T.; Haydon, A.M.; Creese, B.R.; Roberts, K.L.; et al. A phase II study of the heparanase inhibitor PI-88 in patients with advanced melanoma. *Investig. New Drugs* **2008**, *26*, 89–94. [CrossRef]
237. Chhabra, M.; Ferro, V. PI-88 and Related Heparan Sulfate Mimetics. *Adv. Exp. Med. Biol.* **2020**, *1221*, 473–491. [PubMed]
238. Ferro, V.; Liu, L.; Johnstone, K.D.; Wimmer, N.; Karoli, T.; Handley, P.; Rowley, J.; Dredge, K.; Li, C.P.; Hammond, E.; et al. Discovery of PG545: A highly potent and simultaneous inhibitor of angiogenesis, tumor growth, and metastasis. *J. Med. Chem.* **2012**, *55*, 3804–3813. [CrossRef] [PubMed]
239. Weissmann, M.; Bhattacharya, U.; Feld, S.; Hammond, E.; Ilan, N.; Vlodavsky, I. The heparanase inhibitor PG545 is a potent anti-lymphoma drug: Mode of action. *Matrix Biol.* **2019**, *77*, 58–72. [CrossRef]
240. Barash, U.; Cohen-Kaplan, V.; Arvatz, G.; Gingis-Velitski, S.; Levy-Adam, F.; Nativ, O.; Shemesh, R.; Ayalon-Sofer, M.; Ilan, N.; Vlodavsky, I. A novel human heparanase splice variant, T5, endowed with protumorigenic characteristics. *FASEB J.* **2010**, *24*, 1239–1248. [CrossRef] [PubMed]
241. Neagu, M.; Constantin, C.; Caruntu, C.; Dumitru, C.; Surcel, M.; Zurac, S. Inflammation: A key process in skin tumorigenesis. *Oncol. Lett.* **2019**, *17*, 4068–4084. [CrossRef] [PubMed]
242. Neagu, M.; Constantin, C.; Longo, C. Chemokines in the melanoma metastasis biomarkers portrait. *J. Immunoass. Immunochem.* **2015**, *36*, 559–566. [CrossRef]
243. Khaldoyanidi, S.K.; Goncharova, V.; Mueller, B.; Schraufstatter, I.U. Hyaluronan in the healthy and malignant hematopoietic microenvironment. *Adv. Cancer Res.* **2014**, *123*, 149–189.
244. Neagu, M.; Constantin, C.; Popescu, I.D.; Zipeto, D.; Tzanakakis, G.; Nikitovic, D.; Fenga, C.; Stratakis, C.A.; Spandidos, D.A.; Tsatsakis, A.M. Inflammation and Metabolism in Cancer Cell-Mitochondria Key Player. *Front. Oncol.* **2019**, *9*, 348. [CrossRef]
245. Rios de la Rosa, J.M.; Tirella, A.; Gennari, A.; Stratford, I.J.; Tirelli, N. The CD44-Mediated Uptake of Hyaluronic Acid-Based Carriers in Macrophages. *Adv. Heal. Mater.* **2017**, *6*, 4. [CrossRef]
246. Bulman, A.; Neagu, M.; Constantin, C. Immunomics in Skin Cancer-Improvement in Diagnosis, Prognosis and Therapy Monitoring. *Curr. Proteom.* **2013**, *10*, 202–217. [CrossRef] [PubMed]
247. Huang, W.C.; Chen, S.H.; Chiang, W.H.; Huang, C.W.; Lo, C.L.; Chern, C.S.; Chiu, H.C. Tumor Microenvironment-Responsive Nanoparticle Delivery of Chemotherapy for Enhanced Selective Cellular Uptake and Transportation within Tumor. *Biomacromolecules* **2016**, *17*, 3883–3892. [CrossRef]
248. Wang, B.; Zhang, W.; Zhou, X.; Liu, M.; Hou, X.; Cheng, Z.; Chen, D. Development of dual-targeted nano-dandelion based on an oligomeric hyaluronic acid polymer targeting tumor-associated macrophages for combination therapy of non-small cell lung cancer. *Drug Deliv.* **2019**, *26*, 1265–1279. [CrossRef]
249. Parayath, N.N.; Gandham, S.K.; Leslie, F.; Amiji, M.M. Improved anti-tumor efficacy of paclitaxel in combination with MicroRNA-125b-based tumor-associated macrophage repolarization in epithelial ovarian cancer. *Cancer Lett.* **2019**, *461*, 1–9. [CrossRef] [PubMed]
250. Zhang, H.; Zhang, X.; Ren, Y.; Cao, F.; Hou, L.; Zhang, Z. An in situ microenvironmental nano-regulator to inhibit the proliferation and metastasis of 4T1 tumor. *Theranostics* **2019**, *9*, 3580–3594. [CrossRef] [PubMed]
251. Alaniz, L.; Rizzo, M.; Garcia, M.G.; Piccioni, F.; Aquino, J.B.; Malvicini, M.; Atorrasagasti, C.; Bayo, J.; Echeverria, I.; Sarobe, P.; et al. Low molecular weight hyaluronan preconditioning of tumor-pulsed dendritic cells increases their migratory ability and induces immunity against murine colorectal carcinoma. *Cancer Immunol. Immunother.* **2011**, *60*, 1383–1395. [CrossRef] [PubMed]
252. Song, M.; Liu, T.; Shi, C.; Zhang, X.; Chen, X. Bioconjugated Manganese Dioxide Nanoparticles Enhance Chemotherapy Response by Priming Tumor-Associated Macrophages toward M1-like Phenotype and Attenuating Tumor Hypoxia. *ACS Nano* **2016**, *10*, 633–647. [CrossRef]
253. Lee, Y.H.; Yoon, H.Y.; Shin, J.M.; Saravanakumar, G.; Noh, K.H.; Song, K.H.; Jeon, J.H.; Kim, D.W.; Lee, K.M.; Kim, K.; et al. A polymeric conjugate foreignizing tumor cells for targeted immunotherapy in vivo. *J. Control. Release* **2015**, *199*, 98–105. [CrossRef] [PubMed]
254. He, M.; Huang, L.; Hou, X.; Zhong, C.; Bachir, Z.A.; Lan, M.; Chen, R.; Gao, F. Efficient ovalbumin delivery using a novel multifunctional micellar platform for targeted melanoma immunotherapy. *Int. J. Pharm.* **2019**, *560*, 1–10. [CrossRef] [PubMed]

255. Shin, J.M.; Oh, S.J.; Kwon, S.; Deepagan, V.G.; Lee, M.; Song, S.H.; Lee, H.J.; Kim, S.; Song, K.H.; Kim, T.W.; et al. A PEGylated hyaluronic acid conjugate for targeted cancer immunotherapy. *J. Control. Release* **2017**, *267*, 181–190. [CrossRef] [PubMed]
256. Cao, F.; Yan, M.; Liu, Y.; Liu, L.; Ma, G. Photothermally Controlled MHC Class I Restricted CD8(+) T-Cell Responses Elicited by Hyaluronic Acid Decorated Gold Nanoparticles as a Vaccine for Cancer Immunotherapy. *Adv. Healthc. Mater.* **2018**, *7*, e1701439. [CrossRef]
257. Watanabe, K.; Tsuchiya, Y.; Kawaguchi, Y.; Sawada, S.; Ayame, H.; Akiyoshi, K.; Tsubata, T. The use of cationic nanogels to deliver proteins to myeloma cells and primary T lymphocytes that poorly express heparan sulfate. *Biomaterials* **2011**, *32*, 5900–5905. [CrossRef] [PubMed]
258. Georgescu, S.R.; Tampa, M.; Mitran, C.I.; Mitran, M.I.; Caruntu, C.; Caruntu, A.; Lupu, M.; Matei, C.; Constantin, C.; Neagu, M. Tumour Microenvironment in Skin Carcinogenesis. *Adv. Exp. Med. Biol.* **2020**, *1226*, 123–142. [PubMed]
259. Rossi, G.R.; Trindade, E.S.; Souza-Fonseca-Guimaraes, F. Tumor Microenvironment-Associated Extracellular Matrix Components Regulate NK Cell Function. *Front. Immunol.* **2020**, *11*, 73. [CrossRef] [PubMed]
260. Melstrom, L.G.; Salazar, M.D.; Diamond, D.J. The pancreatic cancer microenvironment: A true double agent. *J. Surg. Oncol.* **2017**, *116*, 7–15. [CrossRef] [PubMed]
261. Ancuceanu, R.; Neagu, M. Immune based therapy for melanoma. *Indian J. Med. Res.* **2016**, *143*, 135–144. [CrossRef] [PubMed]
262. Jiang, P.; Li, X.; Thompson, C.B.; Huang, Z.; Araiza, F.; Osgood, R.; Wei, G.; Feldmann, M.; Frost, G.I.; Shepard, H.M. Effective targeting of the tumor microenvironment for cancer therapy. *Anticancer Res.* **2012**, *32*, 1203–1212.
263. Borad, M.J.; Ramanathan, R.K.; Bessudo, A.; LoRusso, P.; Shepard, H.M.; Maneval, D.C.; Jiang, P.; Zhu, J.; Frost, G.I.; Infante, J.R. Targeting hyaluronan (HA) in tumor stroma: A phase I study to evaluate the safety, pharmacokinetics (PK), andpharmacodynamics (PD) of pegylated hyaluronidase (PEGPH20) in patients with solid tumors. *J. Clin. Oncol.* **2012**, *30*, 2579. [CrossRef]
264. Wong, K.M.; Horton, K.J.; Coveler, A.L.; Hingorani, S.R.; Harris, W.P. Targeting the Tumor Stroma: The Biology and Clinical Development of Pegylated Recombinant Human Hyaluronidase (PEGPH20). *Curr. Oncol. Rep.* **2017**, *19*, 47. [CrossRef] [PubMed]
265. Bollyky, P.L.; Wu, R.P.; Falk, B.A.; Lord, J.D.; Long, S.A.; Preisinger, A.; Teng, B.; Holt, G.E.; Standifer, N.E.; Braun, K.R.; et al. ECM components guide IL-10 producing regulatory T-cell (TR1) induction from effector memory T-cell precursors. *Proc. Natl. Acad. Sci. USA* **2011**, *108*, 7938–7943. [CrossRef] [PubMed]
266. Clift, R.; Souratha, J.; Garrovillo, S.A.; Zimmerman, S.; Blouw, B. Remodeling the Tumor Microenvironment Sensitizes Breast Tumors to Anti-Programmed Death-Ligand 1 Immunotherapy. *Cancer Res.* **2019**, *79*, 4149–4159. [CrossRef]
267. Hermano, E.; Lerner, I.; Elkin, M. Heparanase enzyme in chronic inflammatory bowel disease and colon cancer. *Cell. Mol. Life Sci.* **2012**, *69*, 2501–2513. [CrossRef]
268. Mayfosh, A.J.; Baschuk, N.; Hulett, M.D. Leukocyte Heparanase: A Double-Edged Sword in Tumor Progression. *Front. Oncol.* **2019**, *9*, 331. [CrossRef]
269. Weissmann, M.; Arvatz, G.; Horowitz, N.; Feld, S.; Naroditsky, I.; Zhang, Y.; Ng, M.; Hammond, E.; Nevo, E.; Vlodavsky, I.; et al. Heparanase-neutralizing antibodies attenuate lymphoma tumor growth and metastasis. *Proc. Natl. Acad. Sci. USA* **2016**, *113*, 704–709. [CrossRef] [PubMed]
270. Hoellenriegel, J.; Zboralski, D.; Maasch, C.; Rosin, N.Y.; Wierda, W.G.; Keating, M.J.; Kruschinski, A.; Burger, J.A. The Spiegelmer NOX-A12, a novel CXCL12 inhibitor, interferes with chronic lymphocytic leukemia cell motility and causes chemosensitization. *Blood* **2014**, *123*, 1032–1039. [CrossRef]
271. Singha, N.C.; Nekoroski, T.; Zhao, C.; Symons, R.; Jiang, P.; Frost, G.I.; Huang, Z.; Shepard, H.M. Tumor-associated hyaluronan limits efficacy of monoclonal antibody therapy. *Mol. Cancer Ther.* **2015**, *14*, 523–532. [CrossRef] [PubMed]
272. Very, N.; Lefebvre, T.; El Yazidi-Belkoura, I. Drug resistance related to aberrant glycosylation in colorectal cancer. *Oncotarget* **2018**, *9*, 1380–1402. [CrossRef] [PubMed]
273. Khan, S.A.; Mason, R.W.; Kobayashi, H.; Yamaguchi, S.; Tomatsu, S. Advances in glycosaminoglycan detection. *Mol. Genet. Metab.* **2020**, *130*, 101–109. [CrossRef]
274. Pshezhetsky, A.V. Crosstalk between 2 organelles: Lysosomal storage of heparan sulfate causes mitochondrial defects and neuronal death in mucopolysaccharidosis III type C. *Rare Dis.* **2015**, *3*, e1049793. [CrossRef]
275. Ponsiglione, A.M.; Russo, M.; Torino, E. Glycosaminoglycans and Contrast Agents: The Role of Hyaluronic Acid as MRI Contrast Enhancer. *Biomolecules* **2020**, *10*, 1612. [CrossRef]
276. Guzzo, T.; Barile, F.; Marras, C.; Bellini, D.; Mandaliti, W.; Nepravishta, R.; Paci, M.; Topai, A. Stability Evaluation and Degradation Studies of DAC((R)) Hyaluronic-Polylactide Based Hydrogel by DOSY NMR Spectroscopy. *Biomolecules* **2020**, *10*, 1478. [CrossRef]
277. Whitmore, E.K.; Vesenka, G.; Sihler, H.; Guvench, O. Efficient Construction of Atomic-Resolution Models of Non-Sulfated Chondroitin Glycosaminoglycan Using Molecular Dynamics Data. *Biomolecules* **2020**, *10*, 537. [CrossRef] [PubMed]
278. Javadi, J.; Dobra, K.; Hjerpe, A. Multiplex Soluble Biomarker Analysis from Pleural Effusion. *Biomolecules* **2020**, *10*, 1113. [CrossRef] [PubMed]
279. Nikitovic, D.; Kouvidi, K.; Kavasi, R.M.; Berdiaki, A.; Tzanakakis, G.N. Hyaluronan/Hyaladherins-a Promising Axis for Targeted Drug Delivery in Cancer. *Curr. Drug Deliv.* **2016**, *13*, 500–511. [CrossRef] [PubMed]

Review

Heparan Sulfate Glycosaminoglycans: (Un)Expected Allies in Cancer Clinical Management

Isabel Faria-Ramos [1,2], **Juliana Poças** [1,2,3], **Catarina Marques** [1,2,3], **João Santos-Antunes** [1,2,4,5], **Guilherme Macedo** [4,5], **Celso A. Reis** [1,2,3,4] and **Ana Magalhães** [1,2,*]

1. Instituto de Investigação e Inovação em Saúde (i3S), University of Porto, 4200-135 Porto, Portugal; iantunes@ipatimup.pt (I.F.-R.); jpocas@ipatimup.pt (J.P.); cgoncalves@ipatimup.pt (C.M.); joao.claudio.antunes@gmail.com (J.S.-A.); celsor@ipatimup.pt (C.A.R.)
2. Instituto de Patologia e Imunologia Molecular da Universidade do Porto (IPATIMUP), 4200-135 Porto, Portugal
3. Molecular Biology Department, Instituto de Ciências Biomédicas Abel Salazar (ICBAS), University of Porto, 4050-313 Porto, Portugal
4. Pathology Department, Faculdade de Medicina, University of Porto, 4200-319 Porto, Portugal; guilhermemacedo59@gmail.com
5. Gastroenterology Department, Centro Hospitalar S. João, 4200-319 Porto, Portugal
* Correspondence: amagalhaes@ipatimup.pt

Abstract: In an era when cancer glycobiology research is exponentially growing, we are witnessing a progressive translation of the major scientific findings to the clinical practice with the overarching aim of improving cancer patients' management. Many mechanistic cell biology studies have demonstrated that heparan sulfate (HS) glycosaminoglycans are key molecules responsible for several molecular and biochemical processes, impacting extracellular matrix properties and cellular functions. HS can interact with a myriad of different ligands, and therefore, hold a pleiotropic role in regulating the activity of important cellular receptors and downstream signalling pathways. The aberrant expression of HS glycan chains in tumours determines main malignant features, such as cancer cell proliferation, angiogenesis, invasion and metastasis. In this review, we devote particular attention to HS biological activities, its expression profile and modulation in cancer. Moreover, we highlight HS clinical potential to improve both diagnosis and prognosis of cancer, either as HS-based biomarkers or as therapeutic targets.

Keywords: biomarker; cancer; cancer therapy; extracellular vesicles; glycosaminoglycans; heparan sulfate; proteoglycans

1. Introduction

After decades of knowledge about the cellular signalling pathways mediated by glycoconjugates and the impact of the glycan structural characteristics in defining specific cellular responses, researchers are taking advantage of the multiple features of glycosaminoglycans (GAGs) to develop new tools for improving the clinical management of cancer. GAGs are long linear chains of heterogeneous saccharides, comprising one of the major biomolecules class found in all mammalian cells [1]. GAGs have been extensively studied, and their interactions with growth factors, morphogens, chemokines, extracellular matrix (ECM) proteins and their bioactive fragments, receptors, lipoproteins and pathogens are well described [1–5]. This dynamic network orchestrates several essential functions, from critical steps in embryogenesis and early development to ECM (re)modelling and cell signalling regulation in various physiological and pathological contexts, such as metabolic and neurodegenerative diseases, infections and cancer [6,7]. This review focuses on one particular class of GAGs: Heparan sulfate (HS). HS are anionic polysaccharide chains that assemble as disaccharide building blocks of glucuronic acid (GlcA) linked to *N*-acetyl-glucosamine (GlcNAc) and undergo extensive modification through the action of at least four families of sulfotransferases and one epimerase. HS chains are covalently linked to a core protein to form heparan sulfate proteoglycans (HSPGs), which can be expressed at the cell membrane, released into the ECM [8] or secreted in extracellular vesicles (EVs) [9,10]. The HSPGs are the main mediators of cellular interaction with an enormous number of ligands. Over the last decade, new insights have emerged regarding the mechanisms and the biological significance of those interactions [7,11–13], and in this last couple of years, their biomedical potential has been at the forefront in glycobiology translational research [14–16]. HS interfere in many steps of tumour progression, such as cancer cell proliferation, immune response escaping, invasion of neighbour tissues and metastasis [7,9,17]. Moreover, the aberrant expression of different HSPGs and of the key enzymes involved in HS biosynthesis and post-synthesis modifications impact cancer cell behaviour [17,18]. The interplay between researchers and clinicians has been key to identify the major needs in the clinical practice, and therefore, propel a better understanding on the potential of HS with the ultimate goal of improving cancer patients' management. This ladder could not be scaled without the parallel development of powerful analytical equipment and approaches for glycan characterisation [19,20]. These biotechnological advances have contributed to unravel important features regarding the chemical diversity of HS structures, along with the intricate regulation of its biosynthetic pathways.

This review presents the main HS and HSPGs biological functions, from physiological to disease contexts, and summarises the most recent findings on HS as biomarkers and/or as therapeutic targets.

2. Glycosaminoglycans as Main Extracellular Matrix and Glycocalyx Building Blocks

2.1. Glycosaminoglycans and Proteoglycans Diversity

ECM is a well-organised and dynamic macromolecular complex that provides a three-dimensional scaffold for cells and contributes to tissue homeostasis. Generally, the ECM is composed by varied fibrous proteins, polysaccharides and water. However, its major components, and subsequent structural features, are tissue-specific. Its most common constituents include collagens, glycoproteins, such as laminins and fibronectin, proteoglycans (PGs) and GAGs [21,22]. Besides acting as important ECM building blocks, PGs are also major components of the cellular glycocalyx. This cell's surface layer includes a vast group of membrane-attached PGs, secreted GAG chains, glycoproteins and glycolipids being associated with cellular functions in homeostasis, as well as to cell responses to injury and disease [1].

PGs are composed by a core protein with GAG chains covalently attached. GAGs are long and linear polysaccharides composed by repeating disaccharide units and represent an important distinctive structural feature amongst different PGs. According to the disaccharide units that build these chains, GAGs can be classified as HS, chondroitin sulfate (CS), dermatan sulfate (DS) or keratan sulfate (KS) [8]. Hyaluronan (HA) is the exception because it is the only non-sulfated GAG and lacks a covalent bond to a protein core. The different classes of GAGs are schematically represented in Figure 1A.

According to their cellular and subcellular localisation, overall homology and function, PGs can be further classified into five different groups: (i) Intracellular proteoglycans (Serglycin); (ii) Cell surface proteoglycans (syndecans (SDCs), chondroitin sulfate Proteoglycan 4/neuron glia antigen-2 (CSPG4/NG2), betaglycan/TGFβ type III receptor; phosphacan/receptor-type protein tyrosine phosphatase β; glypicans (GPCs)/GPI-anchored proteoglycans); (iii) pericellular and basement membrane proteoglycans (Perlecan; Agrin; Collagens XV and XVIII); (iv) extracellular proteoglycans (Aggrecan; Versican; Neurocan and Brevican); and (v) small leucine-rich proteoglycans (SLRPs) (class I-V), which are abundant ECM glycoconjugates (decorin, biglycan, fibromodulin, luminican, kerotocan, osteoglycan) [24].

The different carriers of HS GAGs at the cellular glycocalyx are shown in Figure 1B.

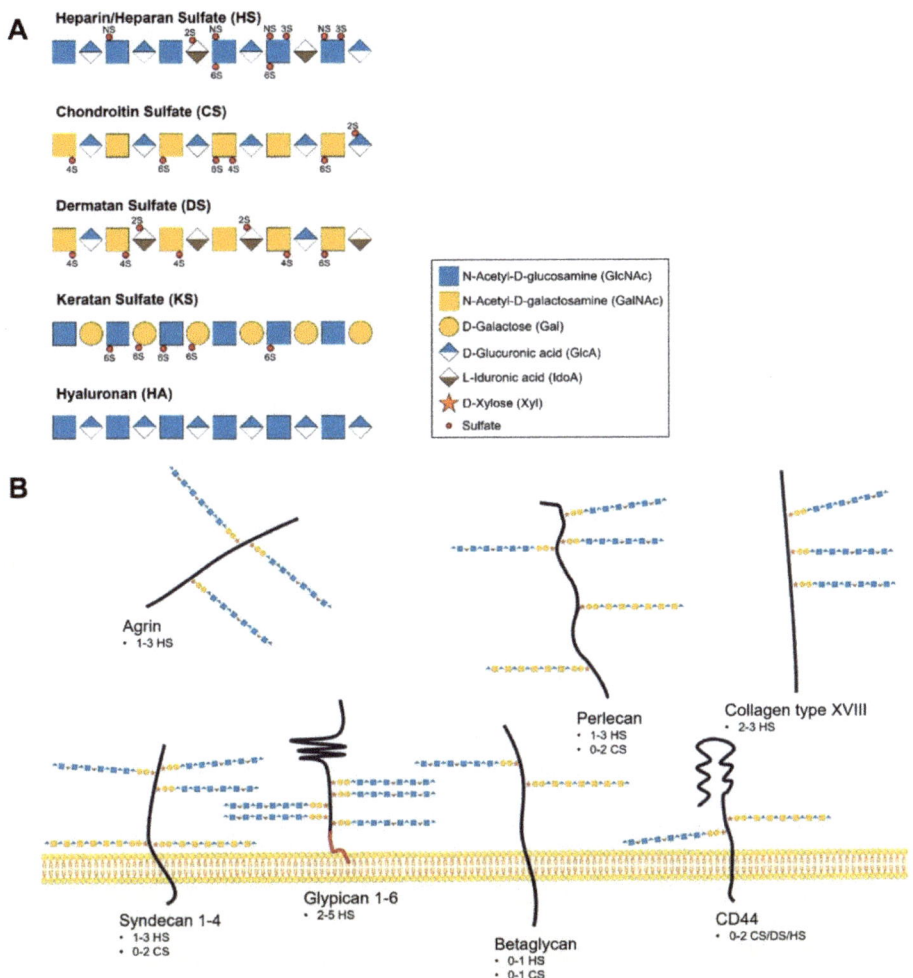

Figure 1. (**A**) Structural composition and classification of glycosaminoglycan chains. Non-reducing termini are to the right of the saccharide's sequences. (**B**) Illustrative representation of major heparan sulfate (HS)-proteoglycans composing the cells' glycocalyx and extracellular matrix (ECM). Below each family of heparan sulfate proteoglycans (HSPGs) is indicated the number and type of glycosaminoglycan (GAG) chains that commonly modify the core protein [8,23].

2.2. Heparan Sulfate Biochemical and Structural Features

HSPGs are composed by a core protein with covalently linked HS chains, whose length ranges between 50–400 disaccharide units [25]. The glycan portion directly attached to the protein is termed tetrasaccharide linker and is composed by a residue of xylose (Xyl) directly attached to the core protein, two galactose (Gal) residues and one GlcA residue. This region is followed by repeating disaccharide units of glucosamine and uronic acid residues. It is the sulfation pattern of these repeating units that generate large structural and functional diversity. The glucosamine residues can either be N-sulfated (GlcNS) or N-acetylated (GlcNAc), both of which can suffer 6-O-sulfation (GlcNS(6S) and GlcNAc(6S)). GlcNS and GlcNS (6S) can also be further 3-O-sulfated (GlcNS(3S) and GlcNS(3,6S)). The uronic acid residues that can either be GlcA or its epimer Iduronic Acid (IdoA), can also be 2-O-

sulfated (IdoA(2S) and GlcA(2S)) [6,25,26]. These sulfation and epimerisation reactions give rise to, at least, 23 different HS disaccharide structures that constitute the sulfated (S)-domains subsequently repeated through the chains. The S-domains are intercalated by N-acetylated (NA)-domains, which are enriched in less modified disaccharides, providing great variability within HS polysaccharides [25].

HS biosynthesis occurs in Golgi apparatus or at the endoplasmic reticulum (ER)-Golgi interface, and is organised in three major events: (i) GAG-protein tetrasaccharide linker assembly, through which HS are covalently attached to particular serine residues in the PG core protein; (ii) HS chains polymerisation; and (iii) structural modifications of the elongated chains [25]. The first two stages include a series of sequential steps catalysed by different glycosyltransferases. It starts with the transfer of a Xyl residue, catalysed by two O-xylosyltransferases (XYLT1 and XYLT2), followed by the addition of a Gal residue, by Galactosyltransferase-I/β4-Galactosyltransferase 7 (β4Gal-T7) and subsequent transient phosphorylation of the Xyl residue mediated by the kinase FAM20B. This last step is essential for the following reactions of assembly, as it enhances the activity of subsequent glycosyltransferases, namely, the Galactosyltransferase-II/β3-Galactosyltransferase 6 (β3Gal-T6), which will then add the second residue of Gal to the nascent polysaccharide chain [27,28]. The biosynthesis of the tetrasaccharide linker (GlcAβ1-3Gal-β1-3Gal-β1-4Xyl-β1-O-Ser) is completed once the Glucuronyltransferase I (GlcAT-I) adds a GlcA residue to the extremity of the chain, in a reaction step that occurs simultaneously with the dephosphorylation of the Xyl residue by the 2-Phosphoxylose phosphatase (XYLP) [29].

Knock-out (KO) cellular glycoengineering showed that abrogation of XYLT2 in CHO cells that do not express XYLT1, abolished HS biosynthesis. Additionally, elimination of *B4galt7* and *B3gat3* (GlcAT-I) gene expression also fully impaired GAGs biosynthesis, while the KO of the genes coding for the enzymes β3Gal-T6 and FAM20B only reduced its synthesis [30]. Koike et al. conducted silencing experiments in HeLa cells and observed great reduction of the levels of HS chains in lower XYLP expressing cells, suggesting that the dephosphorylation of xylose residues is necessary for correct tetrasaccharide linker assembly [29]. However, more recently, it was determined that the KO of *Pxylp1*, performed on CHO cells, did not alter the levels of GAGs [30]. These results indicate that the role of XYLP in the maturation of the tetrasaccharide linker might be dependent of the cellular context.

The above-mentioned enzymatic steps are common to the biosynthesis of heparin/HS and CS/DS GAG chains, while the following events dictate the biosynthesis of a particular type of GAG chains. Focusing on HS, the addition of a GlcNAc residue to the linkage tetrasaccharide initiates the polymerisation of these chains (in detriment of the polymerisation of CS chains). This reaction involves the catalytic activity of two members of the Exostosin (EXT) family, EXT-like proteins 2 and 3 (EXTL2 and EXTL3), and is followed by further elongation promoted by a hetero-oligomeric complex formed by EXT1 and EXT2 that mediates the intercalated transfer of GlcNAc and GlcA residues [31–34].

EXTL3 acts as a highly efficient α1,4-GlcNAc transferase towards mature tetrasaccharide linkers by adding the first GlcNAc to the HS chains [32]. Different in vitro and in vivo models have revealed that KO of *EXTL3* results in the abolition of HS biosynthesis, uncovering the crucial role of this enzyme in initiating the elongation of HS chains [30,35,36]. The regulatory activity of EXTL2 in this step stills raises significant doubt. EXTL2 is characterised as an α1,4-N-acetylhexosaminyltransferase, displaying dual in vitro catalytic activity by adding both GlcNAc and GalNAc residues to linker mimetics. It has been demonstrated that this glycosyltransferase cannot add GlcNAc residues to mature tetrasaccharide linker substrates [33], however it exhibits significant N-acetylglucosamine-transferase activity towards phosphorylated forms of the tetrasaccharide linker. By adding a GlcNAc residue to immature linker structures (GlcAβ1–3Galβ1–3Galβ1-4Xyl(2-O-phosphate)-β1-O-Ser), EXTL2 promotes the synthesis of phosphorylated pentasaccharides (GlcNAcα1-4GlcUAβ1–3Galβ1–3Galβ1-4Xyl(2-O-phosphate)-β1-O-Ser) that neither EXT1 nor EXT2 can further polymerise, ultimately resulting in premature HS chains termination [37].

This is in accordance with the increased HS content reported in EXTL2 KO cell models [30] and EXTL2 deficient mice [37,38].

Once polymerised, HS chains are matured by HS modifying enzymes, including *N*-Deacetylase/*N*-Sulfotransferases (NDST1-4), C5-epimerase and different Sulfotransferases (2OST, 6OSTs, 3OSTs) and sulfatases (Sulf-1 and Sulf-2) [6,39]. HS chain features are not directly encoded by the genome, showing a high level of heterogeneity and large structural diversity in terms of monomer sequence, chain length and sulfation profile, all due to post-translational modifications regulated in the Golgi [25]. Therefore, the resulting HS chains are involved in multiple biological processes, varying over different organs [40,41], stages of development [42–44] and pathologies [45,46]. HS chains sulfation and length are crucial to the roles displayed by HSPGs, as these determine the binding affinity to the respective targets. The HS sulfation degree, in particular, confers high negative charge to GAGs, prompting HSPGs to interact, in a non-covalent ionic manner, with several proteins [12].

3. Heparan Sulfate Biological Activities
3.1. In Physiology

HS are loaded with biological roles, as illustrated in Figure 2A. Acting as mediators in a multitude of regulatory mechanisms, ranging from embryonic development to ECM assembly and regulation of cell signalling [6,47,48]. HS interact with a plethora of molecular partners, including soluble proteins (growth factors, morphogens and chemokines), ECM proteins, bioactive fragments and membrane receptors, such as integrins and receptor tyrosine kinase (RTKs). HS chains also promote pathogen attachment and invasion of specific tissues by binding to numerous microorganisms, including viruses, bacteria, parasites and fungi [2,5,49–51]. Moreover, HSPGs are expressed in all main organ systems having essential roles in several biological activities like metabolism regulation, transcellular transport, cellular communication, ECM support and modulation. The classical role attributed to cell surface HSPGs was to assist as signalling co-receptor for growth factors activity, allowing a correct presentation to their cognate receptors and helping to stabilise gradients, to control the range of signalling and to protect the proteins against degradation [1]. However, it has been increasingly accepted that besides these co-receptor functions, HSPGs stand alone as key regulators of cell behaviour [52].

During embryonic development, HSPGs modulate the morphogen gradients distribution and other extracellular ligands signalling involved in the formation of the different tissue architectures [53]. In this light, the particular interaction of HS with the Hedgehog signalling pathway is very important to a proper embryonic development [54]. Similarly, HSPGs, being the most abundant PGs in basal lamina and cell surface of skeletal muscle, have been shown to regulate fibroblast growth factor (FGF), Wnt and bone morphogenetic protein pathways, fundamental for the development of skeletal structures [55].

More recently, it was revealed that SDCs can regulate calcium channels of the TRPC (transient receptor potential canonical) family, with functional consequences on the actin cytoskeleton, cell adhesion, junctions and migration. Moreover, this interaction was suggested to be evolutionary conserved and relevant for the progression of some diseases [48,56].

HSPGs are also important modulators of metabolism, as illustrated by their role in the liver mediated clearance of triglyceride-rich lipoproteins [57]. Additionally, several SDCs and GPCs have been implicated in the uptake of different forms of lipoproteins [58–60].

Given the many essential cellular and developmental processes in which HS and HSPGs are involved, it is expected that modifications in HSPG expression and structure contribute to a dysregulation in function and lead to pathological scenarios, such as cancer [12,17]. In Section 4, we address several cancer cellular features that are regulated by changes in expression, glycosylation and sulfation profiles of HSPGs, which in turn translate into cancer progression.

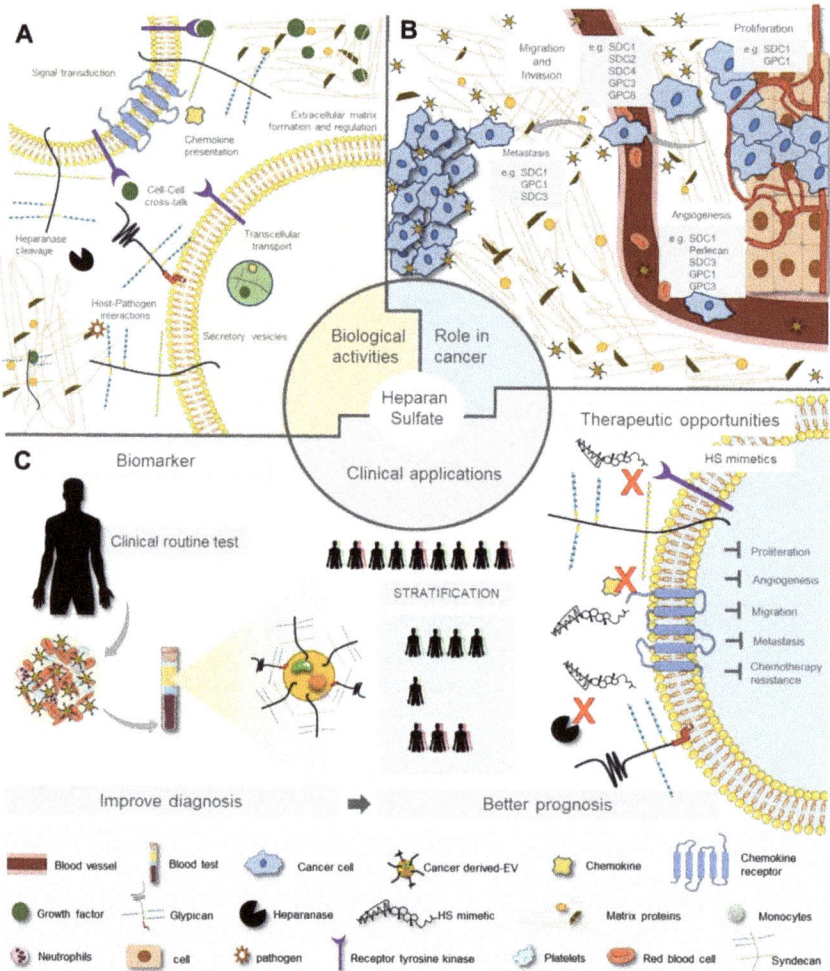

Figure 2. Overview of heparan sulfate proteoglycans functions. (**A**) HSPGs roles in cellular activities, (**B**) HSPGs aberrant expression and functional implications in cancer and (**C**) HSPGs biomedical potential as a biomarker and as a therapeutic target regarding cancer improved diagnosis and prognosis.

3.2. In Inflammation

Inflammation represents a first line protection mechanism for any harmful stimuli. Some of the major events during inflammation are regulated by HS, ranging from immune cells recruitment, adhesion and rolling, to transmigration phenomena [61]. Changes in the expression of HPSGs and HS differ depending on the type of inflammatory stimuli [62]. One of the main roles of HS and HSPGs is to drive the extravasation and migration of inflammatory cells from the vasculature into tissues, where they establish and provide cytokine gradients [63]. Moreover, HSPGs are involved in developing the basement membrane barrier, providing a structure for epithelial support and regulating the transport of solutes [61,64,65]. Perlecan, agrin and collagen XVIII in ECM and basement membrane interact with matrix proteins, such as fibronectin and laminin, and provide support, resistance to mechanical stress and filtration barrier properties [66]. Ultimately, HS chains of HSPGs bind to growth factors that are involved in tissue growth and repair, making them available at sites of tissue remodelling [67].

3.3. In Host-Pathogen Interaction

HS chains provide the gateway for many microorganisms, ranging from normal microbiota to various pathogenic bacteria, viruses and parasites, by mediating adherence to the host cells. This is a crucial step for infection to occur and pathogens exploit the host HSPGs to accomplish it and invade host cells. The previously described structural diversity of the HSPGs offers multiple binding sites and the degree of variability within tissues results in the tissue-specific tropism of some infectious agents [5,51,68,69].

Among the pathogens that bind to host HS chains are parasites like *Plasmodium falciparum* [70]; bacteria, such as *Escherichia coli* [71], *Pseudomonas aeruginosa* [72], *Borrelia burgdorferi* [73] or *Mycobacterium tuberculosis* [74]; and many viruses, amongst which are found Human Papilloma virus, Herpes viruses and Human Immunodeficiency Virus-1 [75,76]. In addition, recently it was reported that severe acute respiratory syndrome coronavirus 2 (SARS-CoV-2) entry in human cells is mediated through the binding to HS chains in a length and sequence-dependent manner [49,77–80]. The interference with the HS-mediated adhesion steps can represent an effective therapeutic approach for these pathogens and can be achieved by the competition with HS mimetics and other highly sulfated polysaccharides [81]. In addition, some pathogens can release GAGs from host cell surfaces and ECM, and use these solubilised GAGs to coat their surface, deceiving and eventually escaping immune detection [82].

4. Heparan Sulfate and Heparan Sulfate Proteoglycans Aberrant Expression in Cancer

4.1. Heparan Sulfate Changes in Cancer

Several hallmarks of cancer, such as continuous growth signalling, abrogation of apoptosis, deregulated metabolism, immune evasion and angiogenesis are boosted through pathological alterations of normal physiological processes [83–85]. There is cumulative evidence that changes in cellular glycosylation are concomitant with the acquisition of cellular features involved in tumour growth and progression and ECM remodelling. The glycosylation alterations described in cancer include the expression of truncated O-glycan structures, increased expression of branched N-glycans, de novo expression of sialylated glycans, altered fucosylation and aberrant PGs expression and modification [86,87].

HS chains are key modulators of cancer cell proliferation events, intervening in altered signalling by interacting with growth factor receptors, promoting their dimerisation and consequent activation, leading to overstimulation of downstream signalling cascades [88]. As an example, in multiple myeloma cells, SDC1 was shown to interact with HGF via HS chains, promoting enhanced activation of Met and consequent activation of the PI3K/protein kinase B and RAS-Raf MAPK pathways, which are related to cell proliferation and survival [89]. In addition, the activation of the Wnt/β-catenin cascade in multiple myeloma is also promoted by HS chains of SDC1, leading to cancer cells proliferation [90].

Besides the altered expression of HSPGs, the abnormal activity of HS biosynthetic and post-synthetic enzymatic machinery, which determines HS chains' length, epimerisation, acetylation and sulfation patterns, is also known as a major event behind HS deregulation in cancer [91]. Comparative studies demonstrated that the expression of genes coding for HS biosynthetic machinery is deregulated in several types of cancer, weighing on its role in carcinogenic events [92]. In colorectal cancer, it was shown the aberrant expression of enzymes that catalyse uronic acid structural changes (epimerisation and 2-O sulfation), and enzymes that impact glucosamine residues sulfation pattern (NDST1 and 2, 6OST isoforms 3B, 5 and 6), depending on the anatomical location and the metastatic nature of the tumours [93,94]. HS modifying enzymes were also shown to present deregulated expression on breast cancer tissue samples [46]. As for glycoenzymes that intervene in HS polymerisation, the analysis performed using estrogen receptor positive and triple negative breast cancer cell lines revealed altered expression of EXT1, EXT2, EXTL2 and EXTL3 [95].

HS 6-O-sulfation levels, determined by the expression of 6OSTs, have been reported as critical for the activation of epidermal growth factor receptor (EGFR) by heparin biding-EGF (HBEGF), and the consequent increase in the expression of angiogenic cytokines on ovarian tumour cells [96]. In lung cancer, 3OST2 hypermethylation and consequent deregulation, was associated with lung tumourigenesis and poor overall patient survival, possibly resulting from the altered HSPGs ability to interact with proteins participating in cell growth and adhesion [97]. Likewise, the hypermethylation in HS 6-O-endosulfatase *Sulf-1* promoter region downregulates its expression in gastric cancer cell lines and tissue samples [98], and the decreased levels of HS 6-O-endosulfatase associate with gastric cancer progression [99]. Conversely, the sulfatase Sulf-2 is described to be overexpressed in hepatocellular carcinomas and associated with worse prognosis [100].

Heparanase (HPSE), a β-D-endoglucuronidase, is the only mammalian enzyme known to cleave HS and is one of the most studied glycosylation-related enzymes in cancer [101,102]. This enzyme is known to be a tumour inducer acting in several signalling pathways modulating angiogenesis, cell proliferation, migration and metastasis [103–105]. HER2- and EGFR-positive breast cancer cells resistant to lapatinib, a tyrosine kinase inhibitor that blocks the activation of the EGFR and HER2 pathways, revealed increased activity of HPSE. This feature was associated with enhanced activation of EGFR, FAK and ERK1/2 signalling pathways and subsequent cell growth. HPSE inhibition, was shown to sensitise these cells to lapatinib and inhibit formation of brain metastases [106]. More recently, vascular endothelial growth factor receptor 3 (VEGFR3)-expressing macrophages and cathepsin release, both playing a significant role in metastasis formation in chemotherapy-treated tumours, were found to promote lymph angiogenesis as a result of VEGF-C upregulation by HPSE [107]. Autophagy is another cellular attribute modulated by HPSE. This catabolic pathway maintains homeostasis in normal cells, while it is completely hijacked in several tumours, promoting cancer cell survival. Autophagy induced by lysosomal HPSE has been shown to enhance tumour development and chemoresistance [108–110]. Although HPSE activity has been mainly described extracellularly or within the cytoplasm, nuclear HPSE has also been reported [111–113]. In melanoma, nuclear HPSE was shown to suppress tumour progression by competing for DNA binding and inhibiting the transcription of genes, such as those coding for ECM-degrading enzymes that promote metastasis formation [112]. In multiple myeloma disease context, HSPE was recently associated with chromatin opening and transcriptional activity concomitant with downregulation of PTEN tumour suppressor activity [111]. Also supporting the role of HPSE in tumour progression and metastasis formation, HPSE has been shown to promote EV secretion by tumour cells, affecting its protein cargo [114,115], and modulating HS structure on recipient cells to facilitate EVs internalisation [9,115,116].

4.2. Role of Heparan Sulfate and Heparan Sulfate Proteoglycans in Cancer Cellular Features and Extracellular Matrix Remodelling

Cancer cells undergo relevant morphological changes, such as the epithelial to mesenchymal transition (EMT), to increase motility capacity. HS chains play a key role in this transition, due to their binding affinity to key growth factors secreted into the tumour microenvironment [117,118]. Particularly, upregulation of SDC4, in lung adenocarcinoma, was shown to stimulate cell's migration and invasion via TGF-β1, accompanied by induction of EMT [119]. Cell proliferation is also a crucial characteristic of malignant transformation. The HSPGs GPC1 and SDC1, overexpressed in a high percentage of breast cancer pathologies, enhance the mitogen effects associated with heparin-binding growth factors like Basic Fibroblast Growth Factor (FGF2), HBEGF and Hepatocyte growth factor (HGF), promoting cell proliferation [120]. Some studies have shown that HS and HSPGs can translocate to the nucleus and contribute to gene expression regulation [121,122]. Although the role of nuclear HSPGs is still not fully uncovered, another important role of nuclear HSPGs is the translocation of specific cargo to the nucleus. Nuclear translocation of SDC1 was shown to regulate tumour signalling by shuttling growth factors to the nucleus and by altering histone acetylation [123].

Tumour progression is accompanied by the development of new blood vessels [124]. HS chains, by binding to angiogenic growth factors, namely, FGFs, platelet-derived growth factors and VEGFs, dictate HSPGs relevant roles in angiogenesis [91,125,126]. It has been shown that HS presence on endothelial-cells' surface can serve as a binding site for the potent antiangiogenic factor endostatin. Several studies have indicated that the HS binding site for endostatin is distinct from that of pro-angiogenic factors, such as FGF. This raises the possibility that endothelial cells modulate their HS cell-surface profile to become either more or less sensitive to angiogenic signals from a growing tumour [127]. As referred earlier, SDC1 overexpression in multiple myeloma was also shown to promote angiogenesis by its ability to physically interact with VEGFR2 and prevent the receptor recycling [128]. Perlecan is also an important player in angiogenesis, since its expression is abnormally high in the basement membranes of highly metastatic human melanoma tumour cells [129], promoting the binding of pro-angiogenic FGF2 to its receptors, and consequently increasing angiogenesis [130]. In addition, SDC3 expression is positively associated with angiogenesis in neuronal and brain tissues [131,132].

Beyond the structural modifications of HS and pattern of sulfation, mentioned in Section 4.1, alterations in the HS levels can compromise the stiffness of the ECM, thus modulating cell adhesion and migration. A steady ECM does not offer the best conditions for cell migration, preventing or delaying cell motility. In this light, the ability of cancer cells to invade the surrounding tissues is modulated by changes in the expression of HS and HSPGs, which mediate several events of cell-matrix interaction, and the secretion of HPSE and metalloproteinases that allow cells to penetrate the basement membrane and ECM to invade surrounding tissues [118,133,134]. In hepatocellular carcinoma, SDC1 and SDC4 are key for migration, invasion and increased motility mediated by chemokine-SDC interactions [135]. SDC1 abnormal expression, for example, is determinant in tumour cell growth, invasion and migration in different types of cancer, such as colorectal, gallbladder and oesophageal carcinomas [136–139]. SDC2 overexpression in breast [133,140], colon [141] and pancreatic [142] tumour cells, is associated with altered cell morphology, focal adhesion formation, spreading, enhanced migration and invasion capabilities, and overall to a more aggressive tumour cell behaviour and disease progression [143]. GPCs are also frequently reported to play a part in cancer progression. GPC1, for instance, when upregulated, increases tumour angiogenesis and metastasis in pancreatic cancer [144,145]. In addition, GPC1 modulates heparin-binding growth factors and plays a role in tumour progression in breast cancer [46,120]. In esophageal squamous cell carcinoma and glioblastomas, GPC1 is also upregulated and associated with tumour angiogenesis and patients' poor prognosis [146–148]. GPC3 is overexpressed in hepatocellular carcinomas tissues [149], and associates with higher invasion and migration [150]. Similar to GPC1, an increased expression of GPC6 has been reported in breast cancer. GPC6 overexpression stimulates cancer invasion through NFAT (nuclear factor of activated T-cells) signalling pathway–previously reported as an inducer of pro-invasion and migration gene expression [151].

Cancer cells ability to penetrate blood vessels is preponderant for metastatic spread and is followed by circulation through the intravascular stream and establishment in other sites [152]. During the process of metastasis formation, the reorganisation of HSPGs in the ECM, creates an opportunity for new partners to bind to tumour stroma. This process also involves interactions between cancer cells and platelets, endothelial cells and host organ cells, being HS implicated in the formation of tumour metastasis in sites, such as the liver, lungs or spleen [91,153]. Moreover, SDC1 expression was shown to decrease in hepatocellular [154] and colorectal [155] carcinomas, resulting in more invasive phenotypes, with higher metastatic potential.

HS can also contribute for the immune system deceiving to either disregard or promote the tumour growth [156]. In breast cancer for example, SDC1 has been suggested to act both as a regulator of cancer stem cell (CSC) phenotype and as a modulator of lymphocytes, in particular of T helper cells, depending on the subtype of the disease [157].

A schematic representation of HSPG functional implications in cancer progression and their clinical potential is illustrated in Figure 2B,C.

4.3. Heparan Sulfate Roles in Cancer Intercellular Communication

Cancer cells communication within the tumour microenvironment is key to defeat stromal challenges, settle and colonise distant sites, leading to metastasis. Despite being the main cause of cancer therapy failure and responsible for the greatest number of cancer-related deaths, metastasis remains poorly understood [158]. It is widely recognised that the process of cancer cell systemic circulation and the development of metastasis requires the participation of several glycoconjugates [86,159].

For years, EVs were thought to be a reservoir of cells undesired material. However, in the last two decades, they have emerged as main players in cellular communication [160,161]. EVs are delimited by a lipid-bilayer and can be classified into different classes, including exosomes, microvesicles and apoptotic bodies. Briefly, EVs are secreted to the environment by all cells and carry bioactive cargo that deliver signals and induce several pathophysiological events in the ECM and recipient cells [162,163]. Regarding cancer, EVs have been demonstrated as important signalling nanoparticles in pre-metastatic niche definition and metastasis [164,165].

The EV cargo includes nucleic acids, proteins, lipids and metabolites [166]. Although several seminal studies have addressed in detail the lipid, protein and nucleic acid contents of EVs, the glycans, and particularly GAGs, remain poorly characterised. However, their biological importance is emerging [160,167,168].

Importantly, HSPGs have been described as key regulators of EVs biogenesis (Figure 3) [162,169]. The biogenesis of EVs depends on the small intracellular adaptor syntenin [170], its interaction with SDC [169] and the endosomal-sorting complex required for transport accessory component ALIX [171–173]. Moreover, HPSE can stimulate intraluminal budding of SDC-syntenin-ALIX complex promoting EVs secretion [174–176]. Recently, it was described that tetraspanin-6 (TSPN6) may act as a negative regulator of exosomes release through the promotion of SDC4 and syntenin degradation. This interaction highlights the importance of the interplay between these membrane glycoproteins to produce exosomes [10].

Furthermore, it has been demonstrated that HSPGs, namely, SDCs and GPCs, are critical internalising receptors of cancer cell-derived EVs and determine their functional activity (Figure 3) [116,160]. Very recently, it was shown that under hypoxia stress, the uptake of EVs is upregulated, through a mechanism dependent on HSPG receptors and lipid raft mediated endocytosis [177].

EVs exhibit several distinctive features, from a longer half-life provided by increased resistance to degradation, therefore offering their cargo a higher stability, to the ability of travelling long distances. Furthermore, EVs can carry multiple cargo possibilities and also exhibit a unique interactive surface area [178], which may establish contact with both cells and components in the ECM microenvironment [179]. HSPGs are herein important mediators with several functions from EV secretion and trafficking to their uptake [60,116,180].

Taking together glycan and EVs functional relevance in cancer development, it is not surprising that glycans in EVs have been implicated in cancer cells proliferation, angiogenesis, therapeutic resistance [181], control of metabolic activity, and immune system evasion mechanisms [160].

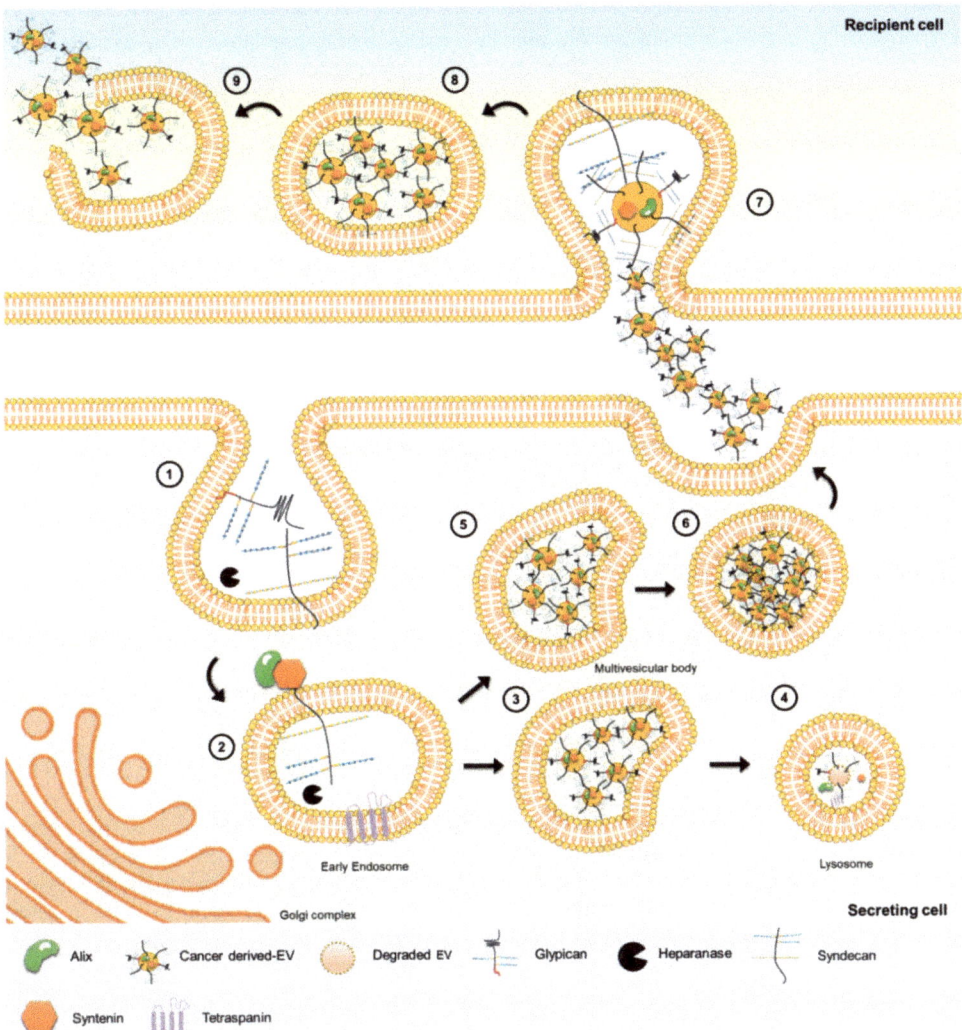

Figure 3. Heparan sulfate proteoglycans regulate EV biogenesis and uptake. (**1**) Cell surface HSPGs can bind multiple ligands through their GAG chains [9]. GAGs can be modified by heparanase activity [174]. (**2**) Syndecan is internalised through endocytosis process, leaving the cytosolic domain clear for syntenin and Alix proteins binding [160,171]. The early endosomes generate the MVBs by inward budding of their membrane. (**3**) EVs, particularly those enriched in tetraspanin-6 (TSPN6), can end on lysosome [10] (**4**) with consequently degradation of their content. (**5**) Alternatively, EVs generated inside of the MVBs can be expelled from the secreting cell, through exocytosis. (**6**) After fusion with cellular membrane EVs are released to the extracellular milieu. (**7**) HSPGs, and specifically GAGs, are important receptors of the cell membrane-EV surface cluster and are directly involved in EV uptake by recipient cell [116]. (**8**) After, the complex is internalised by the recipient cell. (**9**) EV-endosome membrane fusion occurs, and EV content is released to cytoplasmic compartment of the recipient cell, and new biological information is transferred [162].

4.4. Impact of Heparan Sulfate in Cancer Cell Resistance to Therapy

Cancer treatment relies upon four main approaches: surgery, radiation therapy, chemotherapy and immunotherapy. Some individuals will only require one treatment, but most often, a combination of treatments is used to tackle the resistant nature of cancer. Surgery can be used for solid tumours that are located in reachable areas of the body. Nevertheless, many cancers are metastatic or have a high risk for metastasis formation, implying the use of more aggressive treatments, such as radiotherapy and chemotherapy [182].

A great body of evidence indicates that tumour sensitivity to drug treatment is affected by glycosylation, particularly by altered expression of cell-surface HSPGs and/or HPSE [17,87]. GPCs and SDCs are usually implicated in chemo-resistance. GPC3 overexpression has been described to lead to a decrease in the accumulation of drugs associated with atypical multidrug resistance in gastric cancer [183], and high levels of GPC1 expression in patients with oesophageal squamous cell carcinoma are related to chemo-resistance [146]. As previously mentioned, SDC1 overexpression correlates with a malignant phenotype and, in addition, it is also implicated in resistance to cytotoxic or targeted therapeutics in breast cancer and multiple myeloma [157,184–188]. SDC1 levels in pre-chemotherapy breast cancer biopsies correlate with decreased response to treatment with cyclophosphamide and epirubicin [189]. Additionally, the sensitivity of breast cancer cells to trastuzumab is associated with the availability of HS chains on the cell surface and their ability to elicit the antibody response by forming a ternary complex with trastuzumab and HER2 [190].

Cancer cells resistance occurs not only to chemo, but also to radiotherapy. For example, in pancreatic cancer, HPSE was found to be overexpressed throughout the process of ionising radiotherapy, resulting from the downregulation of the transcription factor EGR1, which leads to the upregulation of HPSE and promotes tumour cells invasion [191]. In cervical cancer, HPSE expression was shown to enhance angiogenesis and radiation resistance through the hypoxia-inducible factor 1 (HIF1) pathway [192]. In addition, the interplay between SDC1 and HPSE, through indirect stimulation of HSPG shedding by metalloproteinases, and consequent activation of HS-binding growth factor signalling, was suggested to associate with colorectal cancer cells resistance to chemotherapy [193]. Clinical drugs used for myeloma showed to induce SDC1 shedding, due to the upregulation of HPSE expression. Then, HPSE can be internalised by both tumour cells and macrophages, promoting the transcription of pro-tumourigenic genes. This paradox effect leads to tumour recurrence by recreating a new cancer microenvironment, which induces chemo-resistance [188,194]. On the other hand, as referred to in Section 4.1, HPSE can modulate EVs cargo and enhance its secretion. In breast cancer models, it was shown that in a chemotherapy context, the production of EVs is increased, with upregulation of the levels of annexin A6 and promotion of metastasis [195].

It is important to note that several enzymes are involved in GAGs remodelling, and alterations on their activity can lead to the activation of compensatory routes. The variable levels of HS, CS/DS and HA over each other can lead to unusual GAG profiles, which need to be considered when a GAG-target therapeutic approach is being evaluated, since these biomolecules compete for some common substrates. [196].

5. Heparan Sulfate Clinical Applications–Recent Advances

5.1. Adding Heparan Sulfate to the Equation

The knowledge of the molecular mechanism underlying cancer biology has mightily increased over the last 30 years. The current challenge is to translate this information into benefits for patient care and to convert the new molecular information into therapy and better non-invasive biomarkers.

One of the most obvious cancer-related challenges is the heterogeneity in cancer biology [197]. Again, the challenge is the molecular and cellular heterogeneity of a single tumour, and among tumours from different patients, resulting in the very difficult task to design and select effective therapies without promoting treatment resistance [198,199]. Hereupon, targeting HSPGs and enzymes involved in HS chain editing, as a new anticancer strategy, seems to be a sweet spot of opportunity to deal with these challenges (Figure 2C).

Several different therapies using GAG-based strategies have been reported, and a few HS-specific treatments are on clinical trials [200]. Many new approaches to cancer treatment are emerging, and targeting HS constitutes a novel promising strategy for cancer clinical management [102,201]. These strategies include inhibition of tumour invasion and metastasis using non-anti-coagulant low-molecular-weight heparin (LMWH) analogues and inhibition of tumour progression using HS-mimetics. In animal cancer models, LMWHs were shown to inhibit metastases, diminish primary tumours and increase survival. However, pre-clinical studies have conflicting findings showing an absence of efficacy in reducing disease progression [202].

Likewise, the HPSE contribution to immune regulation is raising clinical interest. Among HS modifying enzymes, HPSE is definitely one of the most investigated as a cancer drug target [105,175,185,188,203,204]. Early observations of the powerful HPSE inhibitory activity of heparin lead the way for the screening of heparin/HS mimetics as HPSE inhibitors [102,205–208].

Noteworthy, substantial research has been devoted to elucidating the roles that EVs play in the regulation of both normal and pathological processes, and recently multiple studies have demonstrated their potential as a source of cancer biomarkers referred to as "liquid biopsies" [160,209]. Biomarkers for early detection of cancer are essential to improve patients' clinical management. Based on several reports showing proof-of-concept results for EV-associated pancreatic cancer biomarkers, GPC1 was the first candidate to enter clinical trials, aiming to evaluate its performance as a biomarker [210]. Consequently, various EV-associated biomarkers have been reported for early detection, diagnosis, treatment monitoring, metastasis burden, prediction and prognosis in cancer patients [211].

Along the next sections, we present some representative examples of HS-related mimetics, therapeutic targets, and biomarkers recently studied, under clinical trials evaluation or already introduced in medical practice.

5.2. Heparan Sulfate-Based Therapeutic Opportunities

Based on promising preclinical data, some HS-based therapies are currently under clinical investigation although none was yet approved [212–214]. Difficulties in interpreting data with HS mimetics are mainly due to their pleiotropic effects [160,206,207].

Currently, leukocyte-based anticancer therapies like Chimeric Antigen Receptor-T (CAR-T) cell therapy, dendritic cell vaccines and viral-therapeutic delivery exploiting HPSE are being developed [102,215]. HPSE is able to change the function of HSPGs in the tumour microenvironment and is a regulator of several cancer hallmarks [216], namely, angiogenesis and the development of metastasis [105,217]. So, targeting HPSE and modulating its activity is a very valuable strategy to overcome several cancer features and improve disease outcome [102,218].

Another possible opportunity for intervention could be the use of the endostatin domain of collagen XVIII. Endostatin exerts an efficient inhibitory effect on tumour angiogenesis and growth [219,220]. Gastric cancer patients with subsequent liver metastasis are extensively studied, and several phase II clinical trials are using recombinant endostatin (either alone or combined with other drugs) as an anti-tumour agent [221–223]. In addition, the effect of endostatin in advanced well-differentiated pancreatic neuroendocrine tumours is being tested [224].

Synthetic peptides that interfere with the interaction between HS and its binding partners, also represent an appealing alternative [225]. This approach allows a specific blocking of interaction with the growth factors, acting on particular pathways. Furthermore, a large number of small molecules can also regulate glycosylation by modulating glycosyltransferases and glycosidases activity [218,226] and are emerging as new therapeutic strategies [226–228].

Recently, another class of enzymes have been capturing attention: the 3-*O*-sulfotransferases (3OSTs). These enzymes produce a rare HS modification, glucosaminyl-3-*O* sulfation, which affects the selective binding of several ligands. Interestingly, these enzymes can either act as pro or anti-tumourigenic according to the cellular specific context [229,230]. An important challenge in this field is to determine how the HS glycan sequence and sulfation pattern drive ligand binding specificity.

Furthermore, the development of specific HSPG targeted therapeutic approaches is also being studied. Antibody-targeting of SDC1 has been used alone or combined with chemotherapy to treat multiple myeloma [231], and a clinical trial using CAR-T cells recognising SDC1 suggests that the treatment is safe, well-tolerable and has potential antitumour activity [232]. In the same line, in hepatocellular carcinoma, the strategies of immunotoxin and CAR-T cells against GPC3 are showing promising results and are under clinical trials [233–235].

5.3. Heparan Sulfate Mimetics Development and Clinical Trials

5.3.1. Heparan Sulfate Mimetics Rational

Heparin derivatives and HS mimetics are drawing great attention for developing new therapeutics for diverse diseases, from inflammation to neurodegenerative disorders and cancer [207]. HS mimetics overcome a problem known for more than half a century in Medicine related to heparin use. Heparin is an anticoagulant drug which has been widely used and remains one of the main drugs for prophylaxis and treatment of thrombosis [236]. Thrombosis is a common complication of cancer patients. The use of heparin has improved the survival rate of cancer patients [237]. The functional roles of heparin seem far more than anticoagulation, since a number of additional beneficial effects have been observed for heparin in other diseases than thrombosis [238]. However, heparin is an animal-derived heterogeneous polysaccharide, therefore the potential risk of contamination and its complicated molecular structure restrict its use. Therefore, and taking in consideration the high structural similarities between heparin and HS, it was tested whether heparin interferes with HS interactions with its ligands mainly through the hamper of angiogenic growth factors, like VEGF and FGF, selectins, and HPSE [239,240]. Indeed, in vitro analyses have shown that heparin is able to inhibit HPSE activity [240]. Since it is important to restrict heparin anticoagulant activity, due to haemorrhagic issues, efforts were made to eliminate the anticoagulation activity by chemical modification.

Roneparstat (SST0001) was the first synthetic product based on this premise. SST0001 is a 100% *N*-acetylated and glycol split heparin synthetic molecule [241]. The SST0001 HPSE inhibitory effect was confirmed on myeloma cell growth, and therefore, it is being evaluated for the treatment of multiple myeloma [242].

The principle of using HS mimetics is to interfere with the interactions between HS and its molecular partners (Figure 2C). There are two types of HS mimetics: (1) Synthetic saccharide-based HS assembled from a backbone sugar structure; and (2) non-sugar scaffold negatively charged with sulfates, sulfonates, carboxylates and/or phosphates [205]. SST0001 and other HS-mimicking compounds main attribution is to inhibit HPSE and compete for HS binding with several growth factors, having impact on cancer by preventing angiogenic and metastatic events [218,243]. In summary, HS mimetics have been shown to enhance antitumour effects, particularly when combined with standard therapies.

5.3.2. Promising Heparan Sulfate Mimicking Molecules

The most promissory HS mimetics that are on clinical trials are: Highly sulfated phosphosulfomannan muparfostat (PI-88) [244], 2,3-*O*-desulfated heparin CX-01 (ODSH) [245], SST0001 [246] and pixatimod (PG545) [214]. The latter was selected from oligosaccharidic HS mimetics of the PG500 series as the best candidate regarding cancer treatment applications [247].

PI-88 is an HPSE inhibitor and also an antagonist of HS-protein interactions. Structurally it is a phospho-mannopentaose obtained through a process of sulfation of a phosphomannan complex produced by yeasts [243,248]. It was the first one entering clinical trials and, generally, phase I/II studies demonstrated a satisfactory pharmacodynamic profile of this mimetic that was also considered safe and well-tolerated, showing minor anticoagulant effects. As main results, PI-88 has proven to be a suitable candidate as an adjuvant for postsurgical hepatocellular carcinoma in phase II clinical trials [249,250]. Subsequently, it reached Phase III in clinical trials regarding large series of liver cancer patients after hepatectomy. In fact, it is currently waiting for approval to enter routine clinical use [250]. Patients with advanced melanoma were recruited for a Phase I and Phase III clinical trials and PI-88 activity was beneficial [251]. Overall, PI-88 showed encouraging results for melanoma, multiple myeloma, prostate and lung cancer treatment [205].

CX-01 is a low anticoagulant 2-O, 3-O desulfated heparin derived from porcine intestinal heparin retaining many anti-inflammatory properties. It has been shown to have a particular potential for acute myeloid leukaemia (AML) treatment [245]. This molecule inhibits leukemic stem cells to concentrate on the bone marrow, therefore enhancing chemotherapy treatments. CX-01 specifically binds chemokine platelet factor 4 (PF4), which is responsible for the negative regulation of megakaryopoiesis [252]. Therefore, by interfering with this process, CX-01 is able to diminish chemotherapy-induced thrombocytopenia. Phase I clinical trial on AML patients demonstrated CX-01 to be well-tolerated and more recently, it entered Phase II and showed promising results in combination with the standard chemotherapy treatment [253].

Preclinical models with SST0001 shown the effective inhibition of myeloma growth in vivo [242]. SST0001 progressed to a Phase I open-label clinical trial design to assess the safety and tolerability profile of this compound in patients with multiple myeloma [254]. Recently, SST0001 was advised for Phase II evaluation [254].

The HS-mimicking molecule with the best clinical evaluation so far is PG545 [255]. PG545 is a fully sulfated glucopyranose tetrasaccharide particularly designed with a hydrophobic 3-cholestanyl group [247,255]. The novel characteristic of this molecule is the stimulation of innate immune cell response to tumours. This stimulation is done via activation of natural killer cells [256]. Clinical trials proved PG545 as a satisfactory alternative in patients with advanced solid malignancies in which standard therapies failed [247]. Currently, PG545 is also being investigated as a potential inhibitor of the SARS-CoV-2 [257].

In addition to the role of HS as an emerging class of molecules for therapeutic strategies, HS glycan chains also constitute important cellular markers. These properties make HS important targets for vaccines development strategies [258].

5.4. Current Heparan Sulfate-Based Biomarkers Landscape

New biomarkers to improve cancer diagnosis are needed, thus improving patient outcomes. In contrast to RNA, DNA and protein synthesis, HS biosynthesis is not a template-driven process. Instead, as mentioned previously, HS are assembled by the activity of a series of enzymes–turning this biosynthetic pathway into a source of very distinct and specific modifications with diverse applications. The type of HSPG glycosylation was shown to affect the ability of immune cells to infiltrate tumour tissues and engage in the immune response [259]. In addition to the critical roles in multiple aspects of tumour biology, described in the previous sections, HSPGs also have value for clinical diagnosis and prognosis in various cancer types. The alteration of GAG abundances is reflected in body fluids (like blood and/or urine), and the HS levels in plasma can, therefore, predict patient's prognosis [260]. For all these attributes, HS and HSPGs are promising new biomarkers in cancer since their recognition by other molecules is based in high affinity and exquisite specificity. It is imperative to discriminate among related isomers the specific glycan-binding partner. The development of more sophisticated equipment and techniques have been critical to surpass these limitations. Recently, using Raman micro-spectroscopy it was possible to determine unique and discrete HS profiles of individual live cells, which can be of value for clinical screening purposes. This method can be utilised for identifying specific molecular signatures of HS and HSPGs as markers of cancer [261].

GPCs have been receiving great attention from the research community as promising biomarkers [262]. GPC1 is overexpressed in several cancer types [263,264]. GPC1 can be detected in the urine of prostate cancer patients [265]. Further, GPC1 was shown to associate with the dissemination levels of glioblastoma [148]. The abundance of GPC1 also positively correlates with disease severity of pancreatic cancer patients, independently on surgical treatment, suggesting that GPC1 is a surgery-independent diagnostic biomarker [263,266]. Recently, GPC3 was pointed as a promising candidate for hepatocellular carcinoma diagnosis and immunotherapy, as previously described [267]. Similarly, GPC6 was identified as a putative biomarker for the metastatic progression of cutaneous melanoma, since it is possible to track higher levels of GPC6 in melanoma samples when compared with normal melanocytes [268].

The biological features of EVs, such as long half-life and physical resistance properties, make EVs a unique source of biomarkers [167]. Particularly, the identification of specific PGs in EVs secreted by cancer cells has demonstrated their potential as biomarkers for minimally invasive diagnosis. Two important examples are GPC1 in pancreatic cancer [266] and SDC1 in glioma [269]. Both HSPGs were detected in EVs isolated from patients' plasma, supporting the concept of a minimally invasive biomarker for patients' stratification. GPC1 was proven to distinguish healthy individuals and patients with a benign pancreatic disease from patients with early- and late-stage pancreatic cancer. Moreover, GPC1-positive EV levels correlated with tumour burden and the patients' survival and were shown as a prognostic marker superior to CA 19-9, the serum biomarker currently employed in pancreatic cancer screening [266]. SDC1 was shown to discriminate between high-grade glioblastoma multiforme and low-grade glioma, and therefore, with the potential to improve the management of brain tumour patients that present high risk of surgery-associated complications [269].

The recent development of EV analysis platforms based on microfluidics technologies holds promise for the development of high sensitivity and high-throughput assays of EVs analysis with clinical diagnosis purpose [270].

6. Conclusions and Future Challenges

The emerging HSPGs biological functions have largely surpassed their classical role as cellular co-receptors and have highlighted HSPGs as main maestros of cancer cell communication and ECM structuring. The HSPGs cellular interactome is vast and is fine-tuned by the biochemical and structural features of the HS chains. HS lack a template for its biosynthesis, and their structural features result from the dynamic cellular GAGosylation pathways that include the sequential, and in some cases competitive, action of specific enzymes, which may associate to form supramolecular complexes [25]. The structural efforts to produce chemically defined HS oligosaccharides [271] have been crucial for the identification of the molecular determinants of enzymatic activity. However, we are still far from fully understanding the complete regulation of HS structural and functional diversity in health and particularly in cancer.

The recent developments in glycosaminoglycanomics, namely, on the analytical techniques for GAGs profiling in cells and clinical samples (tissues and biological fluids), together with the establishment of computational tools for mining GAG-protein interactions and creation of databases, are contributing to significantly improve the knowledge on the human glycosaminoglycome [19,272]. Moreover, the integration of HS structural features and expression with proteomic and transcriptomic analysis will be crucial for further elucidating HS-ligand interactions and unravel HS structure features associated with specific biological functions [273]. Particularly relevant would be the integration of data on relative abundance and structural features of HS for the definition of cancer-specific profiles. Therefore, it is important to apply the most recent GAG analytical approaches to well characterised clinical samples to identify HS signatures that are cancer-specific. To successfully achieve this aim, it is key to incorporate also knowledge on the dysregulation of HS biosynthetic and post-synthetic modification pathways, as well as on the functional redundancy of different PG core protein families in cancer.

The multidimensional roles of HS and HSPGs in different steps of cancer progression have propelled the development of HS-targeted strategies for cancer diagnosis and treatment [200]. Indeed, in this new period of precision oncology, HS GAGs are currently (un)expectedly emerging as allies to improve cancer clinical management by their potential to detect cancer in early stages, allowing an accurate diagnosis, disease monitoring, patients stratification and improve prognosis.

The HSPGs role in regulating EV release, cargo and uptake is well defined [160], but the implications of altered PG expression and glycosylation features in EV biodistribution and metastasis tropism remain to be discovered. It has become highly relevant to understand the impact of HSPG remodelling, both at the level of the glycan structures and core protein expression, in cancer EV-mediated signalling. Understanding the glycosylation modifications involved in EV-cell interaction and cellular uptake is of major relevance for developing therapeutic approaches targeting EV-HSPG interactions as novel cancer treatment strategies.

Author Contributions: Conceptualization I.F.-R., J.P., C.M., and A.M.; writing—original draft preparation I.F.-R., J.P., C.M., J.S.-A. and A.M.; Figures preparation I.F.-R., J.P. and C.M.; writing—review G.M., C.A.R., A.M.; funding acquisition, A.M. All authors have read and agreed to the published version of the manuscript.

Funding: This work was financed by FEDER-Fundo Europeu de Desenvolvimento Regional funds through the COMPETE 2020-Operacional Programme for Competitiveness and Internationalization (POCI), Portugal 2020, and by Portuguese funds through FCT-Fundação para a Ciência e a Tecnologia/Ministério da Ciência, Tecnologia e Inovação in the framework of the project "Institute for Research and Innovation in Health Sciences" (POCI-01-0145-FEDER-007274) and by the grant POCI-01-0145-FEDER-028489 (to A.M.). J.P. and C.M. are funded by FCT PhD scholarships SFRH/BD/137319/2018 and 2020.06412.BD, respectively. The APC was funded by FCT grant POCI-01-0145-FEDER-028489. The authors acknowledge the support of the COST Action CA18103 INNOGLY.

Data Availability Statement: Not applicable.

Conflicts of Interest: The authors declare no conflict of interest. The funders had no role in the design and writing of the manuscript, or in the decision to publish.

References

1. Bishop, J.R.; Schuksz, M.; Esko, J.D. Heparan sulphate proteoglycans fine-tune mammalian physiology. *Nat. Cell Biol.* **2007**, *446*, 1030–1037. [CrossRef] [PubMed]
2. Chang, Y.-C.; Wang, Z.; Flax, L.A.; Xu, D.; Esko, J.D.; Nizet, V.; Baron, M.J. Glycosaminoglycan Binding Facilitates Entry of a Bacterial Pathogen into Central Nervous Systems. *PLoS Pathog.* **2011**, *7*, e1002082. [CrossRef] [PubMed]
3. Esko, J.D.; Linhardt, R.J. Proteins That Bind Sulfated Glycosaminoglycans. In *Essentials of Glycobiology*; Varki, A., Cummings, R.D., Esko, J.D., Stanley, P., Hart, G.W., Aebi, M., Darvill, A.G., Kinoshita, T., Packer, N.H., Eds.; Cold Spring Harbor Laboratory Press: Cold Spring Harbor, NY, USA, 2015; pp. 493–502.
4. Esko, J.D.; Selleck, S.B. Order Out of Chaos: Assembly of Ligand Binding Sites in Heparan Sulfate. *Annu. Rev. Biochem.* **2002**, *71*, 435–471. [CrossRef] [PubMed]
5. Nizet, V.; Esko, J.D. Bacterial and Viral Infections. In *Essentials of Glycobiology*; Varki, A., Cummings, R.D., Esko, J.D., Freeze, H.H., Stanley, P., Bertozzi, C.R., Hart, G.W., Etzler, M.E., Eds.; Cold Spring Harbor Laboratory Press: Cold Spring Harbor, NY, USA, 2009.
6. Li, J.-P.; Kusche-Gullberg, M. Heparan Sulfate: Biosynthesis, Structure, and Function. *Int. Rev. Cell Mol. Biol.* **2016**, *325*, 215–273. [CrossRef] [PubMed]
7. Afratis, N.; Gialeli, C.; Nikitovic, D.; Tsegenidis, T.; Karousou, E.; Theocharis, A.D.; Pavão, M.S.; Tzanakakis, G.N.; Karamanos, N.K. Glycosaminoglycans: Key players in cancer cell biology and treatment. *FEBS J.* **2012**, *279*, 1177–1197. [CrossRef]
8. Lindahl, U.; Couchman, J.; Kimata, K.; Esko, J.D. Proteoglycans and Sulfated Glycosaminoglycans. In *Essentials of Glycobiology*, 3rd ed.; Cold Spring Harbor Laboratory Press: Cold Spring Harbor, NY, USA, 2015.
9. Couchman, J.R.; Multhaupt, H.; Sanderson, R.D. Recent Insights into Cell Surface Heparan Sulphate Proteoglycans and Cancer. *F1000Research* **2016**, *5*, 1541. [CrossRef]

10. Ghossoub, R.; Chéry, M.; Audebert, S.; Leblanc, R.; Egea-Jimenez, A.L.; Lembo, F.; Mammar, S.; Le Dez, F.; Camoin, L.; Borg, J.-P.; et al. Tetraspanin-6 negatively regulates exosome production. *Proc. Natl. Acad. Sci. USA* **2020**, *117*, 5913–5922. [CrossRef]
11. Lawrence, R.; Brown, J.R.; Lorey, F.; Dickson, P.I.; Crawford, B.E.; Esko, J.D. Glycan-based biomarkers for mucopolysaccharidoses. *Mol. Genet. Metab.* **2014**, *111*, 73–83. [CrossRef]
12. Da Costa, D.S.; Reis, R.L.; Pashkuleva, I. Sulfation of Glycosaminoglycans and Its Implications in Human Health and Disorders. *Annu. Rev. Biomed. Eng.* **2017**, *19*, 1–26. [CrossRef]
13. Brown, J.R.; Crawford, B.E.; Esko, J.D. Glycan Antagonists and Inhibitors: A Fount for Drug Discovery. *Crit. Rev. Biochem. Mol. Biol.* **2007**, *42*, 481–515. [CrossRef]
14. Köwitsch, A.; Zhou, G.; Groth, T. Medical application of glycosaminoglycans: A review. *J. Tissue Eng. Regen. Med.* **2018**, *12*, e23–e41. [CrossRef] [PubMed]
15. Lanzi, C.; Yates, E.A.; Cassinelli, G. Editorial: Heparan Sulfate Proteoglycans and Their Endogenous Modifying Enzymes: Cancer Players, Biomarkers and Therapeutic Targets. *Front. Oncol.* **2020**, *10*, 195. [CrossRef]
16. Vitale, D.; Katakam, S.K.; Greve, B.; Jang, B.; Oh, E.; Alaniz, L.; Gotte, M. Proteoglycans and glycosaminoglycans as regulators of cancer stem cell function and therapeutic resistance. *FEBS J.* **2019**, *286*, 2870–2882. [CrossRef] [PubMed]
17. Hassan, N.; Greve, B.; Espinoza-Sánchez, N.A.; Gotte, M. Cell-surface heparan sulfate proteoglycans as multifunctional integrators of signaling in cancer. *Cell. Signal.* **2020**, *77*, 109822. [CrossRef] [PubMed]
18. Raman, K.; Kuberan, B. Chemical Tumor Biology of Heparan Sulfate Proteoglycans. *Curr. Chem. Biol.* **2010**, *4*, 20–31. [CrossRef] [PubMed]
19. Ricard-Blum, S.; Lisacek, F. Glycosaminoglycanomics: Where we are. *Glycoconj. J.* **2016**, *34*, 339–349. [CrossRef] [PubMed]
20. Compagnon, I.; Schindler, B.; Renois-Predelus, G.; Daniel, R. Lasers and ion mobility: New additions to the glycosaminoglycanomics toolkit. *Curr. Opin. Struct. Biol.* **2018**, *50*, 171–180. [CrossRef]
21. Bonnans, C.; Chou, J.; Werb, Z. Remodelling the extracellular matrix in development and disease. *Nat. Rev. Mol. Cell Biol.* **2014**, *15*, 786–801. [CrossRef]
22. Theocharis, A.D.; Skandalis, S.S.; Gialeli, C.; Karamanos, N.K. Extracellular matrix structure. *Adv. Drug Deliv. Rev.* **2016**, *97*, 4–27. [CrossRef]
23. Song, H.H.; Filmus, J. The role of glypicans in mammalian development. *Biochim. Biophys. Acta Gen. Subj.* **2002**, *1573*, 241–246. [CrossRef]
24. Iozzo, R.V.; Schaefer, L. Proteoglycan form and function: A comprehensive nomenclature of proteoglycans. *Matrix Biol. J. Int. Soc. Matrix Biol.* **2015**, *42*, 11–55. [CrossRef] [PubMed]
25. Annaval, T.; Wild, R.; Crétinon, Y.; Sadir, R.; Vivès, R.R.; Lortat-Jacob, H. Heparan Sulfate Proteoglycans Biosynthesis and Post Synthesis Mechanisms Combine Few Enzymes and Few Core Proteins to Generate Extensive Structural and Functional Diversity. *Molecules* **2020**, *25*, 4215. [CrossRef] [PubMed]
26. Rong, J.; Habuchi, H.; Kimata, K.; Lindahl, A.U.; Kusche-Gullberg, M. Substrate Specificity of the Heparan Sulfate Hexuronic Acid 2-O-Sulfotransferase. *Biochemistry* **2001**, *40*, 5548–5555. [CrossRef] [PubMed]
27. Koike, T.; Izumikawa, T.; Tamura, J.-I.; Kitagawa, H. FAM20B is a kinase that phosphorylates xylose in the glycosaminoglycan–protein linkage region. *Biochem. J.* **2009**, *421*, 157–162. [CrossRef] [PubMed]
28. Wen, J.; Xiao, J.; Rahdar, M.; Choudhury, B.P.; Cui, J.; Taylor, G.S.; Esko, J.D.; Dixon, J.E. Xylose phosphorylation functions as a molecular switch to regulate proteoglycan biosynthesis. *Proc. Natl. Acad. Sci. USA* **2014**, *111*, 15723–15728. [CrossRef]
29. Koike, T.; Izumikawa, T.; Sato, B.; Kitagawa, H. Identification of Phosphatase That Dephosphorylates Xylose in the Glycosaminoglycan-Protein Linkage Region of Proteoglycans. *J. Biol. Chem.* **2014**, *289*, 6695–6708. [CrossRef]
30. Chen, Y.-H.; Narimatsu, Y.; Clausen, T.M.; Gomes, C.; Karlsson, R.; Steentoff, C.; Spliid, C.B.; Gustavsson, T.; Salanti, A.; Persson, A.; et al. The GAGOme: A cell-based library of displayed glycosaminoglycans. *Nat. Methods* **2018**, *15*, 881–888. [CrossRef]
31. McCormick, C.; Duncan, G.; Goutsos, K.T.; Tufaro, F. The putative tumor suppressors EXT1 and EXT2 form a stable complex that accumulates in the Golgi apparatus and catalyzes the synthesis of heparan sulfate. *Proc. Natl. Acad. Sci. USA* **2000**, *97*, 668–673. [CrossRef]
32. Kim, B.-T.; Kitagawa, H.; Tamura, J.-I.; Saito, T.; Kusche-Gullberg, M.; Lindahl, U.; Sugahara, K. Human tumor suppressor EXT gene family members EXTL1 and EXTL3 encode 1,4- N-acetylglucosaminyltransferases that likely are involved in heparan sulfate/ heparin biosynthesis. *Proc. Natl. Acad. Sci. USA* **2001**, *98*, 7176–7181. [CrossRef]
33. Kitagawa, H.; Shimakawa, H.; Sugahara, K. The Tumor Suppressor EXT-like GeneEXTL2Encodes an α1, 4-N-Acetylhexosaminyltransferase That TransfersN-Acetylgalactosamine andN-Acetylglucosamine to the Common Glycosaminoglycan-Protein Linkage Region. *J. Biol. Chem.* **1999**, *274*, 13933–13937. [CrossRef]
34. Kim, B.-T.; Kitagawa, H.; Tanaka, J.; Tamura, J.-I.; Sugahara, K. In Vitro Heparan Sulfate Polymerization. *J. Biol. Chem.* **2003**, *278*, 41618–41623. [CrossRef] [PubMed]
35. Takahashi, I.; Noguchi, N.; Nata, K.; Yamada, S.; Kaneiwa, T.; Mizumoto, S.; Ikeda, T.; Sugihara, K.; Asano, M.; Yoshikawa, T.; et al. Important role of heparan sulfate in postnatal islet growth and insulin secretion. *Biochem. Biophys. Res. Commun.* **2009**, *383*, 113–118. [CrossRef] [PubMed]

36. Lee, J.-S.; von der Hardt, S.; Rusch, M.A.; Stringer, S.E.; Stickney, H.L.; Talbot, W.S.; Geisler, R.; Nusslein-Volhard, C.; Selleck, S.B.; Chien, C.-B.; et al. Axon Sorting in the Optic Tract Requires HSPG Synthesis by ext2 (dackel) and extl3 (boxer). *Neuron* **2004**, *44*, 947–960. [CrossRef] [PubMed]
37. Nadanaka, S.; Zhou, S.; Kagiyama, S.; Shoji, N.; Sugahara, K.; Sugihara, K.; Asano, M.; Kitagawa, H. EXTL2, a Member of theEXTFamily of Tumor Suppressors, Controls Glycosaminoglycan Biosynthesis in a Xylose Kinase-dependent Manner. *J. Biol. Chem.* **2013**, *288*, 9321–9333. [CrossRef] [PubMed]
38. Purnomo, E.; Emoto, N.; Nugrahaningsih, D.A.A.; Nakayama, K.; Yagi, K.; Heiden, S.; Nadanaka, S.; Kitagawa, H.; Hirata, K. Glycosaminoglycan Overproduction in the Aorta Increases Aortic Calcification in Murine Chronic Kidney Disease. *J. Am. Hear. Assoc.* **2013**, *2*, e000405. [CrossRef] [PubMed]
39. El Masri, R.; Seffouh, A.; Lortat-Jacob, H.; Vivès, R.R. The "in and out" of glucosamine 6-O-sulfation: The 6th sense of heparan sulfate. *Glycoconj. J.* **2016**, *34*, 285–298. [CrossRef]
40. Maccarana, M.; Sakura, Y.; Tawada, A.; Yoshida, K.; Lindahl, U. Domain Structure of Heparan Sulfates from Bovine Organs. *J. Biol. Chem.* **1996**, *271*, 17804–17810. [CrossRef]
41. Warda, M.; Toida, T.; Zhang, F.; Sun, P.; Munoz, E.; Xie, J.; Linhardt, R.J. Isolation and characterization of heparan sulfate from various murine tissues. *Glycoconj. J.* **2006**, *23*, 555–563. [CrossRef]
42. Nigam, S.K.; Bush, K.T. Growth factor–heparan sulfate "switches" regulating stages of branching morphogenesis. *Pediatr. Nephrol.* **2014**, *29*, 727–735. [CrossRef]
43. Garner, O.B.; Bush, K.T.; Nigam, K.B.; Yamaguchi, Y.; Xu, D.; Esko, J.D.; Nigam, S.K. Stage-dependent regulation of mammary ductal branching by heparan sulfate and HGF-cMet signaling. *Dev. Biol.* **2011**, *355*, 394–403. [CrossRef]
44. Shah, M.M.; Sakurai, H.; Gallegos, T.F.; Sweeney, D.E.; Bush, K.T.; Esko, J.D.; Nigam, S.K. Growth factor-dependent branching of the ureteric bud is modulated by selective 6-O sulfation of heparan sulfate. *Dev. Biol.* **2011**, *356*, 19–27. [CrossRef] [PubMed]
45. Theocharis, A.D.; Vynios, D.H.; Papageorgakopoulou, N.; Skandalis, S.S.; A Theocharis, D. Altered content composition and structure of glycosaminoglycans and proteoglycans in gastric carcinoma. *Int. J. Biochem. Cell Biol.* **2003**, *35*, 376–390. [CrossRef]
46. Fernandez-Vega, I.; García-Suárez, O.; Crespo, A.; Castañón, S.; Menéndez, P.; Astudillo, A.; Quirós, L.M. Specific genes involved in synthesis and editing of heparan sulfate proteoglycans show altered expression patterns in breast cancer. *BMC Cancer* **2013**, *13*, 24. [CrossRef] [PubMed]
47. Xu, D.; Esko, J.D. Demystifying Heparan Sulfate–Protein Interactions. *Annu. Rev. Biochem.* **2014**, *83*, 129–157. [CrossRef] [PubMed]
48. Mitsou, I.; Multhaupt, H.A.; Couchman, J.R. Proteoglycans, ion channels and cell–matrix adhesion. *Biochem. J.* **2017**, *474*, 1965–1979. [CrossRef]
49. Clausen, T.M.; Sandoval, D.R.; Spliid, C.B.; Pihl, J.; Perrett, H.R.; Painter, C.D.; Narayanan, A.; Majowicz, S.A.; Kwong, E.M.; McVicar, R.N.; et al. SARS-CoV-2 Infection Depends on Cellular Heparan Sulfate and ACE2. *Cell* **2020**, *183*, 1043–1057.e15. [CrossRef]
50. Brandhorst, T.T.; Roy, R.; Wüthrich, M.; Nanjappa, S.; Filutowicz, H.; Galles, K.; Tonelli, M.; McCaslin, D.R.; A Satyshur, K.; Klein, B.S. Structure and Function of a Fungal Adhesin that Binds Heparin and Mimics Thrombospondin-1 by Blocking T Cell Activation and Effector Function. *PLoS Pathog.* **2013**, *9*, e1003464. [CrossRef]
51. Liu, J.; Thorp, S.C. Cell surface heparan sulfate and its roles in assisting viral infections. *Med. Res. Rev.* **2002**, *22*, 1–25. [CrossRef]
52. Couchman, J.R.; Gopal, S.; Lim, H.C.; Nørgaard, S.; Multhaupt, H.A. Fell-Muir Lecture: Syndecans: From peripheral coreceptors to mainstream regulators of cell behaviour. *Int. J. Exp. Pathol.* **2014**, *96*, 1–10. [CrossRef]
53. Yamaguchi, Y. Heparan sulfate proteoglycans in the nervous system: Their diverse roles in neurogenesis, axon guidance, and synaptogenesis. *Semin. Cell Dev. Biol.* **2001**, *12*, 99–106. [CrossRef]
54. Koziel, L.; Kunath, M.; Kelly, O.G.; Vortkamp, A. Ext1-Dependent Heparan Sulfate Regulates the Range of Ihh Signaling during Endochondral Ossification. *Dev. Cell* **2004**, *6*, 801–813. [CrossRef] [PubMed]
55. Rodgers, K.D.; Antonio, J.D.S.; Jacenko, O. Heparan sulfate proteoglycans: A GAGgle of skeletal-hematopoietic regulators. *Dev. Dyn.* **2008**, *237*, 2622–2642. [CrossRef] [PubMed]
56. Gopal, S.; Søgaard, P.; Multhaupt, H.A.; Pataki, C.; Okina, E.; Xian, X.; Pedersen, M.E.; Stevens, T.; Griesbeck, O.; Park, P.W.; et al. Transmembrane proteoglycans control stretch-activated channels to set cytosolic calcium levels. *J. Cell Biol.* **2015**, *210*, 1199–1211. [CrossRef] [PubMed]
57. MacArthur, J.M.; Bishop, J.R.; Stanford, K.I.; Wang, L.; Bensadoun, A.; Witztum, J.L.; Esko, J.D. Liver heparan sulfate proteoglycans mediate clearance of triglyceride-rich lipoproteins independently of LDL receptor family members. *J. Clin. Investig.* **2007**, *117*, 153–164. [CrossRef]
58. Christianson, H.C.; Belting, M. Heparan sulfate proteoglycan as a cell-surface endocytosis receptor. *Matrix Biol.* **2014**, *35*, 51–55. [CrossRef]
59. Al-Haideri, M.; Goldberg, I.J.; Galeano, N.F.; Gleeson, A.; Vogel, T.; Gorecki, M.; Sturley, S.L.; Deckelbaum, R.J. Heparan Sulfate Proteoglycan-Mediated Uptake of Apolipoprotein E−Triglyceride-Rich Lipoprotein Particles: A Major Pathway at Physiological Particle Concentrations. *Biochemistry* **1997**, *36*, 12766–12772. [CrossRef]
60. Menard, J.A.; Cerezo-Magaña, M.; Belting, M. Functional role of extracellular vesicles and lipoproteins in the tumour microenvironment. *Philos. Trans. R. Soc. B Biol. Sci.* **2018**, *373*, 20160480. [CrossRef]
61. Collins, L.E.; Troeberg, L. Heparan sulfate as a regulator of inflammation and immunity. *J. Leukoc. Biol.* **2019**, *105*, 81–92. [CrossRef]

62. Reine, T.M.; Kusche-Gullberg, M.; Feta, A.; Jenssen, T.; Kolset, S.O. Heparan sulfate expression is affected by inflammatory stimuli in primary human endothelial cells. *Glycoconj. J.* **2011**, *29*, 67–76. [CrossRef]
63. Massena, S.; Christoffersson, G.; Hjertström, E.; Zcharia, E.; Vlodavsky, I.; Ausmees, N.; Rolny, C.; Li, J.-P.; Phillipson, M. A chemotactic gradient sequestered on endothelial heparan sulfate induces directional intraluminal crawling of neutrophils. *Blood* **2010**, *116*, 1924–1931. [CrossRef]
64. Celie, J.W.A.M. Heparan sulfate proteoglycans in extravasation: Assisting leukocyte guidance. *Front. Biosci.* **2009**, *14*, 4932–4949. [CrossRef] [PubMed]
65. El Masri, R.; Crétinon, Y.; Gout, E.; Vivès, R.R. HS and Inflammation: A Potential Playground for the Sulfs? *Front. Immunol.* **2020**, *11*, 570. [CrossRef] [PubMed]
66. Iozzo, R.V. Basement membrane proteoglycans: From cellar to ceiling. *Nat. Rev. Mol. Cell Biol.* **2005**, *6*, 646–656. [CrossRef]
67. Farach-Carson, M.C.; Warren, C.R.; Harrington, D.A.; Carson, D.D. Border patrol: Insights into the unique role of perlecan/heparan sulfate proteoglycan 2 at cell and tissue borders. *Matrix Biol.* **2014**, *34*, 64–79. [CrossRef] [PubMed]
68. García, B.; Merayo-Lloves, J.; Martin, C.; Alcalde, I.; Quirós, L.M.; Vázquez, F. Surface Proteoglycans as Mediators in Bacterial Pathogens Infections. *Front. Microbiol.* **2016**, *7*, 220. [CrossRef] [PubMed]
69. Maciej-Hulme, M.L.; Skidmore, M.A.; Price, H.P. The role of heparan sulfate in host macrophage infection by Leishmania species. *Biochem. Soc. Trans.* **2018**, *46*, 789–796. [CrossRef]
70. Armistead, J.S.; Wilson, I.B.; van Kuppevelt, T.H.; Dinglasan, R.R. A role for heparan sulfate proteoglycans in Plasmodium falciparum sporozoite invasion of anopheline mosquito salivary glands. *Biochem. J.* **2011**, *438*, 475–483. [CrossRef]
71. Rajan, A.; Robertson, M.J.; Carter, H.E.; Poole, N.M.; Clark, J.; Green, S.I.; Criss, Z.K.; Zhao, B.; Karandikar, U.; Xing, Y.; et al. Enteroaggregative E. coli Adherence to Human Heparan Sulfate Proteoglycans Drives Segment and Host Specific Responses to Infection. *PLoS Pathog.* **2020**, *16*, e1008851. [CrossRef]
72. Paulsson, M.; Su, Y.-C.; Ringwood, T.; Uddén, F.; Riesbeck, K. Pseudomonas aeruginosa uses multiple receptors for adherence to laminin during infection of the respiratory tract and skin wounds. *Sci. Rep.* **2019**, *9*, 18168. [CrossRef]
73. Lin, Y.-P.; Li, L.; Zhang, F.; Linhardt, R.J. Borrelia burgdorferi glycosaminoglycan-binding proteins: A potential target for new therapeutics against Lyme disease. *Microbiol.* **2017**, *163*, 1759–1766. [CrossRef]
74. Menozzi, F.D.; Reddy, V.M.; Cayet, D.; Raze, D.; Debrie, A.-S.; Dehouck, M.-P.; Cecchelli, R.; Locht, C. Mycobacterium tuberculosis heparin-binding haemagglutinin adhesin (HBHA) triggers receptor-mediated transcytosis without altering the integrity of tight junctions. *Microbes Infect.* **2006**, *8*, 1–9. [CrossRef] [PubMed]
75. Agelidis, A.; Shukla, D. Heparanase, Heparan Sulfate and Viral Infection. *Adv. Exp. Med. Biol.* **2020**, *1221*, 759–770. [CrossRef] [PubMed]
76. Connell, B.J.; Lortat-Jacob, H. Human Immunodeficiency Virus and Heparan Sulfate: From Attachment to Entry Inhibition. *Front. Immunol.* **2013**, *4*, 385. [CrossRef] [PubMed]
77. Martino, C.; Kellman, B.P.; Sandoval, D.R.; Clausen, T.M.; Marotz, C.A.; Song, S.J.; Wandro, S.; Zaramela, L.S.; Salido Benitez, R.A.; Zhu, Q.; et al. Bacterial modification of the host glycosaminoglycan heparan sulfate modulates SARS-CoV-2 infectivity. *bioRxiv* **2020**.
78. Liu, L.; Chopra, P.; Li, X.; Wolfert, M.A.; Tompkins, S.M.; Boons, G.J. SARS-CoV-2 spike protein binds heparan sulfate in a length- and sequence-dependent manner. *bioRxiv* **2020**.
79. Lan, J.; Ge, J.; Yu, J.; Shan, S.; Zhou, H.; Fan, S.; Zhang, Q.; Shi, X.; Wang, Q.; Zhang, L.; et al. Structure of the SARS-CoV-2 spike receptor-binding domain bound to the ACE2 receptor. *Nature* **2020**, *581*, 215–220. [CrossRef]
80. Shang, J.; Ye, G.; Shi, K.; Wan, Y.; Luo, C.; Aihara, H.; Geng, Q.; Auerbach, A.; Li, F. Structural basis of receptor recognition by SARS-CoV-2. *Nat. Cell Biol.* **2020**, *581*, 221–224. [CrossRef]
81. Kwon, P.S.; Oh, H.; Kwon, S.-J.; Jin, W.; Zhang, F.; Fraser, K.; Hong, J.J.; Linhardt, R.J.; Dordick, J.S. Sulfated polysaccharides effectively inhibit SARS-CoV-2 in vitro. *Cell Discov.* **2020**, *6*, 1–4. [CrossRef]
82. Park, P.W. Glycosaminoglycans and infection. *Front. Biosci.* **2016**, *21*, 1260–1277. [CrossRef]
83. Hanahan, D.; Coussens, L.M. Accessories to the crime: Functions of cells recruited to the tumor microenvironment. *Cancer Cell* **2012**, *21*, 309–322. [CrossRef]
84. Pickup, M.W.; Mouw, J.K.; Weaver, V.M. The extracellular matrix modulates the hallmarks of cancer. *EMBO Rep.* **2014**, *15*, 1243–1253. [CrossRef] [PubMed]
85. Hanahan, D.; Weinberg, R.A. Hallmarks of Cancer: The Next Generation. *Cell* **2011**, *144*, 646–674. [CrossRef] [PubMed]
86. Pinho, S.S.; Reis, C.A. Glycosylation in cancer: Mechanisms and clinical implications. *Nat. Rev. Cancer* **2015**, *15*, 540–555. [CrossRef]
87. Mereiter, S.; Balmaña, M.; Campos, D.; Gomes, J.; Reis, C.A. Glycosylation in the Era of Cancer-Targeted Therapy: Where Are We Heading? *Cancer Cell* **2019**, *36*, 6–16. [CrossRef] [PubMed]
88. Knelson, E.H.; Nee, J.C.; Blobe, G.C. Heparan sulfate signaling in cancer. *Trends Biochem. Sci.* **2014**, *39*, 277–288. [CrossRef]
89. Derksen, P.W.B.; Keehnen, R.M.J.; Evers, L.M.; van Oers, M.H.J.; Spaargaren, M.; Pals, S.T. Cell surface proteoglycan syndecan-1 mediates hepatocyte growth factor binding and promotes Met signaling in multiple myeloma. *Blood* **2002**, *99*, 1405–1410. [CrossRef]
90. Ren, Z.; van Andel, H.; de Lau, W.; Hartholt, R.B.; Maurice, M.M.; Clevers, H.; Kersten, M.J.; Spaargaren, M.; Pals, S.T. Syndecan-1 promotes Wnt/β-catenin signaling in multiple myeloma by presenting Wnts and R-spondins. *Blood* **2018**, *131*, 982–994. [CrossRef]

91. Nagarajan, A.; Malvi, P.; Wajapeyee, N. Heparan Sulfate and Heparan Sulfate Proteoglycans in Cancer Initiation and Progression. *Front. Endocrinol.* **2018**, *9*, 483. [CrossRef]
92. Suhovskih, A.V.; Domanitskaya, N.V.; Tsidulko, A.Y.; Prudnikova, T.Y.; I Kashuba, V.; Grigorieva, E.V. Tissue-specificity of heparan sulfate biosynthetic machinery in cancer. *Cell Adhes. Migr.* **2015**, *9*, 452–459. [CrossRef]
93. Crespo, A.; García-Suárez, O.; Fernandez-Vega, I.; Solis-Hernandez, M.P.; García, B.; Castañón, S.; Quirós, L.M. Heparan sulfate proteoglycans undergo differential expression alterations in left sided colorectal cancer, depending on their metastatic character. *BMC Cancer* **2018**, *18*, 687. [CrossRef]
94. Fernandez-Vega, I.; García-Suárez, O.; García, B.; Crespo, A.; Astudillo, A.; Quirós, L.M. Heparan sulfate proteoglycans undergo differential expression alterations in right sided colorectal cancer, depending on their metastatic character. *BMC Cancer* **2015**, *15*, 742. [CrossRef] [PubMed]
95. Sembajwe, L.F.; Katta, K.; Grønning, M.; Kusche-Gullberg, M. The exostosin family of glycosyltransferases: mRNA expression profiles and heparan sulphate structure in human breast carcinoma cell lines. *Biosci. Rep.* **2018**, *38*. [CrossRef] [PubMed]
96. Cole, C.L.; Rushton, G.; Jayson, G.C.; Avizienyte, E. Ovarian Cancer Cell Heparan Sulfate 6-O-Sulfotransferases Regulate an Angiogenic Program Induced by Heparin-binding Epidermal Growth Factor (EGF)-like Growth Factor/EGF Receptor Signaling. *J. Biol. Chem.* **2014**, *289*, 10488–10501. [CrossRef] [PubMed]
97. Hwang, J.-A.; Kim, Y.; Hong, S.-H.; Lee, J.; Cho, Y.G.; Han, J.-Y.; Kim, Y.-H.; Han, J.; Shim, Y.M.; Lee, Y.-S.; et al. Epigenetic Inactivation of Heparan Sulfate (Glucosamine) 3-O-Sulfotransferase 2 in Lung Cancer and Its Role in Tumorigenesis. *PLoS ONE* **2013**, *8*, e79634. [CrossRef]
98. Chen, Z.; Fan, J.-Q.; Li, J.; Li, Q.-S.; Yan, Z.; Jia, X.-K.; Liu, W.-D.; Wei, L.-J.; Zhang, F.-Z.; Gao, H.; et al. Promoter hypermethylation correlates with theHsulf-1silencing in human breast and gastric cancer. *Int. J. Cancer* **2008**, *124*, 739–744. [CrossRef]
99. Li, J.; Mo, M.-L.; Chen, Z.; Yang, J.; Li, Q.-S.; Wang, D.-J.; Zhang, H.; Ye, Y.; Xu, J.-P.; Li, H.; et al. HSulf-1 inhibits cell proliferation and invasion in human gastric cancer. *Cancer Sci.* **2011**, *102*, 1815–1821. [CrossRef]
100. Lai, J.; Sandhu, D.S.; Yu, C.; Han, T.; Moser, C.D.; Jackson, K.K.; Guerrero, R.B.; Aderca, I.; Isomoto, H.; Garrity-Park, M.M.; et al. Sulfatase 2 up-regulates glypican 3, promotes fibroblast growth factor signaling, and decreases survival in hepatocellular carcinoma. *Hepatology* **2008**, *47*, 1211–1222. [CrossRef]
101. Nadir, Y.; Brenner, B. Heparanase multiple effects in cancer. *Thromb. Res.* **2014**, *133*, S90–S94. [CrossRef]
102. Vlodavsky, I.; Ilan, N.; Sanderson, R.D. Forty Years of Basic and Translational Heparanase Research. *Adv. Exp. Med. Biol.* **2020**, *1221*, 3–59. [CrossRef]
103. Barash, U.; Cohen-Kaplan, V.; Arvatz, G.; Gingis-Velitski, S.; Levy-Adam, F.; Nativ, O.; Shemesh, R.; Ayalon-Sofer, M.; Ilan, N.; Vlodavsky, I. A novel human heparanase splice variant, T5, endowed with protumorigenic characteristics. *FASEB J.* **2009**, *24*, 1239–1248. [CrossRef]
104. Ilan, N.; Elkin, M.; Vlodavsky, I. Regulation, function and clinical significance of heparanase in cancer metastasis and angiogenesis. *Int. J. Biochem. Cell Biol.* **2006**, *38*, 2018–2039. [CrossRef] [PubMed]
105. Sanderson, R.D.; Iozzo, R.V. Targeting heparanase for cancer therapy at the tumor–matrix interface. *Matrix Biol.* **2012**, *31*, 283–284. [CrossRef] [PubMed]
106. Zhang, L.; Ngo, J.A.; Wetzel, M.D.; Marchetti, D. Heparanase mediates a novel mechanism in lapatinib-resistant brain metastatic breast cancer. *Neoplasia* **2015**, *17*, 101–113. [CrossRef] [PubMed]
107. Alishekevitz, D.; Gingis-Velitski, S.; Kaidar-Person, O.; Gutter-Kapon, L.; Scherer, S.D.; Raviv, Z.; Merquiol, E.; Ben-Nun, Y.; Miller, V.; Rachman-Tzemah, C.; et al. Macrophage-Induced Lymphangiogenesis and Metastasis following Paclitaxel Chemotherapy Is Regulated by VEGFR3. *Cell Rep.* **2016**, *17*, 1344–1356. [CrossRef] [PubMed]
108. Ilan, N.; Shteingauz, A.; Vlodavsky, I. Function from within: Autophagy induction by HPSE/heparanase—new possibilities for intervention. *Autophagy* **2015**, *11*, 2387–2389. [CrossRef] [PubMed]
109. Singh, S.S.; Vats, S.; Chia, A.Y.-Q.; Tan, T.Z.; Deng, S.; Ong, M.S.; Arfuso, F.; Yap, C.T.; Goh, B.C.; Sethi, G.; et al. Dual role of autophagy in hallmarks of cancer. *Oncogene* **2018**, *37*, 1142–1158. [CrossRef]
110. Shteingauz, A.; Boyango, I.; Naroditsky, I.; Hammond, E.; Gruber, M.; Doweck, I.; Ilan, N.; Vlodavsky, I. Heparanase Enhances Tumor Growth and Chemoresistance by Promoting Autophagy. *Cancer Res.* **2015**, *75*, 3946–3957. [CrossRef]
111. Amin, R.; Tripathi, K.; Sanderson, R.D. Nuclear Heparanase Regulates Chromatin Remodeling, Gene Expression and PTEN Tumor Suppressor Function. *Cells* **2020**, *9*, 2038. [CrossRef]
112. Yang, Y.; Gorzelanny, C.; Bauer, A.T.; Halter, N.; Komljenovic, D.; Bäuerle, T.; Borsig, L.; Roblek, M.; Schneider, S.W. Nuclear heparanase-1 activity suppresses melanoma progression via its DNA-binding affinity. *Oncogene* **2015**, *34*, 5832–5842. [CrossRef]
113. Schubert, S.Y.; Ilan, N.; Shushy, M.; Ben-Izhak, O.; Vlodavsky, I.; Goldshmidt, O. Human heparanase nuclear localization and enzymatic activity. *Lab. Investig.* **2004**, *84*, 535–544. [CrossRef]
114. Sanderson, R.D.; Bandari, S.K.; Vlodavsky, I. Proteases and glycosidases on the surface of exosomes: Newly discovered mechanisms for extracellular remodeling. *Matrix Biol.* **2019**, *75*, 160–169. [CrossRef] [PubMed]
115. Thompson, C.A.; Purushothaman, A.; Ramani, V.C.; Vlodavsky, I.; Sanderson, R.D. Heparanase Regulates Secretion, Composition, and Function of Tumor Cell-derived Exosomes. *J. Biol. Chem.* **2013**, *288*, 10093–10099. [CrossRef] [PubMed]
116. Christianson, H.C.; Svensson, K.J.; van Kuppevelt, T.H.; Li, J.-P.; Belting, M. Cancer cell exosomes depend on cell-surface heparan sulfate proteoglycans for their internalization and functional activity. *Proc. Natl. Acad. Sci. USA* **2013**, *110*, 17380–17385. [CrossRef] [PubMed]

117. Wei, J.; Hu, M.; Huang, K.; Lin, S.; Du, H. Roles of Proteoglycans and Glycosaminoglycans in Cancer Development and Progression. *Int. J. Mol. Sci.* **2020**, *21*, 5983. [CrossRef]
118. Brassart-Pasco, S.; Brézillon, S.; Brassart, B.; Ramont, L.; Oudart, J.-B.; Monboisse, J.C. Tumor Microenvironment: Extracellular Matrix Alterations Influence Tumor Progression. *Front. Oncol.* **2020**, *10*, 397. [CrossRef]
119. Toba-Ichihashi, Y.; Yamaoka, T.; Ohmori, T.; Ohba, M. Up-regulation of Syndecan-4 contributes to TGF-β1-induced epithelial to mesenchymal transition in lung adenocarcinoma A549 cells. *Biochem. Biophys. Rep.* **2016**, *5*, 1–7. [CrossRef]
120. Matsuda, K.; Maruyama, H.; Guo, F.; Kleeff, J.; Itakura, J.; Matsumoto, Y.; Lander, A.D.; Korc, M. Glypican-1 is overexpressed in human breast cancer and modulates the mitogenic effects of multiple heparin-binding growth factors in breast cancer cells. *Cancer Res.* **2001**, *61*, 5562–5569.
121. Kovalszky, I.; Hjerpe, A.; Dobra, K. Nuclear translocation of heparan sulfate proteoglycans and their functional significance. *Biochim. Biophys. Acta Gen. Subj.* **2014**, *1840*, 2491–2497. [CrossRef]
122. Stewart, M.D.; Sanderson, R.D. Heparan sulfate in the nucleus and its control of cellular functions. *Matrix Biol.* **2014**, *35*, 56–59. [CrossRef]
123. Stewart, M.D.; Ramani, V.C.; Sanderson, R.D. Shed Syndecan-1 Translocates to the Nucleus of Cells Delivering Growth Factors and Inhibiting Histone Acetylation. *J. Biol. Chem.* **2015**, *290*, 941–949. [CrossRef]
124. Bergers, G.; Benjamin, L.E. Tumorigenesis and the angiogenic switch. *Nat. Rev. Cancer* **2003**, *3*, 401–410. [CrossRef] [PubMed]
125. Fuster, M.M.; Wang, L. Endothelial Heparan Sulfate in Angiogenesis. *Prog. Mol. Biol. Transl. Sci.* **2010**, *93*, 179–212. [CrossRef] [PubMed]
126. Corti, F.; Wang, Y.; Rhodes, J.M.; Atri, D.; Archer-Hartmann, S.; Zhang, J.; Zhuang, Z.W.; Chen, D.; Wang, T.; Wang, Z.; et al. N-terminal syndecan-2 domain selectively enhances 6-O heparan sulfate chains sulfation and promotes VEGFA165-dependent neovascularization. *Nat. Commun.* **2019**, *10*, 1562. [CrossRef] [PubMed]
127. Iozzo, R.V.; Antonio, J.D.S. Heparan sulfate proteoglycans: Heavy hitters in the angiogenesis arena. *J. Clin. Investig.* **2001**, *108*, 349–355. [CrossRef]
128. Lamorte, S.; Ferrero, S.; Aschero, S.; Monitillo, L.; Bussolati, B.; Omede, P.; Ladetto, M.; Camussi, G. Syndecan-1 promotes the angiogenic phenotype of multiple myeloma endothelial cells. *Leukemia* **2011**, *26*, 1081–1090. [CrossRef]
129. Cohen, I.R.; Murdoch, A.D.; Naso, M.F.; Marchetti, D.; Berd, D.; Iozzo, R.V. Abnormal expression of perlecan proteoglycan in metastatic melanomas. *Cancer Res.* **1994**, *54*, 5771–5774.
130. Iozzo, R.V.; Sanderson, R.D. Proteoglycans in cancer biology, tumour microenvironment and angiogenesis. *J. Cell. Mol. Med.* **2011**, *15*, 1013–1031. [CrossRef]
131. Arokiasamy, S.; Balderstone, M.J.M.; de Rossi, G.; Whiteford, J.R. Syndecan-3 in Inflammation and Angiogenesis. *Front. Immunol.* **2020**, *10*, 3031. [CrossRef]
132. De Rossi, G.; Whiteford, J.R. A novel role for syndecan-3 in angiogenesis. *F1000Research* **2013**, *2*, 270. [CrossRef]
133. Lim, H.C.; Multhaupt, H.A.B.; Couchman, J.R. Cell surface heparan sulfate proteoglycans control adhesion and invasion of breast carcinoma cells. *Mol. Cancer* **2015**, *14*, 15–18. [CrossRef]
134. Sanderson, R.D. Heparan sulfate proteoglycans in invasion and metastasis. *Semin. Cell Dev. Biol.* **2001**, *12*, 89–98. [CrossRef] [PubMed]
135. Charni, F.; Friand, V.; Haddad, O.; Hlawaty, H.; Martin, L.; Vassy, R.; Oudar, O.; Gattegno, L.; Charnaux, N.; Sutton, A. Syndecan-1 and syndecan-4 are involved in RANTES/CCL5-induced migration and invasion of human hepatoma cells. *Biochim. Biophys. Acta Gen. Subj.* **2009**, *1790*, 1314–1326. [CrossRef] [PubMed]
136. Reiland, J.; Sanderson, R.D.; Waguespack, M.; Barker, S.A.; Long, R.; Carson, D.D.; Marchetti, D. Heparanase Degrades Syndecan-1 and Perlecan Heparan Sulfate. *J. Biol. Chem.* **2004**, *279*, 8047–8055. [CrossRef] [PubMed]
137. Wang, S.; Zhang, X.; Wang, G.; Cao, B.; Yang, H.; Jin, L.; Cui, M.; Mao, Y. Syndecan-1 suppresses cell growth and migration via blocking JAK1/STAT3 and Ras/Raf/MEK/ERK pathways in human colorectal carcinoma cells. *BMC Cancer* **2019**, *19*, 1160. [CrossRef] [PubMed]
138. Liu, Z.; Jin, H.; Yang, S.; Cao, H.; Zhang, Z.; Wen, B.; Zhou, S. SDC1 knockdown induces epithelial–mesenchymal transition and invasion of gallbladder cancer cells via the ERK/Snail pathway. *J. Int. Med. Res.* **2020**, *48*, 19–32. [CrossRef] [PubMed]
139. Mikami, S.; Ohashi, K.; Usui, Y.; Nemoto, T.; Katsube, K.-I.; Yanagishita, M.; Nakajima, M.; Nakamura, K.; Koike, M. Loss of Syndecan-1 and Increased Expression of Heparanase in Invasive Esophageal Carcinomas. *Jpn. J. Cancer Res.* **2001**, *92*, 1062–1073. [CrossRef] [PubMed]
140. Lim, H.C.; Couchman, J.R. Syndecan-2 regulation of morphology in breast carcinoma cells is dependent on RhoGTPases. *Biochim. Biophys. Acta BBA Gen. Subj.* **2014**, *1840*, 2482–2490. [CrossRef]
141. Park, H.; Kim, Y.; Lim, Y.; Han, I.; Oh, E.-S. Syndecan-2 Mediates Adhesion and Proliferation of Colon Carcinoma Cells. *J. Biol. Chem.* **2002**, *277*, 29730–29736. [CrossRef]
142. De Oliveira, T.; Abiatari, I.; Raulefs, S.; Sauliunaite, D.; Reiser-Erkan, T.M.C.; Kong, B.; Friess, H.; Michalski, C.W.; Kleeff, J. Syndecan-2 promotes perineural invasion and cooperates with K-ras to induce an invasive pancreatic cancer cell phenotype. *Mol. Cancer* **2012**, *11*, 19. [CrossRef]
143. Mytilinaiou, M.; Nikitovic, D.; Berdiaki, A.; Kostouras, A.; Papoutsidakis, A.; Tsatsakis, A.M.; Tzanakakis, G.N. Emerging roles of syndecan 2 in epithelial and mesenchymal cancer progression. *IUBMB Life* **2017**, *69*, 824–833. [CrossRef]

144. Aikawa, T.; Whipple, C.A.; Lopez, M.E.; Gunn, J.; Young, A.; Lander, A.D.; Korc, M. Glypican-1 modulates the angiogenic and metastatic potential of human and mouse cancer cells. *J. Clin. Investig.* **2008**, *118*, 89–99. [CrossRef] [PubMed]
145. Duan, L.; Hu, X.-Q.; Feng, D.-Y.; Lei, S.-Y.; Hu, G.-H. GPC-1 may serve as a predictor of perineural invasion and a prognosticator of survival in pancreatic cancer. *Asian J. Surg.* **2013**, *36*, 7–12. [CrossRef] [PubMed]
146. Hara, H.; Takahashi, T.; Serada, S.; Fujimoto, M.; Ohkawara, T.; Nakatsuka, R.; Harada, E.; Nishigaki, T.; Takahashi, Y.; Nojima, S.; et al. Overexpression of glypican-1 implicates poor prognosis and their chemoresistance in oesophageal squamous cell carcinoma. *Br. J. Cancer* **2016**, *115*, 66–75. [CrossRef]
147. Su, G.; Meyer, K.; Nandini, C.D.; Qiao, D.; Salamat, S.; Friedl, A. Glypican-1 Is Frequently Overexpressed in Human Gliomas and Enhances FGF-2 Signaling in Glioma Cells. *Am. J. Pathol.* **2006**, *168*, 2014–2026. [CrossRef] [PubMed]
148. Saito, T.; Sugiyama, K.; Hama, S.; Yamasaki, F.; Takayasu, T.; Nosaka, R.; Onishi, S.; Muragaki, Y.; Kawamata, T.; Kurisu, K. High Expression of Glypican-1 Predicts Dissemination and Poor Prognosis in Glioblastomas. *World Neurosurg.* **2017**, *105*, 282–288. [CrossRef]
149. Montalbano, M.; Rastellini, C.; McGuire, J.T.; Prajapati, J.; Shirafkan, A.; Vento, R.; Cicalese, L. Role of Glypican-3 in the growth, migration and invasion of primary hepatocytes isolated from patients with hepatocellular carcinoma. *Cell. Oncol.* **2017**, *41*, 169–184. [CrossRef]
150. Zhou, F.; Shang, W.; Yu, X.; Tian, J. Glypican-3: A promising biomarker for hepatocellular carcinoma diagnosis and treatment. *Med. Res. Rev.* **2018**, *38*, 741–767. [CrossRef]
151. Yiu, G.K.; Kaunisto, A.; Chin, Y.R.; Toker, A. NFAT promotes carcinoma invasive migration through glypican-6. *Biochem. J.* **2011**, *440*, 157–166. [CrossRef]
152. Nishida, N.; Yano, H.; Nishida, T.; Kamura, T.; Kojiro, M. Angiogenesis in cancer. *Vasc. Heal. Risk Manag.* **2006**, *2*, 213–219. [CrossRef]
153. Elgundi, Z.; Papanicolaou, M.; Major, G.; Cox, T.R.; Melrose, J.; Whitelock, J.M.; Farrugia, B.L. Cancer Metastasis: The Role of the Extracellular Matrix and the Heparan Sulfate Proteoglycan Perlecan. *Front. Oncol.* **2020**, *9*, 1482. [CrossRef]
154. Matsumoto, A.; Ono, M.; Fujimoto, Y.; Gallo, R.L.; Bernfield, M.; Kohgo, Y. Reduced expression of syndecan-1 in human hepatocellular carcinoma with high metastatic potential. *Int. J. Cancer* **1997**, *74*, 482–491. [CrossRef]
155. Fujiya, M.; Watari, J.; Ashida, T.; Honda, M.; Tanabe, H.; Fujiki, T.; Saitoh, Y.; Kohgo, Y. Reduced Expression of Syndecan-1 Affects Metastatic Potential and Clinical Outcome in Patients with Colorectal Cancer. *Jpn. J. Cancer Res.* **2001**, *92*, 1074–1081. [CrossRef] [PubMed]
156. Saleh, M.E.; Gadalla, R.; Hassan, H.; Afifi, A.; Götte, M.; El-Shinawi, M.; Mohamed, M.M.; Ibrahim, S.A. The immunomodulatory role of tumor Syndecan-1 (CD138) on ex vivo tumor microenvironmental CD4+ T cell polarization in inflammatory and non-inflammatory breast cancer patients. *PLoS ONE* **2019**, *14*, e0217550. [CrossRef] [PubMed]
157. Ibrahim, S.A.; Gadalla, R.; El-Ghonaimy, E.A.; Samir, O.; Mohamed, H.T.; Hassan, H.; Greve, B.; El-Shinawi, M.; Mohamed, M.M.; Gotte, M. Syndecan-1 is a novel molecular marker for triple negative inflammatory breast cancer and modulates the cancer stem cell phenotype via the IL-6/STAT3, Notch and EGFR signaling pathways. *Mol. Cancer* **2017**, *16*, 1–19. [CrossRef] [PubMed]
158. Fares, Y.; Fares, M.Y.; Khachfe, H.H.; Salhab, H.A.; Fares, Y. Molecular principles of metastasis: A hallmark of cancer revisited. *Signal. Transduct. Target. Ther.* **2020**, *5*, 1–17. [CrossRef]
159. Oliveira-Ferrer, L.; Legler, K.; Milde-Langosch, K. Role of protein glycosylation in cancer metastasis. *Semin. Cancer Biol.* **2017**, *44*, 141–152. [CrossRef]
160. Cerezo-Magaña, M.; Bång-Rudenstam, A.; Belting, M. The pleiotropic role of proteoglycans in extracellular vesicle mediated communication in the tumor microenvironment. *Semin. Cancer Biol.* **2020**, *62*, 99–107. [CrossRef]
161. Tkach, M.; Théry, C. Communication by Extracellular Vesicles: Where We Are and Where We Need to Go. *Cell* **2016**, *164*, 1226–1232. [CrossRef]
162. Colombo, M.; Raposo, G.; Théry, C. Biogenesis, secretion, and intercellular interactions of exosomes and other extracellular vesicles. *Annu. Rev. Cell Dev. Biol.* **2014**, *30*, 255–289. [CrossRef]
163. Cocozza, F.; Grisard, E.; Martin-Jaular, L.; Mathieu, M.; Théry, C. SnapShot: Extracellular Vesicles. *Cell* **2020**, *182*, 262–262.e1. [CrossRef]
164. Hoshino, A.; Costa-Silva, B.; Shen, T.-L.; Rodrigues, G.; Hashimoto, A.; Mark, M.T.; Molina, H.; Kohsaka, S.; Di Giannatale, A.; Ceder, S.; et al. Tumour exosome integrins determine organotropic metastasis. *Nature* **2015**, *527*, 329–335. [CrossRef] [PubMed]
165. Adem, B.; Vieira, P.F.; Melo, S.A. Decoding the Biology of Exosomes in Metastasis. *Trends Cancer* **2020**, *6*, 20–30. [CrossRef] [PubMed]
166. Kalluri, R.; le Bleu, V.S. The biology, function, and biomedical applications of exosomes. *Science* **2020**, *367*, eaau6977. [CrossRef] [PubMed]
167. Lane, R.E.; Korbie, D.; Hill, M.M.; Trau, M. Extracellular vesicles as circulating cancer biomarkers: Opportunities and challenges. *Clin. Transl. Med.* **2018**, *7*, 14. [CrossRef] [PubMed]
168. Pang, B.; Zhu, Y.; Ni, J.; Thompson, J.; Malouf, D.; Bucci, J.; Graham, P.; Li, Y. Extracellular vesicles: The next generation of biomarkers for liquid biopsy-based prostate cancer diagnosis. *Theranostics* **2020**, *10*, 2309–2326. [CrossRef]
169. Friand, V.; David, G.; Zimmermann, P. Syntenin and syndecan in the biogenesis of exosomes. *Biol. Cell* **2015**, *107*, 331–341. [CrossRef]

170. Imjeti, N.S.; Menck, K.; Egea-Jimenez, A.L.; Lecointre, C.; Lembo, F.; Bouguenina, H.; Badache, A.; Ghossoub, R.; David, G.; Roche, S.; et al. Syntenin mediates SRC function in exosomal cell-to-cell communication. *Proc. Natl. Acad. Sci. USA* **2017**, *114*, 12495–12500. [CrossRef]
171. Baietti, M.F.; Zhang, Z.; Mortier, E.; Melchior, A.; DeGeest, G.; Geeraerts, A.; Ivarsson, Y.; Depoortere, F.; Coomans, C.; Vermeiren, E.; et al. Syndecan–syntenin–ALIX regulates the biogenesis of exosomes. *Nat. Cell Biol.* **2012**, *14*, 677–685. [CrossRef]
172. Hurley, J.H.; Odorizzi, G. Get on the exosome bus with ALIX. *Nat. Cell Biol.* **2012**, *14*, 654–655. [CrossRef]
173. Ghossoub, R.; Lembo, F.; Rubio, A.; Gaillard, C.B.; Bouchet, J.; Vitale, N.; Slavík, J.; Machala, M.; Zimmermann, P. Syntenin-ALIX exosome biogenesis and budding into multivesicular bodies are controlled by ARF6 and PLD2. *Nat. Commun.* **2014**, *5*, 3477. [CrossRef]
174. David, G.; Zimmermann, P. Heparanase Involvement in Exosome Formation. *Adv. Exp. Med. Biol.* **2020**, *1221*, 285–307. [CrossRef] [PubMed]
175. Sanderson, R.D.; Elkin, M.; Rapraeger, A.C.; Ilan, N.; Vlodavsky, I. Heparanase regulation of cancer, autophagy and inflammation: New mechanisms and targets for therapy. *FEBS J.* **2017**, *284*, 42–55. [CrossRef] [PubMed]
176. Roucourt, B.; Meeussen, S.; Bao, J.; Zimmermann, P.; David, G. Heparanase activates the syndecan-syntenin-ALIX exosome pathway. *Cell Res.* **2015**, *25*, 412–428. [CrossRef] [PubMed]
177. Cerezo-Magana, M.; Christianson, H.C.; van Kuppevelt, T.H.; Forsberg-Nilsson, K.; Belting, M. Hypoxic induction of exosome uptake through proteoglycan dependent endocytosis fuels the lipid droplet phenotype in glioma. *Mol. Cancer Res.* **2020**, *10*, 1158–1541. [CrossRef]
178. Buzás, E.I.; Tóth, E.Á.; Sódar, B.W.; Szabó-Taylor, K.É. Molecular interactions at the surface of extracellular vesicles. *Semin. Immunopathol.* **2018**, *40*, 453–464. [CrossRef]
179. Purushothaman, A.; Bandari, S.K.; Liu, J.; Mobley, J.A.; Brown, E.E.; Sanderson, R.D. Fibronectin on the Surface of Myeloma Cell-derived Exosomes Mediates Exosome-Cell Interactions. *J. Biol. Chem.* **2016**, *291*, 1652–1663. [CrossRef]
180. Nangami, G.; Koumangoye, R.; Goodwin, J.S.; Sakwe, A.M.; Marshall, D.; Higginbotham, J.; Ochieng, J. Fetuin-A associates with histones intracellularly and shuttles them to exosomes to promote focal adhesion assembly resulting in rapid adhesion and spreading in breast carcinoma cells. *Exp. Cell Res.* **2014**, *328*, 388–400. [CrossRef]
181. Berenguer, J.; Lagerweij, T.; Zhao, X.W.; Dusoswa, S.; van der Stoop, P.; Westerman, B.; de Gooijer, M.C.; Zoetemelk, M.; Zomer, A.; Crommentuijn, M.H.W.; et al. Glycosylated extracellular vesicles released by glioblastoma cells are decorated by CCL18 allowing for cellular uptake via chemokine receptor CCR8. *J. Extracell. Vesicles* **2018**, *7*, 1446660. [CrossRef]
182. Arruebo, M.; Vilaboa, N.; Saez, B.; Lambea, J.; Tres, A.; Valladares, M.; González-Fernández, Á. Assessment of the Evolution of Cancer Treatment Therapies. *Cancers* **2011**, *3*, 3279–3330. [CrossRef]
183. Wichert, A.; Stege, A.; Midorikawa, Y.; Holm, P.S.; Lage, H. Glypican-3 is involved in cellular protection against mitoxantrone in gastric carcinoma cells. *Oncogene* **2003**, *23*, 945–955. [CrossRef]
184. Mahtouk, K.; Hose, D.; Raynaud, P.; Hundemer, M.; Jourdan, M.; Jourdan, E.; Pantesco, V.; Baudard, M.; de Vos, J.; Larroque, M.; et al. Heparanase influences expression and shedding of syndecan-1, and its expression by the bone marrow environment is a bad prognostic factor in multiple myeloma. *Blood* **2007**, *109*, 4914–4923. [CrossRef] [PubMed]
185. Ramani, V.C.; Purushothaman, A.; Stewart, M.D.; Thompson, C.A.; Vlodavsky, I.; Au, J.L.-S.; Sanderson, R.D. The heparanase/syndecan-1 axis in cancer: Mechanisms and therapies. *FEBS J.* **2013**, *280*, 2294–2306. [CrossRef]
186. Barbareschi, M.; Aldovini, D.; Cangi, M.G.; Pecciarini, L.; Veronese, S.; Caffo, O.; Lucenti, A.; Palma, P.D. High syndecan-1 expression in breast carcinoma is related to an aggressive phenotype and to poorer prognosis. *Cancer* **2003**, *98*, 474–483. [CrossRef] [PubMed]
187. Nguyen, T.L.; Grizzle, W.E.; Zhang, K.; Hameed, O.; Siegal, G.P.; Wei, S. Syndecan-1 Overexpression Is Associated with Nonluminal Subtypes and Poor Prognosis in Advanced Breast Cancer. *Am. J. Clin. Pathol.* **2013**, *140*, 468–474. [CrossRef]
188. Ramani, V.C.; Zhan, F.; He, J.; Barbieri, P.; Noseda, A.; Tricot, G.; Sanderson, R.D. Targeting heparanase overcomes chemoresistance and diminishes relapse in myeloma. *Oncotarget* **2015**, *7*, 1598–1607. [CrossRef] [PubMed]
189. Götte, M.; Kersting, C.; Ruggiero, M.; Tio, J.; Tulusan, A.H.; Kiesel, L.; Wülfing, P. Predictive value of syndecan-1 expression for the response to neoadjuvant chemotherapy of primary breast cancer. *Anticancer. Res.* **2006**, *26*, 621–627.
190. Suarez, E.R.; Paredes-Gamero, E.J.; del Giglio, A.; Tersariol, I.L.S.; Nader, H.B.; Pinhal, M.A.S. Heparan sulfate mediates trastuzumab effect in breast cancer cells. *BMC Cancer* **2013**, *13*, 444. [CrossRef]
191. Meirovitz, A.; Hermano, E.; Lerner, I.; Zcharia, E.; Pisano, C.; Peretz, T.; Elkin, M. Role of Heparanase in Radiation-Enhanced Invasiveness of Pancreatic Carcinoma. *Cancer Res.* **2011**, *71*, 2772–2780. [CrossRef]
192. Li, J.; Meng, X.; Hu, J.; Zhang, Y.; Dang, Y.; Wei, L.; Shi, M. Heparanase promotes radiation resistance of cervical cancer by upregulating hypoxia inducible factor 1. *Am. J. Cancer Res.* **2017**, *7*, 234–244.
193. Wang, X.; Zuo, D.; Chen, Y.; Li, W.; Liu, R.; He, Y.; Ren, L.; Zhou, L.; Deng, T.; Ying, G.; et al. Shed Syndecan-1 is involved in chemotherapy resistance via the EGFR pathway in colorectal cancer. *Br. J. Cancer* **2014**, *111*, 1965–1976. [CrossRef]
194. Ramani, V.C.; Sanderson, R.D. Chemotherapy stimulates syndecan-1 shedding: A potentially negative effect of treatment that may promote tumor relapse. *Matrix Biol.* **2014**, *35*, 215–222. [CrossRef] [PubMed]
195. Keklikoglou, I.; Cianciaruso, C.; Güç, E.; Squadrito, M.L.; Spring, L.M.; Tazzyman, S.; Lambein, L.; Poissonnier, A.; Ferraro, G.B.; Baer, C.; et al. Chemotherapy elicits pro-metastatic extracellular vesicles in breast cancer models. *Nat. Cell Biol.* **2019**, *21*, 190–202. [CrossRef] [PubMed]

196. Viola, M.; Brüggemann, K.; Karousou, E.; Caon, I.; Caravà, E.; Vigetti, D.; Greve, B.; Stock, C.; de Luca, G.; Passi, A.; et al. MDA-MB-231 breast cancer cell viability, motility and matrix adhesion are regulated by a complex interplay of heparan sulfate, chondroitin−/dermatan sulfate and hyaluronan biosynthesis. *Glycoconj. J.* **2016**, *34*, 411–420. [CrossRef] [PubMed]
197. Alizadeh, A.A.; Aranda, V.V.; Bardelli, A.A.; Blanpain, C.; Bock, C.C.; Borowski, C.C.; Caldas, C.; Califano, A.A.; Doherty, M.M.; Elsner, M.M.; et al. Toward understanding and exploiting tumor heterogeneity. *Nat. Med.* **2015**, *21*, 846–853. [CrossRef] [PubMed]
198. Marusyk, A.; Almendro, V.; Polyak, K. Intra-tumour heterogeneity: A looking glass for cancer? *Nat. Rev. Cancer* **2012**, *12*, 323–334. [CrossRef]
199. Zardavas, D.; Irrthum, A.A.; Swanton, C.; Piccart-Gebhart, M. Clinical management of breast cancer heterogeneity. *Nat. Rev. Clin. Oncol.* **2015**, *12*, 381–394. [CrossRef]
200. Belting, M. Glycosaminoglycans in cancer treatment. *Thromb. Res.* **2014**, *133*, S95–S101. [CrossRef]
201. Espinoza-Sánchez, N.A.; Gotte, M. Role of cell surface proteoglycans in cancer immunotherapy. *Semin. Cancer Biol.* **2020**, *62*, 48–67. [CrossRef]
202. Ripsman, D.; Fergusson, D.; Montroy, J.; Auer, R.C.; Huang, J.W.; Dobriyal, A.; Wesch, N.; Carrier, M.; Lalu, M.M. A systematic review on the efficacy and safety of low molecular weight heparin as an anticancer therapeutic in preclinical animal models. *Thromb. Res.* **2020**, *195*, 103–113. [CrossRef]
203. Ramani, V.C.; Vlodavsky, I.; Ng, M.; Zhang, Y.; Barbieri, P.; Noseda, A.; Sanderson, R.D. Chemotherapy induces expression and release of heparanase leading to changes associated with an aggressive tumor phenotype. *Matrix Biol.* **2016**, *55*, 22–34. [CrossRef]
204. Bandari, S.K.; Purushothaman, A.; Ramani, V.C.; Brinkley, G.J.; Chandrashekar, D.S.; Varambally, S.; Mobley, J.A.; Zhang, Y.; Brown, E.E.; Vlodavsky, I.; et al. Chemotherapy induces secretion of exosomes loaded with heparanase that degrades extracellular matrix and impacts tumor and host cell behavior. *Matrix Biol.* **2018**, *65*, 104–118. [CrossRef] [PubMed]
205. Morla, S. Glycosaminoglycans and Glycosaminoglycan Mimetics in Cancer and Inflammation. *Int. J. Mol. Sci.* **2019**, *20*, 1963. [CrossRef] [PubMed]
206. Moussa, L.; Demarquay, C.; Réthoré, G.; Benadjaoud, M.A.; Siñeriz, F.; Pattappa, G.; Guicheux, J.; Weiss, P.; Barritault, D.; Mathieu, N. Heparan Sulfate Mimetics: A New Way to Optimize Therapeutic Effects of Hydrogel-Embedded Mesenchymal Stromal Cells in Colonic Radiation-Induced Damage. *Sci. Rep.* **2019**, *9*, 164. [CrossRef] [PubMed]
207. Veraldi, N.; Zouggari, N.; de Agostini, A. The Challenge of Modulating Heparan Sulfate Turnover by Multitarget Heparin Derivatives. *Molecules* **2020**, *25*, 390. [CrossRef] [PubMed]
208. Wang, W.; Gopal, S.; Pocock, R.; Xiao, Z.-C. Glycan Mimetics from Natural Products: New Therapeutic Opportunities for Neurodegenerative Disease. *Molecules* **2019**, *24*, 4604. [CrossRef]
209. Whiteside, T.L. The potential of tumor-derived exosomes for noninvasive cancer monitoring. *Expert Rev. Mol. Diagn.* **2015**, *15*, 1293–1310. [CrossRef]
210. Buscail, E.; Alix-Panabières, C.; Quincy, P.; Cauvin, T.; Chauvet, A.; Degrandi, O.; Caumont, C.; Verdon, S.; Lamrissi, I.; Moranvillier, I.; et al. High Clinical Value of Liquid Biopsy to Detect Circulating Tumor Cells and Tumor Exosomes in Pancreatic Ductal Adenocarcinoma Patients Eligible for Up-Front Surgery. *Cancers* **2019**, *11*, 1656. [CrossRef]
211. Zhao, Z.; Fan, J.; Hsu, Y.-M.S.; Lyon, C.J.; Ning, B.; Hu, T.Y. Extracellular vesicles as cancer liquid biopsies: From discovery, validation, to clinical application. *Lab. Chip* **2019**, *19*, 1114–1140. [CrossRef]
212. Dredge, K.; Brennan, T.V.; Hammond, E.; Lickliter, J.D.; Lin, L.; Bampton, D.; Handley, P.; Lankesheer, F.; Morrish, G.; Yang, Y.; et al. A Phase I study of the novel immunomodulatory agent PG545 (pixatimod) in subjects with advanced solid tumours. *Br. J. Cancer* **2018**, *118*, 1035–1041. [CrossRef]
213. Hammond, E.; Haynes, N.M.; Cullinane, C.; Brennan, T.V.; Bampton, D.; Handley, P.; Karoli, T.; Lanksheer, F.; Lin, L.; Yang, Y.; et al. Immunomodulatory activities of pixatimod: Emerging nonclinical and clinical data, and its potential utility in combination with PD-1 inhibitors. *J. Immunother. Cancer* **2018**, *6*, 54. [CrossRef]
214. Weissmann, M.; Bhattacharya, U.; Feld, S.; Hammond, E.; Ilan, N.; Vlodavsky, I. The heparanase inhibitor PG545 is a potent anti-lymphoma drug: Mode of action. *Matrix Biol.* **2019**, *77*, 58–72. [CrossRef] [PubMed]
215. Mayfosh, A.J.; Baschuk, N.; Hulett, M.D. Leukocyte Heparanase: A Double-Edged Sword in Tumor Progression. *Front. Oncol.* **2019**, *9*, 331. [CrossRef] [PubMed]
216. Jayatilleke, K.M.; Hulett, M.D. Heparanase and the hallmarks of cancer. *J. Transl. Med.* **2020**, *18*, 1–25. [CrossRef] [PubMed]
217. Piperigkou, Z.; Mohr, B.; Karamanos, N.; Gotte, M. Shed proteoglycans in tumor stroma. *Cell Tissue Res.* **2016**, *365*, 643–655. [CrossRef] [PubMed]
218. Mohan, C.D.; Hari, S.; Preetham, H.D.; Rangappa, S.; Barash, U.; Ilan, N.; Nayak, S.C.; Gupta, V.K.; Vlodavsky, I. Targeting Heparanase in Cancer: Inhibition by Synthetic, Chemically Modified, and Natural Compounds. *iScience* **2019**, *15*, 360–390. [CrossRef]
219. Kisker, O.; Becker, C.M.; Prox, D.; Fannon, M.; D'Amato, R.; Flynn, E.; E Fogler, W.; Sim, B.K.; Allred, E.N.; Pirie-Shepherd, S.R.; et al. Continuous administration of endostatin by intraperitoneally implanted osmotic pump improves the efficacy and potency of therapy in a mouse xenograft tumor model. *Cancer Res.* **2001**, *61*, 7669–7674.
220. Walia, A.; Yang, J.F.; Huang, Y.-H.; Rosenblatt, M.I.; Chang, J.-H.; Azar, D.T. Endostatin's emerging roles in angiogenesis, lymphangiogenesis, disease, and clinical applications. *Biochim. Biophys. Acta Gen. Subj.* **2015**, *1850*, 2422–2438. [CrossRef]

221. Sato, K.; Akiyama, H.; Kogure, Y.; Suwa, Y.; Momiyama, M.; Ishibe, A.; Endo, I. [A Case of Rhabdomyolysis Related to SOX Therapy for Liver Metastasis of Gastric Cancer]. *Gan Kagaku Ryoho. Cancer Chemother.* **2017**, *44*, 329–331.
222. Wang, M.; Yang, H.; Sui, Y.; Guo, X.; Tan, X.; Li, Y. Endostar continuous intravenous infusion combined with S-1 and oxaliplatin chemotherapy could be effective in treating liver metastasis from gastric cancer. *J. Cancer Res. Ther.* **2018**, *14*, 1148. [CrossRef]
223. Xiao, C.; Qian, J.; Zheng, Y.; Song, F.; Wang, Q.; Jiang, H.; Mao, C.; Xu, N. A phase II study of biweekly oxaliplatin plus S-1 combination chemotherapy as a first-line treatment for patients with metastatic or advanced gastric cancer in China. *Medicine* **2019**, *98*, e15696. [CrossRef]
224. Cheng, Y.-J.; Meng, C.-T.; Ying, H.-Y.; Zhou, J.-F.; Yan, X.-Y.; Gao, X.; Zhou, N.; Bai, C.-M. Effect of Endostar combined with chemotherapy in advanced well-differentiated pancreatic neuroendocrine tumors. *Medicine* **2018**, *97*, e12750. [CrossRef] [PubMed]
225. Billings, P.C.; Pacifici, M. Interactions of signaling proteins, growth factors and other proteins with heparan sulfate: Mechanisms and mysteries. *Connect. Tissue Res.* **2015**, *56*, 272–280. [CrossRef] [PubMed]
226. Wang, H.; Liu, Z.; An, C.; Li, H.; Hu, F.; Dong, S. Self-Assembling Glycopeptide Conjugate as a Versatile Platform for Mimicking Complex Polysaccharides. *Adv. Sci.* **2020**, *7*, 2001264. [CrossRef] [PubMed]
227. Reily, C.; Stewart, T.J.; Renfrow, M.B.; Novak, J. Glycosylation in health and disease. *Nat. Rev. Nephrol.* **2019**, *15*, 346–366. [CrossRef] [PubMed]
228. Coombe, D.R.; Gandhi, N.S. Heparanase: A Challenging Cancer Drug Target. *Front. Oncol.* **2019**, *9*, 1316. [CrossRef] [PubMed]
229. Denys, A.; Allain, F. The Emerging Roles of Heparan Sulfate 3-O-Sulfotransferases in Cancer. *Front. Oncol.* **2019**, *9*, 507. [CrossRef]
230. Hellec, C.; Delos, M.; Carpentier, M.; Denys, A.; Allain, F. The heparan sulfate 3-O-sulfotransferases (HS3ST) 2, 3B and 4 enhance proliferation and survival in breast cancer MDA-MB-231 cells. *PLoS ONE* **2018**, *13*, e0194676. [CrossRef]
231. Jagannath, S.; Heffner, L.T.; Ailawadhi, S.; Munshi, N.C.; Zimmerman, T.M.; Rosenblatt, J.; Lonial, S.; Chanan-Khan, A.; Ruehle, M.; Rharbaoui, F.; et al. Indatuximab Ravtansine (BT062) Monotherapy in Patients with Relapsed and/or Refractory Multiple Myeloma. *Clin. Lymphoma Myeloma Leuk.* **2019**, *19*, 372–380. [CrossRef]
232. Guo, B.; Chen, M.; Han, Q.; Hui, F.; Dai, H.; Zhang, W.; Zhang, Y.; Wang, Y.; Zhu, H.; Han, W. CD138-directed adoptive immunotherapy of chimeric antigen receptor (CAR)-modified T cells for multiple myeloma. *J. Cell. Immunother.* **2016**, *2*, 28–35. [CrossRef]
233. Shimizu, Y.; Suzuki, T.; Yoshikawa, T.; Endo, I.; Nakatsura, T. Next-Generation Cancer Immunotherapy Targeting Glypican-3. *Front. Oncol.* **2019**, *9*, 248. [CrossRef]
234. Liu, X.; Gao, F.; Jiang, L.; Jia, M.; Ao, L.; Lu, M.; Gou, L.; Ho, M.; Jia, S.; Chen, F.; et al. 32A9, a novel human antibody for designing an immunotoxin and CAR-T cells against glypican-3 in hepatocellular carcinoma. *J. Transl. Med.* **2020**, *18*, 1–12. [CrossRef] [PubMed]
235. Li, N.; Gao, W.; Zhang, Y.-F.; Ho, M. Glypicans as Cancer Therapeutic Targets. *Trends Cancer* **2018**, *4*, 741–754. [CrossRef] [PubMed]
236. Casu, B.; Naggi, A.; Torri, G. Re-visiting the structure of heparin. *Carbohydr. Res.* **2015**, *403*, 60–68. [CrossRef] [PubMed]
237. Kakkar, A.K.; Levine, M.N.; Kadziola, Z.; Lemoine, N.R.; Low, V.; Patel, H.K.; Rustin, G.; Thomas, M.D.; Quigley, M.; Williamson, R.C. Low Molecular Weight Heparin, Therapy with Dalteparin, and Survival in Advanced Cancer: The Fragmin Advanced Malignancy Outcome Study (FAMOUS). *J. Clin. Oncol.* **2004**, *22*, 1944–1948. [CrossRef]
238. Ludwig, R.J. Therapeutic use of heparin beyond anticoagulation. *Curr. Drug Discov. Technol.* **2009**, *6*, 281–289. [CrossRef]
239. Pomin, V.H.; Mulloy, B. Current structural biology of the heparin interactome. *Curr. Opin. Struct. Biol.* **2015**, *34*, 17–25. [CrossRef]
240. Gong, F.; Jemth, P.; Galvis, M.L.E.; Vlodavsky, I.; Horner, A.; Lindahl, U.; Li, J.-P. Processing of Macromolecular Heparin by Heparanase. *J. Biol. Chem.* **2003**, *278*, 35152–35158. [CrossRef]
241. Naggi, A.; Casu, B.; Perez, M.; Torri, G.; Cassinelli, G.; Penco, S.; Pisano, C.; Giannini, G.; Ishai-Michaeli, R.; Vlodavsky, I. Modulation of the Heparanase-inhibiting Activity of Heparin through Selective Desulfation, Graded N-Acetylation, and Glycol Splitting. *J. Biol. Chem.* **2005**, *280*, 12103–12113. [CrossRef]
242. Ritchie, J.P.; Ramani, V.C.; Ren, Y.; Naggi, A.; Torri, G.; Casu, B.; Penco, S.; Pisano, C.; Carminati, P.; Tortoreto, M.; et al. SST0001, a Chemically Modified Heparin, Inhibits Myeloma Growth and Angiogenesis via Disruption of the Heparanase/Syndecan-1 Axis. *Clin. Cancer Res.* **2011**, *17*, 1382–1393. [CrossRef]
243. Karoli, T.; Liu, L.; Fairweather, J.K.; Hammond, E.; Li, C.P.; Cochran, S.; Bergefall, K.; Trybala, E.; Addison, R.S.; Ferro, V. Synthesis, Biological Activity, and Preliminary Pharmacokinetic Evaluation of Analogues of a Phosphosulfomannan Angiogenesis Inhibitor (PI-88). *J. Med. Chem.* **2005**, *48*, 8229–8236. [CrossRef]
244. Khachigian, L.M.; Parish, C.R. Phosphomannopentaose sulfate (PI-88): Heparan sulfate mimetic with clinical potential in multiple vascular pathologies. *Cardiovasc. Drug Rev.* **2006**, *22*, 1–6. [CrossRef] [PubMed]
245. Kovacsovics, T.J.; Mims, A.; Salama, M.E.; Pantin, J.; Rao, N.; Kosak, K.M.; Ahorukomeye, P.; Glenn, M.J.; Deininger, M.W.N.; Boucher, K.M.; et al. Combination of the low anticoagulant heparin CX-01 with chemotherapy for the treatment of acute myeloid leukemia. *Blood Adv.* **2018**, *2*, 381–389. [CrossRef] [PubMed]
246. Pala, D.; Rivara, S.; Mor, M.; Milazzo, F.M.; Roscilli, G.; Pavoni, E.; Giannini, G. Kinetic analysis and molecular modeling of the inhibition mechanism of roneparstat (SST0001) on human heparanase. *Glycobiology* **2016**, *26*, 640–654. [CrossRef] [PubMed]
247. Dredge, K.; Hammond, E.; Davis, K.; Li, C.P.; Liu, L.; Johnstone, K.; Handley, P.; Wimmer, N.; Gonda, T.J.; Gautam, A.; et al. The PG500 series: Novel heparan sulfate mimetics as potent angiogenesis and heparanase inhibitors for cancer therapy. *Investig. New Drugs* **2009**, *28*, 276–283. [CrossRef] [PubMed]

248. Ferro, V.; Dredge, K.; Liu, L.; Hammond, E.; Bytheway, I.; Li, C.; Johnstone, K.; Karoli, T.; Davis, K.; Copeman, E.; et al. PI-88 and Novel Heparan Sulfate Mimetics Inhibit Angiogenesis. *Semin. Thromb. Hemost.* **2007**, *33*, 557–568. [CrossRef]
249. Liu, C.-J.; Lee, P.-H.; Lin, D.-Y.; Wu, C.-C.; Jeng, L.-B.; Lin, P.-W.; Mok, K.-T.; Lee, W.-C.; Yeh, H.-Z.; Ho, M.-C.; et al. Heparanase inhibitor PI-88 as adjuvant therapy for hepatocellular carcinoma after curative resection: A randomized phase II trial for safety and optimal dosage. *J. Hepatol.* **2009**, *50*, 958–968. [CrossRef]
250. Liao, B.-Y.; Wang, Z.; Hu, J.; Liu, W.-F.; Shen, Z.-Z.; Zhang, X.; Yu, L.; Fan, J.; Zhou, J. PI-88 inhibits postoperative recurrence of hepatocellular carcinoma via disrupting the surge of heparanase after liver resection. *Tumor Biol.* **2015**, *37*, 2987–2998. [CrossRef]
251. Lewis, K.; Robinson, W.A.; Millward, M.J.; Powell, A.; Price, T.; Thomson, D.B.; Walpole, E.; Haydon, A.M.; Creese, B.R.; Roberts, K.L.; et al. A phase II study of the heparanase inhibitor PI-88 in patients with advanced melanoma. *Investig. New Drugs* **2007**, *26*, 89–94. [CrossRef]
252. Joglekar, M.V.; Diez, P.M.Q.; Marcus, S.; Qi, R.; Espinasse, B.; Wiesner, M.R.; Pempe, E.; Liu, J.; Monroe, D.M.; Arepally, G.M. Disruption of PF4/H multimolecular complex formation with a minimally anticoagulant heparin (ODSH). *Thromb. Haemost.* **2012**, *107*, 717–725. [CrossRef]
253. Winer, E.S.; Stone, R.M. Novel therapy in Acute myeloid leukemia (AML): Moving toward targeted approaches. *Ther. Adv. Hematol.* **2019**, *10*, 2040620719860645. [CrossRef]
254. Galli, M.; Chatterjee, M.; Grasso, M.; Specchia, G.; Magen, H.; Einsele, H.; Celeghini, I.; Barbieri, P.; Paoletti, D.; Pace, S.; et al. Phase I study of the heparanase inhibitor roneparstat: An innovative approach for ultiple myeloma therapy. *Haematology* **2018**, *103*, e469–e472. [CrossRef] [PubMed]
255. Ferro, V.; Liu, L.; Johnstone, K.D.; Wimmer, N.; Karoli, T.; Handley, P.; Rowley, J.; Dredge, K.; Li, C.P.; Hammond, E.; et al. Discovery of PG545: A Highly Potent and Simultaneous Inhibitor of Angiogenesis, Tumor Growth, and Metastasis. *J. Med. Chem.* **2012**, *55*, 3804–3813. [CrossRef] [PubMed]
256. Brennan, T.V.; Lin, L.; Brandstadter, J.D.; Rendell, V.R.; Dredge, K.; Huang, X.; Yang, Y. Heparan sulfate mimetic PG545-mediated antilymphoma effects require TLR9-dependent NK cell activation. *J. Clin. Investig.* **2015**, *126*, 207–219. [CrossRef] [PubMed]
257. Scott, E.; Guimond, C.J.M.-W.; Neha, S.; Gandhi-Julia, A.; Tree, K.R.; Buttigieg, N.C.; Elmore, M.J.; Joanna-Said, K.N.; Yin, X.S.; Alberto, A.; et al. Turnbull. Synthetic Heparan Sulfate Mimetic Pixatimod (PG545) Potently Inhibits SARS-CoV-2 By Disrupting the Spike-ACE2 interaction. *bioRxiv* **2020**.
258. Cagno, V.; Tseligka, E.D.; Jones, S.T.; Tapparel, C. Heparan Sulfate Proteoglycans and Viral Attachment: True Receptors or Adaptation Bias? *Viruses* **2019**, *11*, 596. [CrossRef]
259. El Ghazal, R.; Yin, X.; Johns, S.C.; Swanson, L.; Macal, M.; Ghosh, P.; Zuniga, E.I.; Fuster, M.M. Glycan Sulfation Modulates Dendritic Cell Biology and Tumor Growth. *Neoplasia* **2016**, *18*, 294–306. [CrossRef]
260. Gatto, F.; Volpi, N.; Nilsson, H.; Nookaew, I.; Maruzzo, M.; Roma, A.; Johansson, M.E.; Stierner, U.; Lundstam, S.; Basso, U.; et al. Glycosaminoglycan Profiling in Patients' Plasma and Urine Predicts the Occurrence of Metastatic Clear Cell Renal Cell Carcinoma. *Cell Rep.* **2016**, *15*, 1822–1836. [CrossRef]
261. Brézillon, S.; Untereiner, V.; Mohamed, H.T.; Hodin, J.; Chatron-Colliet, A.; Maquart, F.-X.; Brézillon, S. Probing glycosaminoglycan spectral signatures in live cells and their conditioned media by Raman microspectroscopy. *Analyst* **2017**, *142*, 1333–1341. [CrossRef]
262. Li, N.; Spetz, M.R.; Ho, M. The Role of Glypicans in Cancer Progression and Therapy. *J. Histochem. Cytochem.* **2020**, *68*, 841–862. [CrossRef]
263. Lu, H.; Niu, F.; Liu, F.; Gao, J.; Sun, Y.; Zhao, X. Elevated glypican-1 expression is associated with an unfavorable prognosis in pancreatic ductal adenocarcinoma. *Cancer Med.* **2017**, *6*, 1181–1191. [CrossRef]
264. Zhou, C.-Y.; Dong, Y.-P.; Sun, X.; Sui, X.; Zhu, H.; Zhao, Y.-Q.; Zhang, Y.-Y.; Mason, C.; Zhu, Q.; Han, S.-X. High levels of serum glypican-1 indicate poor prognosis in pancreatic ductal adenocarcinoma. *Cancer Med.* **2018**, *7*, 5525–5533. [CrossRef] [PubMed]
265. Campbell, D.H.; Lund, M.E.; Nocon, A.L.; Cozzi, P.J.; Frydenberg, M.; de Souza, P.; Schiller, B.; Beebe-Dimmer, J.L.; Ruterbusch, J.J.; Walsh, B.J. Detection of glypican-1 (GPC-1) expression in urine cell sediments in prostate cancer. *PLoS ONE* **2018**, *13*, e0196017. [CrossRef] [PubMed]
266. Melo, S.A.; Luecke, L.B.; Kahlert, C.; Fernández, A.F.; Gammon, S.T.; Kaye, J.; LeBleu, V.S.; Mittendorf, E.A.; Weitz, J.; Rahbari, N.N.; et al. Glypican-1 identifies cancer exosomes and detects early pancreatic cancer. *Nat. Cell Biol.* **2015**, *523*, 177–182. [CrossRef] [PubMed]
267. Guo, M.; Zhang, H.; Zheng, J.; Liu, Y. Glypican-3: A New Target for Diagnosis and Treatment of Hepatocellular Carcinoma. *J. Cancer* **2020**, *11*, 2008–2021. [CrossRef]
268. Li, Y.; Li, M.; Shats, I.; Krahn, J.M.; Flake, G.P.; Umbach, D.M.; Li, X.; Li, L. Glypican 6 is a putative biomarker for metastatic progression of cutaneous melanoma. *PLoS ONE* **2019**, *14*, e0218067. [CrossRef]
269. Chandran, V.I.; Welinder, C.; Månsson, A.-S.; Offer, S.; Freyhult, E.; Pernemalm, M.; Lund, S.M.; Pedersen, S.; Lehtiö, J.; Marko-Varga, G.; et al. Ultrasensitive Immunoprofiling of Plasma Extracellular Vesicles Identifies Syndecan-1 as a Potential Tool for Minimally Invasive Diagnosis of Glioma. *Clin. Cancer Res.* **2019**, *25*, 3115–3127. [CrossRef]
270. Huang, G.; Lin, G.; Zhu, Y.; Duan, W.; Jin, D. Emerging technologies for profiling extracellular vesicle heterogeneity. *Lab. Chip* **2020**, *20*, 2423–2437. [CrossRef]
271. Liu, J.; Moon, A.F.; Sheng, J.-Z.; Pedersen, L.C. Understanding the substrate specificity of the heparan sulfate sulfotransferases by an integrated biosynthetic and crystallographic approach. *Curr. Opin. Struct. Biol.* **2012**, *22*, 550–557. [CrossRef]

272. Clerc, O.; Deniaud, M.; Vallet, S.D.; Naba, A.; Rivet, A.; Perez, S.; Thierry-Mieg, N.; Ricard-Blum, S. MatrixDB: Integration of new data with a focus on glycosaminoglycan interactions. *Nucleic Acids Res.* **2019**, *47*, D376–D381. [CrossRef]
273. Vallet, S.D.; Clerc, O.; Ricard-Blum, S. Glycosaminoglycan–Protein Interactions: The First Draft of the Glycosaminoglycan Interactome. *J. Histochem. Cytochem.* **2020**. [CrossRef]

Article

HA and HS Changes in Endothelial Inflammatory Activation

Elena Caravà [1,2], Paola Moretto [2], Ilaria Caon [2], Arianna Parnigoni [2], Alberto Passi [2], Evgenia Karousou [2], Davide Vigetti [2], Jessica Canino [3], Ilaria Canobbio [3] and Manuela Viola [2,*]

1 Quantix Italia S.r.l., 20121 Milano, Italy; carava@quantixitalia.com
2 Department of Medicine and Surgery, University of Insubria, 21100 Varese, Italy; paola.moretto@uninsubria.it (P.M.); i.caon@uninsubria.it (I.C.); a.parnigoni@uninsubria.it (A.P.); alberto.passi@uninsubria.it (A.P.); jenny.karousou@uninsubria.it (E.K.); davide.vigetti@uninsubria.it (D.V.)
3 Department of Biology and Biotechnology, University of Pavia, 27100 Pavia, Italy; jessica.canino@iusspavia.it (J.C.); ilaria.canobbio@unipv.it (I.C.)
* Correspondence: manuela.viola@uninsubria.it; Tel.: +39-0332-397143

Abstract: Cardiovascular diseases are a group of disorders caused by the presence of a combination of risk factors, such as tobacco use, unhealthy diet and obesity, physical inactivity, etc., which cause the modification of the composition of the vessel's matrix and lead to the alteration of blood flow, matched with an inflammation condition. Nevertheless, it is not clear if the inflammation is a permissive condition or a consequent one. In order to investigate the effect of inflammation on the onset of vascular disease, we treated endothelial cells with the cytokine TNF-α that is increased in obese patients and is reported to induce cardiometabolic diseases. The inflammation induced a large change in the extracellular matrix, increasing the pericellular hyaluronan and altering the heparan sulfate Syndecans sets, which seems to be related to layer permeability but does not influence cell proliferation or migration nor induce blood cell recruitment or activation.

Keywords: heparan sulfate; inflammation; Syndecans

1. Introduction

Cardiovascular diseases (CVD) are a group of pathologies of the vascular system that are growing in number and that usually present a build-up of fatty deposits inside the arteries (atherosclerosis) and an increased risk of blood clots and can also be associated with damage to arteries in organs such as the brain, heart, kidneys, and eyes. CVD are triggered by several risk factors, ranging from pathologies such as diabetes, hypercholesterolemia, and metabolic syndrome, to unhealthy habits, e.g., inactivity, smoking, and excessive alcohol. Apart from the medical advice that may include changing lifestyle and diet, the only medical treatment that can help decrease the risk is lowering cholesterol levels and controlling blood pressure and coagulation [1,2]. Despite the high numbers of data about CVD, the exact timing and molecular mechanism at the basis of the pathologies are still unclear, even though they are all inflammation related [3,4]. Some authors proposed that some of the risk factors lead to qualitative changes in the endothelium such as changes in permeability and increase in adhesion molecules expression that attract leucocytes ("response of injury" hypothesis) [5,6], which, in turn, cause the inflammatory condition. Moreover, the localized inflammation causes the thickening of the arterial wall due to the increased deposition of extracellular matrix (ECM) and the newly formed ECM traps lipoprotein and inflammatory/growth factors from the circulation within the vessel wall [3].

In obese patients, the endothelium is characterized by dysfunction associated with a condition of vascular low-grade inflammation in which an excess of TNF-α (tumor necrosis factor) is generated either in small vessels or within the perivascular adipose tissue [2]. TNF-α has also an effect on the release of nitric oxide (NO), which is produced to regulate vascular tone, cardiac contractility, and vascular remodeling [7].

The endothelial layer is composed of several components, the major ones including fibrous proteins such as collagen, fibronectin, elastin, laminin, and glycosaminoglycans (GAGs) chains such as hyaluronan (HA) or proteoglycans (PGs). The PGs and GAGs fill the interstitial space of tissues left free from fibrous protein, forming a well-organized network conferring tissue hydration [8] as well as function as co-receptors and therefore regulating their activity and that of growth factors and cytokines [1,9]. Both HA and heparan sulfate (HS) GAGs are pivotal in the maintenance of endothelial function, and their removal causes a loss of barrier properties comparable to an inflammatory condition [10]; moreover, several data pointed out the Syndecan family of HSPG as central in the endothelial behavior [11]. Syndecans are a family of four transmembrane HSPGs that are expressed in a cell-type-specific manner [11]; Syndecan-1 is present during development and in adult endothelium and cancer cells; Syndecan-2 is expressed in liver, mesenchymal tissue, and neuronal cells; Syndecan-3 is a neuronal type; and Syndecan-4 is ubiquitously distributed [12,13]. The complex mechanism of biosynthesis and modification of the HS chains generate a variability in N-sulfation levels along the polysaccharides. EXT1 and EXT2 are the polymerizing enzymes, while glucosaminyl N-deacetylase/N-sulfotransferase (NDST) is the first modification enzyme that starts to work on the growing heparan sulfate (HS) polysaccharide chain. This enzyme defines the sulphation pattern, which will determine the ability of the HS chain to interact with target molecules [14]. Moreover, the NDST1 has the capacity to bind to EXT2, and EXT1 and EXT2 expressions affect the N-sulfation degree, hinting that the overexpression of all the three enzymes happens simultaneously. The subsequent sulfations occur on the growing chains [15]; it is noteworthy that although there is only a single 2-O-sulfotransferase, there are three 6-O-sulfotransferases (6OST1-3) and seven 3-O-sulfotransferases [16], indicating the great importance of the sulfation pattern on the growth factor binding ability of the HS chains [17]. It produces regions rich in N-acetylated residue (GlcA and GlcNAc) called NA domain, regions rich in N-sulfated residue (IdoA and GlcNS derivates) called NS domain, and sequences that contain alternation of NA and NS. The ligands binding mainly depend upon the distribution of these domains [18,19]. The synthetic GAG machinery can be altered by several events, including the changes in PG core proteins expression or UDP-sugars transporters [20].

A preliminary clinical study in patients with resistant arterial hypertension (CVD risk factor) indicates high Syndecan-4 level as a potential marker for endothelial dysfunction [21]; moreover, the same HSPG has also been shown to regulate focal adhesions junctions, demonstrating the properties of a mechano-transducer effector [22], and importantly, its fragments generated following thrombin cleavage can modulate changes in endothelial barrier resistance [23]. Since one of the major events in atherosclerosis onset is the accumulation in the sub-endothelium of lipids driven by the lipoprotein LDL, the passage of such particles through the endothelial barrier is a critical step. Nevertheless, despite the many studies underscoring the relevance of plasma lipoproteins and the effects of lipids and cholesterol on the cell behavior and extracellular matrix architecture of the tunica intima [3], the data about the events causing the LDL transcytosis and accumulations are still scant. Links regarding Syndecan-4 metabolism/levels and the endothelial barrier system are not yet known.

Following the data reported, we made endothelial cells undergo inflammatory conditions using the TNF-α cytokine and investigated the effect in the main ECM components, i.e., HA and Syndecans, evaluating their effect on cell behavior, with regards to the pro-atherogenic aspect of membrane permeability.

2. Materials and Methods
2.1. Material

HUVEC, human umbilical vein endothelial cells (Gibco, Waltham, MA, USA); M200 culture medium (Gibco, Waltham, MA, USA); DMEM High Glucose w/o Sodium Pyruvate w/ L-Glutamine (Euro Clone, Pero, Italy); TNF-α (Sino Biological Inc., Wayne, NJ, USA); Protease Inhibitor Cocktail (Sigma-Aldrich, St. Louis, MO, USA); Hyaluronate Lyase from Streptomyces hyalurolyticus (Sigma-Aldrich, St. Louis, MO, USA); Transwell system with filter of 0.4 µm pore size, 6.5 mm diameter (Corning, New York, NY, USA); FITC labelled dextran (Mw~250,000, Sigma-Aldrich, St. Louis, MO, USA); heparinases I–II–III (form F. heparinum, Seikagaku, Tokyo, Japan); 2-Aminoacridone, AMAC-(Sigma-Aldrich, St. Louis, MO, USA); Chondroitinase ABC (from Proteus vulgaris, Seikagaku, Tokyo, Japan); 3-hydroxybiphenol (Fluka, Buchs, Switzerland); D-Glucuronic acid (Sigma-Aldrich, St. Louis, MO, USA); prostaglandin E1 and indomethacin (Sigma); all chemicals were purchased by Sigma-Aldrich (St. Louis, MO, USA).

2.2. Methods

2.2.1. Cell Cultures

Human umbilical vein endothelial cells (HUVEC) obtained from Gibco (Waltham, MA, USA), were grown for 4–8 passages in M200 culture medium (Gibco, Waltham, MA, USA) supplemented with 2% fetal bovine serum (FBS). The cultures were maintained in an atmosphere of humidified 95% air, 5% CO_2, at 37 °C. Twenty-four hours before treatments, subconfluent HUVEC were cultured in DMEM with 0.5% FBS. The medium was then changed to M200 with 0.1 µg/mL of TNF-α (Sino Biological Inc., Wayne, NJ, USA) and incubated for 24 or 48 h.

2.2.2. Quantitative RT-PCR

Total RNA samples were extracted from untreated or treated cells with an Absolutely RNA Microprep Kit (Agilent Technologies, Santa Clara, CA, USA). cDNA was generated by using the High-Capacity cDNA synthesis kit (Applied Biosystems, Foster City, CA, USA) and amplified on an Abi Prism 7000 instrument (Applied Biosystems, Foster City, CA, USA) using the Taqman Universal PCR Master Mix (Applied Biosystems). The following human TaqMan gene expression assays were used: HAS2 (Hs00193435_m1), HAS3 (Hs00193436_m1), NOS1 (Hs00167223_m1), NOS2 (Hs01075529_m1), NOS3 (Hs01574659_m1), SYND1 (Hs00174579_m1), SYND2 (Hs00299807_m1), SYND3 (Hs00206320_m1), SYND4 (Hs00161617_m1), NDST1 (Hs00155454_m1), EXT1 (Hs00609162_m1), EXT2 (Hs00181158_m1), and β-actin (Hs99999903_m1) as the reference gene. The relative quantification of gene expression levels was determined by comparing $2^{-\Delta\Delta Ct}$ [3,24] using non-treated cells, or the expression of HAS2 and Syndecan-1 as a normalizers.

2.2.3. Western Blotting

A RIPA buffer (50 mM Tris (pH 7.4), 150 mM NaCl, 1% TRITON X-100, 0.5% sodium deoxycholate, 0.1% SDS) containing Protease Inhibitor Cocktail (Sigma-Aldrich, St. Louis, MO, USA) was used to prepare cell lysates. Proteins were quantified, separated in a 12% SDS polyacrylamide gel electrophoresis, and transferred to nitrocellulose membrane. After incubation in blocking solution, 5% BSA in TBS-T (Tris-Buffered Saline: 0,02 M Tris, 0,136 M NaCl, 0,001 % Tween-20, pH 7.6), the membrane was incubated overnight with a primary antibody at 4 °C. Antibodies used were rabbit polyclonal antibody against Syndecan4 (ABT157, Merck Millipore, Burlington, VT, USA) dilution 1:250, and goat polyclonal antibody against β-actin (#J1805, Santa Cruz Biotechnology, Dallas, TX, USA), dilution 1:1000. The membrane was washed with TBS-T and incubated for 1 h with the secondary antibody. Band visualization was carried out by the chemiluminescence system LiteAblot TURBO (Euro Clone, Pero, Italy). The relative intensities of the protein bands were analyzed with ImageJ software. β-actin levels were used as controls for protein loading.

2.2.4. Cell Transfection

HUVEC were transfected with siRNA against syndecans4 (S12639, Ambion, Carlsbad, CA, USA) using a nucleofector apparatus (Amaxa, Basel, Switzerland) and the Amaxa HUVEC Nucleofector kit (Lonza, Basel, Switzerland) following the manufacturer's instructions. In total, 5 × 100,000 cells were resuspended in 100 µL HUVEC Nucleofector solution and transfected with 40 nM siRNA against syndecans4 and Silencer Negative Control siRNA #1 (AM4611, Ambion, Carlsbad, CA, USA). Syndecans4 silencing efficiency was determined by qRT-PCR. Silenced cells were treated 24 h after the transfection.

2.2.5. Cell Viability Assay

HUVEC metabolic activity was evaluated with the MTT assay. Cells were plated at a density of 6×1000 cells/well in a 96-well plate. After 16–18 h, HUVEC were treated with 0.1 µg/mL of TNF-α. After 4, 16, 24, or 48 h, the cells were washed with PBS, and MTT solution (50 µL of 5 mg/mL) was added to each well for 4 h at 37 °C. Subsequently, the medium was removed, and DMSO (Sigma-Aldrich, St. Louis, MO, USA) was added (200 µL/well) to solubilize the formazan crystals. Optical density was measured at 570 nm with the Tecan microplate reader (Thermo Scientific, Waltham, MA, USA).

2.2.6. Migration Assay

HUVECs were cultured until confluence in 6-well plates and serum-deprived (0.2% FBS) for 16–18 h. Three scratches per well were done with a 20 µL sterile pipette tip. Cells were washed to remove debris and incubated in fresh M200 with or without TNF-α. Images from three different scratch areas in each culture well were obtained using Olympus (Hamburg, Germany) IX51 microscope after 2, 4, 6, and 8 h.

2.2.7. Exclusion Assay

HUVEC pericellular coat was visualized and measured by using a particle exclusion assay. In total, 6×1000 cells/well were seeded in 12-well plate and treated with TNF-α or PBS as control. After 24 h, 500 µL of a suspension of formaldehyde-fixed erythrocytes ($15 \times 1,000,000$ erythrocytes/mL) was added to the wells and allowed to settle for 20 min at 37 °C. Images of the pericellular coat were obtained using phase contrast microscope Olympus IX51. The presence of HA on the pericellular coat was evaluated treating the cultures with 2 U/mL of Hyaluronate Lyase for 1 h at 37 °C before visualization with the particle exclusion assay. Representative cells were photographed at a magnification of ×40; the control experiment was performed with heat inactivated Hyaluronate lyase. ImageJ software was used to quantify the area delimited by red blood cells and the area delimited by the cell membrane to give a coat-to-cell ratio [25].

2.2.8. Permeability Assay

FITC-labelled dextran was used as the representative of hydrophilic molecules to measure the permeability of endothelial cell monolayer [26,27]. HUVEC were plated in the upper part of a Transwell filter with 0.4 µm pore size at a density of 8×1000 cells per well until the formation of a tight monolayer was checked with the microscope. Cells were treated with TNF-α (0.1 µg/mL), and FITC-dextran was added to the top chamber of the Transwell in a final concentration of 1 mg/mL. The culture medium in the upper and in the lower chamber was collected 24 h post-treatment, and fluorescence was measured by fluorimeter (Tecan, Thermo Scientific, Waltham, MA, USA) with an excitation wavelength of 490 nm and an emission wavelength of 520 nm. To evaluate the FITC-dextran passage through the cells monolayer, we calculated the percentage of lower over total fluorescence.

2.2.9. Glycosaminoglycans Purification and Quantification

Glycosaminoglycan from the culture medium or cell membrane was extracted using the protocol described in Viola et al. [28]. Briefly, after sample stimulation, conditioned media were collected as well as trypsin supernatants after cells harvesting (membrane GAGs). Samples were subjected to digestion with proteinase K (20 U/mL, Finnzymes, Espoo, Finland) and precipitation with ethanol (9:1 / ethanol:water).

HE/HS Δ-disaccharides were obtained digesting the pellet with a mix of heparinases I–II–III (form F. heparinum, Seikagaku, Tokyo, Japan) 0.5 U/mL each and then derivatized with AMAC (Sigma-Aldrich St. Louis, MO, USA). HE/HS Δ-disaccharides were analyzed and quantified by HPLC with respect to specific standards.

Intact HS GAGs were purified digesting the pellet with 0.1 U/mL U of Chondroitinase ABC for 5 h at 37 °C.

HS GAGs amount was calculated by means of the uronic acid content, using the van den Hoogen et al. method [29]. Briefly, 40 μL of the HS sample and 200 μL of concentrated sulfuric acid (80% w/w) were added in a 96-well plate. The plate was incubated for 1 h at 80 °C and, after cooling to room temperature, the background absorbance of samples was measured at 540 nm on a microplate reader (Tecan, Thermo Scientific, Waltham, MA, USA). Then, 40 μL of 3-hydroxybiphenol solution (100 μL of 100 mg/mL 3-hydroxybiphenol in DMSO mixed with 4.9 mL 80% (v/v) sulfuric acid) was added. After an overnight incubation, the absorbance was read again at 540 nm. D-Glucuronic acid (Sigma-Aldrich St. Louis, MO, USA) was used for a standard curve.

2.2.10. Data Analysis

Data are presented as mean ± S.E.M. Statistical significance was determined using Student's t test. Statistical significances were $p < 0.05$ for *, $p < 0.01$ for **, and $p < 0.001$ for ***.

3. Results

In the development of atherosclerosis, the endothelial dysfunction is one of the beginning steps or a permissive status of the endothelial layer for the onset of the pathology. Treatment of HUVEC cells with TNF-α can mimic the systemic inflammatory status of the endothelium [30,31].

In order to confirm the inflammatory condition of HUVEC cells, we analyzed nitric oxide synthases (NOSs) expression. Nitric oxide (NO) is important to maintain normal vascular functions and endothelial integrity. As expected, the endothelial isoform NOS3 was the most expressed form in HUVEC (Figure 1A), and the expression levels of NOS3 and NOS1 were significantly decreased after TNF-α stimulation, while NOS2 showed a non-significant tendency to decrease (Figure 1B). These data agree with the literature in which in vitro studies confirm the defect in the NO production in isolated atherosclerotic blood vessels [7,32].

Migration recorded through 8 h and vitality at 24 and 48 h were not affected by the cytokine (Supplementary Figure S1).

The extracellular matrix expressed by endothelial cells is commonly referred to as glycocalyx and has an important role in controlling shear stress from laminar flow through mechano-transduction mechanisms [14] and inflammation, thus controlling cell adhesion, motility, and proliferation [15–17].

The HA production has also an important role in the maintenance of cell homeostasis and in activation of different signal transduction pathways [33]. In HUVEC cells, HAS3 mRNA was the most abundant (Figure 2A), whereas HAS1 messenger was not detected (data not shown). Interestingly, after TNF-α stimulation, HAS2 increased expression while HAS3 was decreased (Figure 2B). In order to evaluate the glycocalyx of the HUVEC, we quantified the glycosaminoglycans from the membrane and from the medium with no significant differences (Supplementary Figure S3). The pericellular coat surrounding the endothelial cells was measured and showed a significant increase after TNF-α stimulation, which was mainly constituted of HA as demonstrated by enzymatic digestion (Figure 2C).

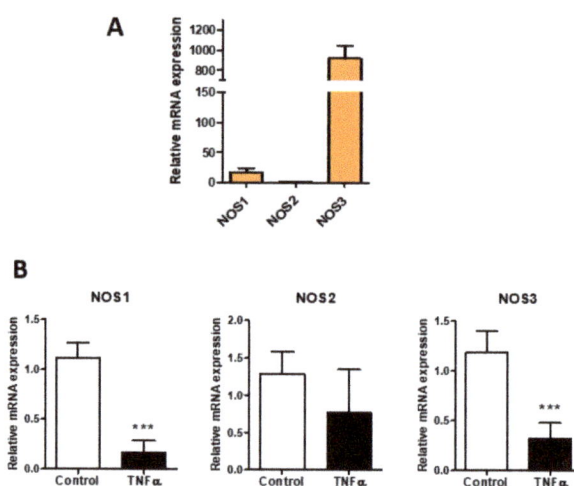

Figure 1. Effect of TNF-α on NO synthetic enzymes in HUVEC. (**A**) relative expression of NOSs (neuronal NOS1, inducible NOS2, and endothelial NOS3) in HUVEC. (**B**) NOSs expression in HUVEC untreated (control) and treated with TNF-α (0.1 μg/mL) for 24 h. Data are mean ± S.E.M. of three independent experiment, *** $p < 0.001$.

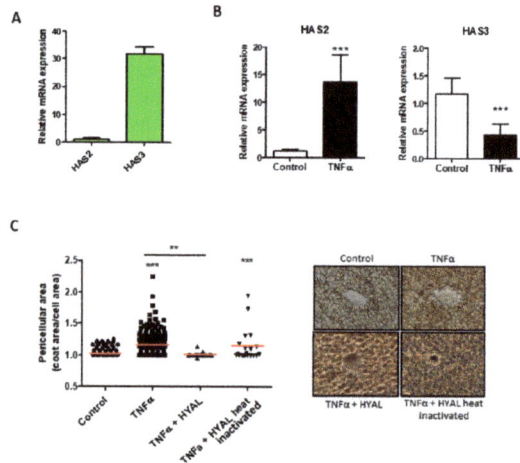

Figure 2. Effect of TNF-α on Hyaluronan synthesis in HUVEC. (**A**) HASs expression profile in HUVEC. The reference gene used for normalization was β-actin and the normalizer HAS2 expression. (**B**) Relative expression of HAS2 and HAS3 after TNF-α stimulation (24 h). The reference gene used for normalization was β-actin and the normalizer untreated samples. Data are mean ± S.E.M. of four independent experiments, *** $p < 0.001$. (**C**). Particle exclusion assay performed on HUVEC untreated (control) and under TNF-α stimulation for 24 h. To clarify the HA composition of the pericellular matrix, we digested HA with 2 U/mL of Hyaluronate Lyase from Streptomyces hyalurolyticus (HYAL) before the addition of erythrocytes. Original magnification 40×. Values represent the measure of the single cell pericellular area, and the red bars are the mean of three independent experiments, *** $p < 0.001$ and ** $p < 0.01$.

Due to the lining of the vessels, ECM is also important in recruitment and activation of immune cells and of platelets from the blood [34–36]. Among all the HSPGs, Syndecans are a family of four transmembrane proteoglycans acting as co-receptors interacting with different molecules including growth factors, matrix components, and cytokines that are present in glycocalyx [11]. In HUVEC, the main Syndecans expressed are the -3 and -4 isoforms (Figure 3A), but only Syndecan-4 increased during TNF-α stimulation from 24 up to 48 h (Figure 3B). The core protein of the proteoglycan was evaluated in the cell extraction and shown in Western blot and turned out to be increased, even if not significantly (Figure 3C). The Western blot shows three different bands positive to antibody recognition. The three different bands at around 27, 37, and 45 kDa can be the proteoglycan with different GAG chains [37], Syndecan-4 bound to growth factor or matrikines, and/or its homo- or hetero-oligomerization forms [38].

As reported, the NDST1 has the capacity to bind to EXT2, and EXT1 and EXT2 expressions affect the N-sulfation degree, suggesting that the overexpression of all the three enzymes happens simultaneously, as shown in Supplementary Figure S3.

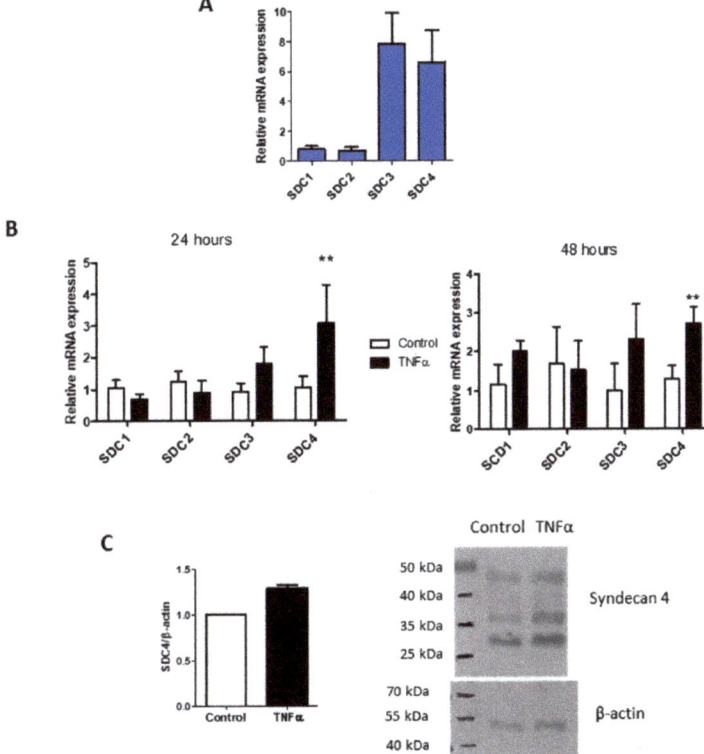

Figure 3. TNF-α influence on Syndecans expression. (**A**) Syndecans expression profile in HUVEC. The reference gene used for normalization was β-actin and the normalizer the Syndecan-1 expression level. (**B**) Syndecans isoforms expressions in HUVEC control and after 24- and 48-h of TNF-α stimulation. The reference gene used for normalization was β-actin and the normalizer untreated samples. Values represent mean ± S.E.M. (n = 3), ** $p < 0.01$. (**C**) Western blot analysis of Syndecan-4 (SDC4) protein in HUVEC control and treated 24 h with TNF-α. Bar chart represents normalized mean ± S.E.M. of two independent experiments and the figure is a representative SDS-PAGE.

The GAG moiety of the HUVEC HSPGs was analyzed by enzymatic digestion followed by HPLC analysis, and the disaccharide percentages are reported in Table 1. The main drastic difference seems related to the increment of N-sulfation on glucosamine residue. The higher amount of N-sulfation correlates well with the increment of expression of the enzyme NDST1, heparan sulfate N-deacetylase/N-sulfotransferase 1 (Supplementary Figure S3), that catalyzes both the N-deacetylation and the N-sulfation of glucosamine.

Table 1. HPLC analysis of the main HS/HE disaccharides. GAGs were isolated from plasma membrane and from culture medium of HUVEC control and TNF-α treated (24 h). To obtain HS/HE disaccharides, we digested GAGs with heparinases. After AMAC derivatization, the disaccharides were analyzed by means of HPLC. Data are expressed as % area of each HS disaccharide/ % area total. The N-sulfation in bold (NS) is catalyzed by NDST1. Values are mean ± SD of three independent experiments ** $p < 0.01$. UA: uronic acid; GlcNAc: N-acetyl; GlcNS: N-sulphonyl glucosamine: S: sulphate group [39].

	GAG medium		GAG membrane	
	Control	TNF-α	Control	TNF-α
ΔUA-2S-β[1→4]-GlcNS-6S	0.7 ± 1.3	0.4 ± 0.7	4.9 ± 6.9	5.5 ± 7.8
ΔUA-β[1→4]-GlcNS-6S	1.2 ± 1.0	2.9 ± 0	0	91 ± 6 **
ΔUA-2S-β[1→4]-GlcNS	0.4 ± 0.7	2.5 ± 4.3	0.9 ± 1.3	0.7 ± 0.5
ΔUA-β[1→4]-GlcNS	3.8 ± 5.0	34 ± 21	34 ± 23	0.2 ± 0.2
ΔUA-β[1→4]-GlcNAc-6S	66 ± 2	48 ± 36	55 ± 21	0
ΔUA-β[1→4]-GlcNAc	50 ± 39	13 ± 19	5.3 ± 6.6	2.3 ± 0.7

The major event in atherosclerosis onset is the accumulation in the sub-endothelium of lipids driven by the lipoprotein LDL [3]; nevertheless, the data about the events causing the LDL particle transcytosis and accumulations within the tunica intima are still scant.

To test whether the endothelial permeability is altered under the inflammatory condition, we incubated a continued layer of HUVEC in a transwell system with FITC-dextran using a permeable membrane with a cut-off unable to let the cells pass. As reported in Figure 4A, the presence of the HUVEC layer (control) blocks the free passage of the fluorescent dextran, and the same cell under the inflammatory condition of TNF-α increases the blocking by a significant, even if small, amount.

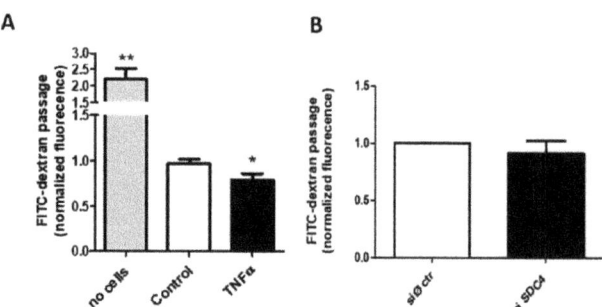

Figure 4. Transwell permeability assay. (**A**) FITC-dextran flow through HUVEC monolayer. Confluent HUVEC cells in the upper chamber of a transwell system +/− TNF-α were added with 1 mg/mL of dextran conjugated with FITC. After 24 h, the medium of the lower chamber was collected and FITC fluorescence was measured, * $p < 0.05$ and ** $p < 0.01$. (**B**). FITC-dextran flow through siRNA control (siØ ctr) or siRNA against SDC4 HUVEC monolayers; data are mean ± S.E.M. and n = 3.

Since Syndecan-4 exerts various effects on the endothelial glycocalyx, with particular regard to TNF-α induced endothelial modifications [40], and has a pivotal role in the dynamics of focal adhesion [41] and in the formation of networks at gap junctions [42], we investigated the HUVEC permeability of SDC4-silenced cells (Figure 4B) (silencing efficiency 80%, data not shown). The abrogation of the proteoglycan does not alter the dextran passage, even considering the high silencing levels, thus indicating a complex metabolism and turnover for the proteoglycan as well as multiple control levels for the layer permeability.

4. Discussion

The atherosclerotic process in which we are interested is a combination of several factors, among which high cholesterol levels, driven by lipoprotein LDL, vasculature inflammation, and oxidative processes are the main components, deeply studied but still not easily correlated [1,3,43]. In particular, it is not yet clear whether the vessel inflammatory condition can influence the LDL passage to the subendothelial space and alter the activation of blood components through the interaction with the glycocalyx. In the in vitro model of inflammation and atherosclerosis using endothelial cells [44] and smooth muscle cells [3], we highlighted the role of the GAG HA in recruiting monocytes/macrophages from blood, promoting their adhesion to the endothelial layer, and altering the ECM composition and thickness of the intima layer.

We determine to use the pro-inflammatory cytokine TNF-α, since it is well known that is produced in the visceral fat of obese and overweight patients that are prone to CVD. Moreover, this chronic low-grade vascular inflammation is hypothesized to be the stimulus by which, within vasculature, reactive oxygen species (ROS) are generated through NAD(P)H oxidase activation and other sources, which in turn reduces NO availability, causing the local endothelial dysfunctions [2]. Therefore, TNF-α can contribute to vascular changes, favouring the development and acceleration of the atherothrombotic process in the clinical condition [2]. In this view, we used the pro-inflammatory cytokine TNF-α treatment on HUVEC cells as a model of vasculature inflammation, investigating the glycocalyx modifications and their effect of the endothelial barrier.

The effects of TNF-α on the NO production shown in Figure 1 confirm the effectiveness of the treatment, and even if the data are only the expression of the synthetic enzymes, this evidence is related to the NO levels [45]. The action of TNF-α on the HA synthesis is reported in Figure 2 and confirms the data previously found with a different cytokine, the IL-1β. Briefly, the synthesis of HA is increased by means of the overexpression of the membrane synthases HAS2, and the GAG is mainly localized around the cells as a pericellular coat that increases the distance between the cells. Among the synthetic HA enzymes, HAS2 is known to be the main producer of the polymer [3]. The HAS3 enzyme decrease in this phase is not yet well understood and deserves other investigations in order to unravel the exact role of the enzyme in the membrane. During the early stage of atherosclerotic lesion formation in Apolipoprotein E (Apoe)-deficient mice, the HAS3 expression is increased and controlled in vascular smooth muscle cells by the cytokine IL-1β [46], and even if data on endothelial cells are unavailable, there are clear indications that HAS3 might be a promising therapeutic target in atherosclerosis.

It is noteworthy that this pericellular coat can increase the adhesion and recruitment of the circulating monocytes [44] and protect cells from apoptotic events [47]. Since HA can also exert its effect on cell proliferation and migration depending on its dimensions [48–50] and organized sovramolecular architecture by soluble factors (e.g., TSG-6) [51], the mere synthesis of the polymer is not predictive of its effect. In our model, in fact, the accumulation of HA in the pericellular coat does not alter cell proliferation and migration as reported in Supplementary Figures S1 and S2, which was in line with the barrier role of the endothelial cells, but nevertheless can favour the adhesion and transmigration of blood cells to the sub-endothelium.

Inflammation is the cause of multiple effects on the endothelium, including changing the glycocalyx composition [10,19,52]. The heparan sulfate proteoglycan belonging to the Syndecans family is largely involved in the onset of different CVD in vasculature, and in particular is reported to be increased in patients with resistant hypertension [21], while syndecan-3/-4 ectodomain fragments, produced by several stimuli, including heparinase or thrombin, decrease endothelial cell–cell adhesive barrier integrity [23] and are involved in the cell-extracellular matrix and cell–cell adhesion mechanisms [22]. HSPGs in endothelial cells include glypicans, located on the cell membrane, anchored by glycosylphosphatidylinositol (GPI), functioning mainly as modulators of growth factor signaling [53], but they seem more critically involved in developmental morphogenesis and positively correlate with the onset of certain types of cancers [54] and less with vascular inflammation.

The TNF-α treatment on HUVEC cells selectively increases the expression of Syndecan-4, and the expression remains high after the incubation for up to 48 h (Figure 3); the increase of syndecan-4 is also evident at the protein level, and is measurable with Western blot, even if the results are not significant. This finding can be either due to the low level of expression of the PG in the cells that do not consent to the change to be significant in the assay sensibility range, or to the turnover rate of Syndecan-4, which is formed by the protein core and the GAG moiety. In our data, in fact, we investigated the composition of the heparan sulfate chains of the membrane-bound PGs as well as those released in the medium and evidenced a difference in the percentage of various disaccharides. The different sulfation of the disaccharides can lead to a modification in the sulfation pattern of the GAG chains, and this can be the real important event for the Syndecan-4 related effects, while the protein core can be maintained at the same amount but continuously replaced.

The total amount of uronic containing GAGs remains invariable (Supplementary Figure S4), indicating a balance between the synthesis of the polymer chains that use the same UDP-sugar precursors as we underlined in a previous paper, involving, in particular HA and HS [20], but a control on the number and dimension of the various chains was impossible. As shown in Table 1 the main changes in disaccharides involve the NS and 6 sulfation. Unfortunately, the small size of the samples and high biological variability among them make it impossible to have statistically significant data, but the trend is very sharp. Together with the changes in the protein core of Syndecan-4 and in the NS sulfation, the synthetic enzymes EXT1 and EXT2, responsible for the polymerization, and NDST1 (N-deacetylase/N-sulfotransferase) are also increased in the TNF-α treated samples (Supplementary Figure S3). The sulfation pattern of the HS chains is frequently found altered in the inflammatory condition of tumours, as reported by several papers that indicate specific HS sulfotransferase as critical for the survival and invasiveness of those cells, for example, the 3-O-sulfotransferase [55] or the 2-O-sulfotransferase [56]. Moreover, the 6-O-sulfation of HS highly influences the polysaccharide structural diversity and is critically involved in the binding of many proteins, in particular growth factors [17].

The NDST1 enzyme is the component that catalyses both the N-deacetylation and the N-sulfation of glucosamine (GlcNAc) residue in the heparan sulfate. This enzyme modifies the GlcNAc-GlcA disaccharide-repeating sugar backbone to make N-sulfated heparosan, a prerequisite substrate for later modifications in heparin biosynthesis, such as 6-O-sulfation [15]. Therefore, the increase in the N-sulfation along the chain consequently also increases the sulfation in the C6, as summarized in Table 1. This modification seems more important in the chains bound to the membrane, i.e., carried on the proteoglycan core. For what concerns the CVD, it was reported that the protein PCSK9, an important drug target because of its crucial role in lipid metabolism, can interact with HS in N-sulfation rich domain [57], eventually linking the high cholesterol load with the vasculature inflammation at the onset of the pathology.

The mechanism of endothelial dysfunction involves the alteration of the layer permeability demonstrated in Figure 4, which is dependent upon the overexpression of Syndecan-4, while the silencing of the protein core of the PG does not change the barrier function of the membrane. The increase in Syndecan-4 after treatment with TNF-α can cause a rearrangement in the adhesion asset of the endothelial layer due to its involvement in the dynamics of focal adhesion and the formation of networks at gap junctions. The different integration of the cells with each other and with the basement membrane can be the molecular mechanism leading to lower permeability of the endothelial layer. Nevertheless, this mechanism needs further investigations to be clarified, in particular regarding the interaction with other membrane components.

5. Conclusions

This research work aims to close small gaps in the sequence events on the onset of vascular lesion at the basis of various cardiovascular diseases, such as atherosclerosis.

The hypothesis we followed is the establishment of a first inflammatory condition due to physical and environmental state (diet and sport exercise habits, health conditions, etc.), and the results we obtained seem to positively correlate with the onset of a pathological state: (i) alteration of the endothelial barrier properties (i.e., membrane permeability); (ii) increase of HA in the pericellular coat and therefore of the monocyte recruitment possibility from blood; (iii) alteration of the sulfation pattern of membrane-bound HS which can cause modifications of the endothelium response to growth factor and cytokines, as well as of the lipid metabolism through the association HS/PCSK9/LDL-receptor; (iv) HAS3 enzymes abundantly decrease in these conditions without affecting the HA amount, which suggests a different role in the cell behaviour.

Concluding, we can assess that inflammation is the leading event of CVD and that it is of pivotal importance to understand if the inflammatory HS carries specific sequences connected to the various events and unravel the role of HAS3 in endothelial cells.

Supplementary Materials: The following are available online at https://www.mdpi.com/article/10.3390/biom11060809/s1, Figure S1: MTT vitality assay. Figure S2: Wound healing assay of HUVEC. Figure S3: Expression levels of Syndecan chains biosynthetic enzymes. Figure S4: GAGs quantification.

Author Contributions: Conceptualization, M.V. and A.P. (Alberto Passi); methodology, E.C., P.M., I.C. (Ilaria Caon); investigations, E.C. and J.C.; writing—original draft preparation, E.C. and M.V.; writing—review and editing: D.V. and I.C. (Ilaria Caon); critical review, E.C., A.P. (Arianna Parnigoni) and I.C. (Ilaria Canobbio).; supervision and project administration, M.V.; funding acquisition, A.P. (Alberto Passi), M.V. and E.K. All authors have read and agreed to the published version of the manuscript.

Funding: This work was supported by PRIN2017 to E.K. (prot. 2017T8CMCY), FAR-University of Insubria, and EU grant RISE-HORIZON 2020 (ID645756) to A.P.

Acknowledgments: E.C. was a PhD student of the "Biotechnology, Biosciences, and Surgical Technology" course at Università degli Studi dell'Insubria; Ar.P. is a PhD student of the "Life Science and Biotechnology" course at Università degli Studi dell'Insubria.

Conflicts of Interest: The author(s) declared no potential conflicts of interest with respect to the research, authorship, and/or publication of this article.

References

1. Viola, M.; Karousou, E.; D'Angelo, M.L.; Moretto, P.; Caon, I.; Luca, G.; Passi, A.; Vigetti, D. Extracellular Matrix in Atherosclerosis: Hyaluronan and Proteoglycans Insights. *Curr. Med. Chem.* **2016**, *23*, 2958–2971. [CrossRef]
2. Virdis, A.; Colucci, R.; Bernardini, N.; Blandizzi, C.; Taddei, S.; Masi, S. Microvascular Endothelial Dysfunction in Human Obesity: Role of TNF-α. *J. Clin. Endocrinol. Metab.* **2019**, *104*, 341–348. [CrossRef] [PubMed]
3. Viola, M.; Bartolini, B.; Vigetti, D.; Karousou, E.; Moretto, P.; Deleonibus, S.; Sawamura, T.; Wight, T.N.; Hascall, V.C.; De Luca, G.; et al. Oxidized low density lipoprotein (LDL) affects hyaluronan synthesis in human aortic smooth muscle cells. *J. Biol. Chem.* **2013**, *288*, 29595–29603. [CrossRef] [PubMed]
4. Viola, J.; Soehnlein, O. Atherosclerosis—A matter of unresolved inflammation. In *Seminars in Immunology*; Academic Press: Cambridge, MA, USA, 2015. [CrossRef]
5. Maiolino, G.; Rossitto, G.; Caielli, P.; Bisogni, V.; Rossi, G.P.; Calò, L.A. The role of oxidized low-density lipoproteins in atherosclerosis: The myths and the facts. *Mediat. Inflamm.* **2013**, *2013*, 714653. [CrossRef] [PubMed]
6. Ross, R. Atherosclerosi—An inflammatory disease. *N. Engl. J. Med.* **1999**, *340*, 115–126. [CrossRef]
7. Matthys, K.E.; Bult, H. Nitric oxide function in atherosclerosis. *Mediat. Inflamm.* **1997**, *6*, 3–21. [CrossRef] [PubMed]
8. Iozzo, R.V.; Schaefer, L. Proteoglycan form and function: A comprehensive nomenclature of proteoglycans. *Matrix Biol.* **2015**, *42*, 11–55. [CrossRef]
9. Bartolini, B.; Caravà, E.; Caon, I.; Parnigoni, A.; Moretto, P.; Passi, A.; Vigetti, D.; Viola, M.; Karousou, E. Heparan Sulfate in the Tumor Microenvironment. *Adv. Exp. Med. Biol.* **2020**, *1245*, 147–161. [CrossRef]

10. Delgadillo, L.F.; Lomakina, E.B.; Kuebel, J.; Waugh, R.E. Changes in endothelial glycocalyx layer protective ability after inflammatory stimulus. *Am. J. Physiol. Cell Physiol.* **2020**. [CrossRef]
11. Götte, M. Syndecans in inflammation. *FASEB J.* **2003**, *17*, 575–591. [CrossRef]
12. Schaefer, L.; Schaefer, R.M. Proteoglycans: From structural compounds to signaling molecules. *Cell Tissue Res.* **2010**, *339*, 237–246. [CrossRef] [PubMed]
13. Couchman, J.R. Transmembrane signaling proteoglycans. *Annu. Rev. Cell Dev. Biol.* **2010**, *26*, 89–114. [CrossRef]
14. Kjellén, L. Glucosaminyl N-deacetylase/N-sulphotransferases in heparan sulphate biosynthesis and biology. *Biochem. Soc. Trans.* **2003**, *31*, 340–342. [CrossRef]
15. Dou, W.; Xu, Y.; Pagadala, V.; Pedersen, L.C.; Liu, J. Role of Deacetylase Activity of N-Deacetylase/N-Sulfotransferase 1 in Forming N-Sulfated Domain in Heparan Sulfate. *J. Biol. Chem.* **2015**, *290*, 20427–20437. [CrossRef]
16. Kreuger, J.; Kjellén, L. Heparan sulfate biosynthesis: Regulation and variability. *J. Histochem. Cytochem.* **2012**, *60*, 898–907. [CrossRef]
17. Kjellén, L.; Lindahl, U. Specificity of glycosaminoglycan-protein interactions. *Curr. Opin. Struct. Biol.* **2018**, *50*, 101–108. [CrossRef]
18. Jackson, R.L.; Busch, S.J.; Cardin, A.D. Glycosaminoglycans: Molecular properties, protein interactions, and role in physiological processes. *Physiol. Rev.* **1991**, *71*, 481–539. [CrossRef]
19. Whitelock, J.M.; Iozzo, R.V. Heparan sulfate: A complex polymer charged with biological activity. *Chem. Rev.* **2005**, *105*, 2745–2764. [CrossRef]
20. Viola, M.; Brüggemann, K.; Karousou, E.; Caon, I.; Caravà, E.; Vigetti, D.; Greve, B.; Stock, C.; De Luca, G.; Passi, A.; et al. MDA-MB-231 breast cancer cell viability, motility and matrix adhesion are regulated by a complex interplay of heparan sulfate, chondroitin-/dermatan sulfate and hyaluronan biosynthesis. *Glycoconj. J.* **2017**, *34*, 411–420. [CrossRef]
21. Lipphardt, M.; Dihazi, H.; Maas, J.H.; Schäfer, A.K.; Amlaz, S.I.; Ratliff, B.B.; Koziolek, M.J.; Wallbach, M. Syndecan-4 as a Marker of Endothelial Dysfunction in Patients with Resistant Hypertension. *J. Clin. Med.* **2020**, *9*, 3051. [CrossRef]
22. Gopal, S.; Multhaupt, H.A.B.; Pocock, R.; Couchman, J.R. Cell-extracellular matrix and cell-cell adhesion are linked by syndecan-4. *Matrix Biol.* **2017**, *60–61*, 57–69. [CrossRef]
23. Jannaway, M.; Yang, X.; Meegan, J.E.; Coleman, D.C.; Yuan, S.Y. Thrombin-cleaved syndecan-3/-4 ectodomain fragments mediate endothelial barrier dysfunction. *PLoS ONE* **2019**, *14*, e0214737. [CrossRef]
24. Livak, K.J.; Schmittgen, T.D. Analysis of relative gene expression data using real-time quantitative PCR and the 2(-Delta Delta C(T)) Method. *Methods* **2001**, *25*, 402–408. [CrossRef] [PubMed]
25. Vigetti, D.; Rizzi, M.; Viola, M.; Karousou, E.; Genasetti, A.; Clerici, M.; Bartolini, B.; Hascall, V.C.; De Luca, G.; Passi, A. The effects of 4-methylumbelliferone on hyaluronan synthesis, MMP2 activity, proliferation, and motility of human aortic smooth muscle cells. *Glycobiology* **2009**, *19*, 537–546. [CrossRef]
26. Lal, B.K.; Varma, S.; Pappas, P.J.; Hobson, R.W.; Durán, W.N. VEGF increases permeability of the endothelial cell monolayer by activation of PKB/akt, endothelial nitric-oxide synthase, and MAP kinase pathways. *Microvasc. Res.* **2001**, *62*, 252–262. [CrossRef]
27. Simoneau, B.; Houle, F.; Huot, J. Regulation of endothelial permeability and transendothelial migration of cancer cells by tropomyosin-1 phosphorylation. *Vasc. Cell* **2012**, *4*, 18. [CrossRef] [PubMed]
28. Viola, M.; Vigetti, D.; Karousou, E.; Bartolini, B.; Genasetti, A.; Rizzi, M.; Clerici, M.; Pallotti, F.; De Luca, G.; Passi, A. New electrophoretic and chromatographic techniques for analysis of heparin and heparan sulfate. *Electrophoresis* **2008**, *29*, 3168–3174. [CrossRef]
29. van den Hoogen, B.M.; van Weeren, P.R.; Lopes-Cardozo, M.; van Golde, L.M.; Barneveld, A.; van de Lest, C.H. A microtiter plate assay for the determination of uronic acids. *Anal. Biochem.* **1998**, *257*, 107–111. [CrossRef] [PubMed]
30. Bilgic Gazioglu, S.; Akan, G.; Atalar, F.; Erten, G. PAI-1 and TNF-α profiles of adipose tissue in obese cardiovascular disease patients. *Int. J. Clin. Exp. Pathol.* **2015**, *8*, 15919–15925.
31. Cawthorn, W.P.; Sethi, J.K. TNF-alpha and adipocyte biology. *FEBS Lett.* **2008**, *582*, 117–131. [CrossRef]
32. Napoli, C.; Ignarro, L.J. Nitric oxide and atherosclerosis. *Nitric Oxide* **2001**, *5*, 88–97. [CrossRef] [PubMed]
33. Caon, I.; Bartolini, B.; Parnigoni, A.; Caravà, E.; Moretto, P.; Viola, M.; Karousou, E.; Vigetti, D.; Passi, A. Revisiting the hallmarks of cancer: The role of hyaluronan. *Semin. Cancer Biol.* **2020**, *62*, 9–19. [CrossRef]
34. Kolářová, H.; Ambrůzová, B.; Svihálková Šindlerová, L.; Klinke, A.; Kubala, L. Modulation of endothelial glycocalyx structure under inflammatory conditions. *Mediat. Inflamm.* **2014**, *2014*, 694312. [CrossRef] [PubMed]
35. Kumar, A.V.; Katakam, S.K.; Urbanowitz, A.K.; Gotte, M. Heparan sulphate as a regulator of leukocyte recruitment in inflammation. *Curr. Protein Pept. Sci.* **2015**, *16*, 77–86. [CrossRef] [PubMed]
36. Ibrahim, S.A.; Hassan, H.; Vilardo, L.; Kumar, S.K.; Kumar, A.V.; Kelsch, R.; Schneider, C.; Kiesel, L.; Eich, H.T.; Zucchi, I.; et al. Syndecan-1 (CD138) modulates triple-negative breast cancer stem cell properties via regulation of LRP-6 and IL-6-mediated STAT3 signaling. *PLoS ONE* **2013**, *8*, e85737. [CrossRef] [PubMed]
37. Liao, W.C.; Yen, H.R.; Chen, C.H.; Chu, Y.H.; Song, Y.C.; Tseng, T.J.; Liu, C.H. CHPF promotes malignancy of breast cancer cells by modifying syndecan-4 and the tumor microenvironment. *Am. J. Cancer Res.* **2021**, *11*, 812–826. [PubMed]
38. Hamon, M.; Mbemba, E.; Charnaux, N.; Slimani, H.; Brule, S.; Saffar, C.; Vassy, R.; Prost, C.; Lievre, N.; Starzec, A.; et al. A syndecan-4/CXCR4 complex expressed on human primary lymphocytes and macrophages and HeLa cell line binds the CXC chemokine stromal cell-derived factor-1 (SDF-1). *Glycobiology* **2004**, *14*, 311–323. [CrossRef] [PubMed]

39. Karousou, E.G.; Viola, M.; Vigetti, D.; Genasetti, A.; Rizzi, M.; Clerici, M.; Bartolini, B.; De Luca, G.; Passi, A. Analysis of glycosaminoglycans by electrophoretic approach. *Curr. Pharm. Anal.* **2008**, *4*, 78–89. [CrossRef]
40. Lipphardt, M.; Song, J.W.; Goligorsky, M.S. Sirtuin 1 and endothelial glycocalyx. *Pflug. Arch.* **2020**, *472*, 991–1002. [CrossRef]
41. Elfenbein, A.; Simons, M. Syndecan-4 signaling at a glance. *J. Cell Sci.* **2013**, *126*, 3799–3804. [CrossRef] [PubMed]
42. Horiguchi, K.; Kouki, T.; Fujiwara, K.; Tsukada, T.; Ly, F.; Kikuchi, M.; Yashiro, T. Expression of the proteoglycan syndecan-4 and the mechanism by which it mediates stress fiber formation in folliculostellate cells in the rat anterior pituitary gland. *J. Endocrinol.* **2012**, *214*, 199–206. [CrossRef]
43. Viola, M.; Vigetti, D.; Karousou, E.; D'Angelo, M.L.; Caon, I.; Moretto, P.; De Luca, G.; Passi, A. Biology and biotechnology of hyaluronan. *Glycoconj. J.* **2015**, *32*, 93–103. [CrossRef]
44. Vigetti, D.; Genasetti, A.; Karousou, E.; Viola, M.; Moretto, P.; Clerici, M.; Deleonibus, S.; De Luca, G.; Hascall, V.; Passi, A. Proinflammatory cytokines induce hyaluronan synthesis and monocyte adhesion in human endothelial cells through hyaluronan synthase 2 (HAS2) and the nuclear factor-kappaB (NF-kappaB) pathway. *J. Biol. Chem.* **2010**, *285*, 24639–24645. [CrossRef]
45. Giaroni, C.; Marchet, S.; Carpanese, E.; Prandoni, V.; Oldrini, R.; Bartolini, B.; Moro, E.; Vigetti, D.; Crema, F.; Lecchini, S.; et al. Role of neuronal and inducible nitric oxide synthases in the guinea pig ileum myenteric plexus during in vitro ischemia and reperfusion. *Neurogastroenterol. Motil.* **2013**, *25*, e114–e126. [CrossRef]
46. Homann, S.; Grandoch, M.; Kiene, L.S.; Podsvyadek, Y.; Feldmann, K.; Rabausch, B.; Nagy, N.; Lehr, S.; Kretschmer, I.; Oberhuber, A.; et al. Hyaluronan synthase 3 promotes plaque inflammation and atheroprogression. *Matrix Biol.* **2018**, *66*, 67–80. [CrossRef]
47. Vigetti, D.; Rizzi, M.; Moretto, P.; Deleonibus, S.; Dreyfuss, J.M.; Karousou, E.; Viola, M.; Clerici, M.; Hascall, V.C.; Ramoni, M.F.; et al. Glycosaminoglycans and Glucose Prevent Apoptosis in 4-Methylumbelliferone-treated Human Aortic Smooth Muscle Cells. *J. Biol. Chem.* **2011**, *286*, 34497–34503. [CrossRef] [PubMed]
48. Viola, M.; Karousou, E.; D'Angelo, M.L.; Caon, I.; De Luca, G.; Passi, A.; Vigetti, D. Regulated Hyaluronan Synthesis by Vascular Cells. *Int. J. Cell Biol.* **2015**, *2015*, 208303. [CrossRef] [PubMed]
49. Karamanos, N.K.; Piperigkou, Z.; Theocharis, A.D.; Watanabe, H.; Franchi, M.; Baud, S.; Brézillon, S.; Götte, M.; Passi, A.; Vigetti, D.; et al. Proteoglycan Chemical Diversity Drives Multifunctional Cell Regulation and Therapeutics. *Chem. Rev.* **2018**, *118*, 9152–9232. [CrossRef]
50. Tavianatou, A.G.; Caon, I.; Franchi, M.; Piperigkou, Z.; Galesso, D.; Karamanos, N.K. Hyaluronan: Molecular size-dependent signaling and biological functions in inflammation and cancer. *FEBS J.* **2019**, *286*, 2883–2908. [CrossRef] [PubMed]
51. Day, A.J.; Milner, C.M. TSG-6: A multifunctional protein with anti-inflammatory and tissue-protective properties. *Matrix Biol.* **2019**, *78–79*, 60–83. [CrossRef]
52. Potje, S.R.; Paula, T.D.; Paulo, M.; Bendhack, L.M. The Role of Glycocalyx and Caveolae in Vascular Homeostasis and Diseases. *Front. Physiol.* **2020**, *11*, 620840. [CrossRef]
53. Wang, S.; Qiu, Y.; Bai, B. The Expression, Regulation, and Biomarker Potential of Glypican-1 in Cancer. *Front. Oncol.* **2019**, *9*, 614. [CrossRef] [PubMed]
54. Li, N.; Spetz, M.R.; Ho, M. The Role of Glypicans in Cancer Progression and Therapy. *J. Histochem. Cytochem.* **2020**, *68*, 841–862. [CrossRef]
55. Vijaya Kumar, A.; Salem Gassar, E.; Spillmann, D.; Stock, C.; Sen, Y.P.; Zhang, T.; Van Kuppevelt, T.H.; Hülsewig, C.; Koszlowski, E.O.; Pavao, M.S.; et al. HS3ST2 modulates breast cancer cell invasiveness via MAP kinase- and Tcf4 (Tcf7l2)-dependent regulation of protease and cadherin expression. *Int. J. Cancer* **2014**, *135*, 2579–2592. [CrossRef] [PubMed]
56. Vijaya Kumar, A.; Brézillon, S.; Untereiner, V.; Sockalingum, G.D.; Kumar Katakam, S.; Mohamed, H.T.; Kemper, B.; Greve, B.; Mohr, B.; Ibrahim, S.A.; et al. HS2ST1-dependent signaling pathways determine breast cancer cell viability, matrix interactions, and invasive behavior. *Cancer Sci.* **2020**, *111*, 2907–2922. [CrossRef] [PubMed]
57. Gustafsen, C.; Olsen, D.; Vilstrup, J.; Lund, S.; Reinhardt, A.; Wellner, N.; Larsen, T.; Andersen, C.B.F.; Weyer, K.; Li, J.P.; et al. Heparan sulfate proteoglycans present PCSK9 to the LDL receptor. *Nat. Commun.* **2017**, *8*, 503. [CrossRef] [PubMed]

Communication

Discovery of Sulfated Small Molecule Inhibitors of Matrix Metalloproteinase-8

Shravan Morla [1,2,†] and Umesh R. Desai [1,2,*]

1. Department of Medicinal Chemistry, Virginia Commonwealth University, Richmond, VA 23298, USA; smorla@scripps.edu
2. Drug Discovery and Development, Institute for Structural Biology, Virginia Commonwealth University, Richmond 23219, VA, USA
* Correspondence: urdesai@vcu.edu; Tel.: +804-828-7575; Fax: +804-827-3664
† Present Address: Department of Molecular Medicine, The Scripps Research Institute, La Jolla, CA 92037, USA.

Received: 12 July 2020; Accepted: 7 August 2020; Published: 9 August 2020

Abstract: Elevated matrix metalloproteinase-8 (MMP-8) activity contributes to the etiology of many diseases, including atherosclerosis, pulmonary fibrosis, and sepsis. Yet, very few small molecule inhibitors of MMP-8 have been identified. We reasoned that the synthetic non-sugar mimetics of glycosaminoglycans may inhibit MMP-8 because natural glycosaminoglycans are known to modulate the functions of various MMPs. The screening a library of 58 synthetic, sulfated mimetics consisting of a dozen scaffolds led to the identification of only two scaffolds, including sulfated benzofurans and sulfated quinazolinones, as promising inhibitors of MMP-8. Interestingly, the sulfated quinazolinones displayed full antagonism of MMP-8 and sulfated benzofuran appeared to show partial antagonism. Of the two, sulfated quinazolinones exhibited a >10-fold selectivity for MMP-8 over MMP-9, a closely related metalloproteinase. Molecular modeling suggested the plausible occupancy of the S_1' pocket on MMP-8 as the distinguishing feature of the interaction. Overall, this work provides the first proof that the sulfated mimetics of glycosaminoglycans could lead to potent, selective, and catalytic activity-tunable, small molecular inhibitors of MMP-8.

Keywords: matrix metalloproteinases; glycosaminoglycans; small molecule inhibitors; drug discovery

1. Introduction

Matrix metalloproteinase-8 (MMP-8), also known as neutrophil collagenase or collagenase-2, is a Zn^{+2}-dependent endopeptidase of the MMP family. It is primarily expressed and secreted by neutrophils as a zymogen (pro-MMP-8), and is activated by the reactive oxygen species released from activated neutrophils, thus rendering MMP-8 as having a key role in both acute and chronic inflammation [1]. Tissue inhibitors of metalloproteinases (TIMPs) regulate the activity of MMP-8 by forming a 1:1 stoichiometric inhibitory complex [2]. However, imbalances in MMP-8 and TIMP levels can lead to increased collagen degradation and pathological remodeling of the extracellular matrix [1]. Although it was initially believed that MMP-8's only function was to degrade collagen, later studies have identified both non-collagenous and non-structural substrates [1], thereby implicating MMP-8 in various pathological processes (see Table 1 for the full listing). In recent years, a wealth of preclinical and clinical data have emerged that correlate high MMP-8 levels with the potential for diagnosing diseases [3]. Likewise, MMP-8 inhibitors have also been suggested as potential therapeutics for treating some of these pathologies [3].

Table 1. Matrix metalloproteinase-8 (MMP-8) in various diseases.

Disease	Role of MMP-8
Atherosclerosis	High MMP-8 in atheromatous and fibrous plaques promote plaque rupture, leading to vascular and cardiac events, including myocardial infarction, ischemic stroke, and abdominal aortic aneurysm [4]. MMP-8 knockdown significantly reduces atherosclerotic events in mouse models [5].
Bacterial meningitis	High MMP-8 levels in the cerebrospinal fluid of children with bacterial meningitis are associated with blood–brain barrier damage and neuronal injury [6].
Cancer	MMP-8 displays apparently contradictory roles in both cancer progression and inhibition, depending on the type of cancer, making it both a target and an anti-target for cancer therapy [7].
Chronic obstructive pulmonary disease (COPD)/emphysema	Increased MMP-8 levels lead to poor pulmonary function and emphysema severity [8].
Coronary artery disease	The plasma MMP-8 levels in patients with coronary artery disease is associated with disease severity [9].
Idiopathic pulmonary fibrosis (IPF)	MMP-8 levels in plasma, bronchoalveolar lavage fluid, and lung macrophages of IPF patients are noted to be high. MMP-8 knockdown protects mice from bleomycin-mediated lung fibrosis [10].
Obesity	MMP-8, which degrades the human insulin receptor, is increased in the serum of obese individuals, and may contribute to insulin resistance. MMP-8 inhibition restores the insulin receptor [11].
Periodontal diseases	Upregulated MMP-8 levels are observed in gingival cervicular fluid, corresponding to 90–95% of all collagenolytic activity. MMP-8 inhibitors cease the progression of periodontitis [12].
Sepsis	The increased gene expression and activity of MMP-8 correlates with disease severity and a worsening clinical outcome [13].
Tuberculosis	MMP-8 dependent tissue destruction is observed in patient lung biopsies [14].
Wound healing	Increased MMP-8 levels in mice prevent tissue repair, leading to impaired wound healing [15].

Most early attempts to develop MMP inhibitors focused on developing peptide-based agents mimicking endogenous substrate sequences aided by a hydroxamate group, which ligates the critical Zn^{+2} co-factor [3,16]. Unfortunately, the high affinity of hydroxamates to Zn^{+2} containing proteins, which include several MMPs, as well as a disintegrin and metalloproteinase (ADAM) and ADAM with thrombospondin motifs (ADAMTS) family members, introduced a significant non-selectivity of inhibition, causing severe side-effects in the clinical trials [3,16]. Later, researchers focused on the development of peptides with a lower affinity to Zn^{+2} by including phosphonates, thiolates, and carboxylates in the structure (see Table S1) [3,16–18]. Although such peptides demonstrated significantly reduced side-effects, they failed to show sufficient efficacy in the clinical trials [3]. This resulted in a rather negative perception of MMPs as drug targets in the community [3].

After nearly a decade of no progress in clinical trials with MMP inhibitors, a better understanding of the MMP biology has led to the insight that the initial clinical trials were perhaps held prematurely [3,16]. Learning from both the fundamental advances made so far and the limitations of the broad-spectrum non-selective MMP inhibitors, researchers have focused on the development of selective small molecule MMP inhibitors. This has led to the development of selective MMP-2, -9, and -13 inhibitors. Several of these molecules are now being tested in clinical trials [3,16].

In contrast to these successes, MMP-8 appears to be a rather challenging target. Although small molecule inhibitors are likely to prevent and treat multiple diseases (Table 1), few reports are available on the development of MMP-8 inhibitors. In fact, only two publications have reported MMP-8 inhibitors in the last decade [19,20]. Thus, the vast chemical space offered by small molecules appears to be almost untouched for MMP-8 inhibitor development.

Natural glycosaminoglycans (GAGs) modulate various functions of MMPs, including cell localization, conformation, and stability [21,22]. As one would expect, based on the extensive literature behind GAG–protein interactions [23,24], GAGs are expected to affect MMP activity, depending on the length of their polymeric chains and the structure of the local sulfated microdomains [21,25]. In fact, distinct GAGs have been found to inhibit MMP-9, which is closely related to MMP-8 [26]. Furthermore, the potency of natural GAGs against MMPs is quite promising. However, the heterogeneity of these GAGs is problematic for drug development. Additionally, polymeric natural GAGs generally display unwanted side-effects; they are impossible to chemically synthesize, and their bio-analytical characterization is a major challenge. These challenges reduce the enthusiasm to pursue natural GAGs as drugs [27].

In contrast, sulfated small molecules that mimic GAGs offer much better prospects. These molecules, referred to as non-saccharide GAG mimetics (NSGMs), have now been shown to bind and functionally modulate multiple GAG-binding proteins [28–34]. For example, distinct NSGMs have been identified as ligands of fibroblast growth factor receptor-1 [28], antithrombin [33,34], neutrophil elastase [29], coagulation factor Xia [30,35], plasmin [31], and viral glycoprotein D [32], each of which are known to bind to heparin. In fact, several NSGMs have been found to display one-to-one correspondence with specific GAG sequences [28,36]. In terms of drug-like properties, NSGMs are fully synthetic, which bodes well for analytical ease and scalability of preparation [27,37]. Their much smaller molecular size (MW 500–2000) in comparison with natural GAGs (MW 15,000–100,000) enables a rational and/or computational design for identifying advanced analogs [28]. Additionally, NSGMs are highly water-soluble owing to the presence of one or more sulfate groups, which enables easier formulation, e.g., direct instillation into the lung. This property could have important value for lung disorders, such as chronic obstructive pulmonary disease (COPD) and idiopathic pulmonary fibrosis (IPF), where MMP-8 inhibition has shown to be beneficial.

Over the past decade, we have developed a library of NSGMs (Figure 1) with 12 different chemical scaffolds (apigenin, benzofuran, catechin, glucoside, gossypetin, inositol, luteolin, morin, phloretin, quercetin, quinazolinone, and resveratrol). The library contains mimetics with different chain lengths (monomer, dimer, and trimer), as well as varying sulfation levels (mono-sulfate through dodeca-sulfate), which affords an excellent potential to mimic the diversity of the chain length and sulfation patterns of GAGs.

The rationale behind the development of the NSGM library was that while the sulfate groups will offer GAG-like protein recognition features, the aromatic scaffold will impart hydrophobic characteristics. These dual ionic and non-ionic forces of recognition are likely to enhance selectivity for the target protein of interest. Using this concept, distinct NSGMs have been developed as promising anticoagulants [35], antifibrinolytics [31], anticancer [39], anti-inflammatory [29], and antiviral agents [32]. More importantly, our work has shown that initial "hits" can be systematically developed through an appropriate hit-to-lead optimization strategy into potent and selective candidates, with favorable pharmacokinetic properties and minimum adverse side-effects [30,36,40,42].

In this work, we screen a library of NSGMs using computational and in vitro techniques to identify promising inhibitors of MMP-8. Our work shows, for the first time, that MMP-8 can be inhibited by targeting electropositive GAG binding sites, and offers at least one promising small molecule inhibitor of MMP-8.

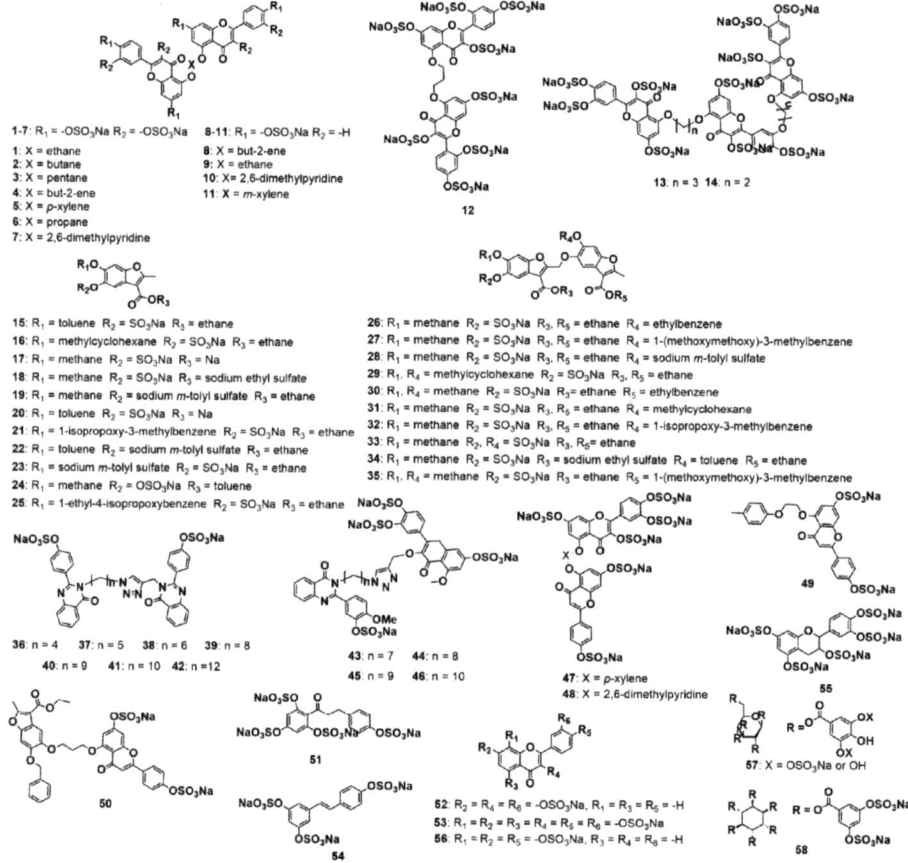

Figure 1. Structures of the 58 non-saccharide glycosaminoglycans (GAGs) mimetics (NSGMs) studied against MMP-8. The synthesis and characterization of these molecules has been previously reported [30,31,38–41].

2. Materials and Methods

2.1. Materials

Human pro-MMP-8 was purchased from R&D Systems (Minneapolis, MN, USA). p-aminomercuric acetate and MMP-8 fluorogenic substrate (DNP-Pro-Leu-Ala-Tyr-Trp-Ala-Arg) were obtained from Sigma-Aldrich (St. Louis, MO). MMP-9 inhibitor screening assay kit was obtained from Abcam (Cambridge, UK). All other reagents were purchased from Fisher Scientific (Waltham, MA, USA).

2.2. Chemistry

The synthetic scheme of the NSGMs was previously reported [30,31,38–41]. All the molecules were characterized by NMR and ultra-performance liquid chromatography (UPLC) – electrospray ionization mass spectrometry (ESI-MS). The purity of each agent was >95%, as analyzed by UPLC-MS.

2.3. Direct Inhibition Studies

NSGMs were screened for the inhibition of MMP-8 using a fluorogenic substrate assay, as described earlier [20]. Briefly, 100 µg/mL pro-MMP-8 was activated by incubating with 1mM p-aminomercuric acetate at 37 °C for 1 hr. Activated MMP-8 (10 nM, final) was then incubated with NSGMs (100 µM, final) in Tris-buffered saline containing 10 mM CaCl$_2$, 1 µM ZnCl$_2$, pH 7.5 for 10 min at 37 °C. Residual MMP-8 activity was measured by adding the fluorogenic substrate (DNP-Pro-Leu-Ala- Tyr-Trp-Ala-Arg, 20 µM, final) and monitoring initial linear rate of increase in fluorescence. IC50s were calculated using the following equation, $Y = Y_0 + \frac{Y_M - Y_0}{1 + 10^{(log[I]_0 - logIC_{50}) \times HS}}$. Y is the ratio of residual enzyme activity in the presence of NSGM to that in its absence, Y_0 and Y_M are the minimal and maximal values of Y, respectively, obtained following regression, and HS is the Hill slope of inhibition.

Screening of sulfated benzofurans and quinazolinones (100 µM, final) against MMP-9 was performed using a commercially available assay kit (Abcam catalog #ab139448) following manufacturer's protocol. All experiments were performed at least twice in duplicates.

2.4. Molecular Modeling Studies

The X-ray crystal structures of MMP-2 (PDB ID: 1CK7), MMP-8 (PDB ID: 5H8X) and MMP-9 (PDB ID: 2OW1) were retrieved from RCSB protein data bank and the electrostatic surface potential (ESP) maps calculated using PyMOL (Molecular Graphics System, Schrödinger, NY). MMP-8 was prepared for docking by removing water molecules, adding polar hydrogens, Kollman charges and solvation parameters using MGL (Molecular Graphics Laboratory) Tools of AutoDock 4.2 (Scripps Research, CA) to generate pdbqt files. Docking was performed in two stages. First, to identify a potential binding site of NSGMs, a blind docking was performed using a grid box of 100 × 100 × 100 Å with 0.375 Å grid spacing to include all the amino acid residues. A short search of 10 runs with 250,000 evaluations per run was performed with Lamarckian Genetic Algorithm (LGA) docking. Later, based on the results of this initial blind docking, a more stringent grid with a radius of 16 Å was defined and the evaluations increased to 25000000. The top 10 binding poses were then individually analyzed, and the pose favoring the best score with least root-mean-square difference was chosen. The interactions of the docked poses were analyzed using PyMOL or LigPlot+ (European Molecular Biology Laboratory, Heidelberg, Germany).

3. Results and Discussions

3.1. GAG Binding Potential of MMP-8

To assess whether MMP-8 is likely to interact with GAGs, we evaluated its electrostatic surface potential (ESP), as discussed in the literature (Figure 2A) [43]. The ESP map of MMP-8 revealed a catalytic site surrounded by electropositive residues that may serve as GAG or NSGM interacting residues. In fact, multiple electropositive microdomains are located close by, which could possibly favor binding to dimeric and/or trimeric GAG-like NSGM molecules. Thus, we reasoned that there is a high probability of identifying an NSGM that inhibits MMP-8.

Interestingly, we noted that the S_1' pocket of MMP-8 is fairly electropositive in nature. The S_1' pocket is the most varied among the different MMPs. In fact, MMPs can be grouped in the order of the depth of their S_1' pockets. MMPs have either "shallow", "intermediate", or "deep" S_1' pockets [3,44]. MMP-8 is categorized as one with an "intermediate" S_1' pocket. Apart from the size of the S_1' pocket, the residues in the S_1' specificity loop are also considerably different among different MMPs [45]. These differences in the S_1' pocket and its specificity loop are currently being exploited to develop selective MMP inhibitors [44]. Thus, to assess if the electropositive nature of the S_1' pocket is common among MMPs in the "intermediate" category, we analyzed the ESPs of MMP-2 and -9, two other "intermediate" S_1' MMPs [3]. Surprisingly, both MMP-2 and -9 displayed an electronegative S_1' pocket in comparison with that of MMP-8 (Figure 2B,C). Considering these differences in the S_1' pocket of MMP-8 with closely related MMPs, we predicted that the inhibition of MMP-8 by GAGs or NSGMs that target the S_1' pocket is likely to yield a good selectivity.

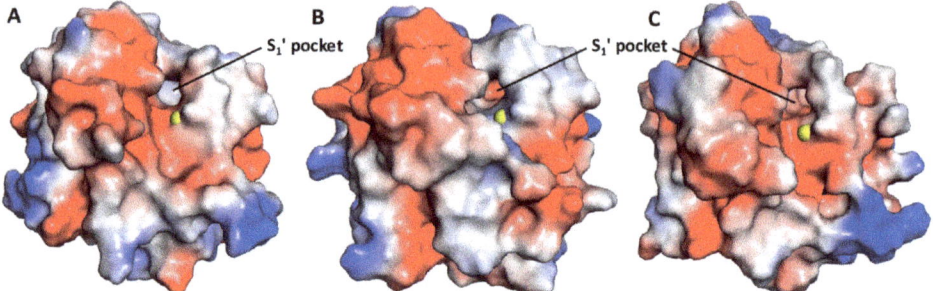

Figure 2. Electrostatic surface potential (ESP) map of MMP-8 (**A**), MMP-9 (**B**), and MMP-2 (**C**) showing electropositive regions (colored blue) that may serve as the site of binding for GAGs and NSGMs. Catalytic Zn^{2+} is shown as a yellow sphere. Note the differences in the nature of ESPs between closely related MMPs, particularly inside the S_1' pocket (blue vs. red), despite their relatively good sequence similarity/homology.

3.2. Structure–Activity Relationships for the Library of NSGMs

To test our hypotheses, we screened our NSGM library against MMP-8 using a fluorogenic substrate hydrolysis assay, as described previously [20]. The screening of the NSGM library resulted in a wide range of inhibitor efficacies (0–100%) against MMP-8 (Figure 3). Interestingly, the inhibition appeared to be related to certain sulfated scaffolds with a majority, e.g., apigenins, catechin, glucoside, inositol, luteolin, morin, phloretin, quercetins, and resveratrol, showing minimal inhibition of MMP-8. This is unusual, because one would expect that the highly sulfated agents, e.g., those carrying more than six sulfate groups, could be expected to target electropositive surfaces more easily. Yet, the results imply that the electropositive regions on MMP-8 are very discriminatory.

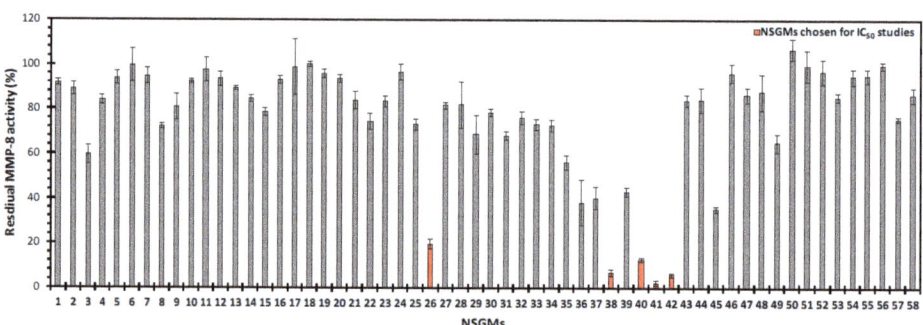

Figure 3. Residual activity of MMP-8 (10 nM) in the presence of each NSGM (100 µM) in Tris-buffered saline containing 10 mM $CaCl_2$, 1 µM $ZnCl_2$, pH 7.5 for 10 min at 37 °C. The residual MMP-8 activity was measured using a fluorogenic substrate. All of the measurements were performed at least in duplicate. Error bars represent ± 1 standard deviation (SD).

Only five NSGMs (**26**, **38**, **40**, **41**, and **42**) demonstrated an excellent MMP-8 inhibition (>80%). These NSGMs belong to the sulfated benzofuran and sulfated quinazolinone scaffolds. When profiled for the quantitative measurement of MMP-8 potency (Figure 4), the five MMP-8 inhibitors were found to exhibit a three-fold range of potency (IC_{50} = 11–34 µM, Table 2).

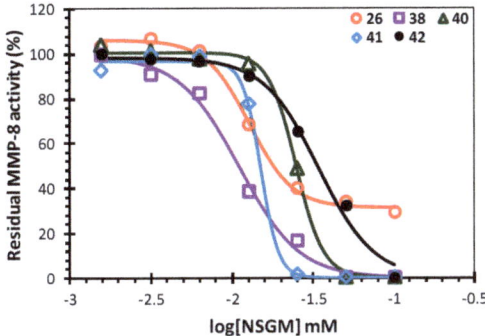

Figure 4. Direct MMP-8 inhibition profiles of the most promising NSGMs, including **26**, **38**, **40**, **41**, and **42**, at pH 7.5 and 37 °C. Solid lines represent the data fitted to the standard sigmoidal dose–response equation to derive IC$_{50}$ and ΔY, which refer to the potency and efficacy of inhibition, respectively.

Table 2. Direct inhibition of MMP-8 by selected NSGMs.

NSGM	IC$_{50}$ (μM) [a]	ΔY (%) [a]
26	13 ± 1 [b]	75 ± 3
38	11 ± 1	99 ± 6
40	25 ± 1	100 ± 2
41	15 ± 1	97 ± 2
42	34 ± 4	98 ± 8

[a] Obtained by regression analysis of the dose-dependence of the MMP-8 activity in Tris-buffered saline containing 10 mM CaCl$_2$, 1 μM ZnCl$_2$, pH 7.5, at 37 °C. IC$_{50}$ and ΔY refer to potency and efficacy of inhibition, respectively.
[b] Error refers to ± 1 SD.

At the level of the MMP-8–GAG system, these results are interesting on several fronts. First, the inhibition profiles support the hypothesis that the family of NSGMs is likely to yield an MMP-8 inhibitor, given its structural diversity. The five NSGMs could be classified as "hits" with moderate potency, which will require secondary drug design efforts to yield "lead(s)". Second, the hit yield, i.e., only 5 out of the 58 studied, is relatively low, which suggests excellent weeding-out by MMP-8. Third, not all NSGMs inhibit MMP-8 fully ($\Delta Y > 90\%$). The lone sulfated benzofuran **26** displays a partial inhibition profile ($\Delta Y = 75\%$), which presents the possibility of regulating the MMP-8 activity ($\Delta Y = 20$–80%), rather than knocking it out completely. Recently, small molecule regulation of soluble enzymes has been discovered and was found to offer beneficial properties [38,46,47]. Similarly, regulation of the MMP-8–NSGM **26** system may offer the benefit of retaining basal levels of the MMP-8 activity that is critical for optimal growth. Fourth, the result that none of the highly sulfated mimetics (NSGMs **1–14**, **57**, or **58**), which carry more than six sulfate groups, were active against MMP-8 (Figure 3) implies significant contributions of the aromatic groups. This supports the hypothesis on dual ionic and non-ionic forces governing recognition.

Among the sulfated benzofurans, the closely related monomers (**15–25**) and dimers (**27–35**) did not inhibit MMP-8 in contrast to **26**, which was one of the three most active NSGMs. In each case, the structural changes were primarily in the terminal substituents (R_1 and R_4, Figure 1), suggesting a stringent size dependence. Among the sulfated quinazolinones **36–42**, agents with linkers of less than six carbons did not inhibit MMP-8. This implies that the NSGM binding site on MMP-8 is likely to contain two sub-sites spaced several angstroms apart.

3.3. Computational Analysis of the Preferred Site of NSGMs Binding to MMP-8

To identify a plausible site of binding of the active sulfated benzofurans and sulfated quinazolinones, we performed genetic algorithm-based molecular docking and scoring studies. Docking was performed using the available MMP-8 crystal structure (PDB ID: 5H8X) and AutoDock 4.2

(Scripps Research). The studies revealed that the sulfate group of NSGM **26** interacts with the catalytic Zn^{2+} and the triad of histidine residues (H197, H201, and H207, Figure 5A,B). This is similar to the previously reported interactions of MMPs with the phosphate groups [48]. Additionally, the oxygen atoms of the furan ring and the linker form a bidentate interaction with S151 (Figure 5B), and the aromatic substituent at the R_4 position contributes towards hydrophobic and cation-Π interactions with residues in the unprimed S_2 and S_3 pockets. In contrast, other sulfated benzofurans did not have any interaction with S151, either because of their sub-optimal chain length (NSGMs **15–25**) or the bulk of their substituents (NSGMs **27–35**). This explained the observed lack of MMP-8 inhibition with these NSGMs.

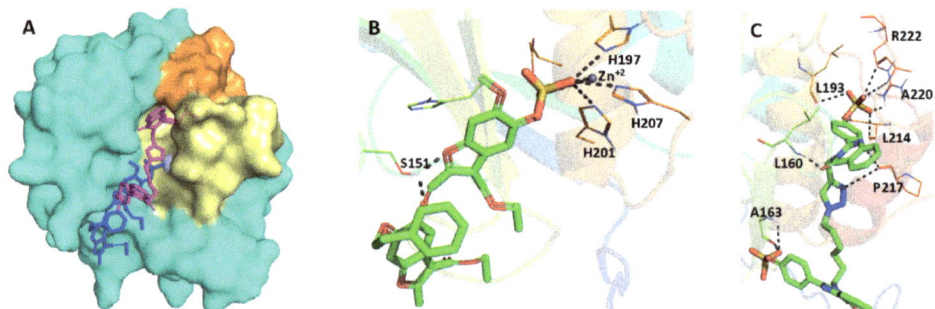

Figure 5. (**A**) Surface representation of MMP-8 (cyan) showing the potential interacting sites of sulfated benzofuran **26** (blue) and sulfated quinazolinone **38** (magenta). The $S_1{'}$ pocket and specificity loop are shown in pale yellow and orange, respectively. (**B,C**) Polar interactions of NSGM **26** (**B**, shown in sticks) and NSGM **38** (**C**, shown in sticks) with MMP-8 (cartoon). Interacting residues of MMP-8 are shown as lines, catalytic Zn^{+2} ion as a sphere, and polar interactions as grey dotted lines.

In the manner of NSGM **26**, sulfated quinazolinone dimers **36–42** also bound near the catalytic site of MMP-8 (Figure 5A,C). For NSGM **38**, one monomeric unit of the quinazolinone was found to insert deep into the $S_1{'}$ pocket (shown as a pale-yellow surface in Figure 5A), which resulted in a strong binding of the sulfate with L193, L214, A220, and R222 (Figure 5C). The insertion of the quinazolinone ring into the $S_1{'}$ pocket positioned the triazole linker close to the catalytic Zn^{+2} ion. However, based on these experiments, NSGM **38** did not make polar interactions with the catalytic Zn^{+2} or the histidine triad. Instead, the nitrogen atom of the triazole ring, along with the nitrogen of the quinazolinone ring form hydrogen bonds with the backbone oxygen of P217. Additional polar, hydrophobic, and cation-Π interactions were observed with multiple residues present in the site of binding (see Table S2). The insertion of sulfated quinazolinone dimers (**36–42**) in the $S_1{'}$ pocket of MMP-8 and their interactions with the residues in the $S_1{'}$ specificity loop is particularly interesting (Figure 5A). Such GAG-like molecules inhibiting MMP-8 by binding in the electropositive $S_1{'}$ pocket have the potential to be selective towards MMP-8. Thus, the identification of NSGM hits in this work is likely to be of significant value for both drug discovery and chemical biology campaigns against MMP-8.

Several other aspects of structure–activity dependence were also observed. For example, when one of the molecules of quinazolinone in the dimers that inhibited MMP-8 (**38**, **40**, **41**, and **42**) was replaced with quercetin, the resulting quercetin–quinazolinone heterodimers (**43–46**) lost their inhibition towards MMP-8 (Figure 3). Although the core scaffold of quinazolinones and quercetins studied here are very similar, i.e., fused bicyclic ring attached to a phenyl group, the presence of three sulfate groups on the quercetins compared with one sulfate on the quinazolinones adds both steric bulk and negative charges. Based on computational modeling, this addition of steric bulk drives the heterodimeric molecules away from the $S_1{'}$ pocket (Figure S1), resulting in a loss of MMP-8 inhibition.

3.4. Sulfated Quinazolinones Do Not Inhibit MMP-9

We next turned our attention to MMP-targeting selectivity studies. Considering that MMP-9 is the most closely related metalloenzyme to MMP-8, and as it belongs to the "intermediate" S_1' pocket category like MMP-8, we screened sulfated benzofurans and sulfated quinazolinones against MMP-9. NSGM **26** also inhibited MMP-9 reasonably well (Figure 6). This was not too unusual, because its inhibition of MMP-8 was predicted to arise from coordination with the catalytic Zn^{2+} and histidine triad (see Figure 5B), which is known to be the origin of a lack of selectivity associated with many inhibitors of MMPs. Another sulfated benzofuran, NSGM **29**, which did not inhibit MMP-8, also showed >80% inhibition of MMP-9. Thus, the future use of sulfated benzofurans as MMP-8 hits is likely to be beset with such non-selectivity concerns.

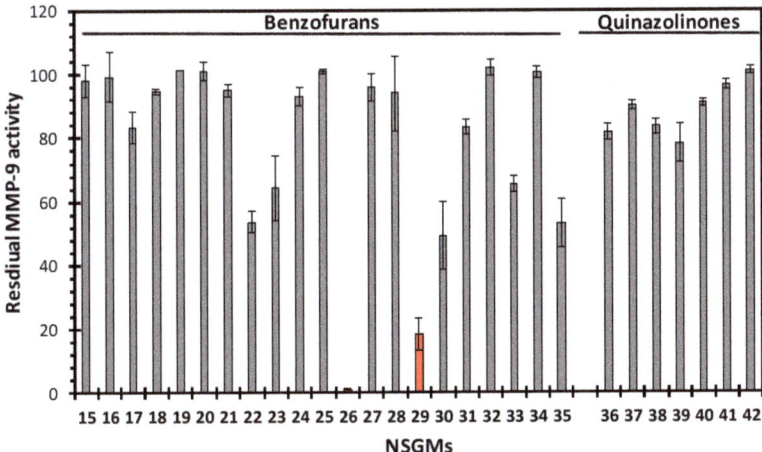

Figure 6. Screening of NSGM **15–42** at 100 µM against MMP-9. Experiments were performed using a commercially available assay kit (Abcam, Cambridge, U.K.) in at least duplicates. Error bars represent ± 1 SD.

In contrast, the sulfated quinazolinones did not inhibit MMP-9 significantly (Figure 6), thus displaying >10-fold selectivity for MMP-8 over MMP-9. Considering that the predicted mode of binding of sulfated quinazolinones does not involve a strong engagement of the catalytic Zn^{2+} or the histidine triad (see Figure 5C), the observation of a lack of MMP-9 activity is encouraging. The sulfate group of **38**, inserted in the S_1' pocket, forms hydrogen and ionic bond interactions with A220 and R222, respectively, of MMP-8. At the corresponding positions in MMP-9, there is neither an arginine nor an alanine, which supports the observed lack of inhibition. In fact, the arginine at 222 is found in only one other MMP (MMP-28), and the alanine at 220 is not present in any other MMPs [45]. Thus, it is possible that NSGM **38** is an excellent hit with a high potential for selectivity of MMP-8. This hypothesis will need to be further experimentally validated through extended selectivity studies.

Overall, our "hit" NSGMs are the first group of GAG-related inhibitors of MMP-8. Given the urgent need and lack of efforts directed towards developing MMP-8 inhibitors, these NSGMs offer a unique starting point for further structure-guided hit-to-lead optimization studies. Two major optimization routes could be considered, namely: (1) Our modeling insights convey that the use of a biphenyl substituent, instead of a phenyl substituent, on the sulfated quinazolinone would lead to a greater reach of a potential inhibitor inside the S_1' pocket. This should enhance both the potency and selectivity. In fact, a peptide-hydroxamate containing a biphenyl substituent was previously shown to inhibit MMP-8 with a low nanomolar potency by occupying the S_1' pocket [49]. However, because of the strong Zn^{+2} chelating nature, metabolic instability, and toxicity associated with hydroxamates, these molecules failed in the clinical trials [3,16]. (2) Increasing the polarity of the linker by changing it from an alkyl-based chain to an ethylene glycol-based chain would likely result in polar interactions with the residues lining the S_2 and S_3 pockets. This should also enhance the potency and perhaps selectivity.

4. Conclusions

The present study shows, for the first time, the possibility of inhibiting MMP-8 by targeting electropositive GAG binding sites, and identifies two new small molecule GAG-mimicking sulfated scaffolds, the sulfated benzofurans and sulfated quinazolinones, as inhibitors of MMP-8. Of these, the sulfated benzofurans appear to be broad-spectrum MMP inhibitors because of their engagement of the catalytic Zn^{2+} and histidines. In contrast, the sulfated quinazolinones bind in the $S_1{'}$ pocket of MMP-8, thereby enhancing the selectivity of inhibition. Considering that several NSGM-based initial hits have been transformed through hit-to-lead optimization strategies into potent and selective candidates [30,36], the potential to transform NSGM **38** into a promising drug-like candidate or a chemical biology tool is high.

Supplementary Materials: The following are available online at http://www.mdpi.com/2218-273X/10/8/1166/s1: Figure S1: Comparison of the mode of binding of **38** and **43** on MMP-8. Figure S2: Mode of binding of the top five NSGM inhibitors (**26, 38, 40, 41,** and **42**) of MMP-8 (shown in ESP map). Table S1: General structure of MMP-8 inhibitors developed to date, Table S2: List of interactions made by NSGMs with MMP-8.

Author Contributions: S.M. performed the library screening, inhibition, and computational studies, and wrote the initial draft of the manuscript; U.R.D. finalized the manuscript, acquired funding, and directed the project. All authors have read and agreed to the published version of the manuscript.

Funding: This work was supported by grants HL090586, HL107152, and CA241951 from the National Institute of Health to U.R.D.

Conflicts of Interest: The authors declare no conflict of interest.

References

1. Van Lint, P.; Libert, C. Matrix metalloproteinase-8: Cleavage can be decisive. *Cytokine Growth Factor Rev.* **2006**, *17*, 217–223. [CrossRef]
2. Jackson, H.W.; Defamie, V.; Waterhouse, P.; Khokha, R. TIMPs: Versatile extracellular regulators in cancer. *Nat. Rev. Cancer* **2016**, *17*, 38–53. [CrossRef] [PubMed]
3. Li, K.; Tay, F.R.; Yiu, C.K. The past, present and future perspectives of matrix metalloproteinase inhibitors. *Pharmacol. Ther.* **2020**, *207*, 107465–107478. [CrossRef] [PubMed]
4. Ye, S. Putative targeting of matrix metalloproteinase-8 in atherosclerosis. *Pharmacol. Ther.* **2015**, *147*, 111–122. [CrossRef] [PubMed]
5. Lenglet, S.; Mach, F.; Montecucco, F. Role of Matrix Metalloproteinase-8 in Atherosclerosis. *Mediat. Inflamm.* **2013**, *2013*, 659282. [CrossRef] [PubMed]
6. Leppert, D.; Leib, S.; Grygar, C.; Miller, K.M.; Schaad, U.B.; Holländer, G.A. Matrix Metalloproteinase (MMP)-8 and MMP-9 in Cerebrospinal Fluid during Bacterial Meningitis: Association with Blood-Brain Barrier Damage and Neurological Sequelae. *Clin. Infect. Dis.* **2000**, *31*, 80–84. [CrossRef]
7. Juurikka, K.; Butler, G.; Salo, T.; Nyberg, P.; Åström, P. Salo The Role of MMP8 in Cancer: A Systematic Review. *Int. J. Mol. Sci.* **2019**, *20*, 4506. [CrossRef]
8. Koo, H.-K.; Hong, Y.; Lim, M.N.; Yim, J.-J.; Kim, W.J. Relationship between plasma matrix metalloproteinase levels, pulmonary function, bronchodilator response, and emphysema severity. *Int. J. Chronic Obstr. Pulm. Dis.* **2016**, *11*, 1129–1137. [CrossRef]
9. Kato, R.; Momiyama, Y.; Ohmori, R.; Taniguchi, H.; Nakamura, H.; Ohsuzu, F. Plasma Matrix Metalloproteinase-8 Concentrations are Associated with the Presence and Severity of Coronary Artery Disease. *Circ. J.* **2005**, *69*, 1035–1040. [CrossRef]

10. Craig, V.J.; Zhang, L.; Hagood, J.S.; Owen, C.A. Matrix Metalloproteinases as Therapeutic Targets for Idiopathic Pulmonary Fibrosis. *Am. J. Respir. Cell Mol. Boil.* **2015**, *53*, 585–600. [CrossRef]
11. Lauhio, A.; Färkkilä, E.; Pietiläinen, K.H.; Åström, P.; Winkelmann, A.; Tervahartiala, T.; Pirilä, E.; Rissanen, A.; Kaprio, J.; Sorsa, T.; et al. Association of MMP-8 with obesity, smoking and insulin resistance. *Eur. J. Clin. Investig.* **2016**, *46*, 757–765. [CrossRef] [PubMed]
12. Sorsa, T.; Gürsoy, U.K.; Nwhator, S.; Hernández, M.; Tervahartiala, T.; Leppilahti, J.; Gursoy, M.; Könönen, E.; Emingil, G.; Pussinen, P.J.; et al. Analysis of matrix metalloproteinases, especially MMP-8, in gingival crevicular fluid, mouthrinse and saliva for monitoring periodontal diseases. *Periodontol. 2000* **2015**, *70*, 142–163. [CrossRef] [PubMed]
13. Solan, P.D.; Dunsmore, K.E.; Denenberg, A.G.; Odoms, K.; Zingarelli, B.; Wong, H.R. A novel role for matrix metalloproteinase-8 in sepsis*. *Crit. Care Med.* **2012**, *40*, 379–387. [CrossRef] [PubMed]
14. Ong, C.W.M.; Elkington, P.T.; Brilha, S.; Ugarte-Gil, C.; Esteban, M.T.T.; Tezera, L.B.; Pabisiak, P.J.; Moores, R.C.; Sathyamoorthy, T.; Patel, V.; et al. Neutrophil-Derived MMP-8 Drives AMPK-Dependent Matrix Destruction in Human Pulmonary Tuberculosis. *PLoS Pathog.* **2015**, *11*, e1004917. [CrossRef]
15. Danielsen, P.L.; Holst, A.V.; Maltesen, H.R.; Bassi, M.R.; Holst, P.; Heinemeier, K.M.; Olsen, J.; Danielsen, C.C.; Poulsen, S.S.; Jørgensen, L.N.; et al. Matrix metalloproteinase-8 overexpression prevents proper tissue repair. *Surgery* **2011**, *150*, 897–906. [CrossRef]
16. Fields, G.B. The Rebirth of Matrix Metalloproteinase Inhibitors: Moving Beyond the Dogma. *Cells* **2019**, *8*, 984. [CrossRef]
17. Bianchini, G.; Aschi, M.; Cavicchio, G.; Crucianelli, M.; Preziuso, S.; Gallina, C.; Nastari, A.; Gavuzzo, E.; Mazza, F. Design, modelling, synthesis and biological evaluation of peptidomimetic phosphinates as inhibitors of matrix metalloproteinases MMP-2 and MMP-8. *Bioorganic Med. Chem.* **2005**, *13*, 4740–4749. [CrossRef]
18. Scozzafava, A.; Supuran, C.T. Protease inhibitors: Synthesis of matrix metalloproteinase and bacterial collagenase inhibitors incorporating 5-amino-2-mercapto-1,3,4-thiadiazole zinc binding functions. *Bioorganic Med. Chem. Lett.* **2002**, *12*, 2667–2672. [CrossRef]
19. Wang, Z.C.; Shen, F.Q.; Yang, M.R.; You, L.X.; Chen, L.Z.; Zhu, H.L.; Lu, Y.D.; Kong, F.L.; Wang, M.H. Dihydropyrazothiazole derivatives as potential MMP-2/MMP-8 inhibitors for cancer therapy. *Bioorganic Med. Chem. Lett.* **2018**, *28*, 3816–3821. [CrossRef]
20. Bhowmick, M.; Tokmina-Roszyk, D.; Onwuha-Ekpete, L.; Harmon, K.; Robichaud, T.; Fuerst, R.; Stawikowska, R.; Steffensen, B.; Roush, W.R.; Wong, H.R.; et al. Second Generation Triple-Helical Peptide Inhibitors of Matrix Metalloproteinases. *J. Med. Chem.* **2017**, *60*, 3814–3827. [CrossRef]
21. Tocchi, A.; Parks, W.C. Functional interactions between matrix metalloproteinases and glycosaminoglycans. *FEBS J.* **2013**, *280*, 2332–2341. [CrossRef] [PubMed]
22. Mannello, F.; Jung, K.; Tonti, G.A.; Canestrari, F. Heparin affects matrix metalloproteinases and tissue inhibitors of metalloproteinases circulating in peripheral blood. *Clin. Biochem.* **2008**, *41*, 1466–1473. [CrossRef] [PubMed]
23. Capila, I.; Linhardt, R.J. Heparin-protein interactions. *Angew. Chem. Int. Ed. Engl.* **2002**, *41*, 391–412. [CrossRef]
24. Gandhi, N.S.; Mancera, R.L. The Structure of Glycosaminoglycans and their Interactions with Proteins. *Chem. Boil. Drug Des.* **2008**, *72*, 455–482. [CrossRef] [PubMed]
25. Ra, H.-J.; Harju-Baker, S.; Zhang, F.; Linhardt, R.J.; Wilson, C.L.; Parks, W.C. Control of Promatrilysin (MMP7) Activation and Substrate-specific Activity by Sulfated Glycosaminoglycans. *J. Boil. Chem.* **2009**, *284*, 27924–27932. [CrossRef]
26. Mannello, F.; Raffetto, J.D. Matrix metalloproteinase activity and glycosaminoglycans in chronic venous disease: The linkage among cell biology, pathology and translational research. *Am. J. Transl. Res.* **2010**, *3*, 149–158.
27. Morla, S. Glycosaminoglycans and Glycosaminoglycan Mimetics in Cancer and Inflammation. *Int. J. Mol. Sci.* **2019**, *20*, 1963. [CrossRef]
28. Nagarajan, B.; Sankaranarayanan, N.V.; Patel, B.B.; Desai, U.R. A molecular dynamics-based algorithm for evaluating the glycosaminoglycan mimicking potential of synthetic, homogenous, sulfated small molecules. *PLoS ONE* **2017**, *12*, e0171619. [CrossRef]

29. Morla, S.; Sankaranarayanan, N.V.; Afosah, D.K.; Kumar, M.; Kummarapurugu, A.B.; Voynow, J.A.; Desai, U.R. On the Process of Discovering Leads That Target the Heparin-Binding Site of Neutrophil Elastase in the Sputum of Cystic Fibrosis Patients. *J. Med. Chem.* **2019**, *62*, 5501–5511. [CrossRef]
30. Al-Horani, R.A.; Abdelfadiel, E.I.; Afosah, D.K.; Morla, S.; Sistla, J.C.; Mohammed, B.; Martin, E.J.; Sakagami, M.; Brophy, D.F.; Desai, U.R. A synthetic heparin mimetic that allosterically inhibits factor XIa and reduces thrombosis in vivo without enhanced risk of bleeding. *J. Thromb. Haemost.* **2019**, *17*, 2110–2122. [CrossRef]
31. Afosah, D.K.; Al-Horani, R.A.; Sankaranarayanan, N.V.; Desai, U.R. Potent, Selective, Allosteric Inhibition of Human Plasmin by Sulfated Non-Saccharide Glycosaminoglycan Mimetics. *J. Med. Chem.* **2017**, *60*, 641–657. [CrossRef] [PubMed]
32. Gangji, R.N.; Sankaranarayanan, N.V.; Elste, J.; Al-Horani, R.A.; Afosah, D.K.; Joshi, R.; Tiwari, V.; Desai, U.R. Inhibition of Herpes Simplex Virus-1 Entry into Human Cells by Nonsaccharide Glycosaminoglycan Mimetics. *ACS Med. Chem. Lett.* **2018**, *9*, 797–802. [CrossRef] [PubMed]
33. Gunnarsson, G.T.; Desai, U.R. Interaction of Designed Sulfated Flavanoids with Antithrombin: Lessons on the Design of Organic Activators. *J. Med. Chem.* **2002**, *45*, 4460–4470. [CrossRef] [PubMed]
34. Al-Horani, R.A.; Liang, A.; Desai, U.R. Designing Nonsaccharide, Allosteric Activators of Antithrombin for Accelerated Inhibition of Factor Xa. *J. Med. Chem.* **2011**, *54*, 6125–6138. [CrossRef]
35. Al-Horani, R.A.; Ponnusamy, P.; Mehta, A.Y.; Gailani, D.; Desai, U.R. Sulfated Pentagalloylglucoside Is a Potent, Allosteric, and Selective Inhibitor of Factor XIa. *J. Med. Chem.* **2013**, *56*, 867–878. [CrossRef] [PubMed]
36. Boothello, R.S.; Patel, N.J.; Sharon, C.; Abdelfadiel, E.I.; Morla, S.; Brophy, D.F.; Lippman, H.R.; Desai, U.R.; Patel, B.B. A Unique Nonsaccharide Mimetic of Heparin Hexasaccharide Inhibits Colon Cancer Stem Cells via p38 MAP Kinase Activation. *Mol. Cancer Ther.* **2018**, *18*, 51–61. [CrossRef] [PubMed]
37. Desai, U.R. The promise of sulfated synthetic small molecules as modulators of glycosaminoglycan function. *Future Med. Chem.* **2013**, *5*, 1363–1366. [CrossRef] [PubMed]
38. Afosah, D.K.; Verespy, S.; Al-Horani, R.A.; Boothello, R.S.; Karuturi, R.; Desai, U.R. A small group of sulfated benzofurans induces steady-state submaximal inhibition of thrombin. *Bioorganic Med. Chem. Lett.* **2018**, *28*, 1101–1105. [CrossRef]
39. Patel, N.J.; Karuturi, R.; Al-Horani, R.A.; Baranwal, S.; Patel, J.; Desai, U.R.; Patel, B.B. Synthetic, Non-saccharide, Glycosaminoglycan Mimetics Selectively Target Colon Cancer Stem Cells. *ACS Chem. Boil.* **2014**, *9*, 1826–1833. [CrossRef]
40. Al-Horani, R.A.; Desai, U.R. Designing Allosteric Inhibitors of Factor XIa. Lessons from the Interactions of Sulfated Pentagalloylglucopyranosides. *J. Med. Chem.* **2014**, *57*, 4805–4818. [CrossRef]
41. Al-Horani, R.A.; Karuturi, R.; White, D.T.; Desai, U.R. Plasmin Regulation through Allosteric, Sulfated, Small Molecules. *Molecules* **2015**, *20*, 608–624. [CrossRef] [PubMed]
42. Al-Horani, R.A.; Gailani, D.; Desai, U.R. Allosteric inhibition of factor XIa. Sulfated non-saccharide glycosaminoglycan mimetics as promising anticoagulants. *Thromb. Res.* **2015**, *136*, 379–387. [CrossRef] [PubMed]
43. Sankaranarayanan, N.V.; Nagarajan, B.; Desai, U.R. So you think computational approaches to understanding glycosaminoglycan–protein interactions are too dry and too rigid? Think again! *Curr. Opin. Struct. Boil.* **2018**, *50*, 91–100. [CrossRef] [PubMed]
44. Gimeno, A.; Beltran-Debon, R.; Mulero, M.; Pujadas, G.; Garcia-Vallve, S. Understanding the variability of the S1' pocket to improve matrix metalloproteinase inhibitor selectivity profiles. *Drug Discov. Today* **2020**, *25*, 38–57. [CrossRef]
45. Overall, C.M.; Kleifeld, O. Towards third generation matrix metalloproteinase inhibitors for cancer therapy. *Br. J. Cancer* **2006**, *94*, 941–946. [CrossRef]
46. Iii, S.V.; Mehta, A.Y.; Afosah, D.; Al-Horani, R.A.; Desai, U.R. Allosteric Partial Inhibition of Monomeric Proteases. Sulfated Coumarins Induce Regulation, not just Inhibition, of Thrombin. *Sci. Rep.* **2016**, *6*, 24043. [CrossRef]
47. Lira, A.L.; Ferreira, R.S.; Torquato, R.J.S.; Oliva, M.L.V.; Schuck, P.; Sousa, A.A. Allosteric inhibition of alpha-thrombin enzymatic activity with ultrasmall gold nanoparticles. *Nanoscale Adv.* **2019**, *1*, 378–388. [CrossRef]

48. Agamennone, M.; Campestre, C.; Preziuso, S.; Consalvi, V.; Crucianelli, M.; Mazza, F.; Politi, V.; Ragno, R.; Tortorella, P.; Gallina, C. Synthesis and evaluation of new tripeptide phosphonate inhibitors of MMP-8 and MMP-2. *Eur. J. Med. Chem.* **2005**, *40*, 271–279. [CrossRef]
49. Whittaker, M.; Floyd, C.D.; Brown, P.; Gearing, A.J.H. Design and therapeutic application of matrix metalloproteinase inhibitors. *Chem. Rev.* **1999**, *99*, 2735–2776. [CrossRef]

© 2020 by the authors. Licensee MDPI, Basel, Switzerland. This article is an open access article distributed under the terms and conditions of the Creative Commons Attribution (CC BY) license (http://creativecommons.org/licenses/by/4.0/).

Article

Perlecan in the Natural and Cell Therapy Repair of Human Adult Articular Cartilage: Can Modifications in This Proteoglycan Be a Novel Therapeutic Approach?

John Garcia [1,2], Helen S. McCarthy [1,2], Jan Herman Kuiper [1,2], James Melrose [3,4,5] and Sally Roberts [1,2,*]

1. School of Pharmacy and Bioengineering, Keele University, Newcastle-under-Lyme, Staffordshire ST5 5BG, UK; john.garcia@nhs.net (J.G.); helen.mccarthy6@nhs.net (H.S.M.); jan.kuiper@nhs.net (J.H.K.)
2. Spinal Studies & Cartilage Research Group, Robert Jones and Agnes Hunt Orthopaedic Hospital NHS Foundation Trust, Oswestry, Shropshire SY10 7AG, UK
3. Raymond Purves Bone and Joint Research Laboratory, Kolling Institute of Medical Research, Northern Sydney Area Local Health District, St. Leonards, NSW 2065, Australia; james.melrose@sydney.edu.au
4. Sydney Medical School, Northern, The University of Sydney, Faculty of Medicine and Health, Royal North Shore Hospital, St. Leonards, NSW 2065, Australia
5. Graduate School of Biomedical Engineering, Faculty of Engineering, University of New South Wales, Sydney, NSW 2052, Australia
* Correspondence: sally.roberts4@nhs.net; Tel.: +44-1-691-404-664

Citation: Garcia, J.; McCarthy, H.S.; Kuiper, J.H.; Melrose, J.; Roberts, S. Perlecan in the Natural and Cell Therapy Repair of Human Adult Articular Cartilage: Can Modifications in This Proteoglycan Be a Novel Therapeutic Approach? *Biomolecules* **2021**, *11*, 92. https://doi.org/10.3390/biom11010092

Received: 24 December 2020
Accepted: 11 January 2021
Published: 13 January 2021

Publisher's Note: MDPI stays neutral with regard to jurisdictional claims in published maps and institutional affiliations.

Copyright: © 2021 by the authors. Licensee MDPI, Basel, Switzerland. This article is an open access article distributed under the terms and conditions of the Creative Commons Attribution (CC BY) license (https://creativecommons.org/licenses/by/4.0/).

Abstract: Articular cartilage is considered to have limited regenerative capacity, which has led to the search for therapies to limit or halt the progression of its destruction. Perlecan, a multifunctional heparan sulphate (HS) proteoglycan, promotes embryonic cartilage development and stabilises the mature tissue. We investigated the immunolocalisation of perlecan and collagen between donor-matched biopsies of human articular cartilage defects ($n = 10 \times 2$) that were repaired either naturally or using autologous cell therapy, and with age-matched normal cartilage. We explored how the removal of HS from perlecan affects human chondrocytes in vitro. Immunohistochemistry showed both a pericellular and diffuse matrix staining pattern for perlecan in both natural and cell therapy repaired cartilage, which related to whether the morphology of the newly formed tissue was hyaline cartilage or fibrocartilage. Immunostaining for perlecan was significantly greater in both these repair tissues compared to normal age-matched controls. The immunolocalisation of collagens type III and VI was also dependent on tissue morphology. Heparanase treatment of chondrocytes in vitro resulted in significantly increased proliferation, while the expression of key chondrogenic surface and genetic markers was unaffected. Perlecan was more prominent in chondrocyte clusters than in individual cells after heparanase treatment. Heparanase treatment could be a means of increasing chondrocyte responsiveness to cartilage injury and perhaps to improve repair of defects.

Keywords: human articular cartilage; perlecan; heparan sulphate; heparanase; cartilage repair; natural repair; chondrocytes

1. Introduction

Articular cartilage can withstand compressive, tensile and shear loading and provides efficient articulation of diarthrodial joints. If left untreated, damaged articular cartilage in a joint can lead to osteoarthritis (OA) and ultimately joint failure [1,2]. Cell-based therapies have been developed to promote cartilage repair and the regeneration of complex articular structure to help patients with damaged or degenerate cartilage [3,4].

It is commonly reported that adult articular cartilage has a limited capacity for self-regeneration [5]; however, a growing body of evidence from in vitro and in vivo models suggests that in some cases, cartilage can undergo some form of natural repair [6,7]. A bovine explant model of cartilage healing showed that both young and mature animals produced an outgrowth of tissue from the artificially damaged sites, but with young tissues generating more hyaline-like cartilage [6]. In humans, magnetic resonance imaging (MRI) observation of the knees of healthy subjects showed that some cartilage defects (tibial and patellar) reduced in size or were completely filled between a baseline scan and a two year follow up [7]. A natural healing response was also seen in some cartilage lesions of subjects with anterior cruciate ligament (ACL) injuries 6–56 months after reconstructive surgery to repair the ligament damage [8]. The mechanisms by which articular cartilage repairs itself is poorly understood, but is believed to involve an interplay between cellular, biochemical and mechanical factors [9–11].

Perlecan, also known as heparan sulphate proteoglycan-2, is a modular, multifunctional proteoglycan with an ability to promote chondrocyte proliferation, differentiation and matrix synthesis through its interactions with a large repertoire of ligands including growth factors, morphogens and extracellular matrix (ECM)-stabilising glycoproteins [12,13]. One of the glycosaminoglycans contained in perlecan, heparan sulphate (HS), is a vital extracellular component. Its cleavage causes matrix remodelling through the release of HS-bound cytokines, growth factors, morphogens, proteases and inhibitory proteins which regulate many cellular pathological and physiological processes [14,15]. Perlecan, through its HS chains, has chondrogenic properties and is able to regulate cell signalling, matrix assembly and new tissue formation [12,16,17]. These attributes make perlecan an important candidate molecule when trying to understand how cartilage repairs itself. Hence, harnessing these attributes could also be beneficial in promoting the repair of damaged articular cartilage in human joints. Heparanase is an endo-β-glucuronidase cleaving the $\beta(1,4)$-glycosidic linkages between GlcN and GlcA in heparan sulphate (HS), and is the only known mammalian enzyme displaying this glycolytic activity [18].

Interactions between perlecan and collagen type VI have been well established [19] and, like perlecan, collagen type VI is believed to be involved in chondrocyte adhesion, integrity and matrix interactions [20,21]. Collagen type III is another minor collagen found in articular cartilage and has been suggested to have a role in reinforcing the cartilage matrix as part of a healing response to matrix damage [22,23].

In the present study, we have immunolocalised perlecan and types III and VI collagen for the first time in donor-matched samples of naturally and cell therapy repaired articular cartilage of the human knee. We have also investigated whether the phenotype and proliferation of cultured human chondrocytes was affected by the removal of cell surface HS. We hypothesise that the distribution of perlecan in repaired adult cartilage mimics its distribution in embryological cartilage.

2. Materials and Methods

2.1. Tissue Samples and Histology

The National Research Ethics Service (11/NW/0875) gave ethical approval and informed written consent was obtained from patients undergoing autologous cell therapy for cartilage defects in their knee (n = 10, aged 29–51 years). This procedure entails harvesting ~200 mg macroscopically healthy cartilage, usually from the trochlea, from which chondrocytes are isolated and culture expanded in monolayer, prior to re-implantation in the defect site, usually on the patella or lateral/medial femoral condyles (LFC/MFC) [3]. At approximately 12 months post-implantation, full-depth cartilage biopsies with subchondral bone (1.8 mm diameter) were obtained using a juvenile bone-marrow biopsy needle from both the harvest site (naturally repaired) and the defect site where the cells had been implanted (cell-treated repair). The location of these was ensured via the use of knee maps [24], where the location of each procedure is recorded at the time of original surgery. Macroscopically healthy cartilage was also obtained from the knees of five cadavers (aged 21–63 years) and four donors undergoing total knee arthroplasty for OA (aged 51–81 years). A description of the donor demographics and tissue samples used in the following experiments can be found in Table 1. Healthy cadaveric cartilage from donors 11–14 was obtained within 24 h of death from the UK Human Tissue Bank with approval by the Trent Research Ethics Committee (UK). Full-depth core biopsies of other samples (from TKR and natural and cell therapy repair patients) of cartilage and underlying bone were snap frozen within 2–4 h of harvesting in liquid nitrogen-cooled hexane and stored at $-196\ °C$ until cryosectioning. Cores were embedded into tissue-freezing medium (Leica) and cryosectioned at 7 μm thickness onto poly-L-lysine-coated slides. Cryosections were then stained with either haematoxylin and eosin (H&E) or toluidine blue for the assessment of general morphology and proteoglycan content of the cartilage, respectively. Collagen fibre organisation and orientation were assessed under polarised light. The quality of the repaired cartilage was assessed and scored semi-quantitatively using both the International Cartilage Repair Society II Histology Score (ICRS II) [25] and the Oswestry Score [26], where a higher score in each system represents better-quality cartilage.

Table 1. Donor demographics and samples.

Donor.	Gender	Age	Surgical Intervention	Tissue Used (Experiments)	Tissue Location	
					Natural Repaired	Cell Therapy Repaired
1	F	42	Follow-up arthroscopy	Naturally and cell repaired cartilage (IHC)	Central Trochlea	LFC
2	M	22	Follow-up arthroscopy	Naturally and cell repaired cartilage (IHC)	Central Trochlea	MFC
3	M	41	Follow-up arthroscopy	Naturally and cell repaired cartilage (IHC)	Central Trochlea	LFC
4	M	29	Follow-up arthroscopy	Naturally and cell repaired cartilage (IHC)	Central Trochlea	Patella
5	M	30	Follow-up arthroscopy	Naturally and cell repaired cartilage (IHC)	Central Trochlea	Patella
6	M	34	Follow-up arthroscopy	Naturally and cell repaired cartilage (IHC)	Central Trochlea	MFC
7	F	36	Follow-up arthroscopy	Naturally and cell repaired cartilage (IHC)	Central Trochlea	Patellar
8	M	51	Follow-up arthroscopy	Naturally and cell repaired cartilage (IHC)	Central Trochlea	MFC
9	M	37	Follow-up arthroscopy	Naturally and cell repaired cartilage (IHC)	Central Trochlea	Patella
10	M	43	Follow-up arthroscopy	Naturally and cell repaired cartilage (IHC)	Central Trochlea	Trochlea
11	Unknown	21	Cadaver	Healthy cartilage (IHC)	MFC	
12	Unknown	30	Cadaver	Healthy cartilage (IHC)	MFC	
13	Unknown	40	Cadaver	Healthy cartilage (IHC)	MFC	
14	Unknown	50	Cadaver	Healthy cartilage (IHC)	MFC	
15	M	63	Cadaver	Healthy cartilage (IHC)	MFC	
16	M	71	TKR	Chondrocytes (heparanase treatment, FC, RT-qPCR)	LFC/MFC	

Table 1. Cont.

Donor.	Gender	Age	Surgical Intervention	Tissue Used (Experiments)	Tissue Location	
					Natural Repaired	Cell Therapy Repaired
17	F	81	TKR	Chondrocytes (heparanase treatment, FC, RT-qPCR)	LFC/MFC	
18	F	51	TKR	Chondrocytes (heparanase treatment, FC, RT-qPCR)	LFC/MFC	
19	M	74	TKR	Chondrocytes (heparanase treatment, FC, RT-qPCR)	LFC/MFC	
20	M	22	Cadaver	Chondrocytes (heparanase treatment, FC, RT-qPCR)	LFC/MFC	

FC = flow cytometry, IHC = immunohistochemistry, TKR = total knee replacement, ACI = autologous chondrocyte implantation, LFC = lateral femoral condyle, and MFC = medial femoral condyle.

2.2. Immunohistochemistry

Cryosections were brought to room temperature and treated with 4800 U/mL hyaluronidase (Sigma, Merck Life Science UK, Dorset, UK) for 2 h and fixed with 4% formaldehyde for 10 min. Slides were washed 3 times in phosphate buffered saline (PBS) between all steps and all steps were performed at room temperature. Goat and horse serum were used to block non-specific binding of the primary mouse and rabbit antibodies, respectively (30 min). Sections were then incubated with mouse monoclonal primary antibodies against perlecan (clone A74, Abcam, Cambridge, UK), collagen type III (clone FH-7A, Abcam) and a polyclonal rabbit antibody to bovine collagen type VI (kindly gifted by Shirley Ayad, University of Manchester, UK) for 60 min, then incubated with biotinylated goat anti-mouse and horse anti-rabbit secondary antibodies (Vectastain Elite ABC kit, Vector Laboratories, Upper Heyford, UK) for monoclonal and polyclonal primary antibodies, respectively, for 30 min. An isotype-matched IgG was used in place of the primary monoclonal antibodies (R&D, Cat No MAB002) as a negative control and normal rabbit serum (Abcam, Cat no ab7487) for the polyclonal, and 0.3% hydrogen peroxide in methanol was used to block endogenous peroxidase activity (30 min). The Vectastain Elite ABC kit (Vector Laboratories) was used to enhance labelling and the ImmPACT® DAB Peroxidase substrate (Vector Laboratories) was used to reveal staining. The sections were dehydrated in serial solutions of 70%, 90% and 100% isopropanol (2 min each) and cleared in xylene (2 × 5 min). The slides were mounted in Pertex (CellPath, Newtown, UK) before imaging.

A semi-quantitative score was developed to assess the immunolocalisation and degree of staining for perlecan in the superficial, mid, and deep zones of the cartilage biopsies. Each zone was scored separately as 0 = no staining, 1 = pericellular staining, 2 = mixture of pericellular and matrix staining, or 3 = matrix staining. Each sample was then given an overall score which was a summation of the scores for the three zones. A high overall score equates to a more widespread matrix immunostaining, whereas a low score equates to more restricted pericellular staining throughout the tissue. Image analysis was performed using FIJI-ImageJ software (Version 1.5), using the Colour Deconvolution and Threshold plugins to establish the levels of perlecan staining as a percentage of the total area of the section.

2.3. Isolation and Culture of Chondrocytes

Chondrocytes were isolated from macroscopically normal cartilage taken from four patients having arthroplasty and one cadaver (Table 1), as previously described [27]. In brief, cartilage tissues were minced and digested for 16 h with collagenase type II (250 IU/mg dry weight, Worthington, New Jersey, USA) at 37 °C. The extracted cells, were seeded at 5000 cell/cm^2 in complete culture media containing Dulbecco's Modified Eagle's Medium/F-12 (DMEM/F-12) with 1% (v/v) penicillin/streptomycin (P/S) and 10% (v/v) foetal calf serum (all Life Technologies, Loughborough, UK). Chondrocytes were passaged at 70–80% confluence and cultured to passage 2 (P2).

2.4. Heparanase Treatment of Chondrocytes and Live Cell Imaging

At P2, chondrocytes were seeded into 12-well plates at or in chamber slides (with 8 chambers) 5200 cells/cm^2, and treated with complete media supplemented with or without 200 ng/mL of recombinant active human heparanase (Bio-Techne, Abingdon, UK; 20 ng of enzyme results in >50% of optical density (OD) reduction as measured by heparan sulphate release from human syndecan-4) for 48 h. The 12-well plate was placed in a Cell-IQ (ChipMan Technologies, Tampere, Finland) live imaging platform to acquire phase contrast images of all wells, every ten minutes, during the 48-h culture. A built-in analysis software in the Cell-IQ was used to determine the number of cells in each image to produce growth curves of cells treated with heparanase, in comparison to control cells with no enzyme. The mean and standard deviation of the cell counts from three fields of view from three separate repeat wells were taken. After 48 h, the cells were harvested and prepared for multichromatic flow cytometry and real-time quantitative polymerase chain reaction (RT-qPCR) analysis.

The cells within the chamber slides were washed three times with PBS, fixed with paraformaldehyde for 10 min and chamber slides were stored at 4 °C until used for immunocytochemistry.

2.5. Immunocytochemistry and Toluidine Blue Staining of Heparanase-Treated Chondrocytes

Chamber slides were brought to room temperate and the PBS replaced with 0.2% Tween 20 for 10 min to permeabilise the cells. After three washes with PBS, the same staining protocol used for immunohistochemistry (see Section 2.2) was followed to reveal the presence of perlecan on the adherent cells, with the addition of a haematoxylin counterstain (diluted 1:3) for 5 s before the slides were mounted in Pertex.

To visualise the presence of glycosaminoglycans, chamber slides were brought to room temperate and the PBS replaced with toluidine blue for 30 s, then washed with distilled water for 5 min. The slides were dehydrated in 70%, 90% and 100% isopropanol (2 min each) and cleared in xylene (2 × 5 min). The slides were mounted in Pertex for imaging.

2.6. Multichromatic Flow Cytometry

A panel of 12 surface markers was used in multichromatic flow cytometry to assess the phenotype of the cells. The harvested cells were blocked with human IgG (Grifols, Cambridge, UK) for 1 h, washed with PBS and incubated for 30 min with antibodies against the mesenchymal stromal/stem cell (MSC) markers CD73, CD90 and CD105 putative chondropotency markers CD151, CD166, FGFR3, CD44 and integrins CD29, CD49a, CD49b CD49c, CD51/CD61 (all BD Biosciences, except for FGFR-3 which was sourced from R&D Systems). The matching isotype controls for each antibody were also prepared according to manufacturer's recommendations. At least 5000 cells were measured per marker via a FACS Canto II cytometer and analysis was performed using the FACS Diva software.

2.7. RNA Extraction and Reverse Transcription Quantitative Polymerase Chain Reaction (RT-qPCR)

To determine the effects of the heparanase treatment on gene expression, RNA was extracted using the RNeasy® mini kit (Qiagen, Manchester, UK) and cDNA was generated using a High-Capacity cDNA Reverse Transcriptase Kit® (Applied Biosystems, Loughborough, UK) according to the manufacturers' protocols. RT-qPCR was performed on a QuantStudio 3 real-time PCR system (Applied Biosystems) using SYBR green QuantiTect primer assays (Qiagen) to assess the gene expression of Sox-9 (*SOX9*), aggrecan (*ACAN*), collagen type II (*COL2A1*), fibroblast growth factor receptor 3 (*FGFR3*), collagen type X (*COL10*) and activin receptor-like kinase (*ALK-1*). Peptidylprolyl Isomerase A (*PPIA*) and TATA-box binding protein (*TBP*) were used as reference genes and the delta-delta C_t method was employed to determine the relative fold change in gene expression levels between heparanase-treated and untreated cells.

2.8. Statistical Analysis

Statistical analysis was performed using GraphPad Prism version 7. The Shapiro–Wilk test was used to determine the normality of data. *T*-tests and Pearson's test were used to compare and correlate histology and immunohistochemistry scores, respectively. A two-way ANOVA with multiple comparisons was used to analyse the growth kinetics of the cells treated with heparanase and a paired, one-sample *t*-test for the fold change in gene expression. A *p*-value ≤ 0.05 was considered statistically significant.

3. Results

3.1. Morphological Structure of Healthy and Repaired Cartilage

The general morphology of the repaired tissue biopsies was very variable, more so for the naturally repaired samples than the cell therapy repaired samples. Overall, donor-matched natural and cell therapy repaired samples showed no distinguishable trend or correlation in terms of tissue morphology (Figure 1). Of the naturally repaired biopsies, 3/10 were predominantly hyaline and 4/10 fibrocartilage, 1/10 was a mixture of hyaline and fibrocartilage and 2/10 were a fibrous morphology. Of the cell therapy repaired biopsies, 7/10 were fibrocartilage and 3/10 were of a mixed hyaline/fibrocartilage morphology with no discernible differences in tissue morphology noted with varying anatomical location of the repair cartilage site. The ICRS overall histology score was not significantly different between naturally repaired and cell therapy repaired samples (mean scores of 5.6 ± 1.9 SD and 5.1 ± 0.8 SD, respectively, *p* = 0.393). Matrix metachromasia was generally better in the cell therapy repaired cartilage samples than in the naturally repaired ones. Cell morphology was marginally better in the cell therapy repaired biopsies, but not significantly different to the naturally repaired biopsies. Vascularisation was observed in 6/10 naturally repaired biopsies, but not in the cell therapy repaired or normal samples.

3.2. Perlecan and Collagen Types III and VI Have a Diffuse Immunolocalisation in Repair Cartilage Tissues

Perlecan was localised in a discrete manner in the pericellular matrix around chondrocytes in healthy cartilage (Figure 2A,B). However, in naturally and cell therapy repaired cartilage staining for perlecan was seen in a pericellular location in some biopsies, diffusely throughout the matrix in others or both patterns within others. Where fibrocartilage was more abundant, perlecan was more diffuse in the cartilage matrix with some strong staining around chondrocytes, which was strikingly different to healthy cartilage as illustrated in Figure 2C,D, showing donors 10 and 9, respectively. In both natural and cell therapy repaired tissues where hyaline cartilage was visible, perlecan was mostly localised in the pericellular regions, but more prominently than in normal cartilage (Figure 2E, showing donor 2). The more elongated cells within fibrocartilaginous repair tissue were generally weak or moderately stained for perlecan, compared to the more rounded chondrocytes in hyaline cartilage (both repair and normal cartilage) which had strong pericellular perlecan immunostaining. Disorganised fibrous tissue was associated with weak matrix perlecan staining. Isotype controls are shown in Figure S1.

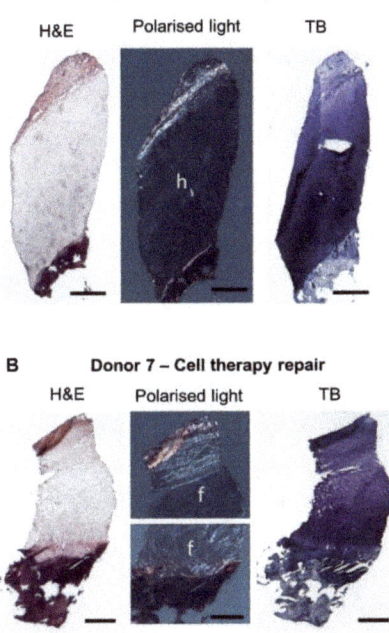

Figure 1. Representative histology images of cartilage repair biopsies from the same donor. Natural repair (**A**) and cell therapy repair (**B**) cryosections were stained with haematoxylin and eosin (H&E) to assess general morphology and toluidine blue (TB) to assess proteoglycan content; both samples demonstrated good to excellent matrix metachromasia. Polarised light was used to assess collagen fibre orientation and determine tissue morphology. The natural repair cartilage demonstrated a mostly hyaline (h) morphology whilst the cell therapy repair cartilage was mostly fibrocartilage (f). Scale bars 500 μm.

The perlecan immunohistochemistry scores were similar between the two repair tissues, with no noticeable trend when comparing individual donor-matched samples (Figure 3A). Image analysis of the percentage of perlecan staining in the tissues showed that naturally repaired and cell therapy repaired cartilage had significantly more staining than the healthy tissues ($p = 0.017$ and $p = 0.018$, respectively, Figure 3B). Interestingly, an increase in the perlecan score significantly correlated with a better-quality cell therapy repair, as defined by the ICRS II 'overall score' parameter ($r = 0.75$, $p = 0.03$, Figure 3C). Perlecan was also strongly localised around small blood vessels that were visible in 6 of the 10 naturally repaired. No blood vessels were observed in either the cell therapy repaired cartilage samples, or the healthy cartilage.

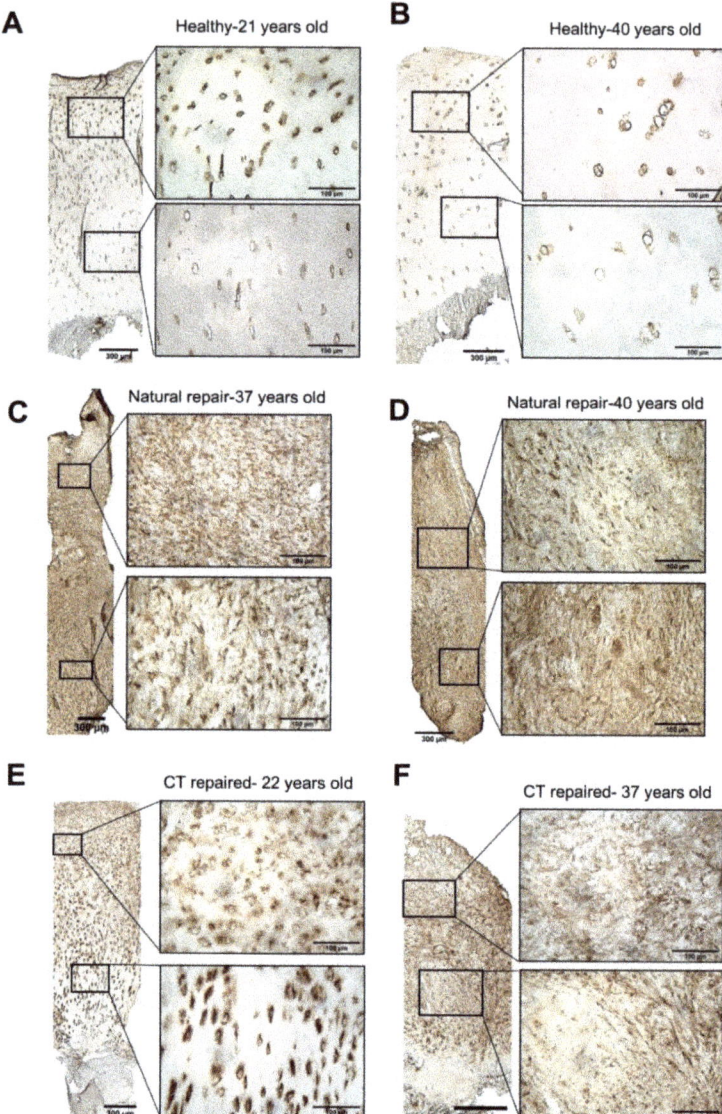

Figure 2. Immunohistochemistry of perlecan. Monoclonal antibodies (A74) were used to detect the presence of perlecan in cryosections of core biopsies. (**A,B**) Heathy cartilage ($n = 5$) from cadavers all showed distinct pericellular staining for perlecan with a typically hyaline morphology. (**C,D**) Naturally repaired cartilage ($n = 10$) from the harvest site of autologous cell therapy donors showed heterogenous staining patterns, some having both widespread matrix and pericellular staining ((**C**), donor 10), whilst in others there was diffuse matrix staining throughout ((**D**), donor 9). (**E,F**) Cell therapy repaired (CT, $n = 9$) cartilage also showed a heterogenous localisation for perlecan, similar to the naturally repaired tissues. The sample depicted in (**E**) (donor 2) shows pericellular staining for perlecan in repair tissue with hyaline cartilage morphology, but not as discretely as in the healthy tissues. The sample depicted in (**F**) (donor 7) shows predominantly matrix immunolocalisation of perlecan. Scale bars show 300 μm for low magnification images and 100 μm for high magnification inserts. Isotype controls found in Supplementary Figure S1A–C.

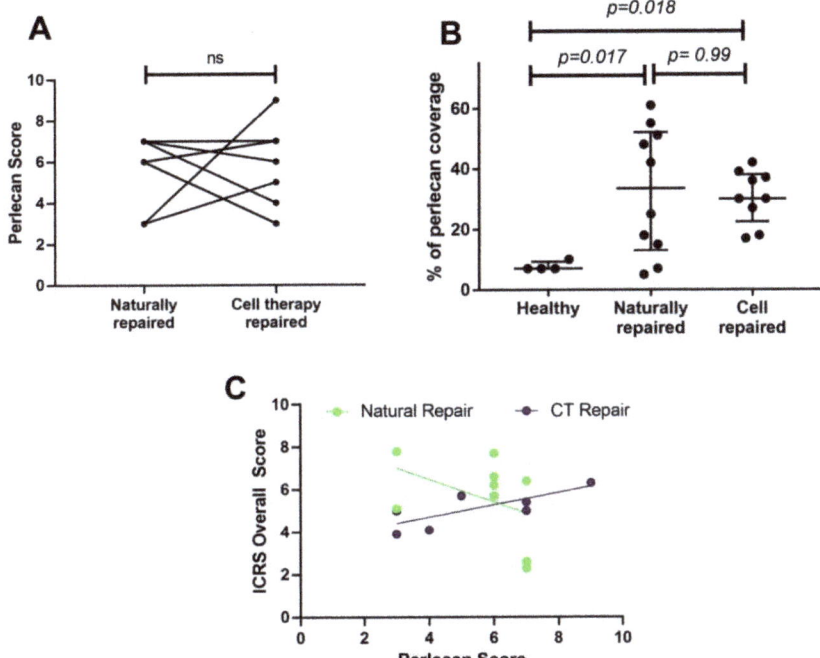

Figure 3. Analysis of tissue morphology and perlecan staining. (**A**) The perlecan immunohistochemistry score gives a general idea of the localisation (pericellular, non-pericellular, mixed) of perlecan in the deep, middle and superficial zones of cartilage. The zones were scored as 0 = no staining, 1 = pericellular staining, 2 = mixture of pericellular and matrix staining, or 3 = matrix staining. The final perlecan score shown here is the summation of the scores for the three zones in each sample. No difference was found between the donor-matched natural and cell therapy repaired tissues. Data show the median with interquartile range. (**B**) Threshold image analysis confirmed a higher percentage of perlecan staining in naturally repaired and CT repaired cartilage than in health cartilage. Perlecan was significantly more prominent in the repair tissues compared to controls. (**C**) Regression analysis showed a positive correlation between the ICRS score and perlecan immunohistochemical score ($r = 0.75$, $p = 0.03$) for cell therapy repaired, but not naturally repaired tissues ($r = -0.4$, $p = 0.25$).

Collagen types III and VI generally exhibited a diffuse staining pattern throughout the interterritorial matrix, covering 94.3 ± 8.9% (range 70–100) and 95.2 ± 7.1% (range 80–100) of the section area, respectively (Figure 4C,D). However, where there was hyaline cartilage present in the repair tissues (Figure 4B), the staining pattern in these regions for both collagen types III and VI was similar to what is typically observed in healthy cartilage (Figure 4A) [28,29], with the pericellular matrix being immunonegative for collagen type III and immunopositive for collagen type VI and the territorial matrix being immunopositive for collagen type III and immunonegative for collagen type VI.

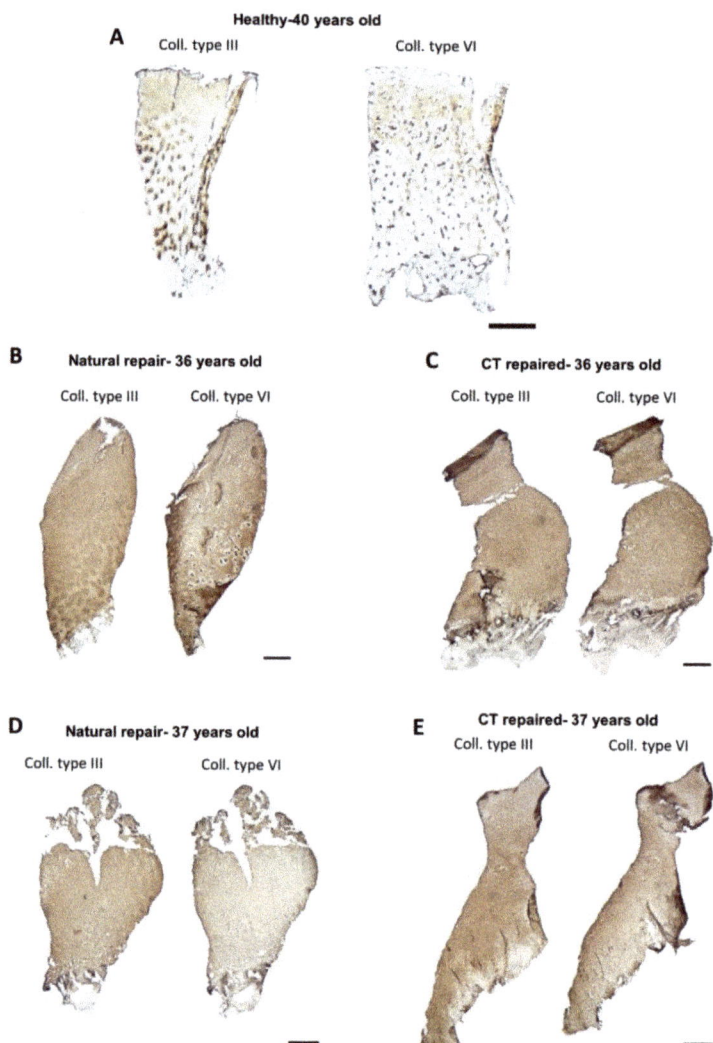

Figure 4. Immunohistochemistry of collagen types III and VI. Monoclonal and polyclonal antibodies were used to detect the presence of collagens type III and VI, respectively, in cryosections of core biopsies. For the repair tissues, two donor-matched samples of natural and cell therapy (CT) repaired cartilage are shown as representative examples (B + C = donor 7, D + E = donor 9). (**A**) Healthy cartilage showing interterritorial staining for collagen type III and pericellular staining for type VI (donor 15). (**B,C**) In this instance of hyaline-like cartilage in naturally repaired cartilage, the collagen type III was localised in the interterritorial region while collagen type VI was localised in the pericellular matrix. (**D,E**) Both collagens type III and VI are diffused in the matrix of fibrocartilage. Scale bar = 500 μm. Isotype controls found in Figure S1D,E.

3.3. Heparanase Increases the Proliferation of Chondrocytes

No discernible difference in morphology was observed in chondrocytes cultured in monolayer which had been treated with 200 ng/mL of heparanase compared to untreated controls after 48 h (Figure 5A). Separate and combined growth plots are shown for the individual donor cell populations tested in Figure 5B. Whilst there is variation between donors, a combined assessment of the cell populations showed that, for the first 20 h, the heparanase-treated and control chondrocytes showed similar growth rates, but diverged from 24 h onwards with treated cells showing significantly higher proliferation rates than untreated control cells between 32 and 48 h (Figure 5B, bottom right plot).

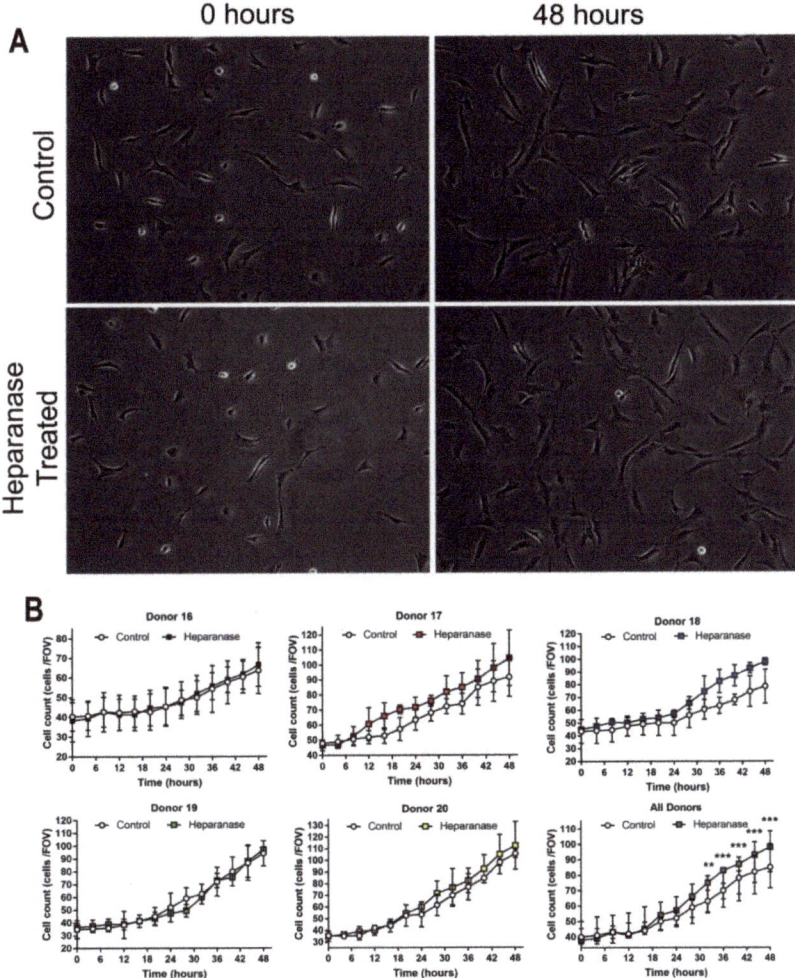

Figure 5. Morphology and growth kinetics of chondrocytes after treatment with heparanase ($n = 5$). (**A**) Phase contrast images were acquired for control and heparanase-treated chondrocytes every 10 min at precise locations for 48 h. The representative images shown are of chondrocytes from donor 20 at t = 0 h and t = 48 h. (**B**) Growth kinetics of chondrocytes were established using a live cell imaging platform and analysis software during the 48 h heparanase treatment period. Individual plots are shown for donors 16 to 20 with mean and SD of cell counts from three FOV from three separate wells. The combined data for all five donors at every time point are also shown (bottom right). FOV = field of view. ** $p < 0.05$, *** $p < 0.01$.

3.4. Stromal/Stem Cell and Chondropotency Markers and Genes Are Not Affected by Heparanase

Flow cytometry demonstrated that the positivity of stromal/stem cell markers, CD73, CD90 and CD105 (Figure 6A), and the chondrogenic markers CD44, CD151, CD166, FGFR3 were unaffected by heparanase treatment (Figure 6B), although CD166 and FGFR3 showed a high level of variability between donors. For the integrins, donor variability was also observed with CD49a, CD49b, CD49c and CD51/61, but not CD29, with no statistical difference between treated cells and controls for any of the integrins (Figure 6C). Chondrocytes from donor 17, the oldest donor, showed a marked heparanase-induced increase in CD166 and a noticeable decrease in CD49a, CD49b and CD49c compared to the cells from the other donors.

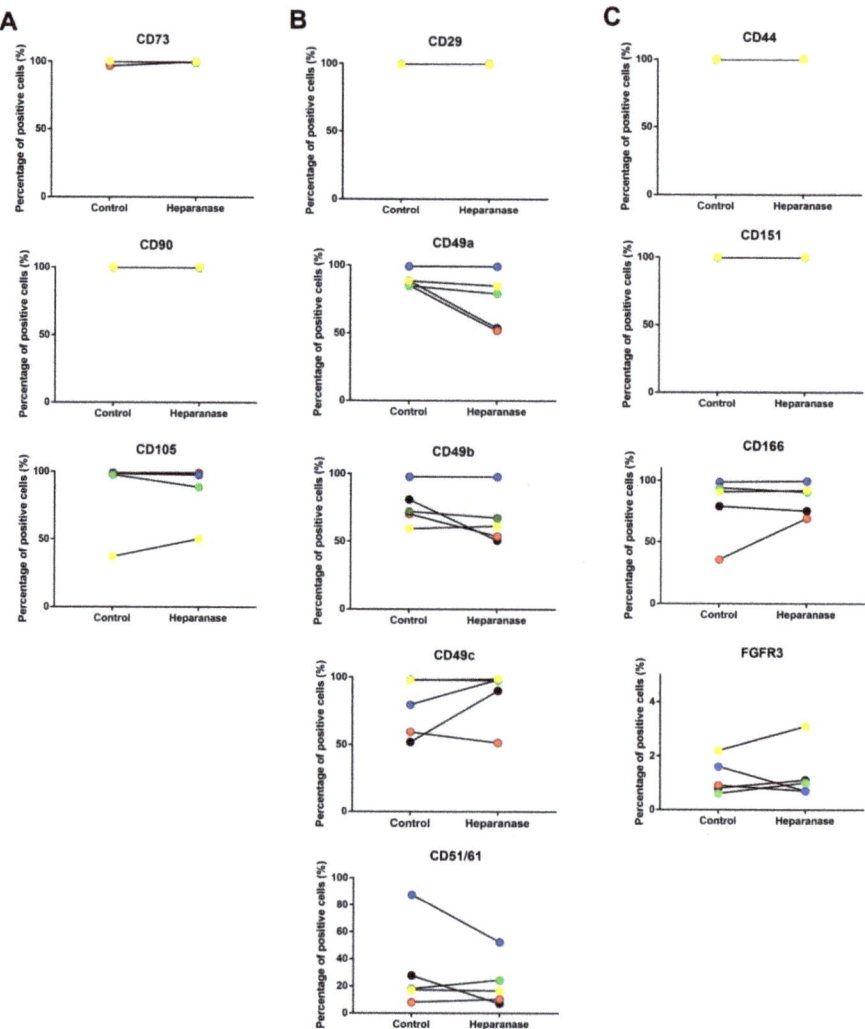

Figure 6. Flow cytometry analysis of the effects of heparanase treatment on surface markers of chondrocytes ($n = 5$). Results are shown as the percentage of positive cells for a particular marker on heparanase-treated chondrocytes and the matching control. Fluorochrome-conjugated antibodies were used to detect (**A**) stem cells markers, (**B**) integrins and (**C**) chondrogenic markers. No significant differences were observed. Matched samples are represented by the same colour dot; donor 16 (black), donor 17 (red), donor 18 (blue), donor 19 (green), and donor 20 (yellow).

Although the relative fold change in chondrogenic gene expression was not statistically significant between the heparinase-treated and untreated chondrocytes, there was a general decrease in SOX9 expression (median= -1.17), and increased expressions for ACAN (median = 1.1), COL2A1 (median =1.2), and FGFR3 (median = 1.1) following heparanase treatment (Figure 7). The relative fold change in expression of the hypertrophic genes COL10 (median = 1.6) and ALK-1 (median= 2.4) was also increased following heparanase treatment, but this was not statistically significant (Figure 7).

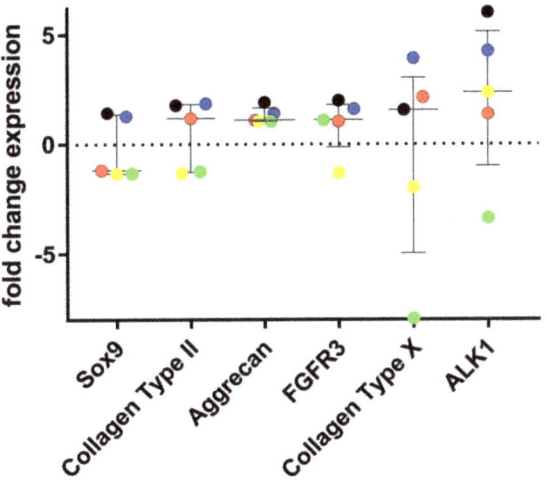

Figure 7. Analysis of the effects of heparanase treatment on gene expression of chondrocytes in monolayer culture (n = 5). Results are presented as log-fold change in the expression of the chondrogenic genes SOX9, collagen type II, aggrecan, FGFR3 and hypertrophy genes collagen type X and ALK1 in chondrocytes that were treated with heparanase compared to the untreated controls. Matched samples are represented by the same colour dot; donor 16 (black), donor 17 (red), donor 18 (blue), donor 19 (green), and donor 20 (yellow). Error bars indicate medians and interquartile ranges.

3.5. Perlecan and Toluidine Blue Staining Is More Prominent in Chondrocyte Clusters

There was immunostaining for perlecan in some cultured cells, some apparently in the cytoplasm and also associated with the cell membrane. This appeared strongest when cells were in clusters, which were more common in cultures without exposure to heparanase (Figure 8A).

Metachromasia with toluidine blue staining for glycosaminoglycans was mostly weak with no consistent difference in pattern between control and heparanase conditions (Figure 8B). However cell, clusters, where present, tended to have stronger toluidine blue staining.

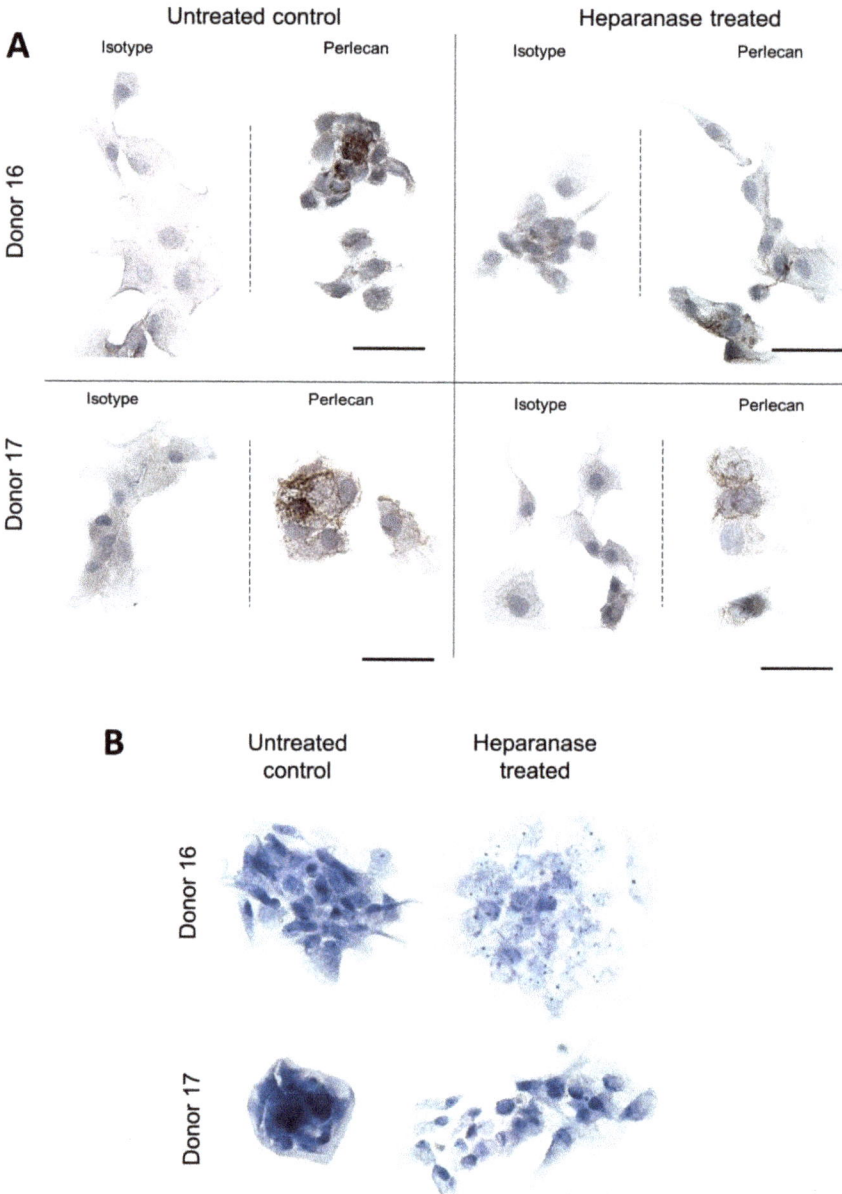

Figure 8. Immunocytochemistry and toluidine blue staining of heparanase-treated chondrocytes (two representative donor examples shown). (**A**) Chondrocytes cultured in chamber slides were treated with heparanase for 48 h and stained for perlecan (n = 4). Chondrocytes in untreated controls tended to remain in clusters that stained more intensely for perlecan compared to heparinase-treated chondrocytes. (**B**) Toluidine blue staining revealed no noticeable trend. However, cell clusters in both controls and heparanase conditions had stronger staining than individual cells. Scale bars = 50 μm.

4. Discussion

Cell-based therapies have shown some degree of success in restoring damaged cartilage [30,31], but no study to date has described the presence of the proteoglycan perlecan in either the natural repair or cell therapy repair of cartilage in humans. Perlecan contributes to processes that are essential to the functioning of chondrocytes such as cell attachment, differentiation and production of extracellular matrix components [12,13], which makes it an ideal candidate molecule to assess in the formation of new cartilage. There is a longstanding biological paradigm that once damaged, articular cartilage cannot heal itself. However, evidence is now mounting to indicate that actually, to a limited extent, articular cartilage does have an innate ability to repair [2,32], although the mechanism and pathways are poorly defined. To our knowledge, this study is the first to assess and compare the differences in perlecan immunolocalisation in matched patient cartilage samples that were repaired either naturally or with autologous cells, while assessing the effects of heparanase on the phenotype of human chondrocytes in vitro.

The variety of tissue morphologies observed in the repair tissues, i.e., fibrous, hyaline, fibrocartilage and a mixture of the two, demonstrates the unpredictable and variable nature of cartilage repair. Some of these differences could be donor dependent, but since the repair tissues have been collected from two different sources (one from the harvest site and the other post-treatment with cell therapy), the repair could have been the result of two different biological mechanisms. Furthermore, the lack of an identifiable pattern of morphology in donor-matched natural and cell therapy repaired tissues could be due to differences in the microenvironment of the location where these defects were found. The high incidence of vascularisation present in the naturally repaired biopsies is of concern, as in its native state, cartilage is avascular. One could hypothesise that there may be a temporary invasion of blood vessels as a means of instigating the repair processes and over time with tissue remodelling and maturation, this vascularisation may disappear. Synovial infiltrates are often vascularised and usually associated with poor cartilage repair [33], but a recent study has provided evidence of the contribution of synovial cells in the repair of cartilage surface injuries in mice [34]. Adhesions identified by MRI (which are likely to be vascularised) have been shown to correlate with better histological features of cartilage repair twelve months after ACI [35].

Perlecan was immunolocalised in the pericellular matrix in healthy cartilage, which was is in line with previous findings [16]. In contrast, the immunolocalisation of perlecan in the repair tissue differed depending upon the type of tissue morphology present, for example, in areas of hyaline cartilage, perlecan appeared to have a more "normal" pericellular appearance whereas in areas of fibrocartilage, it was more associated within the interterritorial extracellular matrix. The latter appears to resemble the disposition of perlecan observed in foetal patella, femoral condyle and tibial plateau tissues [36]. This, combined with the fact that perlecan is a marker of early chondrogenic activity [37], suggests that embryological mechanisms could be contributing to the repair of damaged adult cartilage, either naturally or post-cell therapy. This is further evidenced by the observation that in cell therapy repaired samples, perlecan is associated with better tissue morphology and increased proteoglycan content, more resembling normal, healthy cartilage. (One slight caveat in comparing this immunolocalisation between healthy and surgical samples, however, is that there was some disparity in times between ex-vivo collection and processing; for healthy donors, time to fixation was ~24 h + 2–4 h but for surgical samples it was much quicker (2–4 h).)

Fibrocartilage commonly forms in repair sites following cell therapy, at least in biopsies obtained ~12 months post-treatment [25]. Whilst the aim of cell therapy in the treatment of cartilage defects is the formation of hyaline cartilage, there is evidence that the initial repair tissue which forms is remodelled [38] and does indeed mature towards hyaline cartilage with time post-treatment [39]. The distribution of perlecan seen in our study is perhaps further evidence of this, with the more diffuse and widespread location seen in fibrocartilage resembling that of developing or rudimentary cartilages, some of which subsequently mature to form hyaline cartilage with its definite pericellular staining pattern. The strong vascular localisation of perlecan in the naturally repaired tissues is expected, confirming reports of its role in angiogenesis [40,41].

In healthy articular cartilage, collagen type III has a diffuse localisation in the territorial regions around the chondrocytes, i.e., beyond the pericellular capsule [28]. We found this pattern only in the repair tissues where some hyaline-like cartilage was present, suggesting more matured repair or regeneration. Collagen type III is often associated with collagen type I and is abundant in damaged tissues that are attempting to repair [28,42].

Collagen type VI is a microfibrillar collagen, accounting for approximately 1% of total collagen in adult articular cartilage [43]. Predominantly located in the pericellular matrix (PCM) in developing and mature cartilage, collagen type VI has been demonstrated to be integral for regulating chondrocyte swelling and contributing to the biomechanical integrity of the PCM; indeed, it also binds to the chondrocyte membrane via the RGD sequences [44–47]. During osteoarthritis, however, the localisation of collagen type VI changes to more interterritorial matrix expression, possibly reflecting increased degradation of the collagen fibrils [48,49]. The diffuse pattern of immunolocalisation of collagen type VI in the majority of the repair tissues tested in our study is similar to that of perlecan and indicates an immature PCM in regenerating cartilage. Despite our observations of collagen types III and VI immunlocalisation in repair tissue being similar to those found in OA, it is also possible that they are indicative of an immature, developing cartilage rather than degeneration. Both perlecan and collagen type VI have shown to be pivotal to the biomechanical function of the PCM [50]. As a result, and due to the ability of cartilage to detect and respond to mechanical loading, perlecan in particular could be an active participant in the loading-related aspects of cartilage repair and remodelling. Perlecan's role of "mechanosensing" in tissue maintenance has been demonstrated in bone [51,52], while its ability to influence the elastic modulus of the PCM has been proven in cartilage [50].

Given the unique glycolytic capability of heparanase, this enzyme has been proposed to be a valuable therapeutic target in repair biology [53,54]. The fragments released from the HS by the action of heparanase are often more bioactive than the native molecule [55,56]. For example, when heparanase cleaves HS from perlecan in the basement membrane it releases bound FGF2, which promotes angiogenesis, wound healing and tumour formation [57,58]. In our study, we tested the effects of heparanase on chondrocytes in terms of cell morphology, proliferation, and the expression of surface and genetic markers. Although no noticeable difference in cell morphology was noted, chondrocytes treated with heparanase showed higher proliferation compared to the untreated controls. This finding corroborates a previous study showing a heparanase-induced increase in proliferation and migration of the ATDC5 chondrocyte cell line [59], and supports the theory that the removal of HS encourages an increase in cell proliferation. Further investigations are needed to determine whether this stimulation of chondrocytes by heparanase is reproduced in vivo, and what the pathophysiological implications are, notably in the modulation of tissue repair. One should also consider the source/s of the HS that has been depleted, as perlecan is not the only HS-containing proteoglycan found in cartilage.

The flow cytometry analysis conducted in our experiment produced the first data looking at the effects of heparanase treatment on the expression of a comprehensive panel of surface markers in human chondrocytes. Exposing human chondrocytes to exogenous heparanase did not influence the expression of either surface stem cells markers (CD70, CD90 and CD105), or chondrogenic markers (CD44, CD151 and CD166). Interestingly, another study in mice MSCs has shown similarly that the inhibition of endogenous heparanase has no effect on these stem cells markers [60].

Of the five chondrocyte populations tested for their response to heparanase, three of them showed a marginal increase in FGFR3 as assessed by flow cytometry, while the gene expression of FGFR3 was stable. This is of particular interest in the context of cartilage repair, since signalling through the FGFR3 pathway is essential to chondrocyte function during chondrogenesis. During the embryological development of cartilage rudiments, FGFR1c, FGFR2c, FGFR3c and perlecan are employed by mesenchymal cells to promote the production of extracellular matrix production [17,61,62]. FGF-18 has also been shown to signal through FGFR3 in the cartilaginous development of the human foetal spine [63]. Furthermore, a mouse knockout model revealed that the deletion of domain I in HS improved the symptoms of OA and preserved the expression of FGFR3 with disease progression [64]. We hypothesise that the positivity of FGFR3 on reparative cells in the de novo formation of cartilage could be an essential mediator of natural and CT repaired tissues. Additional work would, however, be needed to investigate this further.

Integrins are a family of cell adhesion receptors that are vital to the interactions between chondrocytes and the cartilage extracellular matrix, that is mediated through the binding of matrix components such as collagen types II and VI, vitronectin and fibronectin [65]. The heparanase treatment of chondrocytes in our study did not affect CD29, which is the $\beta 1$ integrin subunit. CD29 couples with the $\alpha 1$ integrin subunit (CD49a) to form the $\alpha 1 \beta 1$ complex, and facilitates the binding of collagen types II and VI [66,67]. The reduced positivity of CD49a in four of the five cell populations treated with heparanase suggests a possible interaction between HS and the integrins that warrants further characterisation in cartilage repair. The heterogeneity of the expression of integrin subunits CD49b($\alpha 2$), CD49c($\alpha 3$) and the complex CD51/61($\alpha V/\beta 3$) in response to heparanase, may be indicative of the versatility of chondrocytes when interacting with their pericellular environs and extracellular matrix. The marked lower levels in CD49a, CD49b and CD49c detected in chondrocytes from the oldest donor (donor 17) after heparanase treatment could reflect the age-induced decrease in integrins in cartilage that has been previously shown [68].

We tested the effects of heparanase on the expression of key chondrogenic genes and found no significant change in the expression of SOX9, collagen type II and aggrecan. This differed from a previous study that showed an increased gene expression for collagen type II and aggrecan after heparanase treatment, but this was using a more appropriate 3D culture system [59]; even in our monolayer system, heparanase had no inhibitory effect on chondrocytes. The marginal increase in expression of the hypertrophic genes for collagen type X and ALK-1 could be an indirect effect of the increased cell proliferation and is not conducive to the repair of hyaline articular cartilage. This observation should act as a reminder that the mechanisms triggered by the removal of HS would need to be controlled to avoid undesirable matrix formation [69].

The strong immunolocalisation of perlecan in chondrocyte cell clusters suggests that the pericellular matrix of these cells may still be intact, or at least being maintained, in some monolayer cultures with close cell contact. This finding confirms previous studies showing pronounced perlecan staining in clusters found in OA cartilage [70,71]. It was found that domain IV-3 of perlecan was responsible for chondrocyte clustering, by mediating a decrease in ERK1/2 signalling [72]. The presence of perlecan persisted despite after the assumed removal of HS in our study, which could indicate that the heparanase-induced response from chondrocytes is due to the loss of HS from perlecan, and not perlecan itself. Such observations were made in a mouse study where a *Hspg2* exon 3 null strain continued to produce perlecan without the native HS [13].

The present study is not without its limitations. For example, we acknowledge that a bigger sample size would make this study more robust; however, we are confident, based on our experiences, that the tissue morphologies presented here are in line with our previous observations. The naturally and cell therapy repaired tissues that we studied formed at different locations in the joint. This may have limited the direct comparison of the two cartilages, for instance due to differences in biomechanical forces betweeen different regions of the knee joint [73,74]. Regarding the in vitro cell experiments, chondrogenic differentiation may have provided additional insight into the effects of heparanase on chondrocyte function. It is also important to note that the enzymatic activity of heparanase is not specifically targeted to the HS on perlecan and that other HS proteins such as agrin, syndecan 1 and syndecan 4 may also be affected by heparanase [75,76]. This study does not identify an exact pathway or mechanism per se whereby perlecan influences cartilage repair, but it does indicate that it appears to be an integral player and so worthy of further investigation.

5. Conclusions

To conclude, we demonstrate that the HS proteoglycan, perlecan, is clearly present in repair tissue formed both via cell therapy repair of chondral defects and also naturally occurring repair tissue. The localisation of perlecan, as well as type III collagen, which is often found in developing or repairing tissue, is more diffuse for both molecules in the fibrocartilaginous tissue which forms initially, than in the more mature repair tissue. This more mature repair tissue has morphology resembling hyaline cartilage with has more of the typical cell-associated staining pattern seen in adult articular cartilage. The co-localisation of perlecan and collagen type VI and its biomechanical role in the PCM in repair cartilage remains unclear and further research could reveal a key mechanism that incorporates the different loading forces in the articular joint. The strong perlecan staining observed in chondrocyte clusters could be mediated via its domain IV-3 and the suppression of Erk1/2 signalling. We have also shown that heparanase treatment increases the proliferation of chondrocytes, without altering their phenotypical features, at least, as assessed in this study. Taken together, it is plausible to assume that perlecan has an important role in cartilage repair. Further work is required to fully comprehend how heparanase influences different types of repair, and whether this enzyme can be harnessed to enhance the quality of de novo cartilage repair in vivo.

Supplementary Materials: The following are available online at https://www.mdpi.com/2218-273X/11/1/92/s1, Figure S1: Representative negative controls for immunohistochemistry studies.

Author Contributions: Conceptualization, J.G., J.M., S.R., and J.H.K.; Data Acquisition, Data Analysis; J.G., H.S.M., J.M., and J.H.K.; Manuscript Preparation, J.G., H.S.M., J.M., J.H.K., and S.R. All authors have read and agreed to the published version of the manuscript.

Funding: This research was funded by the Medical Research Council (MR/L010453/1 and MR/N02706X/1), Versus Arthritis (grants 18480, 19429, 21156) and NHMRC Project Grant 51267 The role of perlecan in tensional connective tissues.

Institutional Review Board Statement: The study was conducted according to the guidelines of the Declaration of Helsinki, and approved by the National Research Ethics Service—Coventry and Warwickshire (REC reference 11/WM/0175) in 2011.

Informed Consent Statement: Informed consent was obtained from all subjects involved in the study.

Data Availability Statement: Data available on request due to restrictions eg privacy or ethical.

Acknowledgments: We acknowledge the intellectual and clinical contributions of the late James Richardson to this study. We are also grateful to the OsCell team at The John Charnley Laboratory, RJAH Orthopaedic Hospital, UK for the processing of harvested tissues for autologous cell therapy.

Conflicts of Interest: The authors declare no conflict of interest.

References

1. Ding, C.; Garnero, P.; Cicuttini, F.; Scott, F.; Cooley, H.; Jones, G. Knee cartilage defects: Association with early radiographic osteoarthritis, decreased cartilage volume, increased joint surface area and type II collagen breakdown. *Osteoarthr. Cartil.* **2005**, *13*, 198–205. [CrossRef]
2. Davies-Tuck, M.L.; Wluka, A.E.; Wang, Y.; Teichtahl, A.J.; Jones, G.; Ding, C.; Cicuttini, F.M. The natural history of cartilage defects in people with knee osteoarthritis. *Osteoarthr. Cartil.* **2008**, *16*, 337–342. [CrossRef]
3. Brittberg, M.; Lindahl, A.; Nilsson, A.; Ohlsson, C.; Isaksson, O.; Peterson, L. Treatment of deep cartilage defects in the knee with autologous chondrocyte transplantation. *N. Engl. J. Med.* **1994**, *331*, 889–895. [CrossRef]
4. Wakitani, S.; Mitsuoka, T.; Nakamura, N.; Toritsuka, Y.; Nakamura, Y.; Horibe, S. Autologous Bone Marrow Stromal Cell Transplantation for Repair of Full-Thickness Articular Cartilage Defects in Human Patellae: Two Case Reports. *Cell Transplant.* **2004**, *13*, 595–600. [CrossRef]

5. Hunter, W. Of the structure and disease of articulating cartilages. *Philos. Trans.* **1743**, *42*, 514–521.
6. Bos, P.K.; Kops, N.; Verhaar, J.A.N.; van Osch, G.J.V.M. Cellular origin of neocartilage formed at wound edges of articular cartilage in a tissue culture experiment. *Osteoarthr. Cartil.* **2008**, *16*, 204–211. [CrossRef]
7. Ding, C.; Cicuttini, F.; Scott, F.; Boon, C.; Jones, G. Association of prevalent and incident knee cartilage defects with loss of tibial and patellar cartilage: A longitudinal study. *Arthritis Rheum.* **2005**, *52*, 3918–3927. [CrossRef]
8. Nakamura, N.; Horibe, S.; Toritsuka, Y.; Mitsuoka, T.; Natsu-Ume, T.; Yoneda, K.; Hamada, M.; Tanaka, Y.; Boorman, R.S.; Yoshikawa, H.; et al. The location-specific healing response of damaged articular cartilage after ACL reconstruction: Short-term follow-up. *Knee Surg. Sport. Traumatol. Arthrosc.* **2008**, *16*, 843–848. [CrossRef]
9. Dell'accio, F.; Vincent, T.L. Joint surface defects: Clinical course and cellular response in spontaneous and experimental lesions. *Eur. Cell. Mater.* **2010**, *20*, 210–217. [CrossRef]
10. Sherwood, J.C.; Bertrand, J.; Eldridge, S.E.; Dell'accio, F. Cellular and molecular mechanisms of cartilage damage and repair. *Drug Discov. Today* **2014**, *19*, 1172–1177. [CrossRef] [PubMed]
11. Tiku, M.L.; Sabaawy, H.E. Cartilage regeneration for treatment of osteoarthritis: A paradigm for nonsurgical intervention. *Ther. Adv. Musculoskelet. Dis.* **2015**, *7*, 76–87. [CrossRef] [PubMed]
12. Whitelock, J.M.; Melrose, J.; Iozzo, R.V. Diverse cell signaling events modulated by Perlecan. *Biochemistry* **2008**, *47*, 11174–11183. [CrossRef] [PubMed]
13. Smith, S.M.; Melrose, J. Type XI collagen–perlecan–HS interactions stabilise the pericellular matrix of annulus fibrosus cells and chondrocytes providing matrix stabilisation and homeostasis. *J. Mol. Histol.* **2019**, *50*, 285–294. [CrossRef] [PubMed]
14. Chong, K.W.; Chanalaris, A.; Burleigh, A.; Jin, H.; Watt, F.E.; Saklatvala, J.; Vincent, T.L. Fibroblast growth factor 2 drives changes in gene expression following injury to murine cartilage in vitro and in vivo. *Arthritis Rheum.* **2013**, *65*, 2346–2355. [CrossRef] [PubMed]
15. Vincent, T.L.; McLean, C.J.; Full, L.E.; Peston, D.; Saklatvala, J. FGF-2 is bound to perlecan in the pericellular matrix of articular cartilage, where it acts as a chondrocyte mechanotransducer. *Osteoarthr. Cartil.* **2007**, *15*, 752–763. [CrossRef]
16. SundarRaj, N.; Fite, D.; Ledbetter, S.; Chakravarti, S.; Hassell, J.R. Perlecan is a component of cartilage matrix and promotes chondrocyte attachment. *J. Cell Sci.* **1995**, *108*, 2663–2672.
17. Gomes, R.R.; Farach-Carson, M.C.; Carson, D.D. Perlecan Functions in Chondrogenesis: Insights from in vitro and in vivo Models. *Cells Tissues Organs* **2004**, *176*, 79–86. [CrossRef]
18. Rivara, S.; Milazzo, F.M.; Giannini, G. Heparanase: A rainbow pharmacological target associated to multiple pathologies including rare diseases. *Future Med. Chem.* **2016**, *8*, 647–680. [CrossRef]
19. Hayes, A.J.; Shu, C.C.; Lord, M.S.; Little, C.B.; Whitelock, J.M.; Melrose, J. Pericellular colocalisation and interactive properties of type VI collagen and perlecan in the intervertebral disc. *Eur. Cells Mater.* **2016**, *32*, 40–57. [CrossRef]
20. Pfaff, M.; Aumailley, M.; Specks, U.; Knolle, J.; Zerwes, H.G.; Timpl, R. Integrin and Arg-Gly-Asp dependence of cell adhesion to the native and unfolded triple helix of collagen type VI. *Exp. Cell Res.* **1993**, *206*, 167–176. [CrossRef]
21. Wu, J.J.; Eyre, D.R.; Slayter, H.S. Type VI collagen of the intervertebral disc. Biochemical and electron-microscopic characterization of the native protein. *Biochem. J.* **1987**, *248*, 373–381. [CrossRef] [PubMed]
22. Hosseininia, S.; Weis, M.A.; Rai, J.; Kim, L.; Funk, S.; Dahlberg, L.E.; Eyre, D.R. Evidence for enhanced collagen type III deposition focally in the territorial matrix of osteoarthritic hip articular cartilage. *Osteoarthr. Cartil.* **2016**, *24*, 1029–1035. [CrossRef] [PubMed]
23. Wu, J.J.; Weis, M.A.; Kim, L.S.; Eyre, D.R. Type III collagen, a fibril network modifier in articular cartilage. *J. Biol. Chem.* **2010**, *285*, 18537–18544. [CrossRef] [PubMed]
24. Talkhani, I.S.; Richardson, J.B. Knee diagram for the documentation of arthroscopic findings of the knee—Cadaveric study. *Knee* **1999**, *6*, 95–101. [CrossRef]
25. Mainil-Varlet, P.; Van Damme, B.; Nesic, D.; Knutsen, G.; Kandel, R.; Roberts, S. A New Histology Scoring System for the Assessment of the Quality of Human Cartilage Repair: ICRS II. *Am. J. Sports Med.* **2010**, *38*, 880–890. [CrossRef]
26. Roberts, S.; McCall, I.W.; Darby, A.J.; Menage, J.; Evans, H.; Harrison, P.E.; Richardson, J.B. Autologous chondrocyte implantation for cartilage repair: Monitoring its success by magnetic resonance imaging and histology. *Arthritis Res. Ther.* **2003**, *5*, R60. [CrossRef]
27. Garcia, J.; Mennan, C.; McCarthy, H.S.; Roberts, S.; Richardson, J.B.; Wright, K.T. Chondrogenic Potency Analyses of Donor-Matched Chondrocytes and Mesenchymal Stem Cells Derived from Bone Marrow, Infrapatellar Fat Pad, and Subcutaneous Fat. *Stem Cells Int.* **2016**, *2016*, 1–11. [CrossRef]
28. Wotton, S.F.; Duance, V.C. Type III collagen in normal human articular cartilage. *Histochem. J.* **1994**, *26*, 412–416. [CrossRef]
29. Pullig, O.; Weseloh, G.; Swoboda, B. Expression of type VI collagen in normal and osteoarthritic human cartilage. *Osteoarthr. Cartil.* **1999**, *7*, 191–202. [CrossRef]
30. Peterson, L.; Vasiliadis, H.S.; Brittberg, M.; Lindahl, A. Autologous chondrocyte implantation: A long-term follow-up. *Am. J. Sports Med.* **2010**, *38*, 1117–1124. [CrossRef]
31. Richardson, J.B.; Caterson, B.; Evans, E.H.; Ashton, B.A.; Roberts, S. Repair of human articular cartilage after implantation of autologous chondrocytes. *J. Bone Jt. Surg. Br.* **1999**, *81*, 1064–1068. [CrossRef]
32. McCarthy, H.S.; Richardson, J.B.; Parker, J.C.E.; Roberts, S. Evaluating joint morbidity after chondral harvest for autologous chondrocyte implantation (ACI): A Study of ACI-treated ankles and hips with a knee chondral harvest. *Cartilage* **2016**, *7*, 7–15. [CrossRef]

33. Miyamoto, A.; Deie, M.; Yamasaki, T.; Nakamae, A.; Shinomiya, R.; Adachi, N.; Ochi, M. The role of the synovium in repairing cartilage defects. *Knee Surg. Sport. Traumatol. Arthrosc.* **2007**, *15*, 1083–1093. [CrossRef] [PubMed]
34. Roelofs, A.J.; Zupan, J.; Riemen, A.H.K.; Kania, K.; Ansboro, S.; White, N.; Clark, S.M.; Bari, C. De Joint morphogenetic cells in the adult synovium. *Nat. Commun.* **2017**, *8*, 1–14. [CrossRef]
35. McCarthy, H.S.; McCall, I.W.; Williams, J.M.; Mennan, C.; Dugard, M.N.; Richardson, J.B.; Roberts, S. Magnetic Resonance Imaging Parameters at 1 Year Correlate With Clinical Outcomes Up to 17 Years After Autologous Chondrocyte Implantation. *Orthop. J. Sport. Med.* **2018**, *6*, 1–10. [CrossRef]
36. Melrose, J.; Roughley, P.; Knox, S.; Smith, S.; Lord, M.; Whitelock, J. The structure, location, and function of perlecan, a prominent pericellular proteoglycan of fetal, postnatal, and mature hyaline cartilages. *J. Biol. Chem.* **2006**, *281*, 36905–36914. [CrossRef]
37. Smith, S.M.; Shu, C.; Melrose, J. Comparative immunolocalisation of perlecan with collagen II and aggrecan in human foetal, newborn and adult ovine joint tissues demonstrates perlecan as an early developmental chondrogenic marker. *Histochem. Cell Biol.* **2010**, *134*, 251–263. [CrossRef]
38. Roberts, S.; Hollander, A.P.; Caterson, B.; Menage, J.; Richardson, J.B. Matrix turnover in human cartilage repair tissue in autologous chondrocyte implantation. *Arthritis Rheum.* **2001**, *44*, 2586–2598. [CrossRef]
39. Sharma, A.; Rees, D.; Roberts, S.; Kuiper, N.J. A case study: Glycosaminoglycan profiles of autologous chondrocyte implantation (ACI) tissue improve as the tissue matures. *Knee* **2017**, *24*, 149–157. [CrossRef]
40. Lord, M.S.; Chuang, C.Y.; Melrose, J.; Davies, M.J.; Iozzo, R.V.; Whitelock, J.M. The role of vascular-derived perlecan in modulating cell adhesion, proliferation and growth factor signaling. *Matrix Biol.* **2014**, *35*, 112–122. [CrossRef]
41. Ishijima, M.; Suzuki, N.; Hozumi, K.; Matsunobu, T.; Kosaki, K.; Kaneko, H.; Hassell, J.R.; Arikawa-Hirasawa, E.; Yamada, Y. Perlecan modulates VEGF signaling and is essential for vascularization in endochondral bone formation. *Matrix Biol.* **2012**, *31*, 234–245. [CrossRef] [PubMed]
42. Fleischmajer, R.; Timpl, R.; Tuderman, L.; Raisher, L.; Wiestner, M.; Perlish, J.S.; Graves, P.N. Ultrastructural identification of extension aminopropeptides of type I and III collagens in human skin. *Proc. Natl. Acad. Sci. USA* **1981**, *78*, 7360–7364. [CrossRef] [PubMed]
43. Eyre, D.R.; Weis, M.A.; Wu, J.J. Articular cartilage collagen: An irreplaceable framework? *Eur. Cells Mater.* **2006**, *12*, 57–63. [CrossRef] [PubMed]
44. Zelenski, N.A.; Leddy, H.A.; Sanchez-Adams, J.; Zhang, J.; Bonaldo, P.; Liedtke, W.; Guilak, F. Type VI collagen regulates pericellular matrix properties, chondrocyte swelling, and mechanotransduction in mouse articular cartilage. *Arthritis Rheumatol.* **2015**, *67*, 1286–1294. [CrossRef]
45. Hansen, U.; Allen, J.M.; White, R.; Moscibrocki, C.; Bruckner, P.; Bateman, J.F.; Fitzgerald, J. WARP Interacts with Collagen VI-Containing Microfibrils in the Pericellular Matrix of Human Chondrocytes. *PLoS ONE* **2012**, *7*, e52793. [CrossRef]
46. Marcelino, J.; McDevitt, C.A. Attachment of articular cartilage chondrocytes to the tissue form of type VI collagen. *Biochim. Biophys. Acta (BBA)/Protein Struct. Mol.* **1995**, *1249*, 180–188. [CrossRef]
47. Arikawa-Hirasawa, E.; Watanabe, H.; Takami, H.; Hassell, J.R.; Yamada, Y. Perlecan is essential for cartilage and cephalic development. *Nat. Genet.* **1999**, *23*, 354–358. [CrossRef]
48. Hambach, L.; Neureiter, D.; Zeiler, G.; Kirchner, T.; Aigner, T. Severe disturbance of the distribution and expression of type VI collagen chains in osteoarthritic articular cartilage. *Arthritis Rheum.* **1998**, *41*, 986–996. [CrossRef]
49. Söder, S.; Hambach, L.; Lissner, R.; Kirchner, T.; Aigner, T. Ultrastructural localization of type VI collagen in normal adult and osteoarthritic human articular cartilage. *Osteoarthr. Cartil.* **2002**, *10*, 464–470. [CrossRef]
50. Wilusz, R.E.; DeFrate, L.E.; Guilak, F. A biomechanical role for perlecan in the pericellular matrix of articular cartilage. *Matrix Biol.* **2012**, *31*, 320–327. [CrossRef]
51. Wang, B.; Lai, X.; Price, C.; Thompson, W.R.; Li, W.; Quabili, T.R.; Tseng, W.-J.; Liu, X.S.; Zhang, H.; Pan, J.; et al. Perlecan-Containing Pericellular Matrix Regulates Solute Transport and Mechanosensing Within the Osteocyte Lacunar-Canalicular System. *J. Bone Miner. Res.* **2014**, *29*, 878–891. [CrossRef] [PubMed]
52. Pei, S.; Parthasarathy, S.; Parajuli, A.; Martinez, J.; Lv, M.; Jiang, S.; Wu, D.; Wei, S.; Lu, X.L.; Farach-Carson, M.C.; et al. Perlecan/Hspg2 deficiency impairs bone's calcium signaling and associated transcriptome in response to mechanical loading. *Bone* **2020**, *131*, 115078. [CrossRef] [PubMed]
53. Jin, H.; Cui, M. New Advances of Heparanase and Heparanase-2 in Human Diseases. *Arch. Med. Res.* **2018**, *49*, 423–429. [CrossRef] [PubMed]
54. Dao, D.T.; Anez-Bustillos, L.; Adam, R.M.; Puder, M.; Bielenberg, D.R. Heparin-Binding Epidermal Growth Factor–Like Growth Factor as a Critical Mediator of Tissue Repair and Regeneration. *Am. J. Pathol.* **2018**, *188*, 2446–2456. [CrossRef] [PubMed]
55. Kato, M.; Wang, H.; Kainulainen, V.; Fitzgerald, M.L.; Ledbetter, S.; Ornitz, D.M.; Bernfield, M. Physiological degradation converts the soluble syndecan-1 ectodomain from an inhibitor to a potent activator of FGF-2. *Nat. Med.* **1998**, *4*, 691–697. [CrossRef]
56. Sanderson, R.D.; Yang, Y.; Suva, L.J.; Kelly, T. Heparan sulfate proteoglycans and heparanase—Partners in osteolytic tumor growth and metastasis. *Matrix Biol.* **2004**, *23*, 341–352. [CrossRef] [PubMed]
57. Whitelock, J.M.; Murdoch, A.D.; Iozzo, R.V.; Underwood, P.A. The degradation of human endothelial cell-derived perlecan and release of bound basic fibroblast growth factor by stromelysin, collagenase, plasmin, and heparanases. *J. Biol. Chem.* **1996**, *271*, 10079–10086. [CrossRef]

58. Reiland, J.; Kempf, D.; Roy, M.; Denkins, Y.; Marchetti, D. FGF2 binding, signaling, and angiogenesis are modulated by heparanase in metastatic melanoma cells. *Neoplasia* **2006**, *8*, 596–606. [CrossRef]
59. Huegel, J.; Enomoto-Iwamoto, M.; Sgariglia, F.; Koyama, E.; Pacifici, M. Heparanase Stimulates Chondrogenesis and Is Up-Regulated in Human Ectopic Cartilage. *Am. J. Pathol.* **2015**, *185*, 1676–1685. [CrossRef]
60. Cheng, C.C.; Lee, Y.H.; Lin, S.P.; Huangfu, W.C.; Liu, I.H. Cell-autonomous heparanase modulates self-renewal and migration in bone marrow-derived mesenchymal stem cells. *J. Biomed. Sci.* **2014**, *21*, 1–12. [CrossRef]
61. Itoh, N.; Ornitz, D.M. Fibroblast growth factors: From molecular evolution to roles in development, metabolism and disease. *J. Biochem.* **2011**, *149*, 121–130. [CrossRef]
62. Ornitz, D.M.; Marie, P.J. FGF signaling pathways in endochondral and intramembranous bone development and human genetic disease. *Genes Dev.* **2002**, *16*, 1446–1465. [CrossRef] [PubMed]
63. Shu, C.; Smith, S.S.; Little, C.B.; Melrose, J. Comparative immunolocalisation of perlecan, heparan sulphate, fibroblast growth factor-18, and fibroblast growth factor receptor-3 and their prospective roles in chondrogenic and osteogenic development of the human foetal spine. *Eur. Spine J.* **2013**, *22*, 1774–1784. [CrossRef] [PubMed]
64. Shu, C.C.; Jackson, M.T.; Smith, M.M.; Smith, S.M.; Penm, S.; Lord, M.S.; Whitelock, J.M.; Little, C.B.; Melrose, J. Ablation of Perlecan Domain 1 Heparan Sulfate Reduces Progressive Cartilage Degradation, Synovitis, and Osteophyte Size in a Preclinical Model of Posttraumatic Osteoarthritis. *Arthritis Rheumatol.* **2016**, *68*, 868–879. [CrossRef] [PubMed]
65. Loeser, R.F. Integrins and chondrocyte-matrix interactions in articular cartilage. *Matrix Biol.* **2014**, *39*, 11–16. [CrossRef]
66. Woltersdorf, C.; Bonk, M.; Leitinger, B.; Huhtala, M.; Käpylä, J.; Heino, J.; Gil Girol, C.; Niland, S.; Eble, J.A.; Bruckner, P.; et al. The binding capacity of α1β1-, α2β1- and α10β1-integrins depends on non-collagenous surface macromolecules rather than the collagens in cartilage fibrils. *Matrix Biol.* **2017**, *63*, 91–105. [CrossRef]
67. Loeser, R.F.; Sadiev, S.; Tan, L.; Goldring, M.B. Integrin expression by primary and immortalized human chondrocytes: Evidence of a differential role for α1β1 and α2β1 integrins in mediating chondrocyte adhesion to types II and VI collagen. *Osteoarthr. Cartil.* **2000**, *8*, 96–105. [CrossRef]
68. Labat-Robert, J. Cell-matrix interactions in aging: Role of receptors and matricryptins. *Ageing Res. Rev.* **2004**, *3*, 233–247. [CrossRef]
69. Brown, A.J.; Alicknavitch, M.; D'Souza, S.S.; Daikoku, T.; Kirn-Safran, C.B.; Marchetti, D.; Carson, D.D.; Farach-Carson, M.C. Heparanase expression and activity influences chondrogenic and osteogenic processes during endochondral bone formation. *Bone* **2008**, *43*, 689–699. [CrossRef]
70. Tesche, F.; Miosge, N. Perlecan in late stages of osteoarthritis of the human knee joint. *Osteoarthr. Cartil.* **2004**, *12*, 852–862. [CrossRef]
71. Danalache, M.; Erler, A.L.; Wolfgart, J.M.; Schwitalle, M.; Hofmann, U.K. Biochemical changes of the pericellular matrix and spatial chondrocyte organization—Two highly interconnected hallmarks of osteoarthritis. *J. Orthop. Res.* **2020**, *38*, 2170–2180. [CrossRef] [PubMed]
72. Martinez, J.R.; Grindel, B.J.; Hubka, K.M.; Dodge, G.R.; Farach-Carson, M.C. Perlecan/HSPG2: Signaling role of domain IV in chondrocyte clustering with implications for Schwartz-Jampel Syndrome. *J. Cell. Biochem.* **2019**, *120*, 2138–2150. [CrossRef] [PubMed]
73. Goldblatt, J.P.; Richmond, J.C. Anatomy and Biomechanics of the Knee. *Oper. Tech. Sports Med.* **2003**, *11*, 172–186. [CrossRef]
74. Mason, J.J.; Leszko, F.; Johnson, T.; Komistek, R.D. Patellofemoral joint forces. *J. Biomech.* **2008**, *41*, 2337–2348. [CrossRef] [PubMed]
75. Chanalaris, A.; Clarke, H.; Guimond, S.E.; Vincent, T.L.; Turnbull, J.E.; Troeberg, L. Heparan Sulfate Proteoglycan Synthesis Is Dysregulated in Human Osteoarthritic Cartilage. *Am. J. Pathol.* **2019**, *189*, 632–647. [CrossRef]
76. Whitelock, J.; Melrose, J. Heparan sulfate proteoglycans in healthy and diseased systems. *Wiley Interdiscip. Rev. Syst. Biol. Med.* **2011**, *3*, 739–751. [CrossRef]

Communication

Heparin Administered to *Anopheles* in Membrane Feeding Assays Blocks *Plasmodium* Development in the Mosquito

Elena Lantero [1,2], Jessica Fernandes [3], Carlos Raúl Aláez-Versón [4], Joana Gomes [3], Henrique Silveira [3], Fatima Nogueira [3] and Xavier Fernàndez-Busquets [1,2,5,*]

1. Institute for Bioengineering of Catalonia (IBEC), The Barcelona Institute of Science and Technology, Baldiri Reixac 10–12, ES-08028 Barcelona, Spain; elantero@ibecbarcelona.eu
2. Barcelona Institute for Global Health (ISGlobal, Hospital Clínic-Universitat de Barcelona), Rosselló 149-153, ES-08036 Barcelona, Spain
3. Global Health and Tropical Medicine, Instituto de Higiene e Medicina Tropical, Universidade Nova de Lisboa (IHMT-NOVA), Rua da Junqueira 100, 1349-008 Lisbon, Portugal; jessica.j.fernandes94@gmail.com (J.F.); joana.matias.gomes@gmail.com (J.G.); hsilveira@ihmt.unl.pt (H.S.); fnogueira@ihmt.unl.pt (F.N.)
4. BIOIBERICA S.A.U., Polígon Industrial "Mas Puigvert", Ctra. N-II, km. 680.6, ES-08389 Palafolls, Spain; cralaez@bioiberica.com
5. Nanoscience and Nanotechnology Institute (IN2UB, Universitat de Barcelona), Martí i Franquès 1, ES-08028 Barcelona, Spain
* Correspondence: xfernandez_busquets@ub.edu

Received: 2 June 2020; Accepted: 29 July 2020; Published: 1 August 2020

Abstract: Innovative antimalarial strategies are urgently needed given the alarming evolution of resistance to every single drug developed against *Plasmodium* parasites. The sulfated glycosaminoglycan heparin has been delivered in membrane feeding assays together with *Plasmodium berghei*-infected blood to *Anopheles stephensi* mosquitoes. The transition between ookinete and oocyst pathogen stages in the mosquito has been studied in vivo through oocyst counting in dissected insect midguts, whereas ookinete interactions with heparin have been followed ex vivo by flow cytometry. Heparin interferes with the parasite's ookinete–oocyst transition by binding ookinetes, but it does not affect fertilization. Hypersulfated heparin is a more efficient blocker of ookinete development than native heparin, significantly reducing the number of oocysts per midgut when offered to mosquitoes at 5 µg/mL in membrane feeding assays. Direct delivery of heparin to mosquitoes might represent a new antimalarial strategy of rapid implementation, since it would not require clinical trials for its immediate deployment.

Keywords: malaria; heparin; mosquito; *Plasmodium*; *Anopheles*; ookinete; transmission blocking; antimalarial drugs

1. Introduction

The emergence and spread of *Plasmodium falciparum* resistance to most of the existing antimalarial drugs is a key factor that contributes to the global reappearance of malaria [1]. This threat of treatment failure is prompting research oriented to targeting the transmission stages of the pathogen between humans and mosquitoes [2], represented by smaller populations less likely to contain resistant individuals that would benefit from the removal of susceptible parasites [3]. Transmission-blocking vaccines (TBV) aim at stimulating in the human the production of antibodies to actively target and block the parasite development once it is in the mosquito [4]. Among the candidate antigens for TBV strategies are proteins on the surface of the ookinete [5,6], the motile *Plasmodium* stage that forms in the blood bolus and has to traverse the midgut endothelium to progress to the next stage of the parasite, the oocyst.

Among other transmission-blocking strategies that can be envisaged is the interference with parasite–midgut interaction through the inhibitory action of the sulfated glycosaminoglycan (sGAG) heparin and related molecules, which have already shown antimalarial activity against several *Plasmodium* stages in humans. During initial malaria infection in the liver, heparin and heparan sulfate are hepatocyte receptors for sporozoite attachment [7]. In blood stages, heparin antimalarial activity, against which no resistances have been reported so far, unfolds by inhibition of merozoite invasion of the erythrocyte [8]. Chondroitin sulfate proteoglycans in the mosquito midgut and a synthetic polysulfonated polymer that mimics the structure of sGAGs present in the midgut epithelium have been described to bind *Plasmodium* ookinetes during host epithelial cell invasion [9,10], whereas ookinetes and ookinete-secreted proteins possess significant binding to heparin [11,12]. Here, we have explored the potential of heparin against ookinete development.

2. Materials and Methods

All reagents were purchased from Sigma-Aldrich Corporation (St. Louis, MO, USA) unless otherwise specified. Heparin from pig intestinal mucosa was provided by BIOIBERICA (Palafolls, Spain). Oversulfation to generate hypersulfated heparin was done as previously described [13], obtaining a preparation with three sulfate groups/disaccharide (as compared to 1.9–2.0 in native heparin). Briefly, 100 mg of sodium heparin salt were subjected to cation-exchange chromatography to obtain tributylamine salt, lyophilized, and dissolved in 0.8 mL of N,N-dimethylformamide, which contained an excess of pyridine-sulfur trioxide. After 1 h at 40 °C, 1.6 mL of water were added, and the product was precipitated with three volumes of cold ethanol saturated with anhydrous sodium acetate and collected by centrifugation. The product was dissolved in water, extensively dialyzed, and recovered by freeze-drying.

2.1. Animals

Female CD1 mice (*Mus musculus*) from the Instituto de Higiene e Medicina Tropical animal house were used to obtain blood for membrane feeding assays (MFAs) and mosquito infections, with the corresponding license (009511 from 21 April 2019) approved by the Portuguese National Authority Health (DGAV). For ex vivo production of ookinetes, female BALB/c mice (Janvier Labs, Le Genest-Saint-Isle, France) were used, following the protocols reviewed and approved by the Ethical Committee on Clinical Research from the *Hospital Clínic de Barcelona* (Reg. 10100/P2, approved on January 2018). In all cases, for the experimental procedure, mice were anesthetized using 100 mg/kg ketamine (Ketolar) mixed with 10 mg/kg xylazine (Rompun) intraperitoneally (i.p.) administered, and regularly monitored. *Anopheles stephensi* mosquitoes were maintained under standard insectary conditions (26 ± 1 °C, 75% humidity and a 12/12 h light/dark cycle). Adult mosquitoes were fed on 10% glucose solution ad libitum until the day before feeding trials.

2.2. Sugar Feed

Heparin directly dissolved at 5 mg/mL or 50 mg/mL in 10% glucose in H_2O, or 10% glucose in H_2O for the control group, were administered to mosquitoes twice (once each 24 h) in a cotton pad on the top of a net-capped paper cup containing 40–50 *A. stephensi* females. Mosquitoes were allowed to feed for 48 h, and then infected by direct feeding on a CD1 mouse parasitized with *Plasmodium berghei* ANKA-GFP (259cl1; MRA-865 [14]) for 10 min. Non-fed mosquitoes were removed, and fed mosquitoes were placed in the insectary at 21 °C and 75% humidity to allow parasite development. After eight days, mosquitoes were dissected, and the number of GFP-expressing oocysts per midgut was counted manually using an Axioskop fluorescence microscope (Zeiss, Oberkochen, Germany).

The total number of mosquitoes analyzed was 61 in the control group, 63 treated with 5 mg/mL heparin, and 46 treated with 50 mg/mL heparin, distributed in three independent experiments.

2.3. Membrane Blood Feeding Assay

Blood obtained from an intracardiac puncture of a CD1 mouse infected with *P. berghei* ANKA-GFP (259cl1; MRA-865) was treated with 1/10 volume of 3.2% w/v sodium citrate to prevent coagulation. 6.25 µL of a solution prepared by dissolving heparin at 40 mg/mL or 0.4 mg/mL in phosphate buffered saline (PBS) was added to 494 µL of blood:citrate (to obtain final heparin concentrations of 500 and 5 µg/mL, respectively), which was then placed in feeders prepared with two-sided stretched Parafilm® connected to two plastic tubes for water inlet and outlet. The same volumes of PBS and blood:citrate were used for heparin-free controls. Temperature within the multiple cylindrical water-jacked glass was kept at 37 °C by a constant water flow supply. Each feeder was placed on top of a net-covered paper cup containing 40–50 *A. stephensi* females. Mosquitoes were allowed to feed for one hour. Non-fed mosquitoes were removed, and the rest were treated as above. The final number of mosquitoes analyzed for the non-modified heparin assay was 127 in the control group, 106 treated with 5 µg/mL heparin, and 149 treated with 500 µg/mL heparin, distributed in three independent experiments. The final number of mosquitoes analyzed for the hypersulfated heparin assay was 62 in the control group, 91 treated with 5 µg/mL hypersulfated heparin, and 102 treated with 500 µg/mL hypersulfated heparin, distributed in two independent experiments.

2.4. Detection of Heparin-Cy5 in Mosquitoes

Thirty female *A. stephensi* mosquitoes per cage were allowed to feed on 400 µg/mL heparin-Cy5 (Nanocs Inc., New York, NY, USA) on either sugar for 6 h or MFA for 1 h. Some mosquitoes were taken from the cages at 6, 24, 48, and 72 h post-feeding to check Cy5 fluorescence ($\lambda_{ex}/\lambda_{em}$: 650/670 nm) with an Eclipse 80i microscope (Nikon, Tokyo, Japan). Mosquitoes with fluorescent signal were dissected, and their organs were individually observed.

2.5. Ex Vivo Production of Ookinetes and Flow Cytometry Analysis

Eight days before ookinete production, 200 µL of *P. berghei* CTRP-GFP (kindly provided by Dr. Inga Siden-Kiamos [15]) in cryopreservation solution (RBC pellet:Roswell Park Memorial Institute medium (RPMI, Gibco, Dublin, Ireland):30% glycerol in water, 1:1:2) was administered i.p. to a BALB/c mouse. Four days later, this mouse was the donor to infect i.p. with 5×10^7 parasitized red blood cells in 200 µL of PBS a second mouse that one hour before the infection had been pretreated i.p. with phenylhydrazine (120 µL of a 10 mg/mL solution in PBS). For ookinete production, up to 1 mL of blood carrying gametocytes was collected by intracardiac puncture and diluted in 30 mL of ookinete medium: 10.4 g/L of RPMI supplemented with 2% w/v NaHCO$_3$, 0.05% w/v hypoxanthine, 0.02% w/v xanthurenic acid, 50 U/mL penicillin and 50 µg/mL streptomycin, 20% heat-inactivated fetal bovine serum (FBS, Invitrogen, Carlsbad, CA, USA), 25 mM HEPES, pH 7.4. The culture was incubated for 24 h at 21 °C with orbital shaking at 50 rpm (modified from [16]).

To check heparin influence on fertilization, to 487.5 µL of culture in a well of 24-well plates was added 12.5 µL of PBS containing heparin at 20 or 0.2 mg/mL, to provide final heparin concentrations of 500 µg/mL and 5 µg/mL. Samples were taken at two different time points (just after extraction and after 1 h incubation), including a control consisting of PBS only, in three independent experimental replicates. Twenty-four hours later, samples were diluted 1:100 in PBS and analyzed in a LSRFortessa™ flow cytometer (BD Biosciences, San Jose, CA, USA) set up with the five lasers, 20 parameters standard configuration. The GFP positive ookinete population was selected and counted using 488 nm laser excitation and a 525/40 nm emission collection filter. BD FACSDiva software version 6.1.3 (BD Biosciences) was used in data collection, and Flowing Software 2.5.1 (Turku Centre for Biotechnology, Turku, Finland) was used for analysis.

For targeting assays, mature ookinetes were washed twice with PBS and incubated with heparin-Cy5 at 400 µg/mL in ookinete medium without FBS for 1 h. The sample was finally diluted 1:100 in the same medium containing 0.2 µg/mL Hoechst 33342 and events recorded with an Amnis® ImageStream®X Mk II cytometer (Luminex Corporation, Austin, TX, USA) using 375 nm, 488 nm, and 642 nm excitation lasers for Hoechst 33342, GFP and Cy5 signals respectively. Data were analyzed with IDEAS® 6.3 software (Luminex Corporation).

2.6. Statistical Analysis

Oocysts/midgut counts from three independent experiments were plotted and analyzed in GraphPad Prism 6 using an unpaired Mann–Whitney test to determine significant differences. t-Tests with Welch's correction were applied for determining significance in ex vivo ookinete maturation and targeting assays, and degrees of freedom were automatically defined by the software, according to n. In both cases, tests were two-sided.

3. Results

3.1. Characterization of Heparin-Cy5 Binding to Ookinetes

Flow cytometry analysis showed binding of heparin-Cy5 to ookinetes obtained ex vivo from mouse blood infected with the *P. berghei* CTRP-GFP transgenic line (Figure 1a,b), which expresses GFP when reaching ookinete stage. Fluorescence images indicated the binding of heparin-Cy5 to discrete areas on the ookinete (Figure 1c), which suggests clustering of heparin receptors.

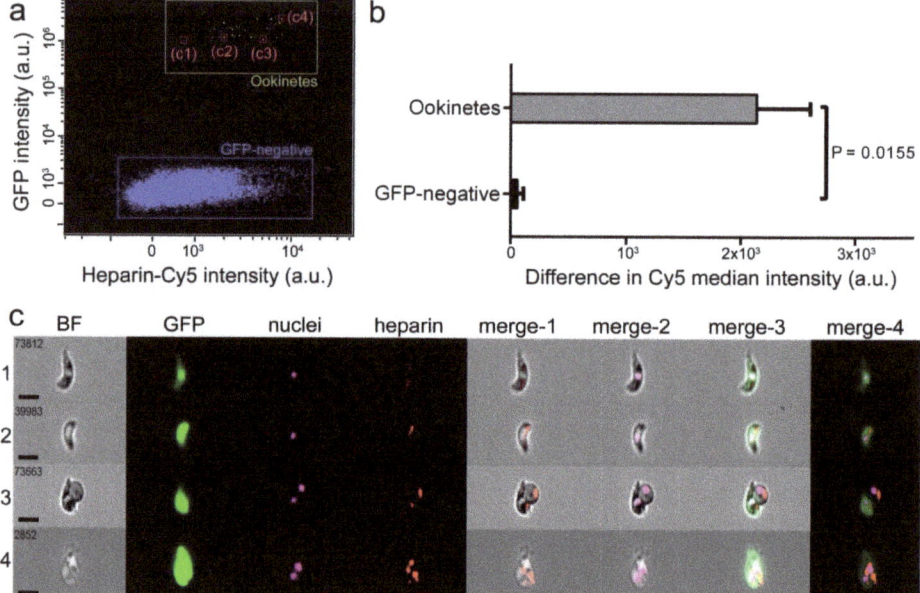

Figure 1. Heparin-Cy5 binding to ookinetes. (**a**) Flow cytometry plot showing heparin-Cy5 signal in green fluorescent protein (GFP)-expressing *P. berghei* ookinetes. GFP-negative events correspond mostly to red blood cells (see Figure S1 for gating strategy). c1 to c4 refer to the individual events reported in panel **c**. (**b**) Difference in Cy5 median intensity between GFP-expressing ookinetes and GFP-negative cells. (**c**) Fluorescence images of the flow cytometry events indicated in panel **a**. The merges of the bright field (BF) image with the fluorescence of heparin, nuclei, GFP, and all three of them, are indicated as merge-1 to merge-4, respectively. Size bars represent 7 µm.

3.2. Effect on Oocyst Development of Heparin Administered to Mosquitoes by Sugar Meal

Heparin-Cy5 fed in the sugar meal to female *A. stephensi* mosquitoes was detected in the midgut of the insects for up to 72 h after administration (Figure 2a–d and Figure S2). Heparin effect on ookinete to oocyst transition was then assessed in live mosquitoes, by offering them heparin by sugar feed during 48 h before infecting them by direct bite to a *P. berghei* ANKA-GFP-parasitized mouse (Figure 2e). Unfed mosquitoes were removed, and eight days later, mosquitoes were dissected, and GFP-expressing oocysts were counted. The prevalence of infection (PI, percentage of mosquitoes with ≥1 oocyst) and the infection intensity (II, number of oocysts per midgut) were not significantly affected when compared to untreated controls up to heparin concentrations in the sugar feed of 50 mg/mL (Figure 2f). It is likely that most of the sugar feed might be pushed out by the blood meal, in which case heparin would not interact with ookinetes. This result led us to explore new strategies to ensure the presence of heparin at the moment of ookinete development in the mosquito midgut by including heparin in a *Plasmodium*-infected blood meal.

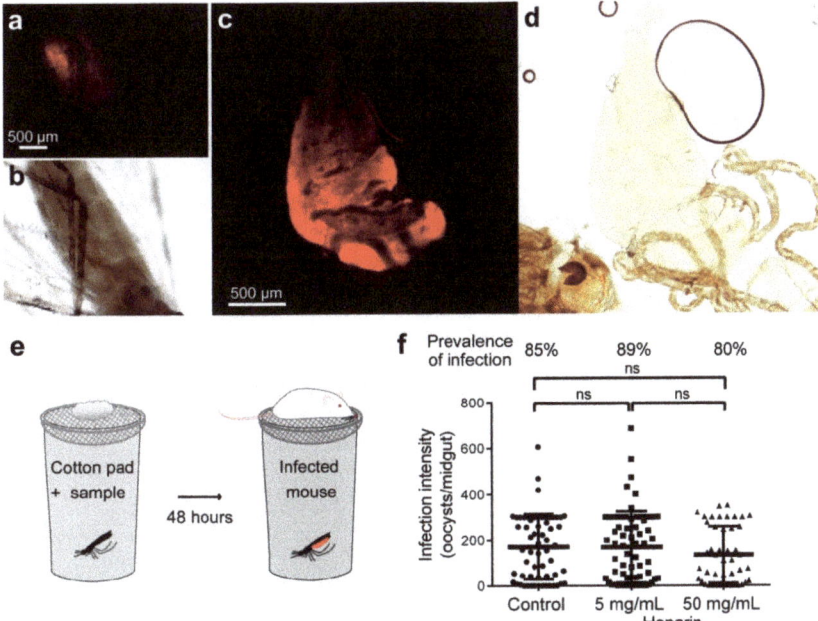

Figure 2. Effect on ookinete development of heparin fed to mosquitoes by sugar meal. (**a,c**) Fluorescence detection in (**a**) intact abdomen and (**c**) dissected midgut of heparin-Cy5 fed to *A. stephensi* female mosquitoes in a sugar meal. (**b,d**) Bright field images of the microscope fields in panels **a** and **c**, respectively. (**e**) Depiction of the method for sugar feed used in mosquito assays. (**f**) Effect on parasite development of heparin delivered by sugar swaps. ns: not significant.

3.3. Effect on Oocyst Development of Heparin Administered to Mosquitoes by Blood Membrane Feeding

Heparin-Cy5 fed to female *A. stephensi* mosquitoes by whole blood MFAs was detected in the midgut of the insect for at least 24 h after administration (Figure 3a–d and Figure S3). Often, Cy5 fluorescence was only faintly observed in the dissected midgut (Figure 3a,b), but it intensified after having pushed out the blood bolus (Figure 3c,d). This might result from light being absorbed or screened by the compacted blood bolus. Heparin activity on ookinete to oocyst transition was then assessed with this method of administration. Heparin was added to the blood of mice infected with *P. berghei* ANKA-GFP, which was then offered to female *A. stephensi* by MFA (Figure 3e). Unfed mosquitoes were removed, and eight days later, mosquitoes were dissected, and oocysts were counted. A significant decrease of PI and II was observed in mosquitoes fed with heparin-containing infected blood samples (Figure 3f,h–k). PI was 38% and 23% for respective MFA heparin concentrations of 5 µg/mL and 500 µg/mL, compared to 52% for the heparin-free control, whereas the mean II for the same samples was, respectively, 24.22 ± 65.10, 0.95 ± 4.12, and 36.38 ± 89.57 oocysts per midgut.

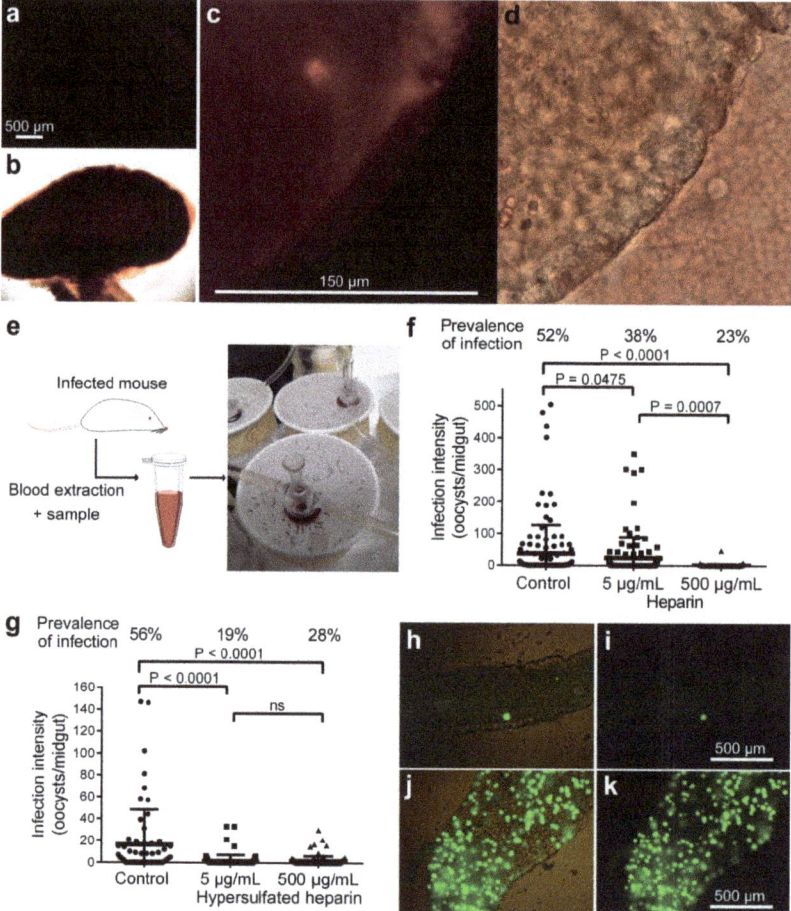

Figure 3. Effect on ookinete development of heparin fed to mosquitoes by membrane feeding assay (MFA). (a–d) Fluorescence detection of heparin-Cy5 fed to *A. stephensi* female mosquitoes by blood feed. (a,b) Whole dissected midgut and (c,d) magnification of the same midgut with the blood bolus pushed away. (b,d) Bright field images of the microscope fields in panels a and c, respectively. (e) Depiction of the MFA method used in mosquito assays. (f) Effect on parasite development of non-modified heparin delivered by MFA. (g) Effect on parasite development of hypersulfated heparin delivered by MFA. ns: not significant. (h–k) Fluorescence images of representative mosquito midguts from the MFA 500 µg/mL non-modified heparin group (h,i) and from the MFA control group (j,k); the fluorescence signal is shown alone (i,k) and merged with bright field images of the midgut contours h,j).

When a modified heparin with higher proportion of sulfated residues in the polysaccharide chain (hypersulfated heparin) was offered to mosquitoes by MFA, a significant decrease in PI and II was observed with as little as 5 µg/mL of heparin (Figure 3g). PI was 19% and 28% for respective hypersulfated heparin concentrations of 5 µg/mL and 500 µg/mL compared to 56% for the control, whereas the mean II for the same samples was, respectively, 1.53 ± 5.56, 1.74 ± 4.61, and 16.29 ± 31.94 oocysts per midgut. Although the extensive dialysis performed at the end of the heparin sulfation process should have removed any residual byproduct, future research has to rule out potential interferences of trace chemicals on the mechanism of oocyst formation. No impact on mosquito viability was observed for any of the heparins studied here (data not shown).

The observation that PI in blood feeding assays was significantly lower than in sugar meal experiments might be explained by the presence of sodium citrate, which is a calcium chelator used to prevent blood coagulation. Since the induction of exflagellation in *Plasmodium* requires calcium [17], sodium citrate could have a synergistic effect with heparin potentiating its inhibitory effect on oocyst formation. Heparin is also a potent calcium chelator which binds ca. one Ca^{2+} ion per average disaccharide [18]. At the high 50 mg/mL heparin concentration of sugar meal assays, the calcium binding capacity of heparin was comparable to that of the sodium citrate amount used in MFAs. Although in these experiments no effect of heparin was seen on ookinete development, the suspected immiscibility of sugar feed and blood meal calls for caution before drawing any conclusions regarding the suspected inhibitory effect of calcium chelators on *Plasmodium* development in the mosquito.

3.4. Effect of Heparin on Fertilization

When blood from *P. berghei*-infected mice was put into culture to obtain ookinetes and heparin was added to the culture either at the moment of blood extraction (t_0) or 1 h later (t_1), no significant effect was observed in the number of ookinetes produced when compared with untreated control cultures (Figure 4). However, the fold-increase in ookinete numbers when heparin was added at t_0 (ookinetes relative to normalized control: 3.32 ± 2.26 and 2.26 ± 1.18 for 5 µg/mL and 500 µg/mL heparin, respectively), though non-significant due to the high dispersion of the results, could indicate that heparin enhances fertilization. This effect might operate through heparin interactions with coagulation factors, which would facilitate gamete motility. Heparin use in MFAs was previously recommended over other anticoagulants such as EDTA, as better infection rates were obtained [19]. Consistently, no effect on ookinete numbers was observed when heparin was added at t_1, when fertilization has already occurred [20]. These results indicated that the inhibitory activity of heparin on *Plasmodium* mosquito stages is not exerted during fertilization or zygote maturation. Although sodium citrate was not used in ex vivo assays, the presence of 500 µg/mL heparin bound a significant amount of calcium, and yet, exflagellation was not affected. However, the potential role of calcium sequestration on this part of the parasite's development deserves further exploration.

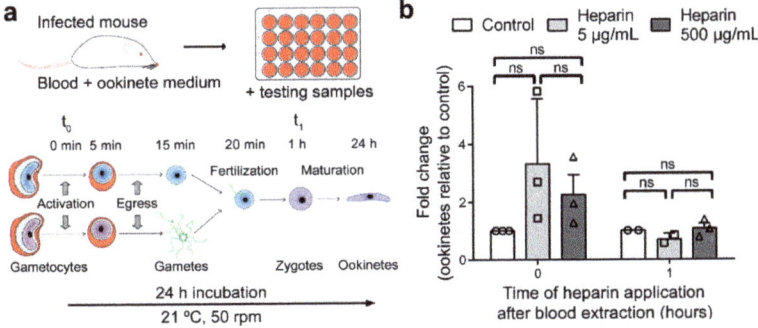

Figure 4. Effect of heparin in the ex vivo development of ookinetes. (**a**) Depiction of the method used for the ex vivo growth of ookinetes. The parasite development scheme has been adapted from Kuehn and Pradel [20]. (**b**) Effect of heparin on ex vivo ookinete maturation analyzed by flow cytometry (see Figure S4 and Table S1). ns: not significant.

4. Discussion

The results presented above validate a potential new antimalarial strategy where heparin binding to ookinetes will prevent the interaction of this *Plasmodium* stage with the mosquito midgut and consequently its development into an oocyst. It has been suggested that chondroitin sulfate is a ligand for the circumsporozoite- and thrombospondin-related anonymous protein-related protein (CTRP) [9], a key molecule for ookinete mobility and parasite development [21]. Characterizing the sGAG ookinete binding domain and sulfation pattern will be important regarding the development of future antimalarials acting on this stage of the parasite's cycle. To start unraveling the relevance of the sulfation pattern in blocking ookinete progression, hypersulfated heparin has been tested here, and has shown interesting potential since the lowest concentration used resulted in a significantly larger inhibition of ookinete development than the same concentration of native heparin. Chemical modifications of heparin or its binding to nanocarriers are strategies that could contribute to increase activity and midgut residence time in view of the potential development of sGAG-based antimalarials as disruptors of the life cycle of *Plasmodium* in the mosquito.

So far, transmission-blocking approaches have focused on the concept of treating humans with vaccines or drugs that will target mosquito stages [22]. A largely unexplored avenue, however, is targeting *Plasmodium* in the insect vector directly [23]. The implementation of antimalarial medicines designed to be delivered directly to mosquitoes might reduce treatment and development costs because the clinical trials otherwise required for therapies to be administered to people could be significantly simplified. Strategies that control malaria using direct action against *Anopheles* are not new but mostly focus on eliminating the vector, either by killing it with pesticides [24] or through the release of sterile males [25]. The administration of drugs to mosquitoes during their blood feed is being used to deliver ivermectin, an endectocide that, at concentrations found in human blood after treatment, is toxic to all *Anopheles* species examined [24]. When *P. falciparum*–infected female *Anopheles gambiae* mosquitoes were exposed to surfaces treated with the antimalarial drug atovaquone, the development of the parasite was completely arrested [26]. Although this strategy is unlikely to work for large hydrophilic molecules like heparin, several other approaches are available for direct drug delivery to mosquitoes, some of which have been used to deliver to dipterans lipid-based [27] and chitosan nanoparticles [28].

The failure of heparin in inhibiting parasite development when delivered in a sugar meal prior to infection of the mosquito indicates that heparin must be present in the midgut simultaneously with ookinetes. This poses a significant obstacle regarding the practical implementation of future antimalarial strategies based on the observations reported here. Heparin is normally present in human plasma in values ranging from 1 mg/L to 2.4 mg/L [29], whereas heparin in the mosquito midgut is active at concentrations >100 times higher. The anticoagulant activity of heparin prevents its administration to people in the amounts required to block ookinete development, although sGAG mimetics [9] or modified heparins having low anticoagulant capacity [30] offer promising perspectives. The use of limited heparin amounts in infected patients for transmission-blocking might actually be beneficial given the described pro-coagulant effects of *Plasmodium*-infected red blood cells [31]. However, to be present in the circulation at the moment of a mosquito bite, the blood residence time of these molecules should be extremely long. An alternative approach could be provided by offering heparin to mosquitoes in an artificial diet simulating vertebrate blood. The available technology is capable of manufacturing artificial blood for mosquito feeding from hemoglobin obtained from citrated rabbit blood [32], outdated bovine blood [33], and other blood-free artificial liquid diets [34–36]. These substitutes mimic in the mosquito the physiological effects of a fresh vertebrate blood meal, supporting ovarian and egg maturation and normal development of offspring into functional adults. Such artificial feedings can substitute direct feeding on mammals and often have prolonged shelf life and do not require refrigeration. For their delivery to mosquitoes, a number of artificial blood feeders are currently under study [37]. This approach would require the presence of attractants in the artificial diet to lure mosquitoes that have already taken a human blood meal and thus potentially carrying ookinetes in their midguts.

The economic landscape of malaria calls for new strategies that take into account the costs of bringing a medicine into the market, which due to expensive clinical trials often prevent promising new drugs from a fast entry into the production pipeline. As a possible approach to solving this problem, the administration of heparin to mosquitoes offers two advantages: first, blocking the life cycle of *Plasmodium* in the mosquito vector through direct drug delivery to the insect, can dramatically shorten product development due to the avoidance of large-scale human tests. Second, applying the three Rs of drug development (rescue, repurpose, reposition) to previously discarded compounds is an interesting strategy to return value to potential treatments in decline or on hold. Heparin is a natural polysaccharide that can be abundantly obtained in large amounts from the intestinal mammalian mucosa and which has a widespread medical use. The results presented here can inspire researchers and entrepreneurs, especially those in malaria endemic regions, to pursue the development of an efficient and economically affordable antimalarial strategy. A chain of heparin production could easily start with the usually discarded mucosae of pigs, goats, sheep, or cows that are consumed for food. In addition, potential strategies to deliver heparin to mosquitoes might involve the use of small containers filled with mosquito attractants, which can boost the economy of many developing regions through either the fabrication of such dispensers made of plastic, glass or aluminum, or the recycling of bottles and cans.

Supplementary Materials: The following are available online at http://www.mdpi.com/2218-273X/10/8/1136/s1, Figure S1. Gating strategy for Figure 1a; Figure S2. Photomicrograph gallery of different time points after heparin-Cy5 administration in sugar feed; Figure S3. Photomicrograph gallery of different time points after heparin-Cy5 administration in MFA; Figure S4. Gating strategy for the data analysis presented in Figure 4b; Table S1. GFP-positive events (n) and corresponding percentages presented in Figure 4b.

Author Contributions: Conceptualization, X.F.-B. and E.L.; methodology, E.L., F.N., H.S., and X.F.-B.; formal analysis, E.L.; investigation, E.L., J.F., J.G., and F.N.; resources, C.R.A.-V., F.N., H.S., and X.F.-B.; writing—original draft preparation, X.F.-B. and E.L.; writing—review and editing, X.F.-B., E.L., H.S., and F.N.; visualization E.L.; supervision, F.N., H.S., and X.F.-B.; project administration, X.F.-B.; funding acquisition, F.N., H.S. and X.F.-B. All authors have read and agreed to the published version of the manuscript.

Funding: X.F.-B. received funding support from (i) Spanish Ministry of Science, Innovation and Universities (http://www.ciencia.gob.es/), grant numbers PCIN-2017-100 and RTI2018-094579-B-I00 (which included FEDER funds), (ii) BIOIBERICA, and (iii) ERA-NET Cofund EURONANOMED (http://euronanomed.net/), grant number 2017-178 (NANOpheles). H.S. and F.N. received funding support from Global Health and Tropical Medicine, Instituto de Higiene e Medicina Tropical, Universidade Nova de Lisboa, GHTM–UID/04413/2020 (https://ghtm.ihmt.unl.pt/).

Acknowledgments: ISGlobal and IBEC are members of the CERCA Programme, *Generalitat de Catalunya*. We acknowledge support from the Spanish Ministry of Science, Innovation and Universities through the "*Centro de Excelencia Severo Ochoa 2019-2023*" Program (CEX2018-000806-S). This research is part of ISGlobal's Program on the Molecular Mechanisms of Malaria, which is partially supported by the *Fundación Ramón Areces*. We are indebted to the Cytometry and Cell Sorting Facility of the *Institut d'Investigacions Biomèdiques August Pi i Sunyer* (IDIBAPS) for technical help. E.L. would like to thank Rafael Oliveira and Helio Rocha for their support and feedback.

Conflicts of Interest: X.F.-B. has received funding from BIOIBERICA and C.R.A.-V. is employed by BIOIBERICA. The funders had no role in the design of the study; in the collection, analyses, or interpretation of data; in the writing of the manuscript, or in the decision to publish the results.

References

1. World Health Organization. *World Malaria Report 2019*; World Health Organization: Geneva, Switzerland, 2019.
2. Sinden, R.; Carter, R.; Drakeley, C.; Leroy, D. The biology of sexual development of *Plasmodium*: The design and implementation of transmission-blocking strategies. *Malar. J.* **2012**, *11*, 70. [CrossRef] [PubMed]
3. Delves, M.; Plouffe, D.; Scheurer, C.; Meister, S.; Wittlin, S.; Winzeler, E.; Sinden, R.E.; Leroy, D. The activities of current antimalarial drugs on the life cycle stages of *Plasmodium*: A comparative study with human and rodent parasites. *PLoS Med.* **2012**, *9*, e1001169. [CrossRef] [PubMed]
4. Kapulu, M.C.; Da, D.F.; Miura, K.; Li, Y.; Blagborough, A.M.; Churcher, T.S.; Nikolaeva, D.; Williams, A.R.; Goodman, A.L.; Sangare, I.; et al. Comparative assessment of transmission-blocking vaccine candidates against *Plasmodium falciparum*. *Sci. Rep.* **2015**, *5*, 11193. [CrossRef]
5. Duffy, P.E.; Kaslow, D.C. A novel malaria protein, Pfs28, and Pfs25 are genetically linked and synergistic as falciparum malaria transmission-blocking vaccines. *Infect. Immun.* **1997**, *65*, 1109–1113. [CrossRef] [PubMed]

6. Kim, T.S.; Kim, H.H.; Moon, S.U.; Lee, S.S.; Shin, E.H.; Oh, C.M.; Kang, Y.J.; Kim, D.K.; Sohn, Y.; Kim, H.; et al. The role of Pvs28 in sporozoite development in *Anopheles sinensis* and its longevity in BALB/c mice. *Exp. Parasitol.* **2011**, *127*, 346–350. [CrossRef]
7. Ancsin, J.B.; Kisilevsky, R. A binding site for highly sulfated heparan sulfate is identified in the N terminus of the circumsporozoite protein: Significance for malarial sporozoite attachment to hepatocytes. *J. Biol. Chem.* **2004**, *279*, 21824–21832. [CrossRef] [PubMed]
8. Boyle, M.J.; Richards, J.S.; Gilson, P.R.; Chai, W.; Beeson, J.G. Interactions with heparin-like molecules during erythrocyte invasion by *Plasmodium falciparum* merozoites. *Blood* **2010**, *115*, 4559–4568. [CrossRef]
9. Mathias, D.K.; Pastrana-Mena, R.; Ranucci, E.; Tao, D.; Ferruti, P.; Ortega, C.; Staples, G.O.; Zaia, J.; Takashima, E.; Tsuboi, T.; et al. A small molecule glycosaminoglycan mimetic blocks *Plasmodium* invasion of the mosquito midgut. *PLoS Pathog.* **2013**, *9*, e1003757. [CrossRef]
10. Dinglasan, R.R.; Alaganan, A.; Ghosh, A.K.; Saito, A.; van Kuppevelt, T.H.; Jacobs-Lorena, M. *Plasmodium falciparum* ookinetes require mosquito midgut chondroitin sulfate proteoglycans for cell invasion. *Proc. Natl. Acad. Sci. USA* **2007**, *104*, 15882–15887. [CrossRef]
11. Marques, J.; Valle-Delgado, J.J.; Urbán, P.; Baró, E.; Prohens, R.; Mayor, A.; Cisteró, P.; Delves, M.; Sinden, R.E.; Grandfils, C.; et al. Adaptation of targeted nanocarriers to changing requirements in antimalarial drug delivery. *Nanomed. NBM* **2017**, *13*, 515–525. [CrossRef]
12. Li, F.; Templeton, T.J.; Popov, V.; Comer, J.E.; Tsuboi, T.; Torii, M.; Vinetz, J.M. *Plasmodium* ookinete-secreted proteins secreted through a common micronemal pathway are targets of blocking malaria transmission. *J. Biol. Chem.* **2004**, *279*, 26635–26644. [CrossRef] [PubMed]
13. Maruyama, T.; Toida, T.; Imanari, T.; Yu, G.; Linhardt, R.J. Conformational changes and anticoagulant activity of chondroitin sulfate following its O-sulfonation. *Carbohydr. Res.* **1998**, *306*, 35–43. [CrossRef]
14. Franke-Fayard, B.; Trueman, H.; Ramesar, J.; Mendoza, J.; van der Keur, M.; van der Linden, R.; Sinden, R.E.; Waters, A.P.; Janse, C.J. A *Plasmodium berghei* reference line that constitutively expresses GFP at a high level throughout the complete life cycle. *Mol. Biochem. Parasitol.* **2004**, *137*, 23–33. [CrossRef] [PubMed]
15. Vlachou, D.; Zimmermann, T.; Cantera, R.; Janse, C.J.; Waters, A.P.; Kafatos, F.C. Real-time, in vivo analysis of malaria ookinete locomotion and mosquito midgut invasion. *Cell. Microbiol.* **2004**, *6*, 671–685. [CrossRef] [PubMed]
16. Blagborough, A.M.; Delves, M.J.; Ramakrishnan, C.; Lal, K.; Butcher, G.; Sinden, R.E. Assessing transmission blockade in *Plasmodium* spp. In *Malaria: Methods and Protocols*; Ménard, R., Ed.; Humana Press: Totowa, NJ, USA, 2013; pp. 577–600.
17. Bansal, A.; Molina-Cruz, A.; Brzostowski, J.; Mu, J.; Miller, L.H. *Plasmodium falciparum* calcium-dependent protein kinase 2 is critical for male gametocyte exflagellation but not essential for asexual proliferation. *mBio* **2017**, *8*. [CrossRef]
18. Grant, D.; Moffat, C.F.; Long, W.F.; Williamson, F.B. Ca^{2+}-heparin interaction investigated polarimetrically. *Biochem. Soc. Trans.* **1991**, *19*, 391S. [CrossRef]
19. Solarte, Y.; Manzano, M.R.; Rocha, L.; Castillo, Z.; James, M.A.; Herrera, S.; Arévalo-Herrera, M. Effects of anticoagulants on *Plasmodium vivax* oocyst development in *Anopheles albimanus* mosquitoes. *Am. J. Trop. Med. Hyg.* **2007**, *77*, 242–245. [CrossRef]
20. Kuehn, A.; Pradel, G. The coming-out of malaria gametocytes. *J. Biomed. Biotechnol.* **2010**, *2010*, 976827. [CrossRef]
21. Dessens, J.T.; Beetsma, A.L.; Dimopoulos, G.; Wengelnik, K.; Crisanti, A.; Kafatos, F.C.; Sinden, R.E. CTRP is essential for mosquito infection by malaria ookinetes. *EMBO J.* **1999**, *18*, 6221–6227. [CrossRef]
22. Wells, T.N.C.; van Huijsduijnen, R.H.; Van Voorhis, W.C. Malaria medicines: A glass half full? *Nat. Rev. Drug Discov.* **2015**, *14*, 424–442. [CrossRef]
23. Paaijmans, K.; Fernàndez-Busquets, X. Antimalarial drug delivery to the mosquito: An option worth exploring? *Future Microbiol.* **2014**, *9*, 579–582. [CrossRef] [PubMed]
24. Chaccour, C.; Kobylinski, K.; Bassat, Q.; Bousema, T.; Drakeley, C.; Alonso, P.; Foy, B. Ivermectin to reduce malaria transmission: A research agenda for a promising new tool for elimination. *Malar. J.* **2013**, *12*, 153. [CrossRef] [PubMed]
25. Andreasen, M.H.; Curtis, C.F. Optimal life stage for radiation sterilization of *Anopheles* males and their fitness for release. *Med. Vet. Entomol.* **2005**, *19*, 238–244. [CrossRef] [PubMed]

26. Paton, D.G.; Childs, L.M.; Itoe, M.A.; Holmdahl, I.E.; Buckee, C.O.; Catteruccia, F. Exposing *Anopheles* mosquitoes to antimalarials blocks *Plasmodium* parasite transmission. *Nature* **2019**, *567*, 239–243. [CrossRef] [PubMed]
27. Whyard, S.; Singh, A.D.; Wong, S. Ingested double-stranded RNAs can act as species-specific insecticides. *Insect Biochem. Mol. Biol.* **2009**, *39*, 824–832. [CrossRef]
28. Zhang, X.; Zhang, J.; Zhu, K.Y. Chitosan/double-stranded RNA nanoparticle-mediated RNA interference to silence chitin synthase genes through larval feeding in the African malaria mosquito (*Anopheles gambiae*). *Insect Mol. Biol.* **2010**, *19*, 683–693. [CrossRef]
29. Engelberg, H. Plasma heparin levels in normal man. *Circulation* **1961**, *23*, 578–581. [CrossRef]
30. Boyle, M.J.; Skidmore, M.; Dickerman, B.; Cooper, L.; Devlin, A.; Yates, E.; Horrocks, P.; Freeman, C.; Chai, W.; Beeson, J.G. Identification of heparin modifications and polysaccharide inhibitors of *Plasmodium falciparum* merozoite invasion that have potential for novel drug development. *Antimicrob. Agents Chemother.* **2017**, *61*, e00709–e00717. [CrossRef]
31. Francischetti, I.M.; Seydel, K.B.; Monteiro, R.Q. Blood coagulation, inflammation, and malaria. *Microcirculation* **2008**, *15*, 81–107. [CrossRef]
32. Dias, L.D.S.; Bauzer, L.G.S.D.; Lima, J.B.P. Artificial blood feeding for Culicidae colony maintenance in laboratories: Does the blood source condition matter? *Rev. Inst. Med. Trop. Sao Paulo* **2018**, *60*, e45. [CrossRef]
33. Haldar, R.; Gupta, D.; Chitranshi, S.; Singh, M.K.; Sachan, S. Artificial blood: A futuristic dimension of modern day transfusion sciences. *Cardiovasc. Hematol. Agents Med. Chem.* **2019**, *17*, 11–16. [CrossRef] [PubMed]
34. Marques, J.; Cardoso, J.C.R.; Felix, R.C.; Power, D.M.; Silveira, H. A blood-free diet to rear anopheline mosquitoes. *J. Vis. Exp.* **2020**, *155*, e60144. [CrossRef] [PubMed]
35. Marques, J.; Cardoso, J.C.R.; Felix, R.C.; Santana, R.A.G.; Guerra, M.D.G.B.; Power, D.; Silveira, H. Fresh-blood-free diet for rearing malaria mosquito vectors. *Sci. Rep.* **2018**, *8*, 17807. [CrossRef] [PubMed]
36. Gonzales, K.K.; Hansen, I.A. Artificial diets for mosquitoes. *Int. J. Environ. Res. Public Health* **2016**, *13*, 1267. [CrossRef]
37. Romano, D.; Stefanini, C.; Canale, A.; Benelli, G. Artificial blood feeders for mosquito and ticks-Where from, where to? *Acta Trop.* **2018**, *183*, 43–56. [CrossRef]

© 2020 by the authors. Licensee MDPI, Basel, Switzerland. This article is an open access article distributed under the terms and conditions of the Creative Commons Attribution (CC BY) license (http://creativecommons.org/licenses/by/4.0/).

Article

Multiplex Soluble Biomarker Analysis from Pleural Effusion

Joman Javadi [1,*], Katalin Dobra [1,2] and Anders Hjerpe [2]

1. Karolinska Institutet, Department of Laboratory Medicine, Division of Pathology, Huddinge University Hospital, SE-14186 Stockholm, Sweden; katalin.dobra@ki.se
2. Karolinska University Hospital, Karolinska University laboratory, Huddinge University Hospital, SE-14186 Stockholm, Sweden; anders.hjerpe@sll.se
* Correspondence: joman.javadi@ki.se; Tel.: +46-762-615-122

Received: 1 June 2020; Accepted: 21 July 2020; Published: 28 July 2020

Abstract: Malignant pleural mesothelioma (MPM) is a highly aggressive and therapy resistant pleural malignancy that is caused by asbestos exposure. MPM is associated with poor prognosis and a short patient survival. The survival time is strongly influenced by the subtype of the tumor. Dyspnea and accumulation of pleural effusion in the pleural cavity are common symptoms of MPM. The diagnostic distinction from other malignancies and reactive conditions is done using histopathology or cytopathology, always supported by immunohistochemistry, and sometimes also by analyses of soluble biomarkers in effusion supernatant. We evaluated the soluble angiogenesis related molecules as possible prognostic and diagnostic biomarkers for MPM by Luminex multiplex assay. Pleural effusion from 42 patients with malignant pleural mesothelioma (MPM), 36 patients with adenocarcinoma (AD) and 40 benign (BE) effusions were analyzed for 10 different analytes that, in previous studies, were associated with angiogenesis, consisting of Angiopoietin-1, HGF, MMP-7, Osteopontin, TIMP-1, Galectin, Mesothelin, NRG1-b1, Syndecan-1 (SDC-1) and VEGF by a Human Premixed Multi-Analyte Luminex kit. We found that shed SDC-1 and MMP-7 levels were significantly lower, whereas Mesothelin and Galectin-1 levels were significantly higher in malignant mesothelioma effusions, compared to adenocarcinoma. Galectin-1, HGF, Mesothelin, MMP-7, Osteopontin, shed SDC-1, NRG1-β1, VEGF and TIMP-1 were significantly higher in malignant pleural mesothelioma effusions compared to benign samples. Moreover, there is a negative correlation between Mesothelin and shed SDC-1 and positive correlation between VEGF, Angiopoietin-1 and shed SDC-1 level in the pleural effusion from malignant cases. Shed SDC-1 and VEGF have a prognostic value in malignant mesothelioma patients. Collectively, our data suggest that MMP-7, shed SDC-1, Mesothelin and Galectin-1 can be diagnostic and VEGF and SDC-1 prognostic markers in MPM patients. Additionally, Galectin-1, HGF, Mesothelin, MMP-7, Osteopontin, shed SDC-1 and TIMP-1 can be diagnostic for malignant cases.

Keywords: malignant pleural mesothelioma; pleural effusion; Luminex; diagnostic biomarkers

1. Introduction

Malignant pleural mesothelioma (MPM) is an aggressive tumor of mesothelial origin that occurs after a long latency period following asbestos exposure often comprising several decades [1–4]. This tumor occurs mostly in the pleura (>75%) and less often in the peritoneum (10–20%) or pericardium (1%) [5,6]. The median survival time of malignant pleural mesothelioma is less than 12 months [7]. This survival is strongly influenced by subtype of the tumor and time of diagnosis [3,8]. The diagnosis of MPM is challenging, which often leads to a late diagnosis and together with tumor aggressiveness results in a poor prognosis. Diagnosis is based on testing of multiple immunohistochemical (IHC) markers of biopsies or cytological samples, and soluble biomarker analysis of pleural effusion to distinguish MPM from other lung malignancies [9]. The optimal treatment is surgery in combination with chemotherapy and radiotherapy, or both [10]. Most patients present with a pleural effusion as first symptom and cytological diagnosis increases the chance for earlier diagnosis and prolongs patient survival [11]. Among soluble biomarkers hyaluronan and mesothelin are the best characterized biomarkers for diagnosis of MPM.

Hyaluronan (HA), a negatively charged glycosaminoglycan of the extracellular matrix (ECM), consists of the repeating disaccharide units (glucuronic acid and N-acetylglucosamine), is an established biomarker used in diagnosis of MPM [12–14]. Hyaluronan is involved in different functions such as cell growth, angiogenesis, differentiation, cell migration, wound healing and the regulation of plasma protein distribution [13,15]. Hyaluronan is a well-known biomarker for MPM [14,16–18].

Mesothelin is a 40-KD cell membrane glycoprotein, which presents on normal mesothelial cells and is highly expressed in different human cancers of the lung, endometrium, stomach and pancreas [19]. Mesothelin is synthesized as a 72-KD protein anchored in the cell membrane. It sheds a 32-KD fragment called N-ERC/mesothelin, leaving a 40-KD fragment named C-ERC/mesothelin. Both fragments can be detected in the effusion supernatant [20]. The role of mesothelin in the development of normal cells is not known, whereas, in cancer cells, mesothelin promotes tumor progression by interacting with other membrane proteins, such as CA125 [21–23]. In recent years, Mesothelin has received considerable attention for immune-based therapies and as a promising biomarker, because of its low expression level in normal cells [24,25].

Syndecan-1 (SDC-1) is a transmembrane heparan sulfate proteoglycan containing glycosaminoglycan (GAG) side chains linked to N-terminal extracellular domain of the core protein. Syndecan-1 acts as a co-receptor for many regulatory proteins, such as growth factors, cytokines and integrins through its GAG, mainly heparan sulfate, chains. This interaction leads to regulation of different cellular processes including proliferation, migration, differentiation and angiogenesis by alteration of different signaling pathways [26,27]. In different cancer types syndecan-1 plays dynamic roles, whereby it can either suppress or promote tumor progression [28–31].

The aim of this study is to evaluate shed SDC-1 in pleural effusions together with angiogenesis related proteins and identify optimal biomarker batteries that allow earlier diagnosis and improve the possibilities to distinguish MPM from metastatic adenocarcinoma and reactive conditions.

2. Materials and Methods

2.1. Study Design

Pleural effusions from 118 patients—42 patients with malignant pleural mesothelioma (MPM), 36 patients with lung adenocarcinoma (AD) and 40 benign (BE) effusions—were collected at the Department of Pathology and Cytology, Karolinska University Hospital Sweden. Malignant mesothelioma effusions comprised the epithelioid and mixed phenotypes, as sarcomatoid mesotheliomas do not exfoliate to the serous effusions. Adenocarcinoma specimens consisted of metastases from lung, breast- gastro-intestinal, ovarian adenocarcinoma and primary tumors of unknown primary, covering the most frequent metastatic tumors to the serosal cavities. Benign effusions were related to inflammation, reactive mesothelial proliferations and heart failure. All cases were diagnosed by cytopathology verified by extensive immunocytochemistry. Mesothelioma cases were also analyzed by established biomarker analyses including Mesothelin and Hyaluronan. All samples were collected before any treatment was given. Samples were centrifuged at 1500 rpm for 5 min directly, and the cell free supernatants were kept at −80 °C without additives. The study was approved by the ethical review board of Stockholm, Sweden (2009/1138–341/3).

2.2. Luminex Assay with Human Premixed Multi-Analyte Kit

Two human premixed multi-analyte kits from the R&D system were used to assess the levels of 10 different biomarkers. First kit (cat: LXSAHM-09) for analyzing Angiopoietin-1, HGF, MMP-7, Osteopontin, TIMP-1, Galectin, Mesothelin, NRG1-b1 and Syndecan-1 simultaneously and second kit (cat: LXSAHM-01) for analyzing VEGF. In total, we analyzed 118 pleural effusions, of which 42 were from MM patients, 36 from AD patients and 40 benign effusions. Effusions were diluted 5-fold, using the dilution buffer included in the kit. All standards and samples were assayed in duplicate.

Analyte specific antibodies are pre-coated onto magnetic microparticles embedded with fluorophores at set ratios for each unique microparticle region. Two spectrally distinct light emitting diodes (LEDs) illuminate the microparticles. One LED excites the dyes inside each microparticle, to identify the region and the second LED excites the Streptavidin-PE to measure the amount of analyte bound to the microparticles. A minimum of 50 beads per region were counted. The median fluorescence intensities were determined on a Luminex® 100/200™ analyzer.

2.3. Enzyme-Linked Immunosorbent Assay (ELISA)

Enzyme-linked immunosorbent assay (ELISA) was performed to measure shed SDC-1, VEGF and Mesothelin levels, following the manufacturer's instructions. ELISA kit for Human shed SDC-1 was from Gen-Probe Diaclone, France (cat. number 950.640.192), Human VEGF was from R&D Systems, UK (cat. number DVE00) and Human N-ERC/Mesothelin Assay kit from IBL. The effusions were diluted 1:5 in kit dilution buffers which also were as blanks. Samples were analyzed in duplicate and the optical densities were determined at 450 nm.

2.4. Statistical Analyses

2.4.1. Analyses of Biomarkers Expression in Pleural Effusions

Three different groupings of patients were used in the analysis of ten different biomarkers, to compare biomarkers expression level between patients with malignant and benign conditions, and also to compare patients with different malignancies comprising malignant mesothelioma (MM) and various adenocarcinomas (AD). Differences in biomarkers expression level between different patient groups were evaluated by performing an unpaired t-test. Correlation analysis between shed SDC-1 and other biomarkers in paired pleural effusions were calculated by the Spearman r test.

2.4.2. Kaplan-Meier Survival

Analyses were performed for patients with malignant mesothelioma, to investigate the correlation of different biomarkers level with the survival of mesothelioma patients. To determine a cut-off value for each biomarker based on the most significant and highest hazard ratio we used the Cutoff Finder online web application. The log-rank test was used to test the significant differences between survival times of two different groups of patients.

2.4.3. Receiver Operating Characteristic (ROC) Analysis

Roc analysis was used to assess the diagnostic utility of each biomarker. ROC curves, areas under the curve (AUC), sensitivity, specificity, likelihood ratio and their 95% confidence intervals (CI) were generated by GraphPad Prism software. All statistical significances were set at a p value equal or lower than 0.05.

2.4.4. Logistic Regression (LR) Analysis

Logistic regression analysis was applied to develop a model for identifying and combine diagnostic biomarkers in pleural effusion for earlier detection of malignant pleural mesothelioma. Angiogenesis-related biomarkers were analyzed using the JMP program. In each model, malignant mesothelioma (MM) and metastatic adenocarcinoma (AD) were set as dependent variable (coded 0 or 1), and the expression values of the angiogenesis-related biomarkers as independent variables. Biomarkers included in a final predictive model were significant at $p \leq 0.05$ and were determined by a stepwise selection procedure.

3. Results

3.1. Correlation between ELISA and Luminex Immunoassays

In order to evaluate the correlation between ELISA and Luminex immunoassays, pleural effusion levels of shed SDC-1, Mesothelin and VEGF were measured independently by both immunoassays. High correlation between immunoassays was observed (Figure 1). We found statistically significant linear correlation between shed SDC-1 ($p = 0.0001$, $r = 0.8$), Mesothelin ($p = 0.0001$, $r = 0.6$) and VEGF ($p = 0.0001$, $r = 0.9$) that measured by ELISA comparing with the Luminex assay, suggesting that Luminex assay can be used in the clinic.

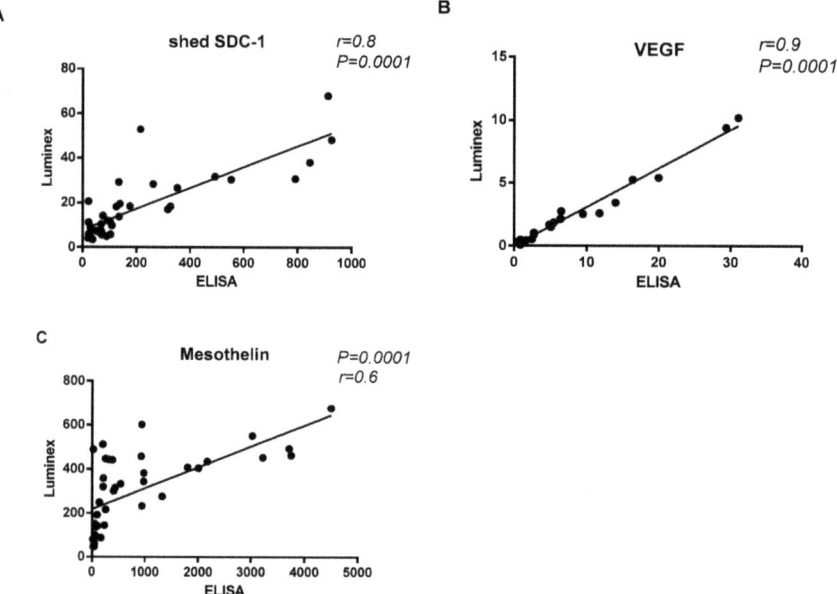

Figure 1. Correlation between enzyme-linked immunosorbent assay (ELISA) and Luminex. (**A**) Correlation between shed Syndecan-1 (SDC-1) level in pleural effusion measured by ELISA and Luminex. (**B**) Correlation between VEGF level in pleural effusion measured by ELISA and Luminex. (**C**) Correlation between Mesothelin level in pleural effusion measured by ELISA and Luminex. Spearman correlation analysis was used to assess the correlation between these two immunoassays.

3.2. Expression Levels of Biomarkers in Malignant Pleural Mesothelioma and Benign Pleural Effusion

In order to describe the diagnostic power of pleural effusion derived Angiopoietin-1, HGF, MMP-7, Osteopontin, TIMP-1, Galectin, Mesothelin, NRG1-b1, Syndecan-1 and VEGF, we compared the expression level of these 10 angiogenesis related biomarkers in pleural effusion from malignant pleural mesothelioma patients ($n = 42$) and benign samples ($n = 40$). The expression levels of Galectin-1, Mesothelin, Osteopontin, shed SDC-1, VEGF, MMP-7, HGF and TIMP-1 were significantly higher, and NRG1-β1 was significantly lower in malignant mesothelioma effusions, compared with the benign effusions (Figure 2). Expression level of these 10 angiogenesis related biomarkers in pleural effusion from metastatic lung adenocarcinoma patients ($n = 38$) showed in Supplementary Figure S1. Differences between malignant mesothelioma cases and benign cases were statistically significant for these nine biomarkers ($p < 0.05$).

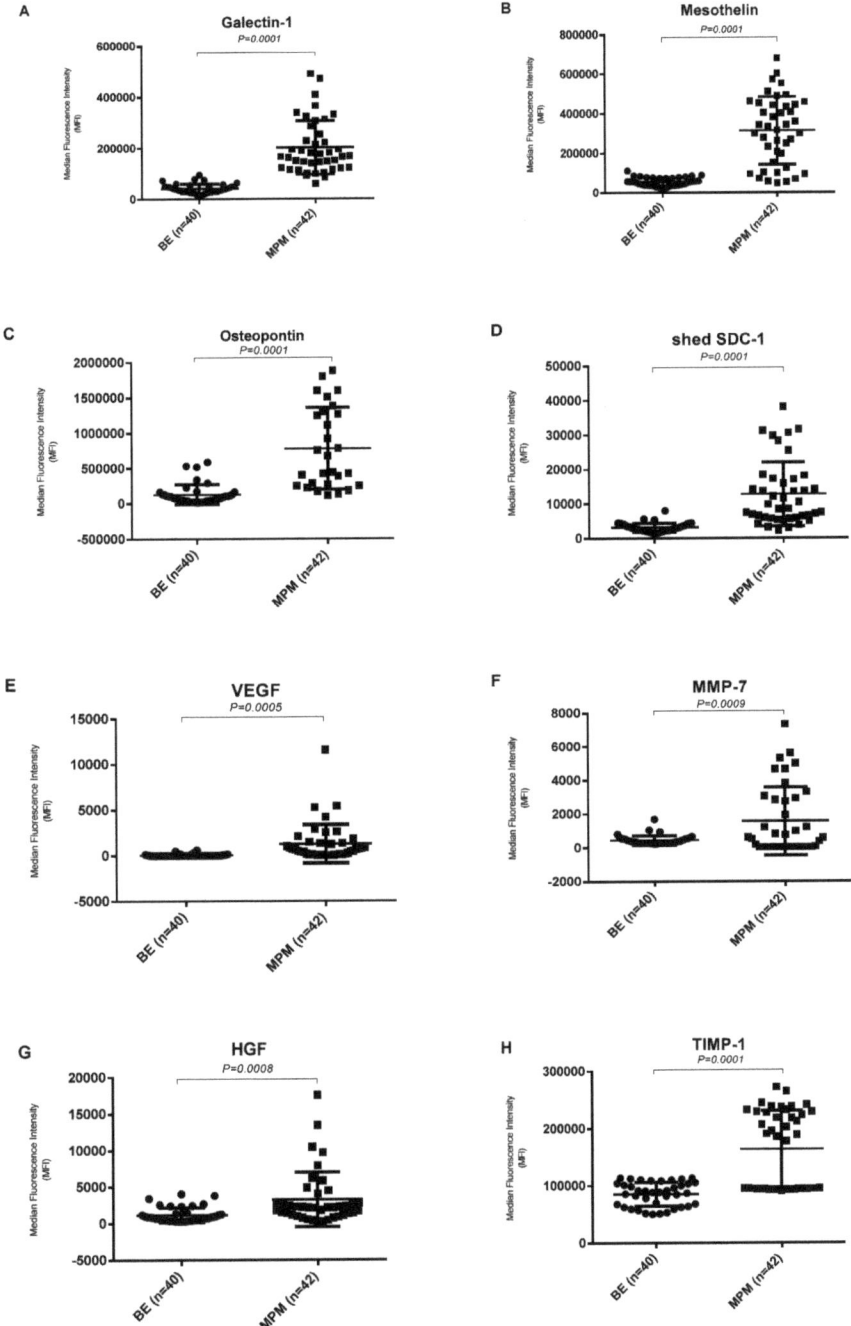

Figure 2. Cont.

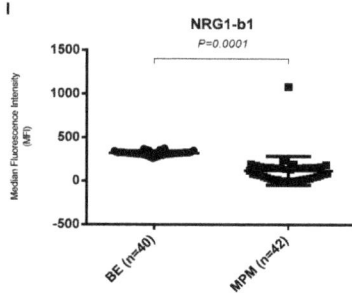

Figure 2. Biomarker levels in pleural effusion of malignant pleural mesothelioma (MPM) patients compared to benign effusions (BE). (**A–H**) Levels of Galectin-1, Mesothelin, Osteopontin, shed SDC-1, VEGF, MMP-7, HGF and TIMP-1 are significantly higher in pleural effusion from malignant pleural mesothelioma (MPM) patients ($n = 42$) comparing with benign (BE) patients ($n = 40$). (**I**) Level of NRG1-β1 is significantly lower Significance was assessed by two-tailed t-test at $p \leq 0.05$.

3.3. Diagnostic Biomarkers for Distinguishing Malignant Mesothelioma from Metastatic Adenocarcinoma

In order to distinguishing malignant mesothelioma from other adenocarcinoma, the expression levels of these 10 angiogenic related biomarkers in pleural effusion were compared between malignant pleural mesothelioma patients and metastatic adenocarcinomas. Of these, expression level of Mesothelin and Galectin-1 were significantly higher in malignant mesothelioma effusions, compared to adenocarcinoma effusions whereas the expression level of shed Syndecan-1 and MMP-7 were significantly lower in malignant mesothelioma effusions compared to the adenocarcinoma effusions (Figure 3). Differences between malignant mesothelioma and adenocarcinoma were statistically significant for these four biomarkers ($p < 0.05$).

Stepwise logistic regression based on biomarker levels showed that MMP-7, Mesothelin and Osteopontin are three variables with a higher predictive value for distinguishing malignant mesothelioma from metastatic adenocarcinoma (Table 1).

Table 1. Parameter estimates of logistic regression model. MMP-7 has higher predictive value for distinguishing malignant pleural mesothelioma from adenocarcinoma, based on estimate and p-value.

Biomarkers	Estimate	Std Error	Prob > ChiSq	p Value
MMP-7	0.00018	0.00012	0.1586	0.00000
Mesothelin	-9.2635×10^{-6}	2.7419×10^{-6}	0.0007	0.00002
Osteopontin	-9.183×10^{-7}	4.3047×10^{-7}	0.0329	0.01

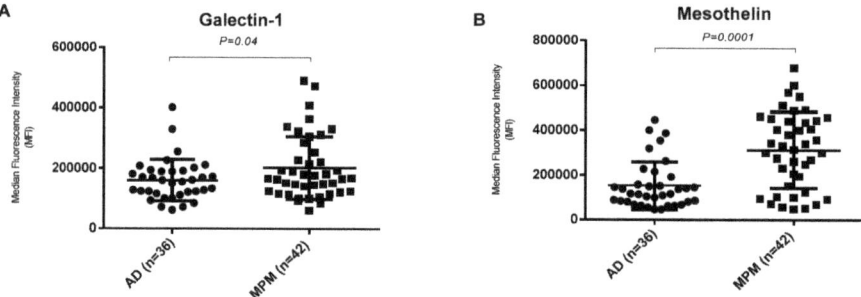

Figure 3. *Cont.*

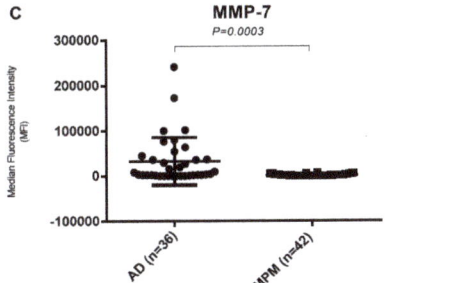
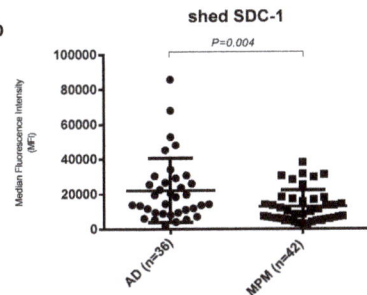

Figure 3. Different biomarkers for distinguishing malignant mesothelioma from metastatic adenocarcinoma patients. (**A,B**) Levels of Galectin-1 and Mesothelin are significantly higher in pleural effusion from malignant pleural mesothelioma (MPM) patients ($n = 42$) compared with adenocarcinoma (AD) patients ($n = 36$). (**C,D**) Levels of MMP-7 and shed SDC-1 are significantly lower in malignant pleural mesothelioma patients ($n = 42$) compared with adenocarcinoma patients ($n = 36$). Significance was assessed by two-tailed t-test at $p \leq 0.05$.

3.4. Shed SDC-1 and VEGF Levels Correlate to Patient Survival in Malignant Mesothelioma

In order to determine the prognostic value of pleural effusion derived Angiopoietin-1, HGF, MMP-7, Osteopontin, TIMP-1, Galectin, Mesothelin, NRG1-b1, SDC-1 and VEGF, we divided patients in two groups depending on the established cut-off value for all analytes. Strikingly, malignant mesothelioma patients with high levels of SDC-1 (>19.28 ng/mL) and VEGF (>0.7 ng/mL) have significantly worse prognosis in comparison to the patients with low levels of SDC-1 (<19.28 ng/mL) and VEGF (<0.7 ng/mL). Median survival time of malignant mesothelioma patients with high VEGF level was significantly shorter (2.6 months) compared to low VEGF level (18 months) ($p = 0.0003$) (Figure 4A). Median survival time of malignant mesothelioma patients with high level of SDC-1 was significantly shorter (2.4 months) compared to patients with low shed SDC-1 level (9.6 months) ($p = 0.03$) (Figure 4B).

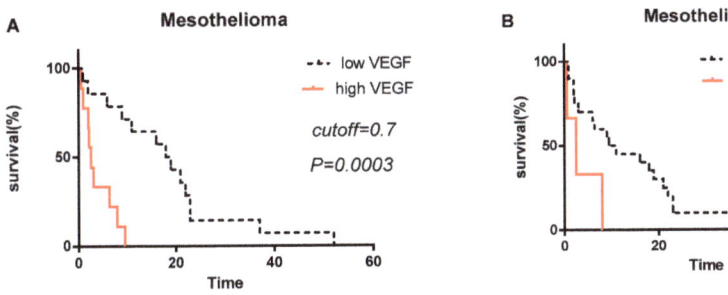

Figure 4. VEGF and shed SDC-1 have prognostic value in malignant mesothelioma patients. Patients were separated into "high" and "low" VEGF and shed SDC-1 level by the online web application Cutoff Finder (VEGF cutoff = 0.7 ng/mL and shed SDC-1 = 19.28 ng/mL). Malignant mesothelioma patients with high level of VEGF (**A**) and shed SDC-1 (**B**) have shorter survival time compared to malignant mesothelioma patients with low VEGF and shed SDC-1 levels. P-values are ≤0.05.

3.5. Correlation between Shed SDC-1 and Other Biomarkers

The correlation between shed SDC-1 and other biomarkers was explored in malignant mesothelioma patients and in all malignant cases (malignant mesothelioma and adenocarcinoma). In all malignant cases, we showed a significant weak-positive correlation between shed SDC-1 and Angiopoietin-1 ($p = 0.004$ and r = 0.3) and a significant weak-negative correlation between shed SDC-1 and Mesothelin ($p = 0.004$ and r = −0.3) (Figure 5). We did not find any significant correlation between expression of shed SDC-1 and other seven angiogenic-related biomarkers (HGF, MMP-7, Osteopontin, TIMP-1, Galectin, VEGF and NRG1-b1).

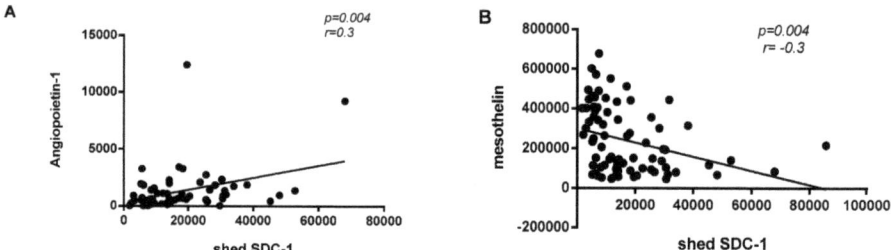

Figure 5. Correlation between shed SDC-1 and other biomarkers in all malignant cases. (**A**) Expression of shed SDC-1 was significantly and positively correlated with Angiopoietin-1 ($p = 0.004$ and $r = 0.3$), whereas, shed SDC-1 expression was negatively correlated with Mesothelin ($p = 0.004$ and $r = -0.3$) (**B**).

In malignant mesothelioma cases, shed SDC-1 expression was significantly and positively correlated with HGF ($p = 0.02$; $r = 0.3$) and NRG1-b1 ($p = 0.001$; $r = 0.4$) (Figure 6). No significant correlations were found between shed SDC-1 and other seven biomarkers (Angiopoietin-1, Mesothelin, MMP-7, Galectin-1, Osteopontin, TIMP-1 and VEGF).

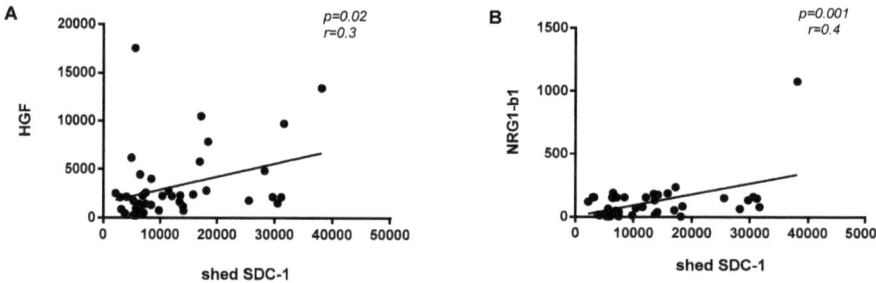

Figure 6. Correlation between shed SDC-1 and other biomarkers in malignant mesothelioma cases. (**A**) Expression of shed SDC-1 was significantly and positively correlated with HGF ($p = 0.02$ and $r = 0.3$) and NRG1-β1 ($p = 0.001$ and $r = 0.4$) (**B**) in malignant mesothelioma cases.

3.6. Diagnostic Value of Individual Biomarkers for Malignant Mesothelioma

Based on ROC curve analyses, Galectin-1, Mesothelin, Osteopontin, NRG1-β1 and shed SDC-1 performed the best diagnostic capacity to distinguish malignant mesothelioma from benign effusions. MMP-7 and angiopoietin-1 have no discriminative capacity to distinguish between malignant mesothelioma and benign effusions (Figure 7). The resulting areas under the curves (AUC) are shown in (Figure 7).

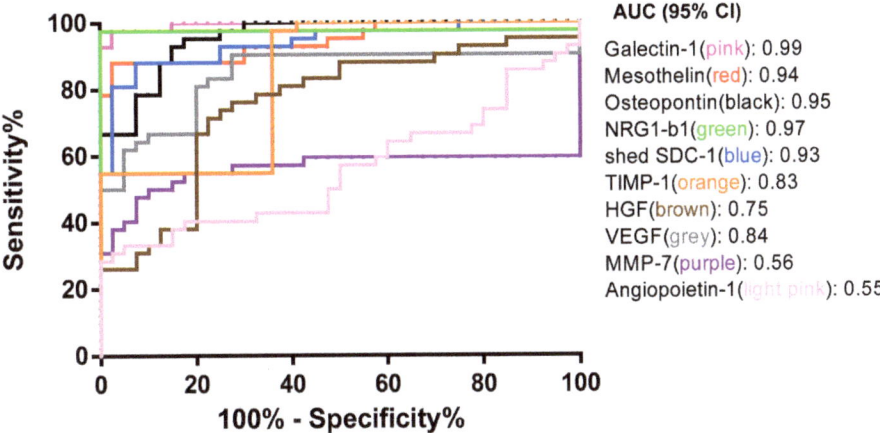

Figure 7. Diagnostic efficiency of individual biomarkers. Receiver operating characteristic (ROC) analysis showing the specificity and sensitivity for each individual angiogenesis related biomarkers. Area under the curve (AUC) values show that Galectin-1 (AUC = 0.99), NRG1-β1 (AUC = 0.97), Osteopontin (AUC = 0.95), Mesothelin (AUC = 0.94) and shed SDC-1 (AUC = 0.93) have the most diagnostic value for malignant pleural mesothelioma.

4. Discussion

Malignant mesothelioma is the locally very aggressive and incurable primary tumor of serous surfaces. Several studies have indicated that clinical outcome of malignant mesothelioma is highly affected by earlier diagnosis of the disease [11,32,33]. Diagnosis of malignant mesothelioma is challenging and requires histological or cytological analysis which can be supported by the analysis of soluble biomarkers in the effusion. Pleural effusion is the first available material for diagnosis, and a possible source for biomarker analysis. Previous studies showed that, Calretinin, Wilms tumor protein 1 (WT-1), HBME-1, D2-40 (podoplanin), Carcinoembryonic antigen (CEA), Napsin-A and Thyroid transcription factor 1 (TTF-1) are immunocytochemical indicators for distinguishing malignant mesothelioma from lung adenocarcinoma [3,34,35]. Calretinin, WT-1 and D2-40 are recommended mesothelioma markers and CEA, Napsin-A and TTF-1 are recommended adenocarcinoma markers by IMIG (International Mesothelioma Interest Group) [36].

In addition, soluble biomarkers are detectable, and can be easily quantified in body fluids before clinical symptoms appear. Several diagnostic biomarkers have been suggested for malignant mesothelioma. Mesothelin, Hyaluronan, Osteopontin and Fibulin-3 are the most promising diagnostic biomarker candidates for malignant mesothelioma [12,37–40]. In this regard hereby we combine multiplex soluble diagnostic and prognostic biomarkers in pleural effusion that can help to establish earlier diagnosis of malignant mesothelioma and to distinguish it from metastatic adenocarcinomas.

To the best of our knowledge, this is the first study that simultaneously combines several soluble diagnostic biomarkers. Here we show that Malignant mesothelioma patients have significantly higher level of Galectin-1, Mesothelin, Osteopontin, VEGF, shed SDC-1, MMP-7, HGF, NRG1-β1 and TIMP-1 compared to benign patients. Pleural effusion Galectin-1, NRG1-β1, Osteopontin, Mesothelin, shed SDC-1, VEGF and TIMP-1 levels are more reliable diagnostic biomarkers than HGF and MMP in pleural effusion. The corresponding AUCs were 0.99, 0.97, 0.95, 0.94, 0.93, 0.84 and 0.83 for Galectin-1, NRG1-β1, Osteopontin, Mesothelin, shed SDC-1, VEGF and TIMP-1, respectively, whereas the AUCs were 0.75 and 0.56 for HGF and MMP-7, respectively. Higher Mesothelin, Osteopontin and VEGF levels were described earlier as individual markers in malignant mesothelioma [41–43]. A recent study showed also higher level of TGF-β 1 in malignant pleural mesothelioma patients compared to lung adenocarcinoma patients [42]. Previously, we have shown that overexpression of membrane-bound

SDC-1 down-regulate TGF-β and TGF-βR1 [44] and the interplay between these two components merits further investigations. High level of shed SDC-1 associates to cancer, infection and inflammation, thus our findings are in line with both previous and recent studies.

In this study, all patients with metastatic adenocarcinoma had significantly higher level of Galectin-1, HGF, Mesothelin, MMP-7, Osteopontin, shed SDC-1, TIMP-1, VEGF and a significantly lower level of NRG1-β1 compared with reactive effusions (Figure S1). Among these angiogenesis related biomarkers, we demonstrated that Galectin-1, Mesothelin, shed SDC-1 and MMP-7 are biomarkers that discriminated best between malignant mesothelioma and metastatic adenocarcinomas. Our data shows that the level of Galectin-1 and mesothelin are significantly higher in MM patients in comparison with metastatic adenocarcinoma patients whereas, shed SDC-1 and MMP-7 levels are significantly decreased in malignant mesothelioma patients. We further combined these biomarkers in several models by using logistic regression method. In our study, a combination of MMP-7, Mesothelin and Osteopontin, showed the best significant model for distinguishing malignant pleural mesothelioma from metastatic adenocarcinoma patients.

Cell surface syndecan-1 expression is essential for the differentiation of various epithelial tumors and it correlates with favorable outcome, whereas decrease or loss of SDC-1 associates with poor survival [45–48]. The elevated SDC-1 level is a result of accelerated shedding or cell decay. Here we show that the shed SDC-1 in contrast to the cell-bound SDC-1 indicates poor prognosis in both malignant pleural mesothelioma and metastatic adenocarcinoma patients. Our results clearly suggest that SDC-1 and VEGF can serve as prognostic biomarkers for malignant pleural mesothelioma. Moreover, inclusion of these soluble factors in the clinical workflow may pave the way for biomarker driven patient selection for antiangiogenic therapy in the future. Though these results are very promising, further studies with larger sample sizes are required to validate these data.

5. Conclusions

Taken together, candidate biomarkers (Galectin-1, Mesothelin, Osteopontin, shed SDC-1, VEGF, MMP-7, HGF, TIMP-1 and NRG1-β1) identified in the current study could be diagnostic biomarkers for distinguishing malignant pleural mesothelioma and metastatic adenocarcinoma from benign patients. Shed SDC-1 and VEGF are prognostic biomarkers for malignant pleural mesothelioma.

Supplementary Materials: The following are available online at http://www.mdpi.com/2218-273X/10/8/1113/s1, Figure S1: Pleural effusion levels of diagnostic biomarkers in metastatic adenocarcinoma patients.

Author Contributions: Conceptualization, A.H. and K.D.; methodology, A.H., K.D. and J.J.; software, A.H. and J.J.; validation, J.J.; investigation, J.J.; writing—original draft preparation, J.J.; writing—review and editing, A.H., K.D. and J.J.; visualization, J.J.; supervision, A.H.; project administration, K.D.; funding acquisition, K.D. All authors have read and agreed to the published version of the manuscript.

Funding: This research was funded by The Swedish Cancer Society, grant number CAN 2018/653; Stockholm County Council, grant number LS 2015-1198; and The Cancer Society in Stockholm, grant number 174073.

Acknowledgments: The authors thank student Hanna Kann, Department of Laboratory Medicine, Division of Pathology, Karolinska Institute, for her skillful help with the Luminex® 100/200™ analyzer.

Conflicts of Interest: The authors declare no conflict of interest. The funders had no role in the design of the study; in the collection, analyses, or interpretation of data; in the writing of the manuscript, or in the decision to publish the results.

References

1. Celsi, F.; Crovella, S.; Moura, R.R.; Schneider, M.; Vita, F.; Finotto, L.; Zabucchi, G.; Zacchi, P.; Borelli, V. Pleural mesothelioma and lung cancer: The role of asbestos exposure and genetic variants in selected iron metabolism and inflammation genes. *J. Toxicol. Environ. Health A* **2019**, *82*, 1088–1102. [CrossRef]
2. Maat, A.; Durko, A.; Thuijs, D.; Bogers, A.; Mahtab, E. Extended pleurectomy decortication for the treatment of malignant pleural mesothelioma. *Multimed. Man. Cardiothorac. Surg.* **2019**, *2019*. [CrossRef]

3. Hjerpe, A.; Abd-Own, S.; Dobra, K. Cytopathologic Diagnosis of Epithelioid and Mixed-Type Malignant Mesothelioma: Ten Years of Clinical Experience in Relation to International Guidelines. *Arch. Pathol. Lab. Med.* **2018**, *142*, 893–901. [CrossRef]
4. Vimercati, L.; Cavone, D.; Caputi, A.; Delfino, M.C.; De Maria, L.; Ferri, G.M.; Serio, G. Malignant mesothelioma in construction workers: The Apulia regional mesothelioma register, Southern Italy. *BMC Res. Notes* **2019**, *12*, 636. [CrossRef] [PubMed]
5. Davidson, B. Expression of cancer-associated molecules in malignant mesothelioma. *Biomark. Insights* **2007**, *2*, 173–184. [CrossRef] [PubMed]
6. Robinson, B.W.S.; Lake, R.A. Medical progress—Advances in malignant mesothelioma. *N. Engl. J. Med.* **2005**, *353*, 1591–1603. [CrossRef]
7. De Reynies, A.; Jaurand, M.C.; Renier, A.; Couchy, G.; Hysi, I.; Elarouci, N.; Galateau-Salle, F.; Copin, M.C.; Hofman, P.; Cazes, A.; et al. Molecular classification of malignant pleural mesothelioma: Identification of a poor prognosis subgroup linked to the epithelial-to-mesenchymal transition. *Clin. Cancer Res.* **2014**, *20*, 1323–1334. [CrossRef]
8. Lo Russo, G.; Tessari, A.; Capece, M.; Galli, G.; de Braud, F.; Garassino, M.C.; Palmieri, D. MicroRNAs for the Diagnosis and Management of Malignant Pleural Mesothelioma: A Literature Review. *Front. Oncol.* **2018**, *8*, 650. [CrossRef]
9. Hjerpe, A.; Ascoli, V.; Bedrossian, C.; Boon, M.; Creaney, J.; Davidson, B.; Dejmek, A.; Dobra, K.; Fassina, A.; Field, A.; et al. Guidelines for cytopathologic diagnosis of epithelioid and mixed type malignant mesothelioma. Complementary statement from the International Mesothelioma Interest Group, also endorsed by the International Academy of Cytology and the Papanicolaou Society of Cytopathology. *Cytojournal* **2015**, *12*, 26. [CrossRef]
10. Sayan, M.; Eren, M.F.; Gupta, A.; Ohri, N.; Kotek, A.; Babalioglu, I.; Oskeroglu Kaplan, S.; Duran, O.; Derinalp Or, O.; Cukurcayir, F.; et al. Current treatment strategies in malignant pleural mesothelioma with a treatment algorithm. *Adv. Respir Med.* **2019**, *87*, 289–297. [CrossRef]
11. Abd Own, S.; Hoijer, J.; Hillerdahl, G.; Dobra, K.; Hjerpe, A. Effusion cytology of malignant mesothelioma enables earlier diagnosis and recognizes patients with better prognosis. *Diagn. Cytopathol.* **2020**. [CrossRef]
12. Creaney, J.; Dick, I.M.; Segal, A.; Musk, A.W.; Robinson, B.W. Pleural effusion hyaluronic acid as a prognostic marker in pleural malignant mesothelioma. *Lung Cancer* **2013**, *82*, 491–498. [CrossRef] [PubMed]
13. Cortes-Dericks, L.; Schmid, R.A. CD44 and its ligand hyaluronan as potential biomarkers in malignant pleural mesothelioma: Evidence and perspectives. *Respir Res.* **2017**, *18*, 58. [CrossRef] [PubMed]
14. Dejmek, A.; Hjerpe, A. The combination of CEA, EMA, and BerEp4 and hyaluronan analysis specifically identifies 79% of all histologically verified mesotheliomas causing an effusion. *Diagn. Cytopathol.* **2005**, *32*, 160–166. [CrossRef]
15. Hirose, Y.; Saijou, E.; Sugano, Y.; Takeshita, F.; Nishimura, S.; Nonaka, H.; Chen, Y.R.; Sekine, K.; Kido, T.; Nakamura, T.; et al. Inhibition of Stabilin-2 elevates circulating hyaluronic acid levels and prevents tumor metastasis. *Proc. Natl. Acad. Sci. USA* **2012**, *109*, 4263–4268. [CrossRef]
16. Hanagiri, T.; Shinohara, S.; Takenaka, M.; Shigematsu, Y.; Yasuda, M.; Shimokawa, H.; Nagata, Y.; Nakagawa, M.; Uramoto, H.; So, T.; et al. Effects of hyaluronic acid and CD44 interaction on the proliferation and invasiveness of malignant pleural mesothelioma. *Tumour Biol.* **2012**, *33*, 2135–2141. [CrossRef]
17. Atagi, S.; Ogawara, M.; Kawahara, M.; Sakatani, M.; Furuse, K.; Ueda, E.; Yamamoto, S. Utility of hyaluronic acid in pleural fluid for differential diagnosis of pleural effusions: Likelihood ratios for malignant mesothelioma. *Jpn. J. Clin. Oncol.* **1997**, *27*, 293–297. [CrossRef]
18. Torronen, K.; Soini, Y.; Paakko, P.; Parkkinen, J.; Sironen, R.; Rilla, K. Mesotheliomas show higher hyaluronan positivity around tumor cells than metastatic pulmonary adenocarcinomas. *Histol. Histopathol.* **2016**, *31*, 1113–1122. [CrossRef]
19. Rump, A.; Morikawa, Y.; Tanaka, M.; Minami, S.; Umesaki, N.; Takeuchi, M.; Miyajima, A. Binding of ovarian cancer antigen CA125/MUC16 to mesothelin mediates cell adhesion. *J. Biol. Chem.* **2004**, *279*, 9190–9198. [CrossRef]
20. Tian, L.; Zeng, R.; Wang, X.; Shen, C.; Lai, Y.; Wang, M.; Che, G. Prognostic significance of soluble mesothelin in malignant pleural mesothelioma: A meta-analysis. *Oncotarget* **2017**, *8*, 46425–46435. [CrossRef]
21. Bera, T.K.; Pastan, I. Mesothelin is not required for normal mouse development or reproduction. *Mol. Cell. Biol.* **2000**, *20*, 2902–2906. [CrossRef] [PubMed]

22. Chen, S.H.; Hung, W.C.; Wang, P.; Paul, C.; Konstantopoulos, K. Mesothelin binding to CA125/MUC16 promotes pancreatic cancer cell motility and invasion via MMP-7 activation. *Sci. Rep.* **2013**, *3*, 1870. [CrossRef]
23. Gubbels, J.A.; Belisle, J.; Onda, M.; Rancourt, C.; Migneault, M.; Ho, M.; Bera, T.K.; Connor, J.; Sathyanarayana, B.K.; Lee, B.; et al. Mesothelin-MUC16 binding is a high affinity, N-glycan dependent interaction that facilitates peritoneal metastasis of ovarian tumors. *Mol. Cancer* **2006**, *5*, 50. [CrossRef] [PubMed]
24. Burt, B.M.; Lee, H.S.; Lenge De Rosen, V.; Hamaji, M.; Groth, S.S.; Wheeler, T.M.; Sugarbaker, D.J. Soluble Mesothelin-Related Peptides to Monitor Recurrence After Resection of Pleural Mesothelioma. *Ann. Thorac. Surg.* **2017**, *104*, 1679–1687. [CrossRef] [PubMed]
25. Sirois, A.R.; Deny, D.A.; Li, Y.; Fall, Y.D.; Moore, S.J. Engineered Fn3 protein has targeted therapeutic effect on mesothelin-expressing cancer cells and increases tumor cell sensitivity to chemotherapy. *Biotechnol. Bioeng.* **2019**. [CrossRef]
26. Heidari-Hamedani, G.; Vives, R.R.; Seffouh, A.; Afratis, N.A.; Oosterhof, A.; van Kuppevelt, T.H.; Karamanos, N.K.; Metintas, M.; Hjerpe, A.; Dobra, K.; et al. Syndecan-1 alters heparan sulfate composition and signaling pathways in malignant mesothelioma. *Cell. Signal.* **2015**, *27*, 2054–2067. [CrossRef]
27. Szatmari, T.; Mundt, F.; Kumar-Singh, A.; Mobus, L.; Otvos, R.; Hjerpe, A.; Dobra, K. Molecular targets and signaling pathways regulated by nuclear translocation of syndecan-1. *BMC Cell. Biol.* **2017**, *18*, 34. [CrossRef]
28. Barbareschi, M.; Maisonneuve, P.; Aldovini, D.; Cangi, M.G.; Pecciarini, L.; Angelo Mauri, F.; Veronese, S.; Caffo, O.; Lucenti, A.; Palma, P.D.; et al. High syndecan-1 expression in breast carcinoma is related to an aggressive phenotype and to poorer prognosis. *Cancer* **2003**, *98*, 474–483. [CrossRef]
29. Sayyad, M.R.; Puchalapalli, M.; Vergara, N.G.; Wangensteen, S.M.; Moore, M.; Mu, L.; Edwards, C.; Anderson, A.; Kall, S.; Sullivan, M.; et al. Syndecan-1 facilitates breast cancer metastasis to the brain. *Breast Cancer Res. Treat.* **2019**, *178*, 35–49. [CrossRef] [PubMed]
30. Wang, S.; Zhang, X.; Wang, G.; Cao, B.; Yang, H.; Jin, L.; Cui, M.; Mao, Y. Syndecan-1 suppresses cell growth and migration via blocking JAK1/STAT3 and Ras/Raf/MEK/ERK pathways in human colorectal carcinoma cells. *BMC Cancer* **2019**, *19*, 1160. [CrossRef]
31. Ren, Z.; van Andel, H.; de Lau, W.; Hartholt, R.B.; Maurice, M.M.; Clevers, H.; Kersten, M.J.; Spaargaren, M.; Pals, S.T. Syndecan-1 promotes Wnt/beta-catenin signaling in multiple myeloma by presenting Wnts and R-spondins. *Blood* **2018**, *131*, 982–994. [CrossRef] [PubMed]
32. Neragi-Miandoab, S. Multimodality approach in management of malignant pleural mesothelioma. *Eur. J. Cardiothorac. Surg.* **2006**, *29*, 14–19. [CrossRef] [PubMed]
33. Negi, Y.; Kuribayashi, K.; Funaguchi, N.; Doi, H.; Mikami, K.; Minami, T.; Takuwa, T.; Yokoi, T.; Hasegawa, S.; Kijima, T. Early-stage Clinical Characterization of Malignant Pleural Mesothelioma. *In Vivo* **2018**, *32*, 1169–1174. [CrossRef] [PubMed]
34. Carella, R.; Deleonardi, G.; D'Errico, A.; Salerno, A.; Egarter-Vigl, E.; Seebacher, C.; Donazzan, G.; Grigioni, W.F. Immunohistochemical panels for differentiating epithelial malignant mesothelioma from lung adenocarcinoma: A study with logistic regression analysis. *Am. J. Surg. Pathol.* **2001**, *25*, 43–50. [CrossRef]
35. Halimi, M.; BeheshtiRouy, S.; Salehi, D.; Rasihashemi, S.Z. The Role of Immunohistochemistry Studies in Distinguishing Malignant Mesothelioma from Metastatic Lung Carcinoma in Malignant Pleural Effusion. *Iran. J. Pathol.* **2019**, *14*, 122–126. [CrossRef]
36. Husain, A.N.; Colby, T.; Ordonez, N.; Krausz, T.; Attanoos, R.; Beasley, M.B.; Borczuk, A.C.; Butnor, K.; Cagle, P.T.; Chirieac, L.R.; et al. Guidelines for pathologic diagnosis of malignant mesothelioma: 2012 update of the consensus statement from the International Mesothelioma Interest Group. *Arch. Pathol. Lab. Med.* **2013**, *137*, 647–667. [CrossRef]
37. Robinson, B.W.; Creaney, J.; Lake, R.; Nowak, A.; Musk, A.W.; de Klerk, N.; Winzell, P.; Hellstrom, K.E.; Hellstrom, I. Mesothelin-family proteins and diagnosis of mesothelioma. *Lancet* **2003**, *362*, 1612–1616. [CrossRef]
38. Ledda, C.; Senia, P.; Rapisarda, V. Biomarkers for Early Diagnosis and Prognosis of Malignant Pleural Mesothelioma: The Quest Goes on. *Cancers* **2018**, *10*. [CrossRef]
39. Pass, H.I.; Levin, S.M.; Harbut, M.R.; Melamed, J.; Chiriboga, L.; Donington, J.; Huflejt, M.; Carbone, M.; Chia, D.; Goodglick, L.; et al. Fibulin-3 as a blood and effusion biomarker for pleural mesothelioma. *N. Engl. J. Med.* **2012**, *367*, 1417–1427. [CrossRef]

40. Hu, Z.D.; Liu, X.F.; Liu, X.C.; Ding, C.M.; Hu, C.J. Diagnostic accuracy of osteopontin for malignant pleural mesothelioma: A systematic review and meta-analysis. *Clin. Chim. Acta* **2014**, *433*, 44–48. [CrossRef]
41. Mundt, F.; Nilsonne, G.; Arslan, S.; Csuros, K.; Hillerdal, G.; Yildirim, H.; Metintas, M.; Dobra, K.; Hjerpe, A. Hyaluronan and N-ERC/mesothelin as key biomarkers in a specific two-step model to predict pleural malignant mesothelioma. *PLoS ONE* **2013**, *8*, e72030. [CrossRef] [PubMed]
42. Stockhammer, P.; Ploenes, T.; Theegarten, D.; Schuler, M.; Maier, S.; Aigner, C.; Hegedus, B. Detection of TGF-beta in pleural effusions for diagnosis and prognostic stratification of malignant pleural mesothelioma. *Lung Cancer* **2019**, *139*, 124–132. [CrossRef] [PubMed]
43. Gao, R.; Wang, F.; Wang, Z.; Wu, Y.; Xu, L.; Qin, Y.; Shi, H.; Tong, Z. Diagnostic value of soluble mesothelin-related peptides in pleural effusion for malignant pleural mesothelioma: An updated meta-analysis. *Medicine* **2019**, *98*, e14979. [CrossRef]
44. Szatmari, T.; Mundt, F.; Heidari-Hamedani, G.; Zong, F.; Ferolla, E.; Alexeyenko, A.; Hjerpe, A.; Dobra, K. Novel genes and pathways modulated by syndecan-1: Implications for the proliferation and cell-cycle regulation of malignant mesothelioma cells. *PLoS ONE* **2012**, *7*, e48091. [CrossRef]
45. Anttonen, A.; Heikkila, P.; Kajanti, M.; Jalkanen, M.; Joensuu, H. High syndecan-1 expression is associated with favourable outcome in squamous cell lung carcinoma treated with radical surgery. *Lung Cancer* **2001**, *32*, 297–305. [CrossRef]
46. Anttonen, A.; Kajanti, M.; Heikkila, P.; Jalkanen, M.; Joensuu, H. Syndecan-1 expression has prognostic significance in head and neck carcinoma. *Br. J. Cancer* **1999**, *79*, 558–564. [CrossRef]
47. Fujiya, M.; Watari, J.; Ashida, T.; Honda, M.; Tanabe, H.; Fujiki, T.; Saitoh, Y.; Kohgo, Y. Reduced expression of syndecan-1 affects metastatic potential and clinical outcome in patients with colorectal cancer. *Jpn J. Cancer Res.* **2001**, *92*, 1074–1081. [CrossRef]
48. Juuti, A.; Nordling, S.; Lundin, J.; Louhimo, J.; Haglund, C. Syndecan-1 expression—A novel prognostic marker in pancreatic cancer. *Oncology* **2005**, *68*, 97–106. [CrossRef]

© 2020 by the authors. Licensee MDPI, Basel, Switzerland. This article is an open access article distributed under the terms and conditions of the Creative Commons Attribution (CC BY) license (http://creativecommons.org/licenses/by/4.0/).

Review

Glycosaminoglycans in Tissue Engineering: A Review

Harkanwalpreet Sodhi [1] and Alyssa Panitch [1,2,*]

1 Department of Biomedical Engineering, University of California Davis, Davis, CA 95616, USA; hssodhi@ucdavis.edu
2 Department of Surgery, University of California Davis, Sacramento, CA 95817, USA
* Correspondence: apanitch@ucdavis.edu

Abstract: Glycosaminoglycans are native components of the extracellular matrix that drive cell behavior and control the microenvironment surrounding cells, making them promising therapeutic targets for a myriad of diseases. Recent studies have shown that recapitulation of cell interactions with the extracellular matrix are key in tissue engineering, where the aim is to mimic and regenerate endogenous tissues. Because of this, incorporation of glycosaminoglycans to drive stem cell fate and promote cell proliferation in engineered tissues has gained increasing attention. This review summarizes the role glycosaminoglycans can play in tissue engineering and the recent advances in their use in these constructs. We also evaluate the general trend of research in this niche and provide insight into its future directions.

Keywords: glycosaminoglycans; tissue engineering; extracellular matrix; chondroitin sulfate; hyaluronic acid; dermatan sulfate; keratan sulfate; heparan sulfate

Citation: Sodhi, H.; Panitch, A. Glycosaminoglycans in Tissue Engineering: A Review. *Biomolecules* **2021**, *11*, 29. https://doi.org/10.3390/biom11010029

Received: 7 December 2020
Accepted: 23 December 2020
Published: 29 December 2020

Publisher's Note: MDPI stays neutral with regard to jurisdictional claims in published maps and institutional affiliations.

Copyright: © 2020 by the authors. Licensee MDPI, Basel, Switzerland. This article is an open access article distributed under the terms and conditions of the Creative Commons Attribution (CC BY) license (https://creativecommons.org/licenses/by/4.0/).

1. Introduction

Glycosaminoglycans (GAGs) are long, unbranched polysaccharide chains made up primarily of repeating disaccharide units. These disaccharide subunits are composed of one hexuronic acid and one amino sugar linked by glycosidic bonds [1] and these variations in disaccharide composition are used to distinguish the major classes of GAGs: Hyaluronic Acid (HA), Chondroitin Sulfate (CS), Dermatan Sulfate (DS), Keratan Sulfate (KS), and Heparan Sulfate (HS). GAGs are sulfated to varying degrees, with the exception of Hyaluronic Acid (HA), which is unsulfated. The different hexuronic acids and amino sugars found in each GAG are summarized in Table 1 and a structural diagram of the repeating disaccharide unit of each GAG is provided in Figure 1. CS, DS, and HS range in molecular mass between 10,000 and 50,000 Daltons and KS and Heparin (a GAG similar to but distinguished from HS) range between 5000 and 15,000 Daltons. In contrast, HA is generally a very high molecular weight GAG, ranging between approximately 100,000 and 10,000,000 Daltons [2]. The presence of the ionizable groups (sulfates and carboxylates on hexuronic acids) confers GAGs with polyionic properties that are responsible for their key abilities such as water retention, cell binding, control of ion fluxes and neuronal signaling [3–5].

Table 1. The hexuronic acid and amino sugar constituents of each glycosaminoglycan.

Glycosaminoglycan	Hexuronic Acid	Hexosamine
Chondroitin Sulfate [1]	glucuronic Acid	N-acetylgalactosamine
Dermatan Sulfate [1]	Iduronic Acid	N-acetylgalactosamine
Keratan Sulfate [1]	galactose	N-acetylglucosamine
Heparan Sulfate [1]	glucuronic Acid	N-acetylglucosamine
Hyaluronic Acid [1]	glucuronic Acid Unsulfated	N-acetyl-D-glucosamine Unsulfated

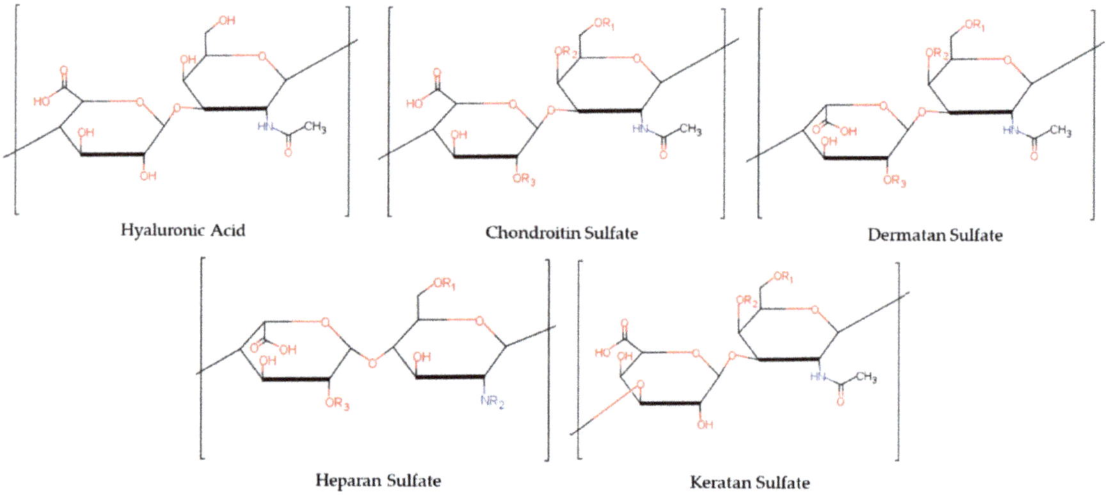

Figure 1. Repeating disaccharide unit of each glycosaminoglycan. "R" indicates a potential sulfation point.

The first reference to GAGs can be found in electronically available published literature dating back to the late 1930s when Karl Meyer summarized GAG chemical properties and biological relevance known at the time. Then, they were referred to as mucopolysaccharides and were classified primary as "containing iduronic acid", with sub-divisions of sulfate-free and sulfate-containing, or "neutral" [6]. Even at this time, knowledge of their general localization within the body was growing. It was known, for example, that "chondroitinsulfuric acid" could be isolated from cartilage, the aorta, and the sclera and also that its presence was decreased in "rachitic" (weak) bones [6]. Fast forward to the 1950s and researchers had identified that cell excretion of mucopolysaccharides could be used to determine the differentiation of fibroblasts in culture [7]. Research in this time period focused on isolating and characterizing new GAGs and elucidating the expression patterns and purpose of GAGs in the body [8], during development [9] and disease [10–12]. Foreshadowing the discovery of the importance of GAGs in tissue remodeling and applications in tissue engineering, in 1958, Bollet et al. analyzed the GAG content of granulation tissue formed when polyvinyl sponges were implanted under the dorsal skin of guinea pigs [13].

The first research using GAGs in tissue engineering scaffolds arose in the 1980s, with scientists investigating hyaluronic acid as a component of scaffolds for regeneration of tissues. Since then, all GAGs, with the exception of Keratan sulfate, have seen increased utilization in tissue engineering constructs for the treatment of a myriad of diseases such as osteoarthritis, neuropathy, and bone defects, to name a few. This review aims to summarize the use of each GAG in the advancement of tissue engineering in the last five years, depicted pictorially in Figure 2, and project how, as GAGs become more thoroughly understood, their utility and ubiquity in the tissue engineering field will expand.

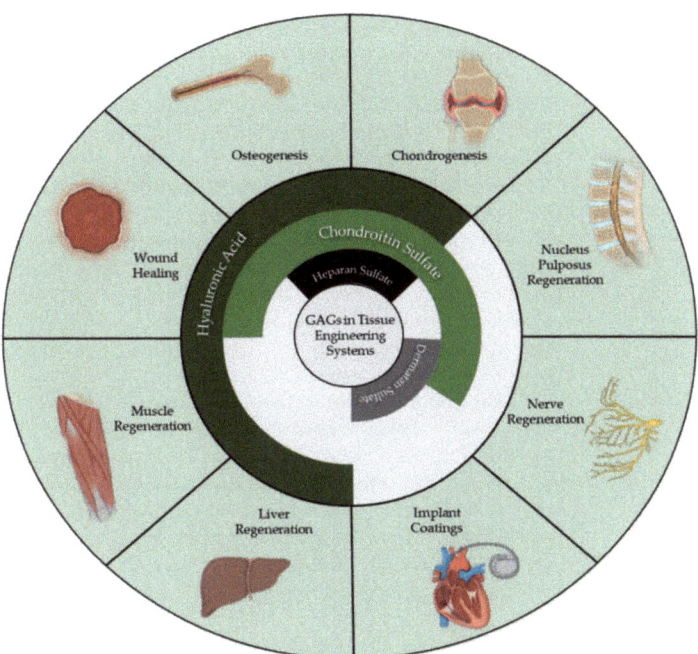

Figure 2. Each GAG and their aforementioned tissue engineering applications. This figure was made using BioRender.

2. Hyaluronic Acid

HA, also referred to as hyaluronan, has a repeating disaccharide unit of D-glucuronic acid and N-acetyl-D-glucosamine (the hexuronic acid and amino sugar, respectively) attached by a beta 1–3 bond, and the disaccharide units are joined by a beta 1–4 bond [1]. [1] It is found in liquid connective tissues such as the synovial fluid of joints and the vitreous humor of the eye where, in conjunction with other charged constituents of the extracellular matrix, it plays a key role in maintaining viscoelasticity via water retention due to both hydrogen bonds and osmotic pressure generated from the high density of anionic groups and accumulation of counter cations. When in water, HA has a gelatin-like consistency. Its viscoelasticity and ability to form matrices that retain water allow it to cushion joints, resist compression and help reduce friction in all joint tissues [3]. It also plays a role in the extracellular matrix of several tissues where it mediates receptor-driven detachment, mitosis, and migration. This control of cell division and cell migration means HA is commonly implicated in tumor development and cancer metastasis [3]. HA is the highest molecular weight GAG [1], presenting with a broad range of molecular masses generally ranging between 10^5 and 10^7 Daltons. This is in stark contrast to other GAGs, which are generally on the order of 103 Daltons [2]. HA's molecular mass plays an important role in its function. Studies have shown that HA fragments of varying lengths may alert the body to trauma and play roles in the progression of wound healing. Degradation of HA increases tissue permeability and HA fragments enhance angiogenesis, promoting tissue healing processes [3]. In contrast, endogenous HA has been shown to promote extracellular matrix secretion, reduce inflammation, and inhibit immune cell migration to maintain homeostasis in healthy tissue [3].

HA has many properties that make it an ideal candidate for tissue engineering scaffolds. It is biodegradable, biocompatible, and resorbable. HA is involved in every step of wound healing in the body [14]. The interplay between its hydrophilicity and control of cell migration allows HA to form a temporary, ideal wound healing environment. Because HA is hygroscopic, it can control the hydration of tissue during healing, allowing for increased flow of nutrients and effluence of waste products [15]. It also stimulates cells via interactions with CD44, RHAMM, and ICAM-1 cell receptors, which allows it to regulate cell adhesion, motility, inflammation, and differentiation [14]. Despite this, for most cell types it does not support sufficient attachment or spreading and requires chemical modification to support cell growth and survival [16].

In its native form, HA is a weak scaffolding material because it is rapidly degraded in vivo by hyaluronidase and is highly soluble, which can cause dissolution. It must, therefore, be chemically modified and crosslinked or combined with another polymer to form stable, structurally integrated scaffolds that support cell adhesion and proliferation [17]. Encouragingly, HA can be crosslinked under basic, acidic, and neutral pH conditions or combined with other natural and synthetic polymers to confer strength, allowing for diverse applications such as treating difficult to heal wounds, burns, and any form of trauma that requires a space-filling scaffold [14].

Additional support for chemical modification of HA comes from the fact that HA itself does not bond to surrounding tissue when it is used to fill defects, and it is often of sufficiently high molecular weight that it does not diffuse into the surrounding tissue to form an integrated seal when crosslinking it in situ. In addition, while viscous HA gels can be injected, injection of unmodified HA has been shown to cause damage and hemorrhaging in some tissues, such as in the heart and liver. With the goal of overcoming all of these shortcomings, Shin, et al. developed a tissue adhesive HA hydrogel functionalized with the adhesive catecholamine motif from mussel foot protein. This gel was shown to reduce apoptosis, increase viability, and enhance the function of human adipose-derived stem cells and hepatocytes. HA-catecholamine laden with hepatocytes was shown to gel and adhere to the liver of athymic mice within minutes. Further, the gel was still present after two weeks and albumin secreted by the transplanted hepatocytes was detectable in the blood stream 3 days after implantation [18], indicating some recapitulation of endogenous tissue behavior.

2.1. Hyaluronic Acid Supports Multiple Crosslinking Mechanisms

While not always supportive of in situ bonding with the surrounding tissue, crosslinking HA is an important tool and can affect the way in which cells interact with HA. While non-crosslinked or loosely crosslinked HA scaffolds do not support the growth of human mesenchymal stem cells (MSCs) due to their low strength and non-adhesive properties, Lou, et al. have developed a viscoelastic HA hydrogel capable of supporting cell adhesion and spreading [19,20]. HA-hydrazine can form transient, non-covalent, hydrazone bonds with HA-aldehyde and/or HA-benzaldehyde. Aldehyde functionalized HA was found to have an order of magnitude faster dissociation and association rate with HA-hydrazine than HA-benzaldehyde. Therefore, increased concentrations of HA-aldehyde results in slower stress-relaxing gels, and increased HA-benzaldehyde results in faster stress-relaxing gels [19]. A hydrogel system composed of both allowed control over gelation time and stress relaxation time. Stress relaxation time, as measured by 50% relaxation of initial stress, was tuned broadly from as slow as 5 h to 20 min by switching from HA-benzaldehyde to HA-aldehyde. Gel storage modulus was also varied from as low as 8.2 ± 1.1 Pa up to 471 ± 31.2 Pa by varying the molecular weight of HA and total HA concentration [19]. These scaffolds have been seeded with human MSCs [19] and human adipose derived stem cells [20] and shown to be cytocompatible and support cell spreading at certain ratios of HA-aldehyde and HA-benzaldehyde. This system is highly adaptable and can also be augmented with other biomolecules. The addition of collagen added cell-binding motifs not present on HA, affected viscoelasticity, and added a fibrillar component to the gels [19]. The addition of cellulose nanocrystals increased network organization and stiffness while also increasing degradation time [21]. The addition of neither collagen nor cellulose nanocrystals adversely affected the ability of HA gels to support cell growth [19–21].

In addition to modulating the viscoelastic properties through creation of transient crosslinks, several techniques have been developed to covalently crosslink HA through reaction of the carboxylate groups on the GAG backbone. In applications related to musculoskeletal tissue engineering, HA based polymer systems have also been shown to trigger endochondral bone formation in vitro and in vivo [22], making it a prime candidate for bone tissue engineering constructs. Poldervaart, et al. have shown that, not only can methacrylated HA support osteogenic differentiation, it can also be 3D printed [23]. 3D bioprinting allows for creation of porous scaffolds with a predefined shape and incorporation of cells and signaling molecules within the constructs in predetermined locations [24]. These photocrosslinkable gels showed long-term stability, lasting up to 14 days in the presence of hyaluronidase at 3% gelatin versus 1–7 days at lower gelatin concentrations. They also exhibited high stiffness, with storage moduli ranging from 170 ± 63 Pa up to as high as 2602 ± 199 Pa and elastic moduli as high as 10.6 kPa depending on concentration of HA (1–3% w/v) [23]. When seeded with bone marrow-derived human MSCs, cell viability at 21 days remained at 64.4% and MSCs showed spontaneous osteogenic differentiation without additional stimuli. While not required, bone morphogenic protein 2 accelerated mineral deposition within these constructs [23]. Similarly, gels made of esterified HA have also been shown to support osteogenic differentiation when seeded with bone marrow concentrate [25]. Bone marrow concentrate allowed MSCs to remain surrounded by their native microenvironment and circumvented the difficult process of pure MSC extraction, while esterified HA provides mechanical support [26,27]. Thus, HA is compatible to various crosslinking modalities.

Current research also explores new hydrogel manufacturing methods that not only overcome the intrinsic weaknesses of HA, but also avoid the use of organic solvents/reagents that cause toxicity and also allow for new types of HA constructs. HA microporous gel systems, where microgels are combined with cells and other additives and then crosslinked to form a macrogel, have been developed to overcome not only nano-porosity issues common with other gels, but also to allow for non-toxic crosslinking after cell seeding [28]. Microporous gels have enhanced cell and tissue integration properties when compared to nano-porous gels [28]. These newer microporous gels improve upon first generation gels that used techniques such as salt leeching [29], gas foaming [30], or using harsh chemicals for creating micropores in gels. Using these older techniques, cells and signaling factors needed to be seeded after pore formation and cell infiltration was limited and slow. Using the newer microbead technique, acrylamide-functionalized HA was formed into crosslinked microspheres using microfluidic droplet generation. The microgels were crosslinked using dithiol matrix metalloprotease sensitive linker peptides. Subsequently, microgels were mixed with human dermal fibroblasts and microbeads crosslinked to one another using light-induced free radical polymerization or carbodiimide chemistry to form a bulk microporous gel. This system allowed for the rapid cell infiltration necessary for endogenous tissue integration without significant scaffold degradation often seen in other hydrogel technologies that use cell remodeling of the matrix to allow for cells to enter [28]. This allows the gel to remain implanted longer while retaining its original mechanical properties, which is important for enhancing cell lineage specification and retention [31].

2.2. Hyaluronic Acid Blends

More control over gel properties and better recapitulation of the extracellular matrix are possible by combining HA with other natural polymers that confer mechanical strength, cell binding motifs, and change the microstructure of the gel. Gelatin is an ideal copolymer for HA as it provides structural support and RGD-integrin binding sites that allow cell adhesion and proliferation, unlike HA alone [32]. Gelatin-HA constructs have been studied extensively for regeneration of articular cartilage [33–35], wound healing [36–38], and even vocal fold repair [39] due to their chondrogenic, angiogenic, and cell adhesive properties, and their tunable viscoelastic properties. Like HA, gelatin can be methacrylated, which allows for photopolymerization of gelatin-gelatin or gelatin-HA crosslinks. Constructs made of methacrylated gelatin and HA have been shown to suppress hypertrophy and increase GAG expression by embedded, human bone marrow stem cells and, when tested in a full thickness osteochondral defect in rabbits, showed good cartilage regeneration [33]. Methacrylated HA and methacrylated gelatin can also be 3D bio-printed and polymerized with embedded cells without affecting their viability or chondrogenic properties, making them a good platform for custom, patient-specific cartilage implants [35]. Feng, et al. have also shown that a slightly different chemistry involving thiolated gelatin and HA-vinylsulphone can form hybrid microgels, generated from crosslinked microbeads, similar to the HA-only microgels mentioned previously. Human bone marrow stem cells encapsulated in HA-gelatin microgels showed high viability and chondrogenic potential. When injected subcutaneously in mice, the cell-laden gels formed smooth, elastic, cartilage-like tissue, and reduced hypertrophy and vascularization over the course of 8 weeks [34].

This ability to drive cell behavior is also augmented by the slow degradation rate of gelatin. While HA alone promotes angiogenesis, decreasing gel degradation rate and providing cell binding sites, via complexation with gelatin accelerates healing and decreases counterproductive inflammatory cell migration at wound sites [36]. When carbohydrazide gelatin was combined with HA-monoaldehyde, they formed an injectable gel that showed no toxicity when tested with human umbilical cord endothelial cells in vitro [36]. Further, gels tested using an ex vivo rat aortic ring assay showed endothelial invasion and microvascular extension into the gel at the aortic ring-gel interface, supporting the hypothesis that HA, which can be angiogenic on its own at the correct concentrations [36,40], can be enhanced with a polymer that presents cell binding sites and slows gel degradation.

In wound healing, where angiogenesis is critical, Ebrahimi, et al. showed that electrospun gelatin-HA constructs could accelerate healing of thermal burns in mice [41]. In contrast to most other natural polymers, gelatin constructs can also be electrospun to generate nanofibrous gels instead of standard monolithic ones. Nanofibrous scaffolds structurally mimic the fibrillar structure of the extracellular matrix, allow for cell adhesion due to the high surface area to volume ratio, allow oxygen to permeate, and allow cell waste to escape, all while inhibiting pathogen infiltration [42], making them excellent candidates for wound healing applications. Similar electrospun constructs have been made using gelatin and HA combined with chitosan, which showed success in a mouse model of wound healing [43] and rabbit models of alkali induced corneal burns [44]. All of these constructs reduced inflammation and improved healing, demonstrating the potential improvement of gelatin-HA construct using a nanofibrous structure. However, nearly all of them, with the exception of acetic acid-based gel systems, use harsh solvents for electrospinning and crosslinking, making their use cumbersome and potentially hazardous. In situations where complex functionalization of the scaffold is required, an easily modifiable polymer such as poly(caprolactone) [45] can be used to electrospin HA instead of gelatin. Poly(caprolactone) (PCL) has been used extensively in biomaterials, especially for electrospinning, but it lacks the cell signaling characteristics and hydrophilicity of HA. PCL electrospun scaffolds doped with HA and epithelial growth factor have been shown to promote cell infiltration while also up-regulating collagen and TGF-ß1 expression in vitro. In vivo, the HA-PCL gels, when doped with endogenous growth factors, showed regeneration of a thicker epidermis layer and formation of an organized dermis layer as well in a rat model of full thickness skin wound healing [42]. Like the HA-Gelatin electrospun constructs, this system also employs harsh solvents, such as chloroform, leaving room for improvement in the electrospinning of nanofibrous HA scaffolds [42].

Less commonly, HA-gelatin solutions have been investigated for regeneration of muscle tissue and as a model system for lung tissue. Gelatin and HA can both be functionalized with tyramine to allow for gelation using horseradish peroxidase and H2O2 [32,46]. C2C12 murine myoblasts seeded on these tyramine crosslinked scaffolds were shown to retain myoblast differentiation and myotube formation, while HA-only and Gelatin-only gels did not. HA gels supported spherical cell morphology due to lack of cell binding sites in HA, and gelatin gels showed dedifferentiation, as the gel collapsed under cell traction forces [32]. Kumar, et al. also showed that tyramine-functionalized HA and gelatin could be spin-coated into membranes and seeded with cells to generate an in vitro model of the alveolar basal epithelium for lung-based research. The films supported attachment, migration, and proliferation of alveolar basal epithelial cell line A549. When laden with growth factors, the membranes also induced some epithelial differentiation in MSCs [37]. Taken together, this research is suggestive of the vast potential of HA blended with gelatin and other bioactive species for tissue regeneration. It also highlights the array of crosslinking and manufacturing modalities that are under investigation to produce fully functional HA-based tissue engineered constructs.

While HA-gelatin blends are promising materials, many other HA blends have been investigated and have also shown promise. Tyramine functionalization of HA has been studied in combination with silk polymers for tissue engineering constructs. Raia, et al. have shown that HA-tyramine and silk fibroin-tyramine can be covalently crosslinked to form tunable hydrogels that begin to approach relevant mechanical properties and overcome some of the inherent weaknesses of HA [47]. In this study, silk fibers formed di-tyrosine bonds via horseradish peroxidase, resulting in highly elastic gels containing crystalline regions of silk. Tyramine-substituted HA, on the other hand, formed weak hydrogels that degraded rapidly. Use of a combination of both biopolymers overcame these weaknesses and resulted in tunable scaffolds. HA concentration in the matrix allowed adjustment of gelation time, degradation rate, and water retention. HA only hydrogels degraded within 6 days, while silk gels retained 70% of their mass on day 6. Hybrid gels allowed for tuning rate of degradation within this range of 1–6 days [47]. Silk-HA gels also achieved 100% strain before breaking, versus 30% in HA-only gels. Silk-only and HA-only gels exhibited storage moduli of 2.27 ± 0.09 KPa and 0.55 ± 0.03 KPa, respectively, while hybrid gels achieved moduli slightly beyond this range, peaking at 3.85 ± 0.08 KPa [47]. Silk gels alone were shown to allow adhesion and promote proliferation of human MSCs and this property was conferred to silk-HA hybrid gels. HA-only hydrogels inhibited MSC growth, showing an unadhered, spherical morphology after one week [47]. Combining silk and HA in this gel construct augments HA with the mechanical strength and degradation properties necessary to support cell growth with fine control over gel mechanical properties.

Similarly, tunable hybrid gels have been developed using HA and agarose. In contrast to HA, agarose has good gelatinizing properties, but exhibits slow degradation, limiting its use in some tissue engineering applications, which often target replacement of the engineered scaffold with host tissue [48]. Chu, et al. have shown that grafting of HA to agarose activated with epichlorohydrin resulted in a scaffold that presented the same cell regulation motifs as HA alone but also supported cell adhesion and proliferation. The gels were shown to stimulate TNF-α secretion in RAW 264 macrophages and upregulate Collagen I and III secretion by 3T3 fibroblasts. Further, when tested in a murine model of full thickness wound healing, agarose-HA gels showed rapid healing when compared to controls over the course of 21 days, showing that HA can facilitate wound healing past 1 week when combined with a slowly-degrading polymer [49].

While this list of HA-polymer blends is not exhaustive, it does demonstrate the enormous potential and versatilely of HA. HA interacts with cell receptors that regulate inflammation, cell differentiation, and cell motility, making it useful for a myriad of tissue engineering applications. However, it forms weak gels alone and does not adhere to tissues or support cell adhesion through integrin receptors thought to be required for tissue regeneration. These weaknesses can be overcome by functionalizing and crosslinking HA or combining it with another polymer, such as silk fibroin, gelatin, collagen, agarose, or polycaprolactone, which can provide strength and cell binding sites. In an appropriate scaffold, HA has been shown to induce chondrogenesis, osteogenesis, and wound healing by driving stem cell behavior. The ubiquitous nature of HA within the body and the ease with which it can be functionalized and combined with other polymers fully supports continued exploration of HA for successful development of tissue engineered products. A summary of HA tissue engineering constructs and their tested behavior can be found in Table 2.

Table 2. Summary of all aforementioned hyaluronic acid hydrogel types and their tested behavior in vitro and in vivo.

HA Type	Copolymer Type	Biological Testing	Biological Outcome
Unmodified HA[46]	Epoxy-Agarose	• Seeded RAW 264 macrophages • Seeded 3T3 fibroblasts • Mouse model of full thickness dermal wound	• Increased TNF-α secretion • Upregulated collagen I and III • Accelerated healing in vivo
Unmodified HA[40]	Polycaprolactone	• Seeded human skin keratinocytes and fibroblasts + epidermal growth factor • Mouse model of full thickness wound	• Upregulated collagen I and III and TGF-ß • Accelerated healing in vivo
Methacrylated HA[22]		• Seeded Human Mesenchymal Stem Cells	• Osteogenic differentiation
Tyramine-HA[43]	Silk Fibroin	• Seeded Human Mesenchymal Stem Cells	• Supports Adhesion and Proliferation
HA-tyramine[37]	Gelatin-Tyramine	• Seeded with C2C12 myoblasts • Spin coated membranes seeded with A549 alveolar epithelial cell[34]	• Induced myotubule formation • Supported epithelial differentiation in the presence of growth factors
HA + Dithiol linker peptide[26]		• Seeded Human Dermal Fibroblasts	• Supports Cell adhesion and proliferation
HA + catecholamine[16]		• Human adipose derived stem cells • Human hepatocytes • Implantation in athymic mice	• Supports Cell Adhesion and proliferation of both stem cells and hepatocytes • Implanted hepatocytes secreted albumin detectable in the blood stream in vivo
Esterified HA[23]		• Bone Marrow Concentrate	• Osteogenic differentiation
HA-hydrazine[17]	HA-aldehyde, HA-aldehyde, and/or collagen	• Seeded with human Mesenchymal stem cells • Doped with cellulose nanocrystals and seeded with 3T3 fibroblasts[17]	• Supports Cell adhesion and proliferation • Increased stiffness and retained support of cell viability
HA-monoaldehyde[33]	Carbohydrazide gelatin	• Seeded with human umbilical cord endothelial cells • Rat aortic ring assay	• No toxicity • Endothelial migration and microvascular extension
Electrospun HA[39]	Electrospun gelatin	• Mouse model of thermal burns	• Improved burn wound healing

3. Chondroitin Sulfate

CS is composed of a repeating disaccharide made up of D-Glucuronic acid, a hexuronic acid, and N-acetyl-d-galactosamine, an amino sugar. It is generally highly sulfated with -SO3 occuring at C4 or C6 on galactosamine [1]. Four subsets for CS exist: A, C, D, E. These subsets are differentiated by the location of the sulfates in the sugar rings. CS type B has subsequently been classified as dermatan sulfate; another GAG discussed later [50]. CS is an integral part of solid connective tissues such as cartilage, bone, skin, ligaments, and tendons [51]. Similar to HA, CS, when bound to a proteoglycan such as aggrecan, plays a key role in retention of water, due to the high density of anionic groups, and resistance to compression making it key in the cushioning and lubrication of joints [50,51].

Chondroitin sulfate-based gel systems have been developed for cartilage [52] and other tissue repair [53]. Similar to HA, chondroitin sulfate has the capacity to induce cell differentiation, making it useful in chondrogenic and osteogenic constructs, however, unmodified and alone, it also lacks the essential mechanical properties necessary for implantation into tissues [54] including cartilage, bone defects, or the nucleus pulposus (NP). Unlike HA, CS promotes cell adhesion and can be used to make non-adhesive polymers adhesive to cells [54]. The bulk of current research, therefore, focuses on adding CS moieties to tissue engineering constructs while mimicking the physical properties of native tissue. This can be done by incorporation of free CS chains into a different bulk material, crosslinking CS to itself or to another polymer [55–58]. In most systems, CS is conjugated with a covalent crosslinker that allows for self-gelation or gelation into a multicomponent matrix. In some systems, CS is entrapped in a matrix and allowed to diffuse in a manner controlled by mesh size and charge interactions [59]. The exact effects of immobilization technique on cell response to CS is still not well understood. However, the wealth of studies incorporating CS is shedding light onto biological activity inherent to CS.

3.1. Sulfation and Sulfation Pattern Suport Biological Activity

In bone tissue engineering, CS is responsible for coordinating osteoblast attachment, cell lineage commitments, and differentiation [60,61]. CS also interacts with growth factors critical for bone regeneration [61]. As such, CS scaffolds have the potential to replace the collagen scaffolds impregnated with Bone Morphogenic Protein 2 (BMP-2) that are currently the medical gold standard for treating critically sized bone defects. BMP-2 is an osteoinductive growth factor. The human recombinant form is approved by the FDA and used clinically with collagen sponges when autograft and allograft are not feasible to repair a bone defect. Andrews, et al. demonstrated extended release of recombinant human BMP-2 from CS scaffolds compared to their collagen sponge counterparts. CS based scaffolds showed very similar total release as compared to collagen gels after 15 days. However, the time to 50% BMP-2 release was 1.5 days for collagen, versus 5 days for CS gels, demonstrating a much more linear release profile, despite comparable total release [62]. Unmodified CS alone cannot, however, form structurally integral scaffolds that support cell growth and be implanted in the body. A common way to overcome this limitation is to methacrylate CS, which allows the polysaccharide chains to be crosslinked via photopolymerization using UV light and a photoinitiator such as 2-hydroxy-4'-(2-hydroxyethoxy)-2-methylpropiophenone [62]. When used to treat a challenging critically sized femoral defect in rats, methacrylated CS scaffolds loaded with BMP-2 induced comparable bone formation to the BMP-2 in collagen sponges as measured by bone volume, strength, and stiffness [62], despite the improved release kinetics of CS based gels. This could be due to the more rapid release of BMP-2 from collagen gels, which showed an initial burst release and demonstrated a collagen deposition pattern characteristic of more mature bone than CS gels [62]. Taken together, these data show the potential of CS-based systems to improve growth factor release kinetics and induce osteogenesis at a level that, at the very least, is equivalent to the current gold standard [62]. This is due in part to the osteogenic interactions between cell surface receptors and CS, but also due to the ability of CS to sequester and release growth factors in a controlled manner via growth factor interactions with the sulfated CS.

This affinity for growth factors and ability to control growth factor presentation to cells also confers CS with the ability to drive neuronal regeneration [63,64]. Besides osteogenic factors like BMP-2, methacrylated CS scaffolds have a strong affinity for fibroblast growth factor 2 (FGF2) and brain derived neurotrophic factor (BDNF), which can be added directly or via impregnation with platelet-rich plasma that contains these growth factors. This affinity for charged growth factors is strong enough that FGF-2 and BDNF release has been shown to be sustainable for 15 days, significantly longer than release from platelet rich plasma alone [63]. The degree of CS sulfation and sulfation patterns affect growth factor binding and release in addition to other cellular responses. Karumbaiah, et al. investigated the effects of disulfated and monosulfated CS on neurotrophic factor binding, neuronal homeostasis and the influence of variably sulfated CS in biomaterials on neural stem cell fate [65]. They found that binding of neurotrophic factors is dependent on CS sulfation and varies between mono and disulfated CS constructs. In addition, they confirmed the cytocompatibility of methacrylated CS gels for neuronally derived cell lines and demonstrated their ability support self-renewal of rat neurospheres. In other studies, when seeded with embryonic chick dorsal root ganglia, CS gels yielded better nerve growth than their HA counterparts [64]. These studies also showed that control over growth factor binding and direction of nerve growth are dependent on the sulfation patterns [64,65]. Therefore, scaffolds containing CS with the appropriate sulfation patterns can potentially be used in combination with growth factors to encourage and direct nerve growth more effectively than commonly used HA scaffolds. Further, sustained controlled release of growth factors utilizing GAGs such as CS may limit systemic exposure and subsequent unintended physiological responses.

Methacrylation of CS also allows for covalent crosslinking to form scaffolds that not only control growth factor presentation, but also drive bone mineralization. Calcium and phosphate are critical components for bone inorganic structure [66–68], so in addition to the need for release of growth factors for bone regeneration, there is a need to support the growth of the ceramic component of these composite tissues. A myriad of methods for incorporating calcium and phosphate ions into biodegradable scaffolds have been explored for bone tissue engineering [69–71]. Hydrogels that provide nucleation points for hydroxy apatite, such as ethylene glycol methacrylate phosphate (EGMP), induce faster apatite growth [71]. Kim, et al. have shown the methacrylated CS can be crosslinked to polyethylene glycol diacrylate (PEGDA) to form a gel which promotes nucleation of hydroxy apatite and sequesters the necessary calcium and phosphate ions, thanks to the charged sulfate groups on CS [72]. PEGDA was selected as a bioinert copolymer that is easy to handle, easily seeded with cells [72] and allows for variation of CS concentration in the system to elucidate the relationship between CS concentration and ion sequestration/deposition. Calcium and phosphate ion concentration in the gel was positivity correlated with CS concentration, and PEGDA-CS gels developed white particulate coatings in the presence of phosphate buffered saline [72], indicating PEGDA + CS is able to provide nucleation points for calcium and phosphate deposition. When embedded with human tonsil-derived MSCs, this gel technology demonstrated acceleration bone mineralization relative to controls and showed ion binding and distribution within negatively charged hydrogel was dependent on CS concentration. Furthermore, the biomineralizing microenvironment induced osteogenesis and deposited calcium and phosphate showed a native hydroxyapatite structure. When tested in a mouse model of critically sized calvarial defect, the cell-laden PEG-CS gels showed 2–3 times higher regeneration volume than controls [72]. Miyamoto, et al. have shown a similar ability of CS to induce hard tissue generation when combined with sodium alginate [73]. Together, these studies highlight the multifunctionality of CS, i.e., growth factor binding and nucleating calcium phosphate deposition, and the important role it plays in tissue regeneration.

3.2. Chondroitin Sulfate Blends

While methacrylated CS forms mechanically robust hydrogels that retain the inherent functionality of CS and have been shown to support chondrogenesis, osteogenesis and neurogenesis in vitro and in vivo, they require photopolymerization, meaning, in clinical applications, these constructs would require specialized tools for delivery and ultra-violet light for polymerization, which could potentially injure surrounding healthy tissue [74]. Tang, et al. approached this issue by developing a hydrogel scaffold comprised of CS functionalized with graphene oxide. Johnson–Claisen rearrangement chemistry allowed graphene oxide (GO) to be functionalization with primary amines. A solution composed of CS and modified GO gelled in situ within 10 min and the incorporation of graphene improved stiffness and toughness drastically (320 and 70%, respectively) over gels made of just CS. They also proved to be highly porous, resistant to degradation, and enabled MSCs to proliferate and deposit collagen matrix [75]. Of note, however, the entrapment of cells and potential chemical remnants from the EDC/NHS chemistry did initially slow cell metabolism [75]. Potentially of value is the conductive nature of GO, which may confer the scaffolds with the ability to induce natural conductive currents to improve tissue regeneration.

New CS crosslinking methods provide new ways to study the interactions of CS with cells in the absence of other extracellular matrix components and, in that regard, are indispensable. However, CS is not the sole constituent of the extracellular matrix in any tissue in the human body, but rather is interlaced with other GAGs and natural polymers. Therefore, combining CS with different polymers presents an opportunity to further control a gel's rheological properties, present additional biological signals, and better mimic native tissue. For example, collagen scaffolds functionalized with CS have been shown to recapitulate the chondrogenic niche, modulate inflammation, and mimic the mechanical properties of native collagen [55]. Corradetti, et al. demonstrated that such constructs support chondrogenic differentiation in rat bone marrow-derived stem cells in vitro and suppressed inflammation in vivo. MSCs grown on CS-collagen constructs aligned with scaffold pores, whereas cells grown on scaffolds containing only collagen showed clustering behavior, demonstrating that the presence of CS in the CS-collagen scaffolds is essential to influence cell-scaffold adhesion and [55], therefore, cytoskeletal organization and differentiation [76,77]. These cells also developed more intracellular vesicles, which have been associated with enhanced intercellular communication [78]. The constructs innately induce chondrogenic differentiation, and even though they didn't support osteogenesis innately, they displayed a synergistic effect with osteogenic media, showing increased expression of osteogenic factors Alp, Spp1, and Bgla compared to controls [55].

CS has also been combined with collagen using genipin as a crosslinker for tissue engineered scaffolds for regeneration of different types of cartilage, such as the nucleus pulposus. Forming a lightly crosslinked, gelatin-like scaffold, type II collagen and CS crosslinked with genipin are biocompatible and support differentiation of adipose-derived stem cells in vitro. When used as an injectable carrier of adipose derived stem cells in a rat model of NP degeneration CS-collagen gels showed increased disc height, water content, proteoglycan and type II collagen synthesis, and partial recovery of NP structure [57].

3.3. Processing Techniques and Manufacturing

As tissue engineering systems become more advanced, research naturally trends towards improving their utility in the clinic. Recent studies using CS in tissue engineering, therefore, explore ways to make cell-seeded CS scaffolds injectable and tailorable to individual patients. Injectable tissue engineering constructs are advantageous as they do not require invasive surgeries to implant. Chen, et al., for example, have developed, an enzymatically crosslinked, injectable, and biodegradable hydrogel system comprised of carboxymethyl pullulan and chondroitin sulfate functionalized with tyramine. These conjugates are crosslinkable under physiological conditions using horseradish peroxidase (HRP) and hydrogen peroxide. Porcine articular chondrocytes embedded in these gels demonstrated proliferation and enhanced cartilage-like extracellular matrix deposition over controls, indicating chondrogenesis [79]. This HRP crosslinking method has the potential to form minimally invasive, injectable hydrogels for a myriad of tissue engineering applications, as the molecular weight ratios, polymer concentrations, and crosslinker concentration can all be modified to fine tune gel properties [80,81]. Li, et al. took this one step further and developed a similar system using oxidized CS and pullulan functionalized with adipic hydrazide that is self-gelling and forms in situ. Similarly, this system demonstrated good biocompatibility and chondrogenic properties [82], and supports the concept of developing in situ gelling CS scaffolds for tissue engineering.

A common way to make cell scaffolds injectable is to make them sheer thinning or thermo responsive via combination with a polymer like Chitosan. Chitosan is broadly used for the synthesis of injectable hydrogels due its biocompatibility [83] and thermosensitive capabilities [84]. CS has been combined with chitosan-poly(hydroxybutyrate-co-valerate) in the form of a nanoparticle for nucleus pulposus regeneration. Similar to CS + collagen systems, this hydrogel system supports viability, adhesion, and chondrogenic differentiation of adipose derived stem cells and shows potential for NP regeneration [85]. Alinejad, et al.'s work provides evidence that the gels made with chitosan and CS can be prepared with weak bases such as sodium hydrogen carbonate and beta-glycerophosphate to form thermosensitive, injectable and biocompatible scaffolds with tunable physical properties. Cytocompatibility of these hydrogels scaffolds was also shown to be good. When evaluated with L929 fibroblasts, they showed high viability and metabolic activity for up to 7 days. This effect was enhanced by the addition of CS relative to controls. [56]. CS can also be linked to a chitosan scaffold if the chitosan is functionalized with hydroxy butyl groups and the CS is oxidized, allowing them to crosslink via the Schiff-base reaction [58]. These injectable gels also show good biocompatibility and support adipose derived stem cells, while not eliciting an immune response [58]. This injectable system, however, differs, in that a pre-gel of oxidized CS and hydroxy butyl chitosan can be injected and subsequently completely gelled by injecting more oxidized CS. The authors see this as applicable in molding processes for custom made tissue engineering constructs that are shaped to the patient [58].

Injectable gels also open the door to 3D bioprinting of tailored, patient-specific constructs. Bioprinting generates 3D scaffolds with reproducible and complex structures and offers the opportunity to generate customized hydrogel scaffolds with a predetermined pattern, shape, and size. Engineered cartilage plugs, for example, can potentially be sized to a patient's joint and shaped exactly to match the defect they aim to repair. In order to 3D print a tissue engineering scaffold, the gel used must have the correct rheological properties to be extruded and undergo rapid gelation upon deposition [86]. Abbadessa, et al. have combined photopolymerizable methacrylated CS with thermosensitive poly(N-(2-hydroxypropyl) methacrylamide-mono/dilactate)-polyethylene glycol triblock copolymer (M15P10). Unlike polymer solutions composed of methacrylated CS alone or M15P10 alone, mixtures containing CS and M15P10 showed strain-softening, thermo-sensitive and shear-thinning properties. The 3D printing of this hydrogel resulted in the generation of constructs with tailorable porosity and embedded chondrogenic cells remained viable and proliferating over a culture period of 6 days [86] confirming the potential of this hydrogel solution for injectable, cell laden tissue engineering constructs.

While 3D printing and injection molding allow engineered tissues to be structurally modified on the macro scale, they do not offer the nanoscale structural control of electrospinning. As mentioned previously, electrospun, nanofibrous scaffolds have many advantages over monolithic hydrogels for some tissue engineering applications, namely for dermal grafts. They structurally mimic the extracellular matrix, allow for cell adhesion, allow oxygen to permeate, and allow cell waste to escape, making them ideal for wound healing [42]. Unmodified CS and the aforementioned hydrogel systems do not allow for electrospinning as do polycaprolactone-based systems [83]. Using acetic acid and water to reduce the use of potentially toxic organic solvents, Sadeghi, et al. have electrospun a gelatin/polyvinyl alcohol/chondroitin sulfate nanofibrous scaffold for skin tissue engineering [87,88]. Results indicated that the gels were not cytotoxic and L929 fibroblasts attach and proliferate on the scaffolds without issue, as assessed via scanning electron microscopy [87], indicating they may be suitable for skin remodeling and regeneration. Further, this suggests that with further work, viable methods for electrospinning biocompatible scaffolds at scale will be realized.

In summary, Chondroitin sulfate-based gel systems have been developed for cartilage and bone repair, and wound healing due to their ability to direct cell attachment, cell lineage commitments, and differentiation [60,61]. Similar to HA, CS lacks the essential mechanical properties necessary for implantation. The bulk of current research, therefore, focuses on adding CS to bulk scaffolds for mechanical support or crosslinking CS. CS, when self-gelled or crosslinked to another organic or inorganic agent, has been shown to promote mineralization and osteogenesis, chondrogenesis, and wound healing in cell-laden tissue engineering constructs. Collectively, these studies demonstrate the key role CS plays in serving as a depot for growth factors to rapidly make them available as necessary for regeneration and engineering of new tissues. They also highlight the importance of highly charged sulfate groups on CS for binding of these factors and aggregation of ions such as the calcium and phosphate required for skeletal and dental bone mineralization. Many advances have also been made in making these constructs injectable and customizable using 3D printing and newer crosslinking modalities, while reducing the use of harsh, cytotoxic chemicals. A summary of CS hydrogel system and their tested behavior in vitro and in vivo can be found in Table 3.

Table 3. Summary of all aforementioned chondroitin sulfate hydrogel types and their tested behavior in vitro and in vivo.

CS Type	Copolymer Type	Biological Testing	Biological Outcome
Unmodified CS[51]	Collagen	• Seeded human mesenchymal stem cells and blood mononuclear cells together • Implantation under mouse dorsal skin	• Bolstered ability of mesenchymal stem cells to reduce inflammation in blood mononuclear cells • Chondrogenic differentiation • Low neutrophil infiltration in vivo
Unmodified CS[53]	Collagen II + Genipin	• Seeded human adipose derived stem cells • Rat model of nucleus pulposus degeneration	• Chondrogenic differentiation (nucleus pulposus specific) • Regeneration of nucleus pulposus in vivo
CS + chitosan nanoparticle[79]	chitosan–Poly(hydroxybutyrate-co-valerate)	• Seeded human adipose derived stem cells • Rat model of nucleus pulposus degeneration	• Chondrogenic differentiation • Regeneration of nucleus pulposus in vivo
Unmodified CS[52]	Chitosan + SHC* + BGP*	• Seeded with L929 Fibroblasts	• Supports Cell adhesion and proliferation
Unmodified CS[54]	Hydroxy–Butyl–Chitosan	• Seeded with Human adipose derived stem cells	• Supports Cell adhesion and proliferation
Unmodified CS[68]	Polyethylene glycol + CS binding peptide + crosslinker peptide	• Embryonic Chick Dorsal Root Ganglia	• Enhanced nerve growth
Methacrylated CS[58]		• Seeded with rate central nervous system neurospheres • Critically sized femoral defect in rats	• Promotes survival and self-renewal of neurospheress • Bone regeneration in constructs containing BMP-2

Table 3. Cont.

CS Type	Copolymer Type	Biological Testing	Biological Outcome
Methacrylated CS[68]	Polyethylene glycol	• Seeded with human mesenchymal stem cells • Mouse model of calvarial defect	• CS-dependent calcium and phosphate sequestration • Osteogenic differentiation and mineral deposition • Bone regeneration
Unmodified CS[69]	Alginate	• Seeded rat bone marrow cells • Implantation in rat dorsal subcutis	• CS-dependent Osteocalcin deposition • CS-dependent Osteogenesis in vitro
CS + Tyramine[75]	Hydroxymethyl Pullulan	• Seeded with porcine articular chondrocytes • Subcutaneous implantation in mice	• Supports Cell adhesion and proliferation • Cartilaginous matrix deposition • Good biocompatibility in vivo
Oxidized CS[76]	Pullulan-adipic hydrazide	• Seeded rabbit articular chondrocytes	• Supports chondrogenesis
Unmodified CS[81]	Polyvinyl alcohol and gelatin	• Seeded with L929 fibroblasts	• Supports Cell adhesion and proliferation
Methacrylated CS[59]	pHPMAlac-PEG triblock polymer*	• Seeded with Chondrogenic ATDC5 cells	• Supports cell survival and proliferation
CS-Graphene Oxide[72]		• Seeded with human mesenchymal stem cells	• Supports cell proliferation and deposition of collagen matrix
Unmodified CS[80]	PDMAEA-Q*	• Tested Adhesion to porcine skin in vitro • Seeded with HEPG2 human liver cancer cells	• Strong adhesion to tissue • Supports cell survival and proliferation

4. Chondroitin Sulfate-Hyaluronic Acid Hybrid Tissue Engineering Systems

More recently, there has been an increase in papers published describing tissue engineering systems that utilize more than one GAG to explore their synergy with respect to directing cell behavior. Fernandes-Cuhna, et al., for example, investigated the ability of an HA+CS construct to support MSCs and accelerate corneal healing in several mouse models of corneal injury. The results showed that a once-daily application of MSCs in HA/CS enhances epithelial cell proliferation and wound healing after injury to the cornea. It also reduced scar formation, neovascularization, and hemorrhage after alkaline corneal burns [89]. Building on single GAG hydrogels, like those formed from methacrylated CS, recent studies show that CS and HA alone can form scaffolds by crosslinking methods including functionalization with tyramine. Tyramine functionalized CS and HA can be covalently bonded to form strong, elastic gels that offer good viability when seeded with MSCs [90]. Similarly, electrospun scaffolds like the gelatin/PVA/CS mentioned earlier can instead be formed using gelatin, HA, and CS. These gels, loaded with sericin, showed several-fold increases in proliferation of human foreskin fibroblast, human keratinocyte and human MSCs, and supported epithelial differentiation in all three cell types. In addition, expression of some dermal proteins was achieved [91]. HA and CS have also been combined with gelatin and silk fibroin for cartilage tissue engineering. This combination was found to induce chondrogenesis of bone marrow MSCs [92]. These experimental systems lend support to the idea that tissue engineering constructs will only improve as the appropriate GAGs for each system are incorporated.

Research using CS or HA alone in tissue engineering has progressed drastically since the inception of tissue engineering in the late 1980s. With this comes the transition to incorporation of both HA and CS into scaffolds to better recapitulate the native extracellular matrix and improve tissue regeneration. Preliminary research combining both suggests they may work to together to improve regeneration of the cornea, articular cartilage, or skin following trauma.

5. Dermatan Sulfate

DS is found in the cornea, where it maintains optical clarity, and in the sclera, where it helps to maintain the eye's overall shape [93]. Further, it is found in blood vessel walls, heart valves, and the umbilical cord during pregnancy where it plays a key role in regulation of the extracellular matrix [93]. Its composition is very similar to that of CS, as demonstrated by its former name CS type B, however I-iduronic acid a C5 epimer of glucuronic acid, substitutes for hexuronic acid found in CS [50]. Sulfation is found on C4 or C6 of the galactosamine ring and sulfation levels increase with age [93].

DS has been implicated in the development of many pathologies, such as cancer metastasis [94], connective tissue diseases [95], and inhibited neuron regeneration [96]. Research focusing on DS in tissue engineering is sparse, with the bulk of research focusing on discovery of its functions and some research focusing on DS, modified DS, and DS proteoglycans as therapeutics or as a targeting mechanism for drug delivery [97–101]. This lack of exploration can likely be attributed to two key factors: the recent reclassification of DS from chondroitin sulfate B, and the extreme complexity of DS synthesis and physiological interactions. DS interactions are based upon the composition and sulfate functionalization patterns of the chain allowing for a high diversity of patterns and potential interactions similar to those seen for CS. For example, xyloside-primed dermatan sulfate from breast carcinoma cells has cytotoxic effects and this behavior is only exhibited by DS of a defined disaccharide composition [102]. This is the first example of cytotoxic effects of dermatan sulfate and highlights the complexity of cell interactions with sulfated GAGs. There has, however, been some research focusing on the use of DS in tissue engineering. DS proteoglycans are key moderators of fibrinogenesis and K.M., et al. have shown that this behavior can be recapitulated in vitro when combining DS with collagen scaffolds. Collagen fibril formation was shown to be dependent on DS concentration, with low concentrations resulting in disorganized fibrils and higher concentrations resulting in more organized, but less dense fibrils [103]. A more unique use for DS in tissue engineering may be in surface modification of implantable devices. DS, when combined with chitosan in a multi-layer coating on polyethylene terephthalate surface show high surface wettability and inhibited biofilm formation, two important factors in implantable devices such as vascular prosthetics [104].

It has also been shown recently that mouse embryonic stem cells undergo neuronal differentiation via activation of signal-regulated kinase 1/2 and human neural stem cells undergo neuronal differentiation and neuronal migration in the presence of DS [105]. This lends some promise to the use of DS to drive stem cell differentiation in neuronal tissue engineering constructs similar to the use of HA and CS as mentioned previously.

6. Heparan Sulfate and Heparin

Heparan sulfate's dominant repeating disaccharide unit is composed of glucuronic acid linked to N-acetylglucosamine [1]. Heparan sulfate is considered the most complex GAG and medical uses of this GAG are currently few and far between. Heparin, on the other hand has seen a myriad of medical applications. It is composed primarily of iduronic acid-N-sulfoglucosamine disaccharide units and is heavily sulfated. Many naturally occurring GAGs display a hybrid structure that blurs the line between HS and heparin. It has been proposed that the name heparin be only applied to GAGs containing more N-sulfate groups than N-acetyl groups. This falls in line with the generally accepted distinction that heparin is more highly sulfated than HS [106].

Pan, et al. have combined CS-chitosan scaffolds with heparin-gelatin microspheres to utilize the growth factor sequestering properties of heparin. These gels were formed from oxidized CS and carboxymethyl chitosan using the schiff's base reaction similar to the gels mentioned previously. Doping with these microspheres accelerated gelation, slowed weight loss, increased water uptake, and increased the compressive modulus over controls. Adipose-derived stem cells showed good viability as they did with the CS-chitosan gels, but had the added benefit of controlled release of incorporated growth factors such as insulin-like growth factor 1, while gels without heparin-gelatin microbeads exhibited burst release. These gels also showed the same injectability of CS-chitosan only gels for non-invasive tissue engineering therapies [107].

Tissue engineering research using heparan sulfate and, in some cases heparan sulfate mimetics [108], has recently increased as it became clear that HS can be administered to injury sites to support bone healing [109] and angiogenesis [110] and might, therefore, confer benefits to tissue engineering constructs. Lee, et al. recently investigated the binding affinity of a myriad of growth factors including TGF-β1, BMP-2, FGF-2, PDGF-BB, and VEGF165 and found it binds them all but with varying affinities, that may depend on the sulfation pattern and composition of the HS used. Further, in a mouse model of osteochondral defect, HA gels loaded with HS and no growth factors or stem cells showed recovery to normal or near normal as measured using the International Cartilage Regeneration and Joint Preservation Society cartilage injury evaluation scoring system. Gels containing HS were also the only to support regeneration of bone and cartilage, while HA only gels did not support bone regeneration [111]. Sefkow-Werner, et al. also noted that HS as part of a gel construct including bone morphogenic protein 2 and cyclic RGD worked synergistically with the growth factor and cell adhesion molecule in eliciting osteogenic differentiation and promoting enhanced and sustained signaling [112]. Possibly more importantly, they developed a streptavidin-based system that allows for tunable amounts of each ligand to be immobilized in a gel to investigate how their relative densities affect cell behavior with the potential to further the use of GAGs to improve tissue engineering constructs. While work with HS and heparin in tissue engineering is nascent with the exception of controlled release of growth factors, as more is learned about these important extracellular matrix components we anticipate that, as with other GAGs, their uses will increase.

7. Keratan Sulfate

Keratan sulfate is the exception to the usual hexuronic acid plus amino sugar composition of GAGs and is instead composed of galactose and acetylated glucosamine [1]. KS is a widely distributed GAG, even more so than those previously mentioned. It is found in the weight bearing connective tissues and epithelial tissues, as well as in the central and peripheral nervous systems, where it plays a key role in control of ion fluxes between neurons [4]. Cells' ability to respond to biochemical stimuli is contingent on the ability to control and sense ion fluxes and KS plays a key role in the regulation of this, and further, the pathophysiology neuronal disorders such as epilepsy [4]. Control of charges and ion gradients such as these also play a role in adhesion, proliferation, and differentiation of cells, and even wound healing [113]. This importance in chemical and ionic signaling is further highlighted by the fact that the brain is the second most KS rich organ following the eyes, where, as part of a KS proteoglycan, it plays a role in neurogenesis, demarcation of brain areas, direction of neuronal growth, and repair processes [5].

Despite its ubiquity, Keratan sulfate's interactions and uses in tissue engineering are the least understood out of all of the GAGs [114] and its applications in tissue engineering to date are nonexistent. The majority of current research focuses on discovery of KS' role in regenerative neural processes [114], airway/lung inflammation [115,116], and infection [117], emphysema [118], corneal dystrophy [119], cancer malignancy [120], and specialized functions in other species [121,122]. An emphasis has also been placed on analyzing sources for isolatable KS [123], and KS as a contaminant in CS purification [124–126].

8. Summary and Future Directions

GAGs are used in tissue engineering constructs to recapitulate the ECM and, thereby, drive stem cell differentiation or retention of phenotype of implanted cells. This allows them to be used as implants for regeneration of damaged tissue. Recent research in this field focuses on tissue engineering constructs for wound healing in skin and cornea, restoring damaged cartilage, such as articular cartilage and the NP, restoring bone, and neuronal regeneration. Research in the last five years generally focuses on three key areas: (1) overcoming the physical limitations of GAGs alone, by developing scaffolds that mimic the rheological properties of native tissues that can be doped with CS to present its moieties; (2) exploration of growth factor and ion sequestering by GAGs in TECs and how this affects their ability to promote cell differentiation and tissue regeneration with one or multiple GAGs; (3) advancement of hydrogel crosslinking technologies to reduce cytotoxicity of components and reagents and confer new, useful properties, such as sheer thinning/thermos-responsiveness for 3D printing, or to allow for GAG-only scaffolds that do not require polymers such as chitosan, collagen, PEGDA, etc. Area (1) has been investigated quite heavily to date, but new potential applications are still emerging, and (2) and (3) leave much room for exploration. We have seen that incorporation of gel formations with one GAG to drive cell fate and tissue regeneration has been heavily explored in some tissues, such as cartilage and bone with some emerging research in combining multiple GAGs into a construct. Moreover, each GAG has been investigated to a different extent. The general trends in tissue engineering using GAGs and where each GAG stands in the process are summarized in Figure 3. Moving forward, we expect to see more tissue engineering constructs that incorporate multiple GAGs to elucidate their synergistic effects on stem cell fate and their composite potential for tissue regeneration, whether the form is as functionalized, crosslinked GAGs alone or GAGs immobilized or crosslinked into a gel composed of another polymer.

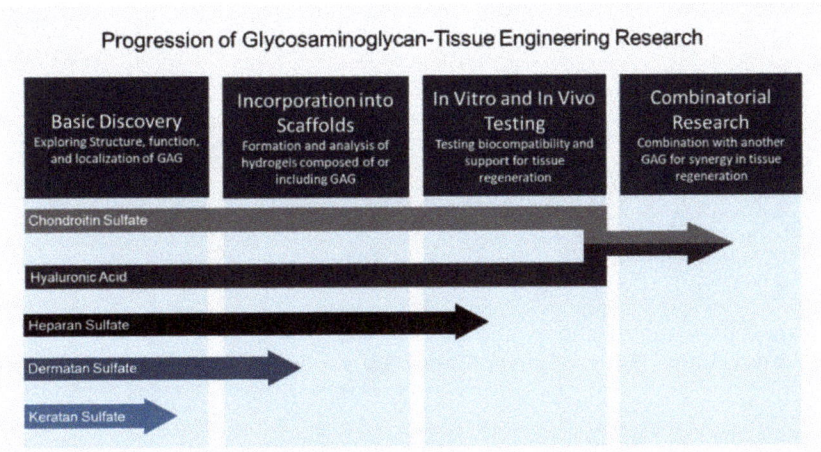

Figure 3. The four generalized steps of incorporating glycosaminoglycans into tissue engineering and the relative progress of each in this process.

All tissues contain a complex mix of different GAGs with different compositions, as described previously. It follows, then, that future research will increasingly focus on different "versions" of the same GAG and combinations of different GAGs in different ratios to recapitulate native tissue. This will be especially true as the wealth of knowledge regarding each GAG individually grows. Moving forward, we expect to see an increasing number of tissue engineering constructs with two or more GAGs and insight into their interplay. We also expect increased application of GAG based or GAG+ polymer gels in a wider variety of biological systems.

Author Contributions: All authors contributed equally to this work. All authors have read and agreed to the published version of the manuscript.

Funding: This research was funded by: National Institutes of Health: T32 HL086350.

Data Availability Statement: No new data was created or analyzed in this study. Data sharing is not applicable in this article.

Conflicts of Interest: The authors declare no conflict of interest.

References

1. Blanco, A.; Blanco, G. (Eds.) *Carbohydrates, in Medical Biochemistry*; Academic Press: Cambridge, MA, USA, 2017; Chapter 4; pp. 73–97.
2. Fraser, J.R.E.; Laurent, T.C.; Laurent, U.B.G. Hyaluronan: Its nature, distribution, functions and turnover. *J. Intern. Med.* **1997**, *242*, 27–33. [CrossRef] [PubMed]
3. Necas, J.B.L.B.P.; Bartosikova, L.; Brauner, P.; Kolar, J.J.V.M. Hyaluronic acid (hyaluronan): A review. *Vet. Med.* **2008**, *53*, 397–411. [CrossRef]
4. Caterson, B.; Melrose, J. Keratan sulfate, a complex glycosaminoglycan with unique functional capability. *Glycobiology* **2018**, *28*, 182–206. [CrossRef] [PubMed]
5. Crowder, K.M.; Gunther, J.M.; Jones, T.A.; Hale, B.D.; Zhang, H.Z.; Peterson, M.R.; Scheller, R.H.; Chavkin, C.; Bajjalieh, S.M. Abnormal neurotransmission in mice lacking synaptic vesicle protein 2A (SV2A). *Proc. Natl. Acad. Sci. USA* **1999**, *96*, 15268–15273. [CrossRef]
6. Meyer, K. The chemistry and biology of mucopolysaccharides and glycoproteins. In *Cold Spring Harbor Symposia on Quantitative Biology*; Cold Spring Harbor Laboratory Press: Cold Spring Harbor, NY, USA, 1938.
7. Grossfeld, H.; Meyer, K.; Godman, G. Differentiation of Fibroblasts in Tissue Culture, as Determined by Mucopolysaccharide Production. *Proc. Soc. Exp. Biol. Med.* **1955**, *88*, 31–35. [CrossRef]
8. Fessler, J.H. Water and Mucopolysaccharide as Structural Components of Connective Tissue. *Nat. Cell Biol.* **1957**, *179*, 426–427. [CrossRef]
9. Anseth, A. Glycosaminoglycans in the developing corneal stroma. *Exp. Eye Res.* **1961**, *1*, 116–121. [CrossRef]
10. Taylor, H.E. The role of mucopolysaccharides in the pathogenesis of intimal fibrosis and atherosclerosis of the human aorta. *Am. J. Pathol.* **1953**, *29*, 871–883.
11. Altshuler, C.H.; Angevine, D.M. Acid mucopolysaccharide in degenerative disease of connective tissue, with special reference to serous inflammation. *Am. J. Pathol.* **1951**, *27*, 141–156.
12. Stidworthy, G.; Masters, Y.F.; Shetlar, M. The effect of aging on mucopolysaccharide composition of human costal cartilage as measured by hexosamine and uronic acid content. *J. Gerontol.* **1958**, *13*, 10–13. [CrossRef]
13. Bollet, A.J.; Goodwin, J.F.; Simpson, W.F.; Anderson, D.V. Mucopolysaccharide, protein and desoxyribosenucleic acid concentration of granulation tissue induced by polyvinyl sponges. *Proc. Soc. Exp. Biol. Med.* **1958**, *99*, 418–421. [CrossRef] [PubMed]
14. Collins, M.N.; Birkinshaw, C. Hyaluronic acid based scaffolds for tissue engineering—A review. *Carbohydr. Polym.* **2013**, *92*, 1262–1279. [CrossRef] [PubMed]
15. Voigt, J.; Driver, V.R. Hyaluronic acid derivatives and their healing effect on burns, epithelial surgical wounds, and chronic wounds: A systematic review and meta-analysis of randomized controlled trials. *Wound Repair Regen.* **2012**, *20*, 317–331. [CrossRef] [PubMed]
16. Yamanlar, S.; Sant, S.; Boudou, T.; Picart, C.; Khademhosseini, A. Surface functionalization of hyaluronic acid hydrogels by polyelectrolyte multilayer films. *Biomaterials* **2011**, *32*, 5590–5599. [CrossRef]
17. Khunmanee, S.; Jeong, Y.; Park, H. Crosslinking method of hyaluronic-based hydrogel for biomedical applications. *J. Tissue Eng.* **2017**, *8*. [CrossRef]
18. Shin, J.; Lee, J.S.; Lee, C.; Park, H.-J.; Yang, K.; Jin, Y.; Ryu, J.H.; Hong, K.S.; Moon, S.; Chung, H.; et al. Tissue Reconstruction: Tissue Adhesive Catechol-Modified Hyaluronic Acid Hydrogel for Effective, Minimally Invasive Cell Therapy (Adv. Funct. Mater. 25/2015). *Adv. Funct. Mater.* **2015**, *25*, 3798. [CrossRef]

19. Lou, J.; Stowers, R.; Nam, S.; Xia, Y.; Chaudhuri, O. Stress relaxing hyaluronic acid-collagen hydrogels promote cell spreading, fiber remodeling, and focal adhesion formation in 3D cell culture. *Biomaterials* **2018**, *154*, 213–222. [CrossRef]
20. Yin, F.; Lin, L.; Zhan, S. Preparation and properties of cellulose nanocrystals, gelatin, hyaluronic acid composite hydrogel as wound dressing. *J. Biomater. Sci. Polym. Ed.* **2019**, *30*, 190–201. [CrossRef]
21. Domingues, R.M.; Silva, M.; Gershovich, P.; Betta, S.; Babo, P.; Caridade, S.G.; Mano, J.F.; Motta, A.; Reis, R.L.; Gomes, M.E. Development of Injectable Hyaluronic Acid/Cellulose Nanocrystals Bionanocomposite Hydrogels for Tissue Engineering Applications. *Bioconjugate Chem.* **2015**, *26*, 1571–1581. [CrossRef]
22. Solchaga, L.A.; Yoo, J.U.; Lundberg, M.; Dennis, J.E.; Huibregtse, B.A.; Goldberg, V.M.; I Caplan, A. Hyaluronan-based polymers in the treatment of osteochondral defects. *J. Orthop. Res.* **2000**, *18*, 773–780. [CrossRef]
23. Poldervaart, M.T.; Goversen, B.; De Ruijter, M.; Abbadessa, A.; Melchels, F.P.; Öner, F.C.; Dhert, W.J.; Vermonden, T.; Alblas, J. 3D bioprinting of methacrylated hyaluronic acid (MeHA) hydrogel with intrinsic osteogenicity. *PLoS ONE* **2017**, *12*, e0177628. [CrossRef]
24. Fedorovich, N.E.; Schuurman, W.; Wijnberg, H.M.; Prins, H.-J.; Van Weeren, P.R.; Malda, J.; Alblas, J.; Dhert, W.J. Biofabrication of Osteochondral Tissue Equivalents by Printing Topologically Defined, Cell-Laden Hydrogel Scaffolds. *Tissue Eng. Part C Methods* **2012**, *18*, 33–44. [CrossRef] [PubMed]
25. Cavallo, C.; Desando, G.; Ferrari, A.; Zini, N.; Mariani, E.; Grigolo, B. Hyaluronan scaffold supports osteogenic differentiation of bone marrow concentrate cells. *J. Biol. Regul. Homeost. Agents* **2016**, *30*, 409–420.
26. Bianco, P.; Riminucci, M.; Gronthos, S.; Robey, P.G. Bone Marrow Stromal Stem Cells: Nature, Biology, and Potential Applications. *Stem Cells* **2001**, *19*, 180–192. [CrossRef] [PubMed]
27. Block, J.E. The role and effectiveness of bone marrow in osseous regeneration. *Med. Hypotheses* **2005**, *65*, 740–747. [CrossRef] [PubMed]
28. Sideris, E.; Griffin, D.R.; Ding, Y.; Li, S.; Weaver, W.M.; Di Carlo, D.; Hsiai, T.; Segura, T. Particle Hydrogels Based on Hyaluronic Acid Building Blocks. *ACS Biomater. Sci. Eng.* **2016**, *2*, 2034–2041. [CrossRef]
29. Chiu, Y.-C.; Larson, J.C.; Isom, A.; Brey, E.M. Generation of Porous Poly(Ethylene Glycol) Hydrogels by Salt Leaching. *Tissue Eng. Part C Methods* **2009**, *16*, 905–912. [CrossRef]
30. Nam, Y.S.; Yoon, J.J.; Park, T.G. A novel fabrication method of macroporous biodegradable polymer scaffolds using gas foaming salt as a porogen additive. *J. Biomed. Mater. Res.* **2000**, *53*, 1–7. [CrossRef]
31. Hung, B.P.; Harvestine, J.N.; Saiz, A.M.; Gonzalez-Fernandez, T.; Sahar, D.E.; Weiss, M.L.; Leach, J.K. Defining hydrogel properties to instruct lineage- and cell-specific mesenchymal differentiation. *Biomaterials* **2019**, *189*, 1–10. [CrossRef]
32. Poveda-Reyes, S.; Moulisova, V.; Sanmartín-Masiá, E.; Quintanilla-Sierra, L.; Salmeron-Sanchez, M.; Ferrer, G.G. Gelatin-Hyaluronic Acid Hydrogels with Tuned Stiffness to Counterbalance Cellular Forces and Promote Cell Differentiation. *Macromol. Biosci.* **2016**, *16*, 1311–1324. [CrossRef]
33. Lin, H.; Beck, A.M.; Shimomura, K.; Sohn, J.; Fritch, M.R.; Deng, Y.; Kilroy, E.J.; Tang, Y.; Alexander, P.G.; Tuan, R.S. Optimization of photocrosslinked gelatin/hyaluronic acid hybrid scaffold for the repair of cartilage defect. *J. Tissue Eng. Regen. Med.* **2019**, *13*, 1418–1429. [CrossRef] [PubMed]
34. Feng, Q.; Li, Q.; Wen, H.; Chen, J.; Liang, M.; Huang, H.; Lan, D.; Dong, H.; Cao, X. Injection and Self-Assembly of Bioinspired Stem Cell-Laden Gelatin/Hyaluronic Acid Hybrid Microgels Promote Cartilage Repair In Vivo. *Adv. Funct. Mater.* **2019**, *29*, 1906690. [CrossRef]
35. Lam, T.; Dehne, T.; Krüger, J.P.; Hondke, S.; Endres, M.; Thomas, A.; Lauster, R.; Sittinger, M.; Kloke, L. Photopolymerizable gelatin and hyaluronic acid for stereolithographic 3D bioprinting of tissue-engineered cartilage. *J. Biomed. Mater. Res. Part B Appl. Biomater.* **2019**, *107*, 2649–2657. [CrossRef] [PubMed]
36. Hozumi, T.; Kageyama, T.; Ohta, S.; Fukuda, J.; Ito, T. Injectable Hydrogel with Slow Degradability Composed of Gelatin and Hyaluronic Acid Cross-Linked by Schiff's Base Formation. *Biomacromolecules* **2018**, *19*, 288–297. [CrossRef] [PubMed]
37. Kumar, P.; Ciftci, S.; Barthes, J.; Knopf-Marques, H.; Muller, C.B.; DeBry, C.; Vrana, N.E.; Ghaemmaghami, A. A composite Gelatin/hyaluronic acid hydrogel as an ECM mimic for developing mesenchymal stem cell-derived epithelial tissue patches. *J. Tissue Eng. Regen. Med.* **2020**, *14*, 45–57. [CrossRef] [PubMed]
38. Luo, J.-W.; Liu, C.; Wu, J.-H.; Zhao, D.-H.; Lin, L.-X.; Fan, H.-M.; Sun, Y.-L. In situ forming gelatin/hyaluronic acid hydrogel for tissue sealing and hemostasis. *J. Biomed. Mater. Res. Part B Appl. Biomater.* **2020**, *108*, 790–797. [CrossRef]
39. Kazemirad, S.; Heris, H.K.; Mongeau, L. Viscoelasticity of hyaluronic acid-gelatin hydrogels for vocal fold tissue engineering. *J. Biomed. Mater. Res. Part B: Appl. Biomaterials* **2016**, *104*, 283–290. [CrossRef]
40. Slevin, M.; Kumar, S.; Gaffney, J. Angiogenic Oligosaccharides of Hyaluronan Induce Multiple Signaling Pathways Affecting Vascular Endothelial Cell Mitogenic and Wound Healing Responses. *J. Biol. Chem.* **2002**, *277*, 41046–41059. [CrossRef]
41. Ebrahimi-Hosseinzadeh, B.; Pedram, M.; Hatamian-Zarmi, A.; Salahshour-Kordestani, S.; Rasti, M.; Mokhtari-Hosseini, Z.B.; Mir-Derikvand, M. In vivo evaluation of gelatin/hyaluronic acid nanofiber as Burn-wound healing and its comparison with ChitoHeal gel. *Fibers Polym.* **2016**, *17*, 820–826. [CrossRef]
42. Wang, Z.; Qian, Y.; Li, L.; Pan, L.; Njunge, L.W.; Dong, L.; Yang, L. Evaluation of emulsion electrospun polycaprolactone/hyaluronan/epidermal growth factor nanofibrous scaffolds for wound healing. *J. Biomater. Appl.* **2016**, *30*, 686–698. [CrossRef]

43. Bazmandeh, A.Z.; Mirzaei, E.; Fadaie, M.; Shirian, S.; Ghasemi, Y. Dual spinneret electrospun nanofibrous/gel structure of chitosan-gelatin/chitosan-hyaluronic acid as a wound dressing: In-vitro and in-vivo studies. *Int. J. Biol. Macromol.* **2020**, *162*, 359–373. [CrossRef] [PubMed]
44. Xu, W.; Wang, Z.; Liu, Y.; Wang, L.; Jiang, Z.; Li, T.; Zhang, W.; Liang, Y. Carboxymethyl chitosan/gelatin/hyaluronic acid blended-membranes as epithelia transplanting scaffold for corneal wound healing. *Carbohydr. Polym.* **2018**, *192*, 240–250. [CrossRef] [PubMed]
45. Yew, C.H.T.; Azari, P.; Choi, J.R.; Muhamad, F.; Murphy, B.P. Electrospun Polycaprolactone Nanofibers as a Reaction Membrane for Lateral Flow Assay. *Polymers* **2018**, *10*, 1387. [CrossRef] [PubMed]
46. Sanmartín-Masiá, E.; Poveda-Reyes, S.; Ferrer, G.G. Extracellular matrix–inspired gelatin/hyaluronic acid injectable hydrogels. *Int. J. Polym. Mater.* **2016**, *66*, 280–288. [CrossRef]
47. Raia, N.R.; Partlow, B.P.; McGill, M.; Kimmerling, E.P.; Ghezzi, C.E.; Kaplan, D.L. Enzymatically crosslinked silk-hyaluronic acid hydrogels. *Biomaterials* **2017**, *131*, 58–67. [CrossRef]
48. Zarrintaj, P.; Manouchehri, S.; Ahmadi, Z.; Saeb, M.; Urbanska, A.M.; Kaplan, D.L.; Mozafari, M. Agarose-based biomaterials for tissue engineering. *Carbohydr. Polym.* **2018**, *187*, 66–84. [CrossRef]
49. Chu, B.; Zhang, A.; Huang, J.; Peng, X.; You, L.; Wu, C.; Tang, S. Preparation and biological evaluation of a novel agarose-grafting-hyaluronan scaffold for accelerated wound regeneration. *Biomed. Mater.* **2020**, *15*, 045009. [CrossRef]
50. Foot, M.; Mulholland, M. Classification of chondroitin sulfate A, chondroitin sulfate C, glucosamine hydrochloride and glucosamine 6 sulfate using chemometric techniques. *J. Pharm. Biomed. Anal.* **2005**, *38*, 397–407. [CrossRef]
51. Henrotin, Y.; Mathy, M.; Sanchez, C.; Lambert, C. Chondroitin sulfate in the treatment of osteoarthritis: From in vitro studies to clinical recommen-dations. *Ther. Adv. Musculoskelet. Dis.* **2010**, *2*, 335–348. [CrossRef]
52. Hwang, N.S.; Varghese, S.; Lee, H.J.; Theprungsirikul, P.; Canver, A.; Sharma, B.; Elisseeff, J. Response of zonal chondrocytes to extracellular matrix-hydrogels. *FEBS Lett.* **2007**, *581*, 4172–4178. [CrossRef]
53. Liu, Y.; Cai, S.; Shu, X.Z.; Shelby, J.; Prestwich, G.D. Release of basic fibroblast growth factor from a crosslinked glycosaminoglycan hydrogel promotes wound healing. *Wound Repair Regen.* **2007**, *15*, 245–251. [CrossRef] [PubMed]
54. Aravamudhan, A.; Ramos, D.M.; Nada, A.A.; Kumbar, S.G. Natural polymers: Polysaccharides and their derivatives for biomedical applications. In *Natural and Synthetic Biomedical Polymers*; Kumbar, S.G., Laurencin, C.T., Deng, M., Eds.; Elsevier: Oxford, UK, 2014; Chapter 4; pp. 67–89.
55. Corradetti, B.; Taraballi, F.; Minardi, S.; Van Eps, J.L.; Cabrera, F.; Francis, L.W.; Gazze, S.A.; Ferrari, M.; Weiner, B.K.; Tasciotti, E. Chondroitin Sulfate Immobilized on a Biomimetic Scaffold Modulates Inflammation While Driving Chondrogenesis. *Stem Cells Transl. Med.* **2016**, *5*, 670–682. [CrossRef] [PubMed]
56. Alinejad, Y.; Adoungotchodo, A.; Hui, E.; Zehtabi, F.; Lerouge, S. An injectable chitosan/chondroitin sulfate hydrogel with tunable mechanical properties for cell therapy/tissue engineering. *Int. J. Biol. Macromol.* **2018**, *113*, 132–141. [CrossRef] [PubMed]
57. Zhou, X.; Wang, J.; Fang, W.; Tao, Y.; Zhao, T.; Xia, K.; Liang, C.; Hua, J.; Li, F.; Chen, Q. Genipin cross-linked type II collagen/chondroitin sulfate composite hydrogel-like cell delivery system induces differentiation of adipose-derived stem cells and regenerates degenerated nucleus pulposus. *Acta Biomater.* **2018**, *71*, 496–509. [CrossRef]
58. Li, C.; Wang, K.; Zhou, X.; Li, T.; Xu, Y.; Qiang, L.; Peng, M.; Xu, Y.; Xie, L.; He, C.; et al. Controllable fabrication of hydroxybutyl chitosan/oxidized chondroitin sulfate hydrogels by 3D bioprinting technique for cartilage tissue engineering. *Biomed. Mater.* **2019**, *14*, 025006. [CrossRef]
59. Piai, J.F.; Rubira, A.F.; Muniz, E.C. Self-assembly of a swollen chitosan/chondroitin sulfate hydrogel by outward diffusion of the chondroitin sulfate chains. *Acta Biomater.* **2009**, *5*, 2601–2609. [CrossRef] [PubMed]
60. Stanford, C.M.; Solursh, M.; Keller, J.C. Significant role of adhesion properties of primary osteoblast-like cells in early adhesion events for chondroitin sulfate and dermatan sulfate surface molecules. *J. Biomed. Mater. Res.* **1999**, *47*, 345–352. [CrossRef]
61. Hempel, U.; Matthäus, C.; Preissler, C.; Moller, S.; Hintze, V.; Dieter, P. Artificial Matrices with High-Sulfated Glycosaminoglycans and Collagen Are Anti-Inflammatory and Pro-Osteogenic for Human Mesenchymal Stromal Cells. *J. Cell. Biochem.* **2014**, *115*, 1561–1571. [CrossRef]
62. Andrews, S.; Cheng, A.; Stevens, H.; Logun, M.T.; Webb, R.; Jordan, E.; Xia, B.; Karumbaiah, L.; Guldberg, R.E.; Stice, S.L. Chondroitin Sulfate Glycosaminoglycan Scaffolds for Cell and Recombinant Protein-Based Bone Regeneration. *Stem Cells Transl. Med.* **2019**, *8*, 575–585. [CrossRef]
63. Birdwhistell, K.E.; Karumbaiah, L.; Franklin, S.P. Sustained Release of Transforming Growth Factor-β1 from Platelet-Rich Chondroitin Sulfate Glycosaminoglycan Gels. *J. Knee Surg.* **2018**, *31*, 410–415. [CrossRef]
64. Conovaloff, A.; Panitch, A. Characterization of a chondroitin sulfate hydrogel for nerve root regeneration. *J. Neural Eng.* **2011**, *8*. [CrossRef] [PubMed]
65. Karumbaiah, L.; Enam, S.F.; Brown, A.C.; Saxena, T.; Betancur, M.I.; Barker, T.H.; Bellamkonda, R. Chondroitin Sulfate Glycosaminoglycan Hydrogels Create Endogenous Niches for Neural Stem Cells. *Bioconjugate Chem.* **2015**, *26*, 2336–2349. [CrossRef] [PubMed]
66. Müller, P.; Bulnheim, U.; Diener, A.; Lüthen, F.; Teller, M.; Klinkenberg, E.-D.; Neumann, H.-G.; Nebe, B.; Liebold, A.; Steinhoff, G.; et al. Calcium phosphate surfaces promote osteogenic differentiation of mesenchymal stem cells. *J. Cell. Mol. Med.* **2008**, *12*, 281–291. [CrossRef] [PubMed]

67. Shih, Y.-R.V.; Hwang, Y.; Phadke, A.; Kang, H.; Hwang, N.S.; Caro, E.J.; Nguyen, S.; Siu, M.; Theodorakis, E.A.; Gianneschi, N.C.; et al. Calcium phosphate-bearing matrices induce osteogenic differentiation of stem cells through adenosine signaling. *Proc. Natl. Acad. Sci. USA* **2014**, *111*, 990–995. [CrossRef]
68. Boonrungsiman, S.; Gentleman, E.; Carzaniga, R.; Evans, N.D.; McComb, D.W.; Porter, A.E.; Stevens, M.M. The role of intracellular calcium phosphate in osteoblast-mediated bone apatite formation. *Proc. Natl. Acad. Sci. USA* **2012**, *109*, 14170–14175. [CrossRef]
69. Jun, S.-H.; Lee, E.-J.; Jang, T.-S.; Kim, H.-E.; Jang, J.-H.; Koh, Y.-H. Bone morphogenic protein-2 (BMP-2) loaded hybrid coating on porous hydroxyapatite scaffolds for bone tissue engineering. *J. Mater. Sci. Mater. Med.* **2013**, *24*, 773–782. [CrossRef]
70. Kweon, H. A novel degradable polycaprolactone networks for tissue engineering. *Biomaterials* **2003**, *24*, 801–808. [CrossRef]
71. Nuttelman, C.R.; Benoit, D.S.; Tripodi, M.C.; Anseth, K.S. The effect of ethylene glycol methacrylate phosphate in PEG hydrogels on mineralization and viability of encapsulated hMSCs. *Biomaterials* **2006**, *27*, 1377–1386. [CrossRef]
72. Kim, H.D.; Lee, E.A.; An, Y.-H.; Kim, S.L.; Lee, S.S.; Yu, S.J.; Jang, H.L.; Nam, K.T.; Im, S.G.; Hwang, N.S. Chondroitin Sulfate-Based Biomineralizing Surface Hydrogels for Bone Tissue Engineering. *ACS Appl. Mater. Interfaces* **2017**, *9*, 21639–21650. [CrossRef]
73. Miyamoto, A.; Yoshikawa, M.; Maeda, H. Hard Tissue-Forming Ability and Ultra-Micro Structure of Newly Developed Sponges as Scaffolds Made with Sodium Alginate Gel and Chondroitin Sulfate. *J. Biomed. Sci. Eng.* **2018**, *11*, 289–306. [CrossRef]
74. Sharma, B.; Fermanian, S.; Gibson, M.; Unterman, S.; Herzka, D.A.; Cascio, B.; Coburn, J.; Hui, A.Y.; Marcus, N.; Gold, G.; et al. Human Cartilage Repair with a Photoreactive Adhesive-Hydrogel Composite. *Sci. Transl. Med.* **2013**, *5*, 167ra6. [CrossRef] [PubMed]
75. Tang, C.; Holt, B.D.; Wright, Z.M.; Arnold, A.M.; Moy, A.C.; Sydlik, S.A. Injectable amine functionalized graphene and chondroitin sulfate hydrogel with potential for cartilage regeneration. *J. Mater. Chem. B* **2019**, *7*, 2442–2453. [CrossRef] [PubMed]
76. Huebsch, N.; Arany, P.R.; Mao, A.S.; E Shvartsman, D.; Ali, O.A.; Bencherif, S.A.; Rivera-Feliciano, J.; Mooney, D.J. Harnessing traction-mediated manipulation of the cell/matrix interface to control stem-cell fate. *Nat. Mater.* **2010**, *9*, 518–526. [CrossRef]
77. Kato, M.; Mrksich, M. Using model substrates to study the dependence of focal adhesion formation on the affinity of integrin-ligand complexes. *Biochemistry* **2004**, *43*, 2699–2707. [CrossRef] [PubMed]
78. Robbins, P.D.; Morelli, A.E. Regulation of immune responses by extracellular vesicles. *Nat. Rev. Immunol.* **2014**, *14*, 195–208. [CrossRef] [PubMed]
79. Chen, F.; Yu, S.; Liu, B.; Ni, Y.; Yu, C.; Su, Y.; Zhu, X.; Yu, X.; Zhou, Y.; Chunyang, Y. An Injectable Enzymatically Crosslinked Carboxymethylated Pullulan/Chondroitin Sulfate Hydrogel for Cartilage Tissue Engineering. *Sci. Rep.* **2016**, *6*, 20014. [CrossRef] [PubMed]
80. Autissier, A.; Le Visage, C.; Pouzet, C.; Chaubet, F.; Letourneur, D. Fabrication of porous polysaccharide-based scaffolds using a combined freeze-drying/cross-linking process. *Acta Biomater.* **2010**, *6*, 3640–3648. [CrossRef]
81. Bae, H.; Ahari, A.F.; Shin, H.; Nichol, J.W.; Hutson, C.B.; Masaeli, M.; Kim, S.-H.; Aubin, H.; Yamanlar, S.; Khademhosseini, A. Cell-laden microengineered pullulan methacrylate hydrogels promote cell proliferation and 3D cluster formation. *Soft Matter* **2011**, *7*, 1903–1911. [CrossRef]
82. Li, T.; Song, X.; Weng, C.; Wang, X.; Sun, L.; Gong, X.; Yang, L.; Chen, C. Self-crosslinking and injectable chondroitin sulfate/pullulan hydrogel for cartilage tissue engineering. *Appl. Mater. Today* **2018**, *10*, 173–183. [CrossRef]
83. Berger, J.; Reist, M.; Mayer, J.; Felt, O.; Peppas, N.; Gurny, R. Structure and interactions in covalently and ionically crosslinked chitosan hydrogels for biomedical applications. *Eur. J. Pharm. Biopharm.* **2004**, *57*, 19–34. [CrossRef]
84. Chenite, A. Rheological characterisation of thermogelling chitosan/glycerol-phosphate solutions. *Carbohydr. Polym.* **2001**, *46*, 39–47. [CrossRef]
85. Nair, M.B.; Baranwal, G.; Vijayan, P.; Keyan, K.S.; Jayakumar, R. Composite hydrogel of chitosan-poly(hydroxybutyrate-co-valerate) with chondroitin sulfate nano-particles for nucleus pulposus tissue engineering. *Colloids Surf. B Biointerfaces* **2015**, *136*, 84–92. [CrossRef] [PubMed]
86. Abbadessa, A.; Blokzijl, M.; Mouser, V.; Marica, P.; Malda, J.; Hennink, W.; Vermonden, T. A thermo-responsive and photo-polymerizable chondroitin sulfate-based hydrogel for 3D printing applications. *Carbohydr. Polym.* **2016**, *149*, 163–174. [CrossRef] [PubMed]
87. Sadeghi, A.; Pezeshki-Modaress, M.; Zandi, M. Electrospun polyvinyl alcohol/gelatin/chondroitin sulfate nano-fibrous scaffold: Fabrication and in vitro evaluation. *Int. J. Biol. Macromol.* **2018**, *114*, 1248–1256. [CrossRef] [PubMed]
88. Sadeghi, A.; Zandi, M.; Pezeshki-Modaress, M.; Rajabi, S. Tough, hybrid chondroitin sulfate nanofibers as a promising scaffold for skin tissue engineering. *Int. J. Biol. Macromol.* **2019**, *132*, 63–75. [CrossRef] [PubMed]
89. Fernandes-Cunha, G.M.; Na, K.; Putra, I.; Lee, H.J.; Hull, S.; Cheng, Y.; Blanco, I.J.; Eslani, M.; Djalilian, A.R.; Myung, D. Corneal Wound Healing Effects of Mesenchymal Stem Cell Secretome Delivered Within a Viscoelastic Gel Carrier. *Stem Cells Transl. Med.* **2019**, *8*, 478–489. [CrossRef]
90. Ni, Y.; Tang, Z.; Cao, W.; Lin, H.; Fan, Y.; Guo, L.; Zhang, X. Tough and elastic hydrogel of hyaluronic acid and chondroitin sulfate as potential cell scaffold materials. *Int. J. Biol. Macromol.* **2015**, *74*, 367–375. [CrossRef]
91. Bhowmick, S.; Scharnweber, D.; Koul, V. Co-cultivation of keratinocyte-human mesenchymal stem cell (hMSC) on sericin loaded electrospun nanofibrous composite scaffold (cationic gelatin/hyaluronan/chondroitin sulfate) stimulates epithelial differentiation in hMSCs: In vitro study. *Biomaterials* **2016**, *88*, 83–96. [CrossRef]

92. Sawatjui, N.; Damrongrungruang, T.; Leeanansaksiri, W.; Jearanaikoon, P.; Hongeng, S.; Limpaiboon, T. Silk fibroin/gelatin–chondroitin sulfate–hyaluronic acid effectively enhances in vitro chondrogen-esis of bone marrow mesenchymal stem cells. *Mater. Sci. Eng. C* **2015**, *52*, 90–96. [CrossRef]
93. Huffman, F.G. Uronic acids. In *Encyclopedia of Food Sciences and Nutrition*, 2nd ed.; Caballero, B., Ed.; Academic Press: Oxford, UK, 2003; pp. 5890–5896.
94. Listik, E.; Xavier, E.G.; Pinhal, M.A.S.; Toma, L. Dermatan sulfate epimerase 1 expression and mislocalization may interfere with dermatan sulfate synthesis and breast cancer cell growth. *Carbohydr. Res.* **2020**, *488*, 107906. [CrossRef]
95. Mizumoto, S.; Kosho, T.; Yamada, S.; Sugahara, K. Pathophysiological Significance of Dermatan Sulfate Proteoglycans Revealed by Human Genetic Disorders. *Pharmaceuticals* **2017**, *10*, 34. [CrossRef] [PubMed]
96. Rezaei, S.; Bakhtiyari, S.; Assadollahi, K.; Heidarizadi, S.; Moayeri, A.; Azizi, M. Evaluating Chondroitin Sulfate and Dermatan Sulfate Expression in Glial Scar to Determine Appro-priate Intervention Time in Rats. *Basic Clin. Neurosci.* **2020**, *11*, 31–40. [PubMed]
97. Walimbe, T.; Panitch, A. Proteoglycans in Biomedicine: Resurgence of an Underexploited Class of ECM Molecules. *Front. Pharmacol.* **2020**, *10*. [CrossRef] [PubMed]
98. Dehghani, T.; Thai, P.N.; Sodhi, H.; Ren, L.; Sirish, P.; Nader, C.E.; Timofeyev, V.; Overton, J.L.; Li, X.; Lam, K.S.; et al. Selectin-Targeting Glycosaminoglycan-Peptide Conjugate Limits Neutrophil Mediated Cardiac Reperfusion Injury. *Cardiovasc. Res.* **2020**. [CrossRef] [PubMed]
99. Li, S.; Zhang, F.; Yu, Y.; Zhangb, Q. A dermatan sulfate-functionalized biomimetic nanocarrier for melanoma targeted chemotherapy. *Carbohydr. Polym.* **2020**, *235*, 115983. [CrossRef] [PubMed]
100. Blachman, A.; Funez, F.; Birocco, A.M.; Saavedra, S.L.; Lázaro-Martinez, J.M.; Camperi, S.A.; Glisoni, R.; Sosnik, A.; Calabrese, G.C. Targeted anti-inflammatory peptide delivery in injured endothelial cells using dermatan sul-fate/chitosan nanomaterials. *Carbohydr. Polym.* **2020**, *230*, 115610. [CrossRef]
101. Rasente, R.Y.; Imperiale, J.C.; Algarra, M.; Gualco, L.; Oberkersch, R.; Sosnik, A.; Calabrese, G.C. Dermatan sulfate/chitosan polyelectrolyte complex with potential application in the treatment and diagnosis of vascular disease. *Carbohydr. Polym.* **2016**, *144*, 362–370. [CrossRef]
102. Persson, A.; Tykesson, E.; Westergren-Thorsson, G.; Malmström, A.; Ellervik, U.; Mani, K. Xyloside-primed Chondroitin Sulfate/Dermatan Sulfate from Breast Carcinoma Cells with a Defined Disaccharide Composition Has Cytotoxic Effects in Vitro. *J. Biol. Chem.* **2016**, *291*, 14871–14882. [CrossRef]
103. Jyothsna, K.M.; Sarkar, P.; Jha, K.K.; Raghunathan, V.; Bhat, R. Differential levels of dermatan sulfate generate distinct Collagen I gel architectures. *bioRxiv* **2020**. [CrossRef]
104. Hayder, J.; Chaouch, M.A.; Amira, N.; Ben Mansour, M.; Majdoub, H.; Chaubet, F.; Maaroufi, R.M. Co-immobilization of chitosan and dermatan sulfate from Raja montagui skin on polyethylene ter-ephthalate surfaces: Characterization and antibiofilm activity. *Int. J. Polym. Mater. Polym. Biomater.* **2018**, *67*, 277–287. [CrossRef]
105. Ogura, C.; Hirano, K.; Mizumoto, S.; Yamada, S.; Nishihara, S. Dermatan sulphate promotes neuronal differentiation in mouse and human stem cells. *J. Biochem.* **2020**. [CrossRef] [PubMed]
106. Gallagher, J.T.; Walker, A. Molecular Distinctions between Heparan-Sulfate and Heparin—Analysis of Sulfation Patterns Indicates That Heparan-Sulfate and Heparin Are Separate Families of N-Sulfated Polysaccharides. *Biochem. J.* **1985**, *230*, 665–674. [CrossRef] [PubMed]
107. Pan, Y.; Xiao, C.; Tan, H.; Yuan, G.; Li, J.; Li, S.; Jia, Y.; Xiong, D.; Hu, X.; Niu, X. Covalently injectable chitosan/chondroitin sulfate hydrogel integrated gelatin/heparin microspheres for soft tissue engineering. *Int. J. Polym. Mater.* **2019**, 1–9. [CrossRef]
108. Mammadov, B.; Sever, M.; Gecer, M.; Zor, F.; Ozturk, S.; Akgun, H.; Ulas, U.H.; Orhan, Z.; Guler, M.O.; Tekinay, A.B. Sciatic Nerve Regeneration Induced by Glycosaminoglycan and Laminin Mimetic Peptide Nan-ofiber Gels. *RSC Adv.* **2016**, *6*, 110535–110547. [CrossRef]
109. Murali, S.; Rai, B.; Dombrowski, C.; Lee, J.; Lim, Z.; Bramono, D.; Ling, L.; Bell, T.; Hinkley, S.; Nathan, S.; et al. Affinity-selected heparan sulfate for bone repair. *Biomaterials* **2013**, *34*, 5594–5605. [CrossRef]
110. Wang, C.; Poon, S.; Murali, S.; Koo, C.Y.; Bell, T.J.; Hinkley, S.F.; Yeong, H.; Bhakoo, K.; Nurcombe, V.; Cool, S. Engineering a vascular endothelial growth factor 165-binding heparan sulfate for vascular therapy. *Biomaterials* **2014**, *35*, 6776–6786. [CrossRef]
111. Lee, J.H.; Luo, X.; Ren, X.; Tan, T.C.; Smith, R.A.; Swaminathan, K.; Sekar, S.; Bhakoo, K.; Nurcombe, V.; Hui, J.H.P.; et al. A Heparan Sulfate Device for the Regeneration of Osteochondral Defects. *Tissue Eng. Part A* **2018**, *25*, 352–363. [CrossRef]
112. Sefkow-Werner, J.; Machillot, P.; Sales, A.; Castro-Ramirez, E.; Degardin, M.; Boturyn, D.; Cavalcanti-Adam, E.A.; Albiges-Rizo, C.; Picart, C.; Migliorini, E. Heparan sulfate co-immobilized with cRGD ligands and BMP2 on biomimetic platforms promotes BMP2-mediated osteogenic differentiation. *Acta Biomater.* **2020**, *114*, 90–103. [CrossRef]
113. Casella, A.; Panitch, A.; Leach, J.K. Endogenous Electric Signaling as a Blueprint for Conductive Materials in Tissue Engineering. *Bioelectricity* **2020**. [CrossRef]
114. Melrose, J. Keratan sulfate (KS)-proteoglycans and neuronal regulation in health and disease: The importance ofKS-glycodynamics and interactive capability with neuroregulatory ligands. *J. Neurochem.* **2019**, *149*, 170–194. [CrossRef]
115. Gonzalez-Gil, A.; Porell, R.N.; Fernandes, S.M.; Wei, Y.; Yu, H.; Carroll, D.J.; McBride, R.; Paulson, J.C.; Tiemeyer, M.; Aoki, K.; et al. Sialylated keratan sulfate proteoglycans are Siglec-8 ligands in human airways. *Glycobiology* **2018**, *28*, 786–801. [CrossRef] [PubMed]

116. Kumagai, T.; Kiwamoto, T.; Brummet, M.E.; Wu, F.; Aoki, K.; Zhu, Z.; Bochner, B.S.; Tiemeyer, M. Airway glycomic and allergic inflammatory consequences resulting from keratan sulfate galactose 6-O-sulfotransferase (CHST1) deficiency. *Glycobiology* **2018**, *28*, 406–417. [CrossRef] [PubMed]
117. Carpenter, J.; Kesimer, M. Membrane-bound mucins of the airway mucosal surfaces are densely decorated with keratan sulfate: Revisiting their role in the Lung's innate defense. *Glycobiology* **2020**. [CrossRef] [PubMed]
118. Gao, C.; Fujinawa, R.; Yoshida, T.; Ueno, M.; Ota, F.; Kizuka, Y.; Hirayama, T.; Korekane, H.; Kitazume, S.; Maeno, T.; et al. A keratan sulfate disaccharide prevents inflammation and the progression of emphysema in murine models. *Am. J. Physiol. Cell. Mol. Physiol.* **2017**, *312*, L268–L276. [CrossRef]
119. Zheng, T.; Zhao, C.; Zhao, B.; Liu, H.; Wang, S.; Wang, L.; Liu, P. Impairment of the autophagy-lysosomal pathway and activation of pyroptosis in macular corneal dystrophy. *Cell Death Discov.* **2020**, *6*, 1–13. [CrossRef] [PubMed]
120. Leiphrakpam, P.D.; Patil, P.P.; Remmers, N.; Swanson, B.; Grandgenett, P.M.; Qiu, F.; Yu, F.; Radhakrishnan, P. Role of keratan sulfate expression in human pancreatic cancer malignancy. *Sci. Rep.* **2019**, *9*, 1–10. [CrossRef] [PubMed]
121. Hadley, J.A.; Horvat-Gordon, M.; Kim, W.K.; Praul, C.A.; Burns, D.; Leach, R.M., Jr. Bone sialoprotein keratan sulfate proteoglycan (BSP-KSPG) and FGF-23 are important physiolog-ical components of medullary bone. *Comp. Biochem. Physiol. Part A Mol. Integr. Physiol.* **2016**, *194*, 1–7. [CrossRef]
122. Melrose, J. Mucin-like glycopolymer gels in electrosensory tissues generate cues which direct electrolocation in am-phibians and neuronal activation in mammals. *Neural Regen. Res.* **2019**, *14*, 1191–1195. [CrossRef] t
123. Fu, L.; Sun, X.; He, W.; Cai, C.; Onishi, A.; Zhang, F.; Linhardt, R.J.; Liu, Z. Keratan sulfate glycosaminoglycan from chicken egg white. *Glycobiology* **2016**, *26*, 693–700. [CrossRef]
124. Restaino, O.F.; Finamore, R.; Diana, P.; Marseglia, M.; Vitiello, M.; Casillo, A.; Bedini, E.; Parrilli, M.; Corsaro, M.M.; Trifuoggi, M.; et al. A multi-analytical approach to better assess the keratan sulfate contamination in animal origin chondroitin sulfate. *Anal. Chim. Acta* **2017**, *958*, 59–70. [CrossRef]
125. Da Cunha, A.L.; de Oliveira, L.G.; Maia, L.F.; de Oliveira, L.F.C.; Michelacci, Y.M.; de Aguiar, J.A.K. Pharmaceutical grade chondroitin sulfate: Structural analysis and identification of contaminants in different commercial preparations. *Carbohydr. Polym.* **2015**, *134*, 300–308. [CrossRef]
126. Bottelli, S.; Grillo, G.; Barindelli, E.; Nencioni, A.; Di Maria, A.; Fossati, T. Validated high-performance anion-exchange chromatography with pulsed amperometric detection method for the determination of residual keratan sulfate and other glucosamine impurities in sodium chondroitin sulfate. *J. Chromatogr. A* **2017**, *1505*, 43–49. [CrossRef]

Article

Supramolecular Structuring of Hyaluronan-Lactose-Modified Chitosan Matrix: Towards High-Performance Biopolymers with Excellent Biodegradation

Riccardo Ladiè [1], Cesare Cosentino [1], Irene Tagliaro [2], Carlo Antonini [2], Giulio Bianchini [3] and Sabrina Bertini [1,*]

1. Istituto di Ricerche Chimiche e Biochimiche G. Ronzoni, Carbohydrate Science Department, 20133 Milan, Italy; ladie@ronzoni.it (R.L.); cosentino@ronzoni.it (C.C.)
2. Department of Materials Science, University of Milano-Bicocca, 20125 Milan, Italy; i.tagliaro@campus.unimib.it (I.T.); carlo.antonini@unimib.it (C.A.)
3. JoinTherapeutics Srl, 35122 Padova, Italy; giulio.bianchini@jointherapeutics.com
* Correspondence: bertini@ronzoni.it; Tel.: +39-02-70641627

Citation: Ladiè, R.; Cosentino, C.; Tagliaro, I.; Antonini, C.; Bianchini, G.; Bertini, S. Supramolecular Structuring of Hyaluronan-Lactose-Modified Chitosan Matrix: Towards High-Performance Biopolymers with Excellent Biodegradation. *Biomolecules* **2021**, *11*, 389. https://doi.org/10.3390/biom11030389

Academic Editor: Dragana Nikitovic

Received: 28 January 2021
Accepted: 26 February 2021
Published: 5 March 2021

Publisher's Note: MDPI stays neutral with regard to jurisdictional claims in published maps and institutional affiliations.

Copyright: © 2021 by the authors. Licensee MDPI, Basel, Switzerland. This article is an open access article distributed under the terms and conditions of the Creative Commons Attribution (CC BY) license (https://creativecommons.org/licenses/by/4.0/).

Abstract: Non-covalent interactions in supramolecular chemistry provide useful systems to understand biological processes, and self-assembly systems are suitable assets to build-up innovative products for biomedical applications. In this field, polyelectrolyte complexes are interesting, especially when polysaccharides are involved, due to their non-toxicity and bio-absorbability. In this work, we investigated a polyelectrolyte formed by hyaluronic acid (HA), a negatively charged linear polysaccharide, with Chitlac (Ch), a positively charged lactose-modified chitosan. The aim of the study was the investigation of a novel Ch–HA polyelectrolyte complex, to understand the interaction between the two polysaccharides and the stability towards enzymatic activity. By means of gel permeation chromatography–triple detector array (GPC–TDA), nuclear magnetic resonance (NMR), dynamic viscosity, Zeta Potential and scanning electron microscopy (SEM), the polyelectrolyte complex properties were identified and compared to individual polysaccharides. The complex showed monodisperse molecular weight distribution, high viscosity, negative charge, and could be degraded by specific enzymes, such as hyaluronidase and lysozyme. The results suggest a close interaction between the two polysaccharides in the complex, which could be considered a self-assembly system.

Keywords: chitosan; hyaluronic acid; lactose modified chitosan; NMR; molecular weight distribution; SEM; rheology; hyaluronidase

1. Introduction

Supramolecular chemistry relies on non-covalent interactions, like hydrogen bonds, hydrophobic and Van der Waals forces, and metal–ligand coordination. These interactions control many self-assembly processes, such as biological processes, and can be used to design innovative products for biomedical applications. Polysaccharides, due to their promising biomedical and biological applications, such as tissue engineering, biosensor and wound healing, are a particularly interesting class of molecules: intra- and inter-chain interactions, together with ion pairs, originate primary, secondary, tertiary, and quaternary structures, leading to supramolecular architectures [1,2]. Such architectures are suitable for a variety of applications, ranging from drug delivery to enhance the bioavailability of poorly soluble drugs [3]. Hydrophobized polysaccharides were synthetized to produce supramolecular structures in water, and their interactions with soluble proteins or other molecular assemblies, such as monolayers, black lipid membranes, liposomes and oil-in-water emulsion were studied [4].

Among the extensive number of polysaccharides available as candidates for biopolymer engineering, glycosaminoglycans (GAGs) are highly promising candidates [5]. GAGs are linear polysaccharides, which consist of repeating disaccharide units, usually include a uronic acid component (such as glucuronic acid), and a hexosamine component (such as N-acetyl-D-glucosamine).

In particular, hyaluronic acid (HA) is a linear polysaccharide with a poly-repeating disaccharide structure [(1→3)- β-D-GlcNAc-(1→4)- β-D-GlcA], which can be found ubiquitously in all vertebrate tissues extracellular matrix (ECM) [6,7]. Glucuronic acid residues contain a carboxyl group, which confers a negative charge on HA. Despite its relatively simple chemical composition, HA is involved in several biological functions, such as morphogenesis, tissue remodelling, inflammation, and tumours development and metastasis; in addition, HA contributes directly to the maintenance of tissue homeostasis and biomechanics [7]. Biocompatibility, biodegradability, high viscoelasticity, and immunoneutrality make HA an attractive polymer for biomedical and pharmaceutical applications. Frequently, HA acts as a space filler, applied to treat joint diseases such as in osteoarthritis and in eye surgery as replacement fluid, for drug delivery, and tissue engineering applications. HA is availed in viscosupplementation therapy, with the therapeutic goal to restore the viscoelasticity of synovial fluid. Indeed, HA is continuously secreted in the articular cavity and is one of the most common components of synovial liquid, giving rise to its characteristic viscoelastic properties. These rheological properties are essential for the lubrication and shock protection of healthy joints, allowing the protection of cartilage and soft tissues. In the event of traumatic and degenerative diseases, the amount of HA is lowered and synovial fluid experiences a viscosity drop, impairing joint functionality, and causing pain [8–10]. For applications, unmodified HA and chemically modified or cross-linked HA are used [11,12]; the derivatization of HA increases its mechanical properties and stability, which allows it to be used as a biomaterial [13–15].

Chitosan is also a linear polysaccharide, derived from deacetylation of chitin, the structural component of fungal cell walls, and exoskeleton of arthropods. It is composed of β-1→4 glucosamine units, with some residual interspersed N-acetyl-glucosamine residues; it is soluble in acidic solution, with a positive charge density, dependent on pH and percentage of deacetylation. Chitosan is well-known for its numerous and interesting biological properties—it is biocompatible, bioresorbable, and bioactive. Availability, safety for medical use, and biodegradability make chitosan very interesting for tissue engineering and biomaterials products. Chitosan can be further improved from the bioactive features and the physical–chemical behaviour. In particular, Chitlac (Ch) is a compound obtained by the modification of chitosan with lactitol moieties [16], in which the oligosaccharide pendant groups alter its solubility at physiological pH. The physico-chemical and biological properties of Ch are already reported in the literature [17].

Both HA and Ch are hydrolyzed by enzymes; linear HA can be easily degraded by enzymes, such as bovine testicular hyaluronidase, an endo-β-N-D-acetylhexosaminidase that hydrolyzes HA at the β (1→4)-N-acetylglucosaminide bonds [18]. Hyaluronidases simultaneously display both hydrolytic and transglycosylation activities—the optimal conditions for the hydrolysis of HA by hyaluronidase are pH 4.0 and the presence of NaCl, whereas for transglycosylation they are pH 7.0 and the absence of NaCl [19]. In the case of Ch, there is no specific enzyme; nonetheless, previous studies report that degradation can be performed by lysozyme, which is, for example, present at concentrations ranging from 4 to 13 mg/L in serum and in tears (450 to 1230 mg/L) [20].

A polyelectrolyte complex is formed when polymers with opposite charges are combined in solution. A strong complex is obtained if the anions and cations in the polymers contain strong acids and bases, or if the polyions attain their fully ionized forms and vice versa. The polyelectrolyte complexes are recently gaining attention as supramolecular carriers for controlled release of drugs and proteins [21], which is widely used in many applications such as membranes, medical prosthetics, environmental sensors, and protein separation systems [22,23]. These complexes prepared from natural polymers, such as polysaccharides, have the additional advantage of being non-toxic and bioabsorbable [24]. For example, biomaterials, constituted of a Chondroitin sulphate-Chitosan complex, have interesting biological properties, such as wound-healing acceleration and cellular assistance for skin and cartilage recovery [25]. Some applications of Chitosan and HA complexes include ophthalmic surgery, arthritis treatment, scaffolds for wound healing, tissue engineering, and the use as a component in implant materials. In particular, different authors showed that the potential of Chitosan/HA complexes coacervates in the biomaterials field [26–31]. A recent study about self-supporting multi-layered film containing a Chitosan and HA polyelectrolyte complex, showed high selectivity during the separation of water from the ethanol–water mixtures for membrane technology applications [32].

Moreover, recent studies highlighted promising results for biological and medical applications of the HA/Ch complex, such as in the treatment of osteoarthritis [33,34]; in vivo osteoarthritis treatments with a viscosupplementation containing Ch and HA showed a decrease in morphological and histolopathological cartilage damage and synovial membrane inflammation, in comparison to the treatment performed with HA alone [33]. A combination of Ch and HA-attenuated macrophage-induced inflammation, inhibited metalloproteinases expression, and exhibited anti-oxidative effects, providing interesting insights into the biological effects of mixture of these polysaccharides for the development of osteoarthritis treatments [34]. However, the literature does not provide exhaustive chemical-physical characterization of HA/Ch for a better understanding of the observed biological performances.

As such, in this work we investigated the properties of the polyelectrolyte complex to understand the interaction between HA and Ch. Specifically, the molecular weight of the supramolecular product through size-exclusion chromatography with a triple detector array (HP-SEC-TDA) was determined. The viscosity property and the molecular mobility was elucidated through dynamic viscosity and NMR diffusion ordered spectroscopy (DOSY) experiments, respectively. Finally, the stability toward enzymatic actions was evaluated.

2. Materials and Methods

2.1. Materials

Ch was provided by Join Therapeutics S.r.l. (Padova, Italy). Sodium Hyaluronate were purchased from HTL Biotechnology. Sodium azide, sodium nitrate, sodium dihydrogen phosphate monohydrate, sodium hydrogen phosphate dihydrate, trimethylsilyl-3-propionic acid (TSP), Hyaluronidase from bovine testes (400–1000 u/mg) and Lysozyme from hen egg white (93,300 u/mg) were purchased from SigmaAldrich (Milan, Italy). Deionized water (conductivity less than 0.1 µS) was prepared with an inverse osmosis system (Culligan, Milan, Italy). PolyCAL TM Pullulan std-57k (Malvern Instruments LtD, Malvern, United Kingdom). The reagent grades were $\geq$ 98%.

2.2. Molecular Weight Distribution by HP-SEC-TDA

HP-SEC-TDA was used extensively to obtain molecular weight distribution of HA, Ch, and the complex. This method does not require any chromatographic calibration and is considered to be suitable to analyze polysaccharides. Measurements for molecular weight distribution were performed on a Viscotek 305 HPLC system (Malvern Instruments LtD, Malvern, UK). The array exploits simultaneous action of refraction index detector (RI), viscometer, and Right Angle Laser Light-scattering (RALS) detector, using a method adapted from Bertini et al. [35]. To prepare the solutions for HP-SEC-TDA measurements, 10–20 mg was dissolved in a mobile phase volume, to obtain a sample concentration of ~3 mg/mL. The Ch solution was stirred for 3 h and then diluted in a mobile phase to a concentration of ~1 mg/mL. The HA solution was stirred overnight and then diluted in the mobile phase to a concentration of ~0.5 mg/mL, respectively. For the HA/Ch complex, 75 mg of Ch were dissolved in 10 mL of phosphate-buffer saline solution pH 6.9 (PBS), and stirred at room temperature to obtain a completely solubilized solution. Afterwards, 125 mg of HA were added to the solution and the complex was stirred for 12 h. Prior to the analysis of the solution, the HA/Ch complex was diluted in a mobile phase to reach 1 mg/mL sample concentration. The analyses were performed at 40 °C, using 2 x TSKPWXL columns in series (Tosoh Bioscience, 7 mm 7.8 × 30 cm). A total of 0.1 M $NaNO_3$ for HA, 0.2 M $NaNO_3$ for Ch, and an HA/Ch complex, both containing 0.05% NaN_3, prefiltered using 0.22 mm filter, were used at a flow rate of 0.6 mL/min. Chromatographic profiles were elaborated using the OmniSEC software version 4.6.2. RI increments, referred to as dn/dc, were determined to enable conversion of RI values into a concentration for Ch and HA/Ch [36]. dn/dc values equal to 0.129 mL/g and 0.119 mL/g were observed for Ch and HA/Ch, respectively. For HA, the dn/dc value of 0.155 was taken from the literature [37].

2.3. NMR Analysis

The Ch and HA spectra were obtained with a Bruker AVANCE IIIHD 500 MHz spectrometer (Bruker, Karlsruhe, Germany) equipped with a 5 mm BBO probe, at 343 K. Spectra were processed with BrukerTopspin software version 4.0.6.

About 30 mg of sample were dissolved in 3 mL of deuterium oxide (D2O) with 0.002 %TSP and subsequently 0.6 mL were transferred into a 5 mm NMR tube. The HA sample was stirred overnight to ensure a complete solubilization before transfer in an NMR tube (Bruker, Karlsruhe, Germany) for the analysis. ^1H NMR spectra were acquired with pre-saturation of residual HDO, using 64 scans, 12 s relaxation delay, and a number of time-domain points equal to 32k. For the Ch sample, the ^1H-^{13}C HSQC spectra were acquired using 16 scans, 5s relaxation delay, $^1J_{C-H}$ 150 Hz.

The HA/Ch complex spectra were acquired with a Bruker AVANCE NEO 500MHz spectrometer (Bruker, Karlsruhe, Germany) equipped with a 5 mm TCI cryogenic probe at 303K for ^1H NMR and Diffusion Order NMR Spectroscopy (DOSY). Spectra were processed with Bruker Topspin software version 4.0.6. Ch and HA were dissolved separately in a PBS deuterium volume (mL), which allowed us to reach 7.5 mg/mL and 12.5 mg/mL, respectively. Then, 500 µL of both solutions were taken and mixed, it was vortexed to obtain a homogeneous solution.

Spectra were acquired with pre-saturation of residual HOD using 64 scans, 5 s relaxation delay, and a number of time-domain points equal to 65 k.

DOSY experiments were acquired using the 2D-stimulated echo sequence with bipolar gradient pulse for diffusion. The gradient pulse (δ) and the diffusion time (Δ) were set to 5 ms and 300 ms, respectively. The 2D DOSY experiments were run with gradients varied linearly from 5 to 95% in 32 steps, with 16 scans per step. The diffusion coefficients D were extracted using the Bruker Dynamics Center 2.5 (Bruker, Karlsruhe, Germany).

2.4. Zeta Potential Analysis

Zeta Potential (Zp) of HA, Ch, and the HA/Ch complex were measured using the Zetasizer Nano ZS (Malvern, Worcestershire, United Kingdom), with a fixed scattering angle of 173° and a 633-nm helium–neon laser. Data were analyzed using the Zetasizer software version 7.11 (Malvern Instruments LtD, Malvern, UK). For the analysis, Ch and HA solutions were diluted in PBS to reach the concentration of 7.5 and 12.5 mg/mL, respectively. In the case of the HA/Ch complex, 75 mg of Ch were dissolved in 10 mL of PBS and stirred at room temperature, to obtain a completely solubilized solution. Then, about 125 mg of HA were added to the solution and the complex was stirred for 12 h. Disposable plastic cuvettes DTS1070 were used for the Zp analysis.

2.5. Rheological Properties

The rheological properties of HA, Ch, and the complex were studied using a Modular Compact Rheometer MCR 92 (Anton Paar GmbH, Graz, Austria), equipped with a 50-mm-diameter cone–plate geometry, with a cone angle of 1°. For all tests, the temperature and the gap between the plates were kept constant 20 °C and 0.98 mm, respectively.

Viscosity measurements were performed in rotation mode, they were investigated in the range of 0.1–100 s^{-1} and ten points per decade were acquired.

The sample viscoelastic behaviour was investigated in the oscillation mode, to determine the storage modulus $G'(\omega)$ and the loss modulus $G''(\omega)$. First, preliminary tests were conducted to determine the upper amplitude limit of the linear viscoelastic region (LVE), testing the samples over an extended strain field (0.01–100%). Second, after the LVE was determined (2% for HA and HA/Ch and 20% for Ch), the samples were tested by performing a frequency sweep test over the 0.628 rad/s–628 rad/s (i.e., 0.1–100 Hz) frequencies, at a constant strain. Data were elaborated with RheoCompass™ software.

For the analysis, the same solutions as for Zp were tested. Before the analysis, solutions were sonicated to reduce air bubbles.

2.6. SEM Analysis

Scanning Electron Microscopy (SEM) analysis was performed with Zeiss Gemini 500 Field-Emission SEM (Carl Zeiss Microscopy, Oberkochen, Germany) at 5 kV. Polysaccharide samples of 50 mg were freeze-dried in a 100 mL Falcon tube. The dried samples where broken with tweezers, deposited on SEM adhesive tape and sputtered with gold to enhance conductivity.

2.7. Enzymatic Degradation Procedure

CH (solution A): 75 mg of sample were dissolved in 10 mL of PBS and stirred to obtain a completely solubilized solution. HA (solution B): 62 mg of HA sample were dissolved in 5 mL of PBS and stirred for 16 h. HA/Ch complex (solution C): 62 mg HA were dissolved in 5 mL of solution A. The final concentration of the polysaccharide (HA/Ch) was about 20 mg/mL (7.5 mg/mL CH + 12.5 mg/mL HA). For enzyme solution preparations, about 15 mg of hyaluronidase were dissolved in about 1.5 mL of deionized water and stirred for 1 h (10 mg/mL). Lysozyme: 75 mg of lysozyme was dissolved in about 1 mL of deionized H_2O and stirred for 1 h (75 mg/mL). The solutions were stirred at 38 °C in an oil bath for about 2 h, before adding the enzyme solution. Different aliquots of solutions were collected at different times (15 min, 30 min, 1 h, 2 h, 4 h, 6 h, 24 h) and heated at 100 °C for 5 min, using a thermo-shaker, (BioSan, Riga, Latvia) to denature the enzyme. The solutions were diluted to different concentrations for HP-SEC-TDA analysis. The Ch solutions were diluted to 2 mg/mL with the mobile phase, whereas the solutions containing HA or HA/Ch were diluted to 0.5 mg/mL (calculated on the total amount of polysaccharide in solution) with the mobile phase. Before the HP-SEC-TDA analysis, all solutions were filtrated to remove the precipitated enzyme (LLG-Syringe filter, CA pore size 0.20 µm, Ø 13 mm).

3. Results
3.1. Molecular Weight Distribution

Chromatographic profiles of HA and Ch are shown in Figure 1a,b, respectively. The samples had an elution time between 10 and 15 mL, with a large a-symmetric bell-shape chromatographic peak, caused by a high polydisperse index.

Values of weight-average molecular weight (Mw), number-average molecular weight (Mn), and molecular-weight dispersity (Đ; Mw/Mn) are reported in Table 1. In addition, values of the intrinsic viscosity value (η), hydrodynamic radius (Rh), and Recovery %, derived from TDA, are also reported. To estimate the Recovery %, the OmniSEC software computes actual concentration in each chromatographic slice, based on the *dn/dc* value, and a detector constant for the RI. The values of *a* and *log k*, corresponding to the slope and intercept constants of the Mark-Houwink curve, respectively, are also reported. The results refer to the mean values of duplicate injections.

For the HA samples, a recovery value of 90% indicated that no material remained adsorbed in the columns, considering that the water content in the sample was about 10% *w/w*; for Ch, the values were close to 100%, due to the low water content in the sample. The Đ value for HA was lower than that for Ch, indicating low molecular weight dispersion. However, the values were relatively high, indicating the presence of chains of different length.

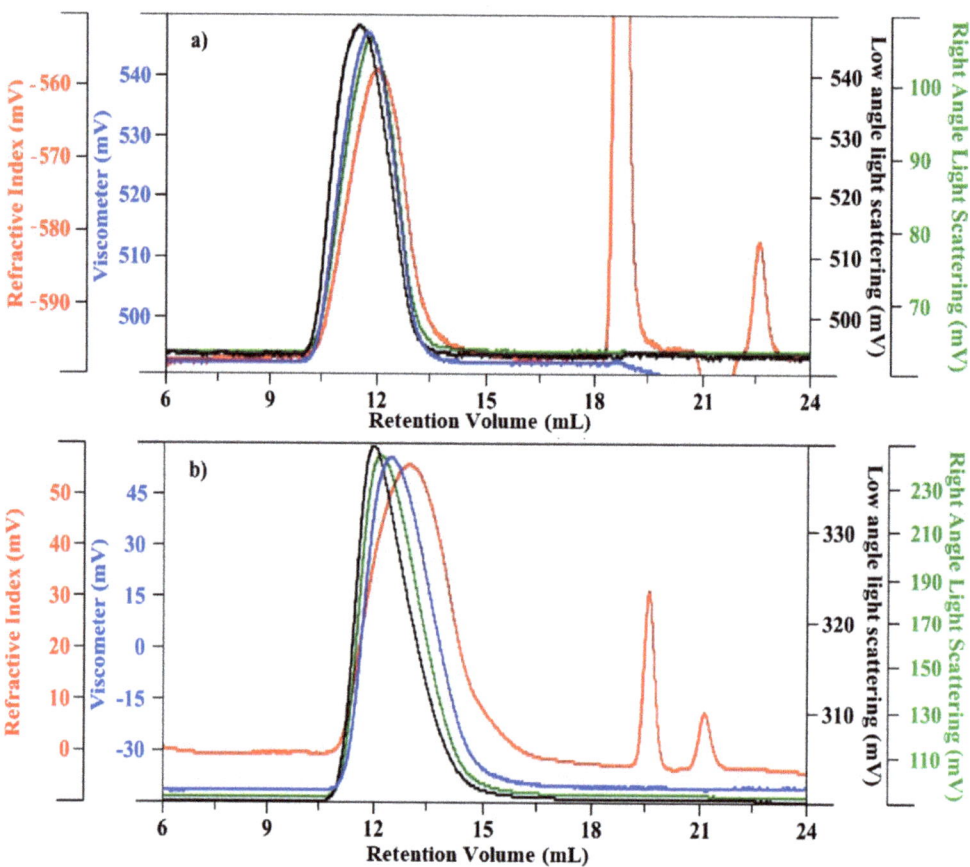

Figure 1. Chromatographic profile (red—refractive index; black—low laser light scattering; green—right angle light scattering; blue—viscometer). (**a**) HA, and (**b**) Ch.

Table 1. Main results of the HP-SEC-TDA analysis for the HA and Ch samples.

Sample	Mw (kDa)	Mn (kDa)	Đ	a	log K	[η] (dL/g)	Rh (nm)	Recovery%
HA	992	660	1.5	0.69	−2.9	17.4	63	90
Ch	1020	554	1.8	0.68	−3.6	3.0	34	96

The HA/Ch complex (0.75:1.25 w/w) molecular weight was also determined by the same technique—results are visualized in Figure 2 As expected, due to very similar molecular weight, the two polysaccharides were eluted together, so it is important to evaluate the dn/dc of the complex. From analysis on several dilutions of a HA/Ch complex, the refractive index increment (dn/dc) was determined as 0.118, through a linear regression.

Using 0.118 as dn/dc, the molecular weight analysis results are reported in Table 2. Molecular weight, dispersity, and hydrodynamic radius of the complex (Table 2) remained quite similar to those of the original polymers (Table 1). Generally, through gel permeation chromatography analysis, the average molecular weight of polymers increased due to cross-linked reactions or derivatizations. In the HA/Ch complex, only weak interactions between the two polysaccharides (hydrogen bonds and Van der Waals forces) are present, which are influenced by high salt concentration of mobile phase.

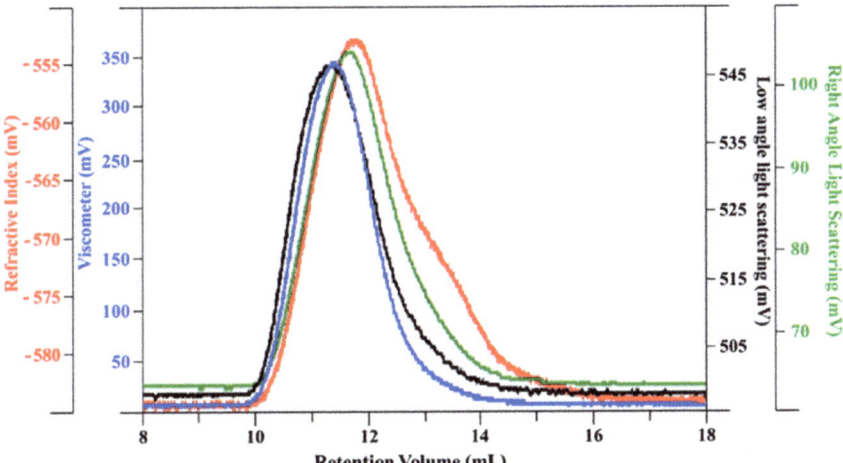

Figure 2. Chromatographic profile (red—refractive index; black—low laser light scattering; green—right angle light scattering; blue—viscometer) of the HA/Ch complex.

Table 2. HP-SEC-TDA HA/Ch complex results.

Sample	Mw (kDa)	Mn (kDa)	Đ	a	log K	[η] (dL/g)	Rh (nm)
HA/Ch complex	1099	619	1.8	1.28	−6.8	7.9	49

3.2. NMR Characterization

Most of the properties of chitosan and its derivates, such as solubility and biodegradability, depend on the proportion between acetylated and non-acetylated glucosamine units, corresponding to degree of acetylation (DA), and eventually substitution degree (DS). NMR spectroscopy is one of the most accurate methods for determining the degree of acetylation for chitosan [38] and DS for Ch compounds [39].

Figure 3 shows the ^{1}H NMR spectrum of a Ch sample. The characteristic peaks are located at 3.32 and 3.40 ppm (black arrows Figure 3)—such signals represent the H2 of substituted deacetylate units (H2-NHR) and the hydrogens of the CH$_2$ group involved in the amide bond (-NH-CH$_2$-), respectively. The H2 of unsubstituted deacetylate units (H2-GlcN) appears at about 3.21 ppm. The peaks in the anomeric region, A (5.00 ppm) and B (4.88 ppm), in Figure 3 are attributed to H1 of the substituted and unsubstituted deacetylate units, respectively. These peaks and the CH$_3$ signal of the acetyl group at 2.06 ppm (C, Figure 3) were used for the DA and DS calculation. In the anomeric region, the two sharp peaks were attributed to the anomeric protons of β-galactose side chains, at 4.57 ppm [17].

By assigning 300 to the integral of the signal related to the methyl group of acetyl (about 2.06 ppm), the percent N-acetylation (%DA) and substitution degree (%DS) were evaluated, using Equations (1) and (2).

$$\%DA = \frac{\frac{C}{3}}{A + B + \frac{C}{3}} \cdot 100 \tag{1}$$

$$\%DS = \frac{A}{A + B} \cdot (100 - \%DA) \tag{2}$$

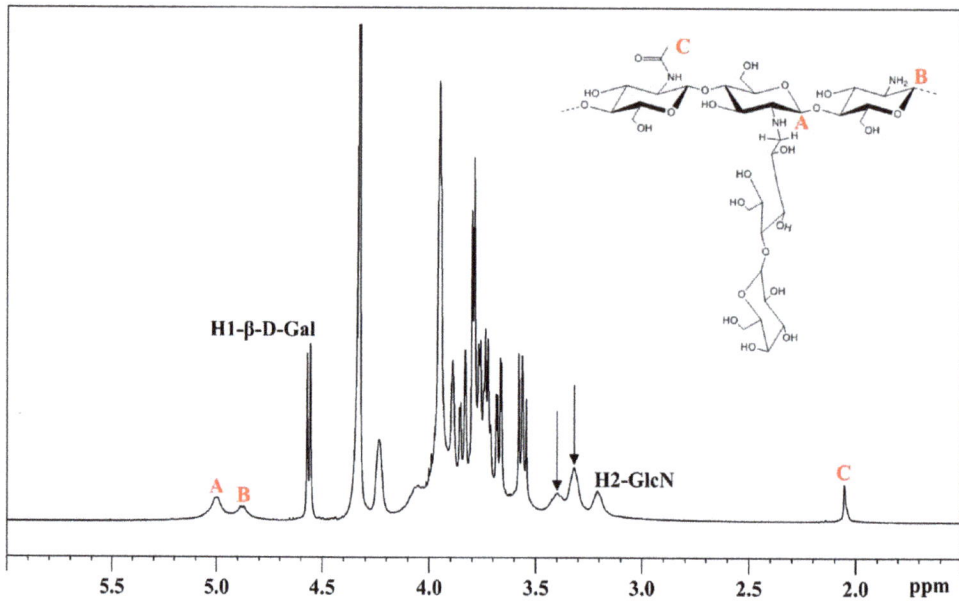

Figure 3. ^{1}H NMR spectrum of Ch in D_2O at 343 K.

From the integral values it was seen that in the Ch sample, *DA* and *DS* were 7.5% and 59%, respectively.

To confirm the peak attributions and to verify that the integrated signals involved in quantification corresponding to the unique function units, the ^{1}H-^{13}C HSQC spectrum was acquired (Figure 4). The 2D spectrum confirmed that the anomeric and methyl peaks in the ^{1}H spectrum corresponded to one peak in the ^{13}C dimension, and that ^{13}C chemical shifts were consistent with the attributed signals of product [17].

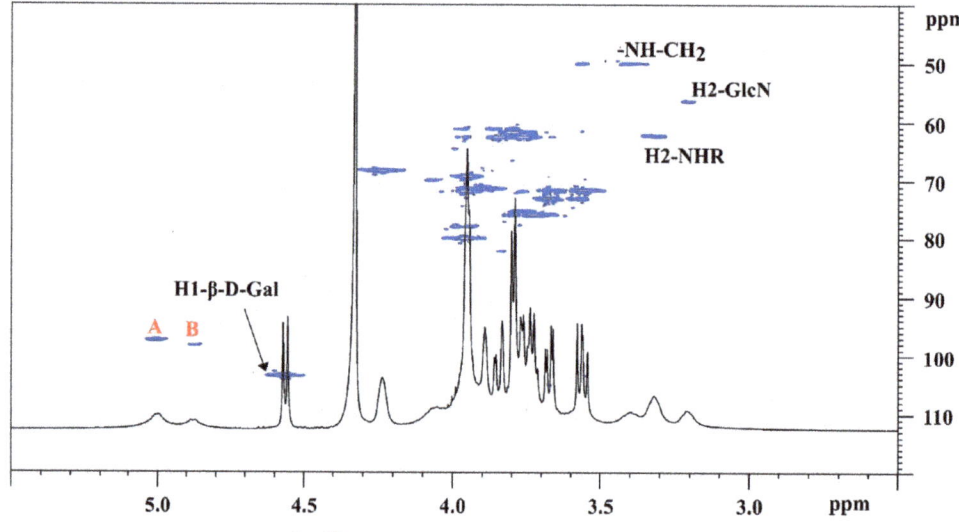

Figure 4. ^{1}H-^{13}C HSQC NMR spectrum of CH in D_2O at 343 K.

Pulsed field gradient (PFG) diffusion ordered spectroscopy (DOSY) is the translational diffusion of dissolved molecules. In addition to the overall molecular size and shape, the diffusion coefficient magnitude provides direct information on molecular dynamics, including intermolecular interactions [40], aggregation, conformational changes [41,42], and viscosity [43]. DOSY processing is a particularly suitable technique for complex samples, since it provides a direct correlation of translational diffusion to the chemical shift in the second dimension. Therefore, a prior separation of complex components is not required [44,45]. A DOSY experiment is represented in a 2D spectrum, with chemical shift along one axis and the diffusion coefficients along the other [46,47].

In this study, DOSY was used to evaluate the interaction between HA and Ch in the complex. The outcomes of the DOSY analysis for all samples are visualized in a 2D map in Figure 5a.

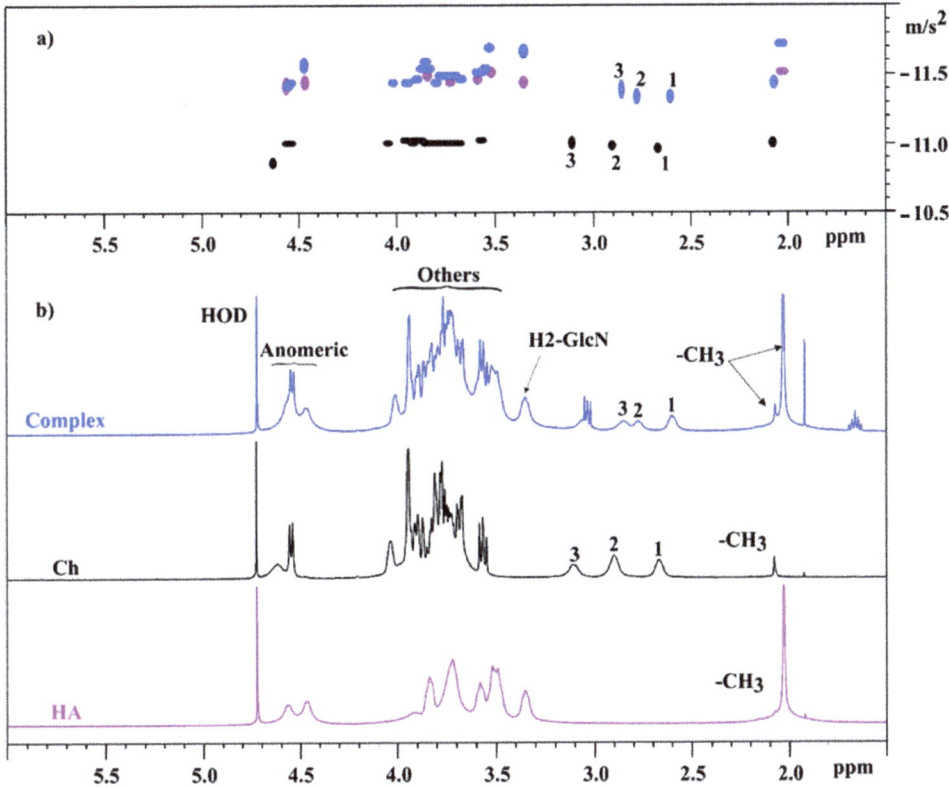

Figure 5. (a) DOSY and (b) ^{1}H spectra of HA (purple), Ch (black), and complex (blue).

For the HA and Ch samples, there is only one population with the same diffusion coefficient, due to a unique molecular weight. For the HA/Ch complex, the results showed different diffusive fronts, due to different hydrodynamic radius of (5a) the HA resonances, and (5b) the resonances due to Ch moiety. The different chemical shift for 1, 2, and 3 signals were related to different small pH of solutions in the presence of HA. Usually, chemical shift of main chain protons, involved in the amino bond (H2) are highly sensitive to the protonation equilibrium [17]. The D values, shown in Table 3, were determined for the peaks corresponding to the components of the complex.

Table 3. DOSY results of HA, Ch, and HA/Ch complex (D = H$_2$O 2.25 × 10^{-9} m^2/s; D = TSP: 6.50 × 10^{-10} m^2/s).

Peak	HA		Ch		HA/Ch Complex			
	^{1}H ppm	D (m^2/s)	^{1}H ppm	D (m^2/s)	^{1}H ppm		D (m^2/s)	
CH$_3$	2.03	1.48 × 10^{-12}	2.07	6.89 × 10^{-12}	2.03	2.07	6.35 × 10^{-13}	1.86 × 10^{-12}
1	-	-	2.73	7.05 × 10^{-12}		2.60		2.64 × 10^{-12}
2	-	-	2.96	7.11 × 10^{-12}		2.77		2.89 × 10^{-12}
3	-	-	3.14	7.53 × 10^{-12}		2.85		3.05 × 10^{-12}
H2-GlcA	3.36	1.61 × 10^{-12}	-	-		3.35		7.51 × 10^{-13}
Anomeric	4.40–4.60	1.72 × 10^{-12}	4.50–4.55	6.60 × 10^{-12}	4.40–4.60		1.86 × 10^{-12}	
others	3.51–3.84	1.51 × 10^{-12}	3.54–4.04	6.65 × 10^{-12}	3.54–4.04		1.60 × 10^{-12}	

The average D value of HA was lower than Ch, despite having similar Mw values. In this case, DOSY was particularly successful in distinguishing among different molar diffusivities, due to the different hydrodynamic radii, confirming the HP-SEC-TDA results. In the HA/Ch complex, the D values of the HA component remained almost constant, meanwhile the Ch diffusion coefficients increased, with values analogous to HA.

3.3. Rheological Properties

Rheology as reported by Ambrosio et al. [48], is a useful tool to explore the relationships between the mechanical behavior and chemical properties of HA, or other biopolymer solutions. In our study, the flow behavior and viscoelastic measurement were evaluated to understand the difference in the rheological properties of HA and Ch solutions with their complex.

The variation of the viscosity as a function of the shear rate were acquired; in Figure 6a the viscosity curves, with a shear rate (γ) ranging from 0.1 s^{-1} to 100 s^{-1}, are reported.

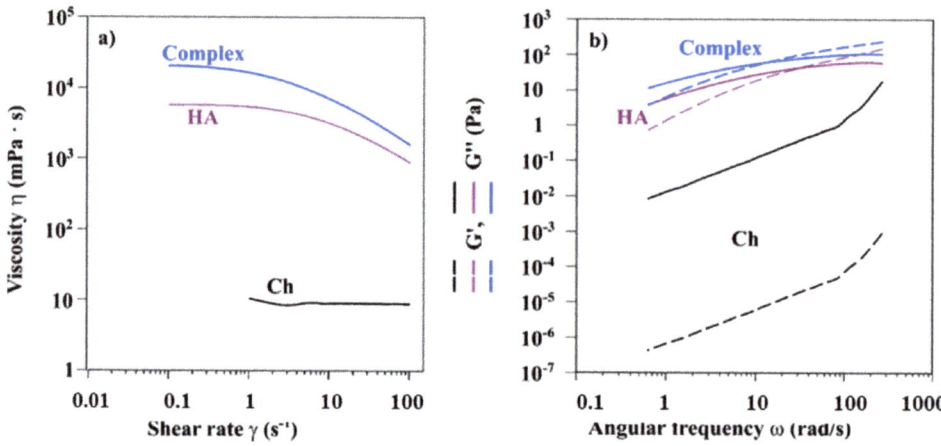

Figure 6. (a) Viscosity curves of the Ch, HA, and HA/Ch complex. Due to low viscosity, data for Ch were only acquired in the range (1–100) s^{-1} (b). Frequency sweep test of the Ch (γ = 20%), HA, and the HA/Ch complex (γ = 2%).

HA and Ch (black and purple curves in Figure 6a, respectively) showed different behaviours. Ch viscosity remained constant at a value of 8.8 mPa s over the entire tested shear rate range, thus showing ideal viscous flow behaviour. Such behaviour was mainly related to the sample concentration and substitution degree of chitosan—according to previous results [49]. The pseudoplasticity was inversely proportional to DS and directly correlated to the concentration of sample. For HA, viscosity was shear-dependent, viscosity decreased as γ increased, in agreement with a pseudoplastic behaviour [48]. For the complex, shear-thinning behaviour was also observed. Note that the viscosity values were higher than that for pure HA.

The rheological tests were also performed in the oscillation mode, to evaluate both the storage modulus (G′) and the loss modulus (G″). The strain sweep test (data not shown) was initially performed to evaluate the LVE zone, where the intrinsic sample structural properties are independent of the applied strain:2% (HA and Complex) and 20% (Ch) strain value was selected for subsequent frequency sweeps tests. In Figure 6b, values of G′ and G″ as function of the angular frequency for the Ch, HA, and the complex are reported. In the investigated frequency range (from 0.628 rad/s to 628 rad/s), the mechanical spectrum of Ch showed that G″ was greater than G′, with both moduli strongly dependent on the frequency, as typically a liquid-like (viscous) behavior dominates over the solid-like (elastic) character. Differently, HA and HA/Ch presented a "weak gel" or viscoelastic behavior with G′ and G″, which became less dependent on the frequency and the crossover point, when G′ = G″, was observed. This suggests that at high frequencies, when G′ > G″, the material show a predominant solid-like behavior.

The angular frequency values, ω_c, and the corresponding G′ = G″ values to the crossover point data are reported in Table 4.

Table 4. Rheological measurements—evaluation of G′ and G″ crossover.

Sample	Angular Frequency ω_c (rad/s)	Crossover Point G′ = G″ (Pa)
Ch	n.d. [1]	n.d. [1]
HA	30	44
HA/Ch	13	63

[1] value not detected in the frequency range investigated.

3.4. Zeta Potential

Zp, a parameter typically obtained by model-dependent transformation of the measured electrophoretic mobility, is a parameter used to estimate the magnitude of the electrostatic repulsion or attraction between particles. By measuring the surface charge, the stability of nanosuspensions, particles, and polymers in solution can be determined [50]. A large positive or negative Zp value, with values >+25 mV or <−25 mV, indicate good stability due to electrostatic repulsion of individual particles; a lowZp value can result in particle aggregation and flocculation due to the van der Waals attractive forces [51–53]. The Zp is frequently used to understand polysaccharide–protein complexation [54].

In this study, the Zp values of solutions of the Ch, HA, and HA/Ch complex were determined; the results are reported in Table 5. The Ch and HA samples were studied at concentrations of 7.5 and 12.5 mg/mL, respectively, which reflect the concentrations of the individual samples in the complex.

Table 5. Zp values for HA, Ch, and complex.

Sample	Concentration (mg/mL)	Zp (mV)
HA	12.5	−33.74
Ch	7.5	0.03
HA/Ch complex	20.0	−23.28

HA showed a negative Zp value, which was consistent with its anionic nature to the presence of carboxylic groups. Low zeta potential values provided an electrostatic repulsion that led to the formation of suitable stability in solution. Else, the Zp value of Ch was zero, indicating a slightly positive surface charge, close to zero, in agreement with the chemical structure. When the two polysaccharides were mixed, the Zp of the complex was negative and homogeneous. The value of about −23 mV meant that the complex in solution was stable without any macro-aggregation phenomena that could induce precipitation.

3.5. SEM

SEM analysis was performed to assess the morphological features of the HA/Ch complex [55], after a freeze-drying process. In the first step of freeze-drying, the water-based solutions (or suspensions) were first frozen—ice nucleated and concentrated the solute (or suspended matter) in the regions between the growing crystals. In the second step, ice was removed via direct sublimation at low pressure, avoiding the intermediate liquid phase, and leaving a porous morphology. The process is widely used to create porous materials using biopolymer [56–58]. The final morphology depends on the specificmaterial, and is controlled by many parameters of the freezing process [59], including sample size and shape, cooling rate, cooling temperature, process time, etc.

Figure 7 illustrates the specific characteristics of the HA/Ch complex that were compared to the single components of Ch and HA.

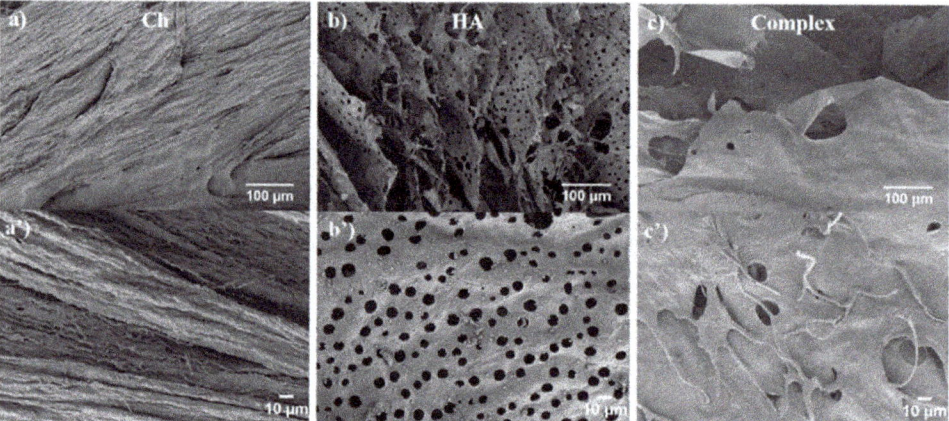

Figure 7. SEM analysis of (**a**) Ch (magnification 150×); (**a'**) Ch (magnification 500×); (**b**) HA (magnification 150×); (**b'**) HA (magnification 500×); (**c**) Complex (magnification 150×); and (**c'**) Complex (magnification 500×).

Ch (Figure 7a,a') showed a compact structure with a fibrous appearance, causing a rough texture on the Ch surface. HA presents even more clearly a lamellar structure (Figure 7b,b'), which is characteristic of an ice-templated process—the structure was oriented parallel to the temperature gradient (i.e., from the skin to the sample centre), along which the ice crystal grew. HA walls were characterized by a pore regular pattern, with a characteristic diameter of ~10 μm. Such pores might also be related to ice-templating, with holes forming due to the secondary crystal growth, perpendicular to the primary crystal growth direction.

The HA/Ch composite (Figure 7c,c') showed a compact and smooth surface, where fibers were visible on the top of the surface. The material was made of a lamellar structure with the holes between 10 and 20 μm as diameter. Similar to Zhang et al. [60], who investigated the structure of porous chitosan/HA/sodium glycerophosphate hydrogel systems, the material appeared homogeneous overall, and phase separation at the microscale was not observed. As such, we could confirm good compatibility between Ch and HA in the complex.

3.6. Enzymatic Degradation

The HA/Ch enzymatic degradation with lysozyme and hyaluronidase was studied. Such tests provide information on the polysaccharide stability in a physiological environment. The samples incubated at different times were analyzed. The Mw values vs. time, determined by chromatographic elaboration, are shown in Figure 8.

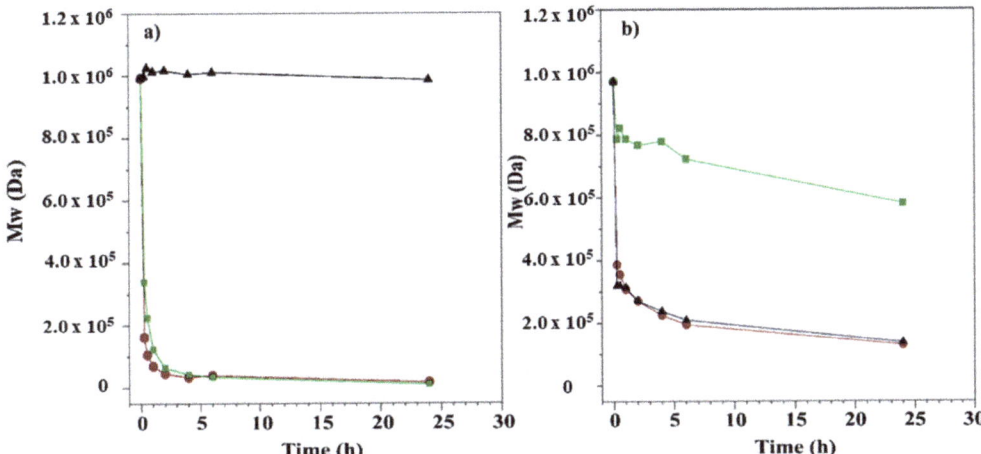

Figure 8. (a) HA and (b) Ch kinetic degradation with HAse (green curves), Lys (black curves), and a mix of enzymes (red curves) over 24 h.

The HA treatment (Figure 8a, green line) resulted in a substantial reduction of Mw indicating depolymerization, and such depolymerization took place very quickly during the initial 3 h, this indicated that the enzymatic hydrolysis was an endo-action. After 3 h, degradation slowed down, probably due to the inhibition of the enzyme activity by the end products, and the reaction was completed after 24 h, at the end of which a reduction of the Mw of 98% was observed (from 992 kDa to 14 kDa). In the case of Lys (Figure 8a, black line), the enzyme did not have any degradation effect, and Mw remained constant for the incubation time (24 h). As expected, the presence of Lys in complex with HAse (Figure 8a, red line) had a mild effect on the HA hydrolysis, showing a slightly different initial degradation rate, but the end point was the same.

For Ch, in the enzymatic degradation with Lys (Figure 8b, black line), the Mw decreased slowly and, after 24 h of incubation, the value was reduced by 86% (from 970 kDa to 137 kDa); with HAse (Figure 8b, green line) a small reduction of Mw was observed, at about 20%, which was in agreement with the lack of a specific substrate for this enzyme. The decrease of Mw was the same with the complex of enzymes than with Lys alone—as expected, the "driving force" for the degradation of Ch was Lys but the presence of hyaluronidase seemed to induce an effect on the first part of hydrolysis process.

To study how the interactions between HA and Ch could affect the enzymatic activity, the complex was hydrolyzed. The results with HAse and Lys individually and finally with enzymatic complex are shown in Table 6. All chromatograms of solutions until 6 h of incubation present only one peak, derived from HA and Ch, due to the very similar molecular weight. Consequently, the Mw and [η] values are related to the complex, imposing the dn/dc of 0.118 (see HP-SEC-TDA results section).

After 6 h, the Mw was reduced by about 70% with HAse, and 20% and 92% with Lys and with HAse + Lys, respectively. Therefore, the enzyme activity was not affected by the combination of the two polysaccharides.

After 24 h of incubation, the chromatograms present two distinct peaks (Figure 9), one related to the HA component and the other one attributed to Ch, so Mw and [η] for each polysaccharide could be determined, using the specific dn/dc values. The results are presented in Table 7.

Table 6. HP-SEC-TDA results of the HA/Ch complex degradation.

	HAse		Lys		HAse + Lys	
Time (h)	Mw (kDa)	[η] dL/g	Mw (kDa)	[η] dL/g	Mw (kDa)	[η] dL/g
0	1099	10.0	1099	10.0	1099	10.0
0.25	406	2.0	902	7.3	192	1.6
0.30	377	1.7	875	7.3	167	1.2
1	370	1.5	884	7.4	150	1.0
2	352	1.4	904	7.6	128	0.9
4	345	1.3	878	7.4	104	0.7
6	331	1.3	884	7.7	95	0.6

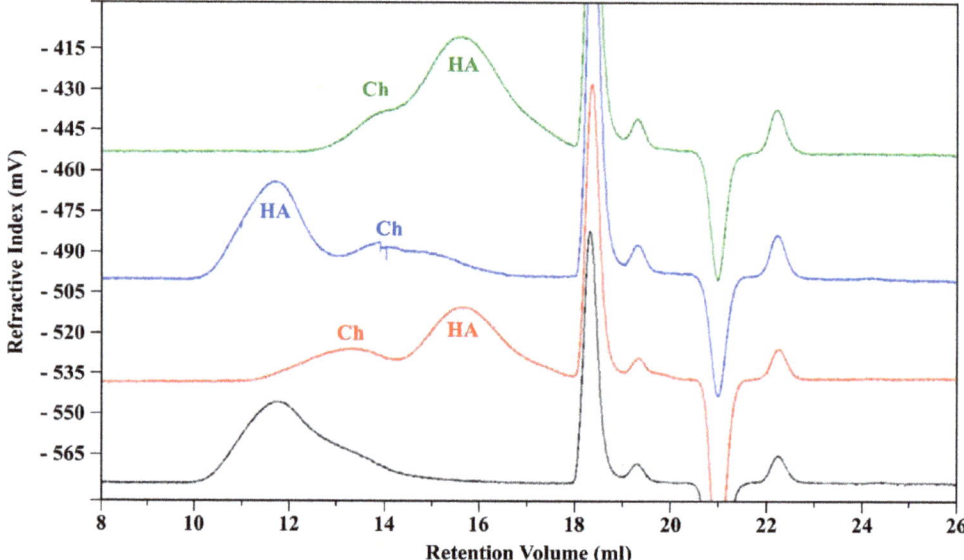

Figure 9. RI signal from the HP-SEC-TDA analysis of the HA/Ch complex at t = 0 (black) and after degradation t = 24 h with HAse (red), Lys (blue), and a mix of enzymes (green).

Table 7. HP-SEC-TDA results of the HA/Ch complex degradation after 24 h.

		HAse		Lys		HAse + Lys	
		Mw (kDa)	[h] dL/g	Mw (kDa)	[h] dL/g	Mw (kDa)	[η] dL/g
HA		16	0.4	955	15	19	0.3
Ch		620	2.3	155	1.6	220	1.1

The Mw values were very similar to the results obtained for the solution aliquots containing only HA or Ch, at 24 h, suggesting that the co-presence of both polysaccharides did not affect the enzymatic action.

4. Discussion

The polyelectrolyte complex in the aqueous medium interacts and self-assembles by electrostatic interactions. Therefore, the development of new stable formulations, constituting polyelectrolyte, requires an in-depth characterisation of the interactions to obtain biodegradable and biocompatible products with unique properties, without losing the polymers inherent characteristics. In the present work, efficacies analytical approaches were set up to characterize the HA/Ch complex and to evaluate its stability toward enzymes.

The interaction between HA and Ch was affected by the structural peculiarity of each polysaccharide. With regards to HA, the chemical structure was well-defined, meanwhile for Ch, the DS and DA were variable. Generally, the proportion of glucosamine residues in chitosan had a significant influence on their various properties. In particular, the distribution of N-acetyl groups along the biopolymer chain might influence the solubility of the polymer and the inter-chain interactions, due to the hydrogen bonds and the hydrophobic character of the acetyl group. Moreover, the derivatization of chitosan with alkyl groups induced high mobility side chains, causing them to be highly hydrated. It follows that the DS and DA of Ch are important parameters for the property of the polysaccharide in solution and for the interaction with HA. A schematic representation of how Ch and HA interact to form a complex is given in Figure S1.

A quantitative ^{1}H NMR method was applied to determine the DA and DS of Ch. The Ch sample with DA and the DS values of 7.5% and 59%, respectively, is soluble in water at neutral pH and it presents a low viscosity, as compared to the same molecular weight of HA, as shown by the HP-SEC-TDA results (Table 1) and the viscosity curves (Figure 6a). HA was approximately 6 times more viscous in terms of intrinsic viscosity than Ch, and the hydrodynamic radius for the negative charge polysaccharide was double, as compared to Ch, although the molecular weights were quite similar. This behavior was confirmed by the DOSY NMR experiment (Table 3). Indeed, the average D values of HA were lower than those of Ch, indicating a lower mobility due to higher viscosity. When the two polysaccharides were mixed, anaverage molecular weight distribution between HA and Ch was obtained, which maintained a high intrinsic viscosity value (Figure 2, Table 2).

The homogeneity of the complex was demonstrated by the Zp measurements (Table 5) —one Zp value was observed and through the comparison of the results, the HA/Ch complex Zp was about −23 mV, lower than that obtained for the HA alone (−34 mV), indicating that the negative charges of HA were partially reduced in the presence of Ch, due to the phenomena of secondary interaction between the polysaccharides. The strict interaction between HA and Ch was also demonstrated by DOSY:the D values of HA and Ch in the complex were similar (Table 3).

It clearly appears that by mixing the two polysaccharides, an interesting modulation of the viscoelastic properties could be easily obtained (Figure 6b). The mechanical spectra of the solutions prepared using only HA or Ch represent two different behaviors—the HA solution had a relevant viscoelastic character, whereas the Ch sample presented a more evident liquid/elastic character. Finally, the HA/Ch solution showed the highest values of both elastic and viscous moduli, with a shift toward higher frequency of the crossover point (Table 4).

SEM analysis on the Ch/HA composite supported the data that suggest an intimate mixing of the components, revealing characteristic features of the individual polysaccharide, such as lamellae, fibers, and pores.

We demonstrated that HA and Ch, solubilized together, can generate secondary interactions that modified the chemico-physical properties of polysaccharides. To verify if these phenomena could affect the biocompatibility, the stability toward enzymatic actions was evaluated, comparing the results obtained from a single component preparation and the complex of HA/Ch. While the HA was fully degraded by Hyaluronidase and was not hydrolyzed by Lysozyme, Ch was mildly degraded by Hyaluronidase, and was fully hydrolyzed by Lysozyme. Using the Hyaluronidase and Lysozyme complex with each polysaccharide, the Mw values were the same for those obtained with the specific enzymes. This meant that there was no cooperative effect in the degradation process. Analyzing the HA/Ch complex, HA hydrolysis was not modified by Ch, wheras Ch appeared more resistant to enzyme hydrolysis in presence of HA. Our hypothesis suggests that the penetration of the enzymes to the network HA/Ch, and in particular, the availability of Ch molecules to the active site of the enzymes were slightly hindered; however, but the hydrolysis processes was maintained.

5. Conclusions

In this study, we investigated the supramolecular structuring of a HA/Ch complex using diverse analytical methods. The HA/Ch solution was characterized in terms of physicochemical properties. The supramolecular complex showed interesting rheological properties and promising preliminary evidence for biocompatibility—monodisperse molecular weight distribution, high viscosity, negative charge, degradation by specific enzyme, such as Hyaluronidase and Lysozyme. Due to the wide range of applications in biomedicine and biotechnology, the development of such polyelectrolyte complexes is of scientific and technological interest. Such an analytical approach would facilitate the development of new formulations, demonstrating the interactions with the different components.

Supplementary Materials: The following are available online at https://www.mdpi.com/2218-273X/11/3/389/s1, Figure S1: Schematic interaction between HA and Ch in Phosphate buffer saline.

Author Contributions: Conceptualization, S.B. and G.B.; investigation, R.L., I.T., C.A., and C.C.; writing—original draft preparation S.B. and G.B.; funding acquisition, G.B. All authors have read and agreed to the published version of the manuscript.

Funding: The research was partially funded by G. Ronzoni foundation.

Data Availability Statement: The data that support the findings of this study are available from the corresponding author upon reasonable request.

Acknowledgments: C.A. acknowledges partial support from the Italian Ministry for University and Research through the Rita Levi Montalcini fellowship for young researchers. C.A. and I.T. acknowledge technical support by Tiziano Catelani for SEM imaging. S.B., R.L., and C.C. acknowledge technical support by Regina Fratiello for HP-SEC-TDA analyses.

Conflicts of Interest: The authors declare no conflict of interest.

References

1. Aravamudhan, A.; Ramos, D.M.; Nada, A.A.; Kumbar, S.G. *Natural Polymers: Polysaccharides and Their Derivatives for Biomedical Applications*; Elsevier Inc.: Amsterdam, The Netherlands, 2014; ISBN 9780123969835.
2. Jaedig, F.; Schefer, L.; Diener, M.; Adamcik, J.; Sa, A. Primary, Secondary, Tertiary and Quaternary Structure Levels in Linear Polysaccharides: From Random Coil, to Single Helix to Supramolecular Assembly. *Biomacromolecules* **2019**, *20*, 1731–1739. [CrossRef]
3. Cho, E.; Jung, S. Supramolecular Complexation of Carbohydrates for the Bioavailability Enhancement of Poorly Soluble Drugs. *Molecules* **2015**, *20*, 19620–19646. [CrossRef]
4. Akiyoshi, K.; Sunamoto, J. Supramolecular assembly of hydrophobized polysaccharides. *Supramol. Sci.* **1996**, *3*, 157–163. [CrossRef]
5. Scott, R.A.; Panitch, A. Glycosaminoglycans in biomedicine. *Wiley Interdiscip. Rev. Nanomed. Nanobiotechnol.* **2013**, *5*, 388–398. [CrossRef] [PubMed]
6. Necas, J.; Bartosikova, L.; Brauner, P.; Kolar, J. Hyaluronic acid (hyaluronan): A review. *Vet. Med.* **2008**, *53*, 397–411. [CrossRef]
7. Dicker, K.T.; Gurski, L.A.; Pradhan-Bhatt, S.; Witt, R.L.; Farach-Carson, M.C.; Jia, X. Hyaluronan: A simple polysaccharide with diverse biological functions. *Acta Biomater.* **2014**, *10*, 1558–1570. [CrossRef]
8. Cowman, M.K.; Schmidt, T.A.; Raghavan, P.; Stecco, A. Viscoelastic Properties of Hyaluronan in Physiological Conditions. *F1000Research* **2015**, *4*, 622. [CrossRef]
9. Swann, D.A.; Radin, E.L.; Nazimiec, M.; Weisser, P.A.; Curran, N.; Lewinnek, G. Role of hyaluronic acid in joint lubrication. *Ann. Rheum. Dis.* **1974**, *33*, 318–326. [CrossRef]
10. Radin, E.L.; Paul, I.L.; Swann, D.A.; Schottstaedt, E.S. Lubrication of synovial membrane. *Ann. Rheum. Dis.* **1971**, *30*, 322–325. [CrossRef] [PubMed]
11. Salwowska, N.; Bebenek, K.A.; Zadło, D.A.; Wcisło-Dziadecka, D.L. Physiochemical properties and application of hyaluronic acid: A systematic review. *J. Cosmet. Dermatol.* **2016**, *15*, 520–526. [CrossRef]
12. Tiwari, S.; Bahadur, P. Modified hyaluronic acid based materials for biomedical applications. *Int. J. Biol. Macromol.* **2019**, *121*, 556–571. [CrossRef]
13. Lapčík, L.; Lapčík, L.; De Smedt, S.; Demeester, J.; Chabreček, P. Hyaluronan: Preparation, structure, properties, and applications. *Chem. Rev.* **1998**, *98*. [CrossRef]

14. Volpi, N.; Schiller, J.; Stern, R. Role, Metabolism, Chemical Modifications and Applications of Hyaluronan. *Curr. Med. Chem.* **2009**, *16*, 1718–1745. [CrossRef]
15. Segura, T.; Anderson, B.C.; Chung, P.H.; Webber, R.E.; Shull, K.R.; Shea, L.D. Crosslinked hyaluronic acid hydrogels: A strategy to functionalize and pattern. *Biomaterials* **2005**, *26*, 359–371. [CrossRef]
16. Yalpani, M.; Hall, L.D. Some chemical and analytical aspects of polysaccharide modifications. III. Formation of branched-chain, soluble chitosan derivatives. *Macromolecules* **1984**, *17*, 272–281. [CrossRef]
17. D'Amelio, N.; Esteban, C.; Coslovi, A.; Feruglio, L.; Uggeri, F.; Villegas, M.; Benegas, J.; Paoletti, S.; Donati, I. Insight into the Molecular Properties of Chitlac, a Chitosan Derivative for Tissue Engineering. *J. Phys. Chem. B* **2013**, *117*, 13578–13587. [CrossRef]
18. Laffleur, F.; Netsomboon, K.; Erman, L.; Partenhauser, A. Evaluation of modified hyaluronic acid in terms of rheology, enzymatic degradation and mucoadhesion. *Int. J. Biol. Macromol.* **2018**, *123*, 1204–1210. [CrossRef] [PubMed]
19. Saitoh, H.; Takagaki, K.; Majimas, M.; Nakamura, T.; Matsuki, A.; Kasai, M.; Narita, H.; Endo, M. Enzymic Reconstruction of Glycosaminoglycan Oligosaccharide Chains Using the Transglycosylation Reaction of Bovine Testicular Hyaluronidase. *J. Biol. Chem.* **1995**, *270*, 3741–3747. [CrossRef] [PubMed]
20. Ren, D.; Yi, H.; Ma, X. The enzymatic degradation and swelling properties of chitosan matrices with different degrees of N-acetylation. *Carbohydr. Res.* **2005**, *340*, 2403–2410. [CrossRef] [PubMed]
21. Deb, S.; Abueva, C.; Kim, B.; Taek, B. Chitosan—hyaluronic acid polyelectrolyte complex scaffold crosslinked with genipin for immobilization and controlled release of BMP-2. *Carbohydr. Polym.* **2015**, *115*, 160–169. [CrossRef]
22. Tsao, C.T.; Chang, C.H.; Lin, Y.Y.; Wu, M.F.; Wang, J.L.; Young, T.H.; Han, J.L.; Hsieh, K.H. Evaluation of chitosan/γ-poly(glutamic acid) polyelectrolyte complex for wound dressing materials. *Carbohydr. Polym.* **2011**, *84*, 812–819. [CrossRef]
23. Lehmann, P.; Symitz, C.; Brezesinski, G.; Kraβ, H.; Kurth, D.G. Langmuir and Langmuir-Blodgett Films of Metallosupramolecular Polyelectrolyte-Amphiphile Complexes. *Langmuir* **2005**, *21*, 5901–5906. [CrossRef]
24. Ji, D.; Kuo, T.; Wu, H.; Yang, J.; Lee, S. A novel injectable chitosan / polyglutamate polyelectrolyte complex hydrogel with hydroxyapatite for soft-tissue augmentation. *Carbohydr. Polym.* **2012**, *89*, 1123–1130. [CrossRef] [PubMed]
25. Denuzière, A.; Ferrier, D.; Domard, A. Interactions between chitosan and glycosaminoglycans (chondroitin sulfate and hyaluronic acid): Physicochemical and biological studies. *Ann. Pharm. Françaises* **2000**, *58*, 47–53.
26. Coueta, F.; Rajana, N. Diego Mantovani Macromolecular Biomaterials for Scaffold-Based Vascular Tissue Engineering. *Macromol. Biosci.* **2007**, *7*, 701–718. [CrossRef]
27. Florczyk, S.J.; Wang, K.; Jana, S.; Wood, D.L.; Sytsma, S.K.; Sham, J.G.; Kievit, F.M.; Zhang, M. Biomaterials Porous chitosan-hyaluronic acid scaffolds as a mimic of glioblastoma microenvironment ECM. *Biomaterials* **2013**, *34*, 10143–10150. [CrossRef]
28. Muzzarelli, C.; Stanic, V.; Gobbi, L.; Tosi, G.; Muzzarelli, R.A.A. Spray-drying of solutions containing chitosan together with polyuronans and characterisation of the microspheres. *Carbohydr. Polym.* **2004**, *57*, 73–82. [CrossRef]
29. Kayitmazer, A.B.; Koksal, A.F.; Kilic Iyilik, E. Complex coacervation of hyaluronic acid and chitosan: Effects of pH, ionic strength, charge density, chain length and the charge ratio. *Soft Matter* **2015**, *11*, 8605–8612. [CrossRef]
30. Furlani, F.; Donati, I.; Marsich, E.; Sacco, P. Characterization of Chitosan/Hyaluronan Complex Coacervates Assembled by Varying Polymers Weight Ratio and Chitosan Physical-Chemical Composition. *Colloids Interfaces* **2020**, *4*, 12. [CrossRef]
31. Lalevée, G.; David, L.; Montembault, A.; Blanchard, K.; Meadows, J.; Malaise, S.; Crépet, A.; Grillo, I.; Morfin, I.; Delair, T.; et al. Highly stretchable hydrogels from complex coacervation of natural polyelectrolytes. *Soft Matter* **2017**, *13*, 6594–6605. [CrossRef]
32. Kononova, S.V.; Kruchinina, E.V.; Petrova, V.A.; Baklagina, Y.G.; Klechkovskaya, V.V.; Orekhov, A.S.; Vlasova, E.N.; Popova, E.N.; Gubanova, G.N.; Skorik, Y.A. Pervaporation membranes of a simplex type with polyelectrolyte layers of chitosan and sodium hyaluronate. *Carbohydr. Polym.* **2019**, *209*, 10–19. [CrossRef]
33. Salamanna, F.; Giavaresi, G.; Parrilli, A.; Martini, L.; Nicoli Aldini, N.; Abatangelo, G.; Frizziero, A.; Fini, M. Effects of intra-articular hyaluronic acid associated to Chitlac (arty-duo®) in a rat knee osteoarthritis model. *J. Orthop. Res.* **2019**, *37*, 867–876. [CrossRef]
34. Tarricone, E.; Mattiuzzo, E.; Belluzzi, E.; Elia, R.; Benetti, A.; Venerando, R.; Vindigni, V.; Ruggieri, P.; Brun, P. Anti-Inflammatory Performance of Lactose-Modified Osteoarthritis Model. *Cells* **2020**, *9*, 1328. [CrossRef]
35. Bertini, S.; Bisio, A.; Torri, G.; Bensi, D. Molecular Weight Determination of Heparin and Dermatan Sulfate by Size Exclusion Chromatography with a Triple Detector Array. *Biomacromolecules* **2005**, *6*, 168–173. [CrossRef]
36. Alekseeva, A.; Raman, R.; Eisele, G.; Clark, T.; Fisher, A.; Larry, S.; Jiang, X.; Torri, G.; Sasisekharan, R.; Bertini, S.; et al. In-depth structural characterization of pentosan polysulfate sodium complex drug using orthogonal analytical tools. *Carbohydr. Polym.* **2020**, *234*, 115913. [CrossRef] [PubMed]
37. Harding, A.; Theisen, C.; Deacon, M.P.; Harding, S.E. *Refractive Increment Data-Book*, 1st ed.; Nottingham University Press: Nottingham, UK, 1999; ISBN 1897676298.
38. Fernandez-Megia, E.; Novoa-Carballal, R.; Quinoà, E.; Riguera, R. Optimal routine conditions for the determination of the degree of acetylation of chitosan by 1 H-NMR. *Carbohydr. Polym.* **2005**, *61*, 155–161. [CrossRef]
39. Donati, I.; Stredanska, S.; Silvestrini, G.; Vetere, A.; Marcon, P.; Marsich, E. The aggregation of pig articular chondrocyte and synthesis of extracellular matrix by a lactose-modified chitosan. *Biomaterials* **2005**, *26*, 987–998. [CrossRef]
40. Johnson, C.S., Jr. Diffusion ordered nuclear magnetic resonance spectroscopy: Principles and applications. *Prog. Nucl. Magn. Reson. Spectrosc.* **1999**, *34*, 203–256. [CrossRef]

41. Pinheiro, J.P.; Domingos, R.; Lopez, R.; Brayner, R.; Fiévet, F.; Wilkinson, K. Determination of diffusion coefficients of nanoparticles and humic substances using scanning stripping chronopotentiometry (SSCP). *Colloids Surf. A Physiochem. Eng. Asp.* **2007**, *295*, 200–208. [CrossRef]
42. Viel, S.; Mannina, L.; Segre, A. Detection of a π—π complex by diffusion-ordered spectroscopy (DOSY). *Tetrahedron Lett.* **2002**, *43*, 2515–2519. [CrossRef]
43. Li, W.; Kagan, G.; Hopson, R.; Williard, P.G. Measurement of Solution Viscosity via Diffusion-Ordered NMR Spectroscopy (DOSY). *J. Chem. Educ.* **2011**, *88*, 1331–1335. [CrossRef]
44. Price, K.E.; Lucas, L.H.; Larive, C.K. Analytical applications of NMR diffusion measurements. *Anal. Bioanal. Chem.* **2004**, *378*, 1405–1407. [CrossRef] [PubMed]
45. Cobas, J.C.; Groves, P.; Martin-Pastor, M.; De Capua, A. New Applications, Processing Methods and Pulse Sequences Using Diffusion NMR. *Curr. Anal. Chem.* **2005**, *1*, 289–305. [CrossRef]
46. Peuravuori, J. NMR Spectroscopy Study of Freshwater Humic Material in Light of Supramolecular Assembly. *Environ. Sci. Technol.* **2005**, *39*, 5541–5549. [CrossRef] [PubMed]
47. Jerschow, A.; Müller, N. Diffusion-separated nuclear magnetic resonance spectroscopy of polymer mixtures. *Macromolecules* **1998**, *31*, 6573–6578. [CrossRef]
48. Ambrosio, L.; Borzacchiello, A.; Netti, P.A.; Nicolais, L. Rheological study on hyaluronic acid and its derivative solutions. *J. Macromol. Sci. Part A Pure Appl. Chem.* **2007**, *36*, 991–1000. [CrossRef]
49. Yang, T.; Chou, C.; Li, C. Preparation, water solubility and rheological property of the N-alkylated mono or disaccharide chitosan derivatives. *Food Res. Int.* **2002**, *35*, 707–713. [CrossRef]
50. Jiang, J.; Oberdörster, G.; Biswas, P. Characterization of size, surface charge, and agglomeration state of nanoparticle dispersions for toxicological studies. *J. Nanoparticle Res.* **2009**, *11*, 77–89. [CrossRef]
51. Hunter, R.J. Interaction between Colloidal Particles. In *Zeta Potential in Colloid Science*; Rowell, R.H., Ottewill, R.L., Eds.; Academic Press: Sydney, Australia, 1981; pp. 363–369.
52. Freitas, C.; Müller, R.H. Effect of light and temperature on zeta potential and physical stability in solid lipid nanoparticle (SLN^{TM}) dispersions. *Int. J. Pharm.* **1998**, *168*, 221–229. [CrossRef]
53. Shah, R.; Eldridge, D.; Palombo, E.; Harding, I. Optimisation and stability assessment of solid lipid nanoparticles using particle size and zeta potential. *J. Phys. Sci.* **2014**, *25*, 59–75.
54. Bertini, S.; Fareed, J.; Madaschi, L.; Risi, G.; Torri, G.; Naggi, A. Characterization of PF4-Heparin Complexes by Photon Correlation Spectroscopy and Zeta Potential. *Clin. Appl. Thromb.* **2017**, *23*, 725–734. [CrossRef] [PubMed]
55. Deville, S. Ice-templating, freeze casting: Beyond materials processing. *J. Mater. Res.* **2013**, *28*, 2202–2219. [CrossRef]
56. Cooney, M.J.; Lau, C.; Windmeisser, M.; Liaw, B.Y.; Klotzbach, T.; Minteer, S.D. Design of chitosan gel pore structure: Towards enzyme catalyzed flow-through electrodes. *J. Mater. Chem.* **2008**, *18*, 667–674. [CrossRef]
57. Orsolini, P.; Antonini, C.; Stojanovic, A.; Malfait, W.J.; Caseri, W.R.; Zimmermann, T. Superhydrophobicity of nanofibrillated cellulose materials through polysiloxane nanofilaments. *Cellulose* **2018**, *25*, 1127–1146. [CrossRef]
58. Antonini, C.; Wu, T.; Zimmermann, T.; Kherbeche, A.; Thoraval, M.-J.; Nystrom, G.; Geiger, T. Ultra-Porous Nanocellulose Foams: A Facile and Scalable Fabrication Approach. *Nanomaterials* **2019**, *9*, 1142. [CrossRef]
59. Pawelec, K.M.; Husmann, A.; Best, S.M.; Cameron, R.E. Ice-templated structures for biomedical tissue repair: From physics to final scaffolds. *Appl. Phys. Rev.* **2014**, *1*, 021301. [CrossRef]
60. Zhang, W.; Jin, X.; Li, H.; Zhang, R.; Wu, C. Injectable and body temperature sensitive hydrogels based on chitosan and hyaluronic acid for pH sensitive drug release. *Carbohydr. Polym.* **2018**, *186*, 82–90. [CrossRef] [PubMed]

Article

Stability Evaluation and Degradation Studies of DAC® Hyaluronic-Polylactide Based Hydrogel by DOSY NMR Spectroscopy

Tatiana Guzzo [1], Fabio Barile [1], Cecilia Marras [1], Davide Bellini [2], Walter Mandaliti [3], Ridvan Nepravishta [3], Maurizio Paci [3] and Alessandra Topai [1,*]

1. Colosseum Combinatorial Chemistry Centre for Technology S.r.l (C4T), Via della Ricerca Scientifica snc, 00133 Rome, Italy; tatiana.guzzo@c4t.it (T.G.); fabaril@gmail.com (F.B.); cm.ceciliamarras@gmail.com (C.M.)
2. Novagenit Srl, 38017 Mezzolombardo (TN), Italy; d.bellini@novagenit.com
3. Department of Chemical Science and Technology, University of Rome, Tor Vergata, 00133 Rome, Italy; w.mandaliti@alice.it (W.M.); nepravishta@gmail.com (R.N.); paci@uniroma2.it (M.P.)
* Correspondence: alessandra.topai@c4t.it

Received: 6 October 2020; Accepted: 21 October 2020; Published: 24 October 2020

Abstract: The stability and the degradation of polymers in physiological conditions are very important issues in biomedical applications. The copolymer of hyaluronic acid and poly-D,L-lactic acid (made available in a product called DAC®) produces a hydrogel which retains the hydrophobic character of the poly-D,L-lactide sidechains and the hydrophilic character of a hyaluronic acid backbone. This hydrogel is a suitable device for the coating of orthopedic implants with structured surfaces. In fact, this gel creates a temporary barrier to bacterial adhesion by inhibiting colonization, thus preventing the formation of the biofilm and the onset of an infection. Reabsorbed in about 72 h after the implant, this hydrogel does not hinder bone growth processes. In the need to assess stability and degradation of both the hyaluronan backbone and of the polylactic chains along time and temperature, we identified NMR spectroscopy as a privileged technique for the characterization of the released species, and we applied diffusion-ordered NMR spectroscopy (DOSY-NMR) for the investigation of molecular weight dispersion. Our diffusion studies of DAC® in physiological conditions provided a full understanding of the product degradation by overcoming the limitations observed in applying classical chromatography approaches by gel permeation UV.

Keywords: DAC® HA-PLA copolymers; biopolymer degradation; polymer stability; DOSY NMR

1. Introduction

Important new advances have been reported about new materials and biomaterials [1]. In the field of biomaterials, polymers and copolymers have found large application in modern medicine [2]. Biodegradability of polymeric biomaterials constituted a significant advantage, being that these materials are able to be broken down and removed when they have exerted their function [2]. Clinically, there are a wide range of applications of degradable polymers, such as surgical sutures and implants. Desired physical, chemical, biological, biomechanical, and degradation properties can be selected and tuned in order to fit all the requirements of the functional demand.

Fortunately, novel materials are developed constantly to meet new challenges providing a growing number of natural and synthetic degradable polymers, investigated for biomedical applications. Particularly, degradable polymers are of interest since these biomaterials are able to be molecularly broken down and eliminated or resorbed without physical removal or surgical revision [3].

In the design of biodegradable biomaterials, great importance should be given to their properties [3]. In fact, these materials should not produce a sustained inflammatory response and have a degradation time coinciding with their function. Moreover, they should have appropriate mechanical properties for the scope, produce non-toxic degradation products that can be readily resorbed or excreted, and include appropriate permeability and processability for the designed application [3].

Orthopedic and traumatology (O&T) are up to 38% of the worldwide leading markets of implanted biomaterials [4], involving each year millions of new patients with an increasing trend. Infections related to implanted medical devices depend on the bacterial capability to establish highly structured multilayered biofilms on artificial surfaces and represent the most devastating complication in O&T, with millions of cases. In many cases, a slow release of antibiotic [5] can help the success of the intervention.

The aim of the present work is to assess the stability and the molecular degradation of a disposable coating of the implanted biomaterial Defensive Antibacterial Coating (DAC). Produced in the form of a powder, this device must be hydrated with water for its preparations, and is injectable alone or in solution, which is associated with an antibiotic to obtain a formulation in a hydrogel. The specific indication is for the prevention of peri-implant infection. This device, based on a novel resorbable hydrogel, would act as a resorbable barrier delivering local antibiofilm and antibacterial compounds. The active drugs will be mixed at the same time as the application of the hydrogel during surgery, allowing the correct choice for any given patient, reducing costs and improving storage life and versatility of use. This hydrogel is a derivative of hyaluronic acid (HA) which is present in large quantity in synovial fluid and vitreous humor, which contributes to these tissues' viscoelastic properties and which plays an important structural role in articular cartilage. HA is a natural linear polymer with disaccharide repeating units, namely, [(1→4)-beta-d-GlcpA-(1→3)-beta-d-GlcpNAc]. Natively, HA has been extracted from a variety of animal tissues in the past. Due to the limited sources and a risk of viral infection, the extraction technique has been replaced by microbial fermentation with a high purification efficiency, low production cost, and low risk rate of cross-species viral infection [6–9]. Owing to its relevantly unique physico-chemical properties and its important biological activities as a drug carrier [10–12], HA has been used in a wide variety of applications, such as food, biomedicine, biomaterials, and cosmetics. The gels obtainable by using cross-linked HA chains by chemical derivatization led to several chemical modifications. The cross-linking has been proposed to achieve chemical and mechanical HA robustness [13–15]. In particular, the conjugation with poly-D,L-lactic acid (PLA) [4,16–19] (Figure 1) gives an ideal product for the predesigned pharmacological use.

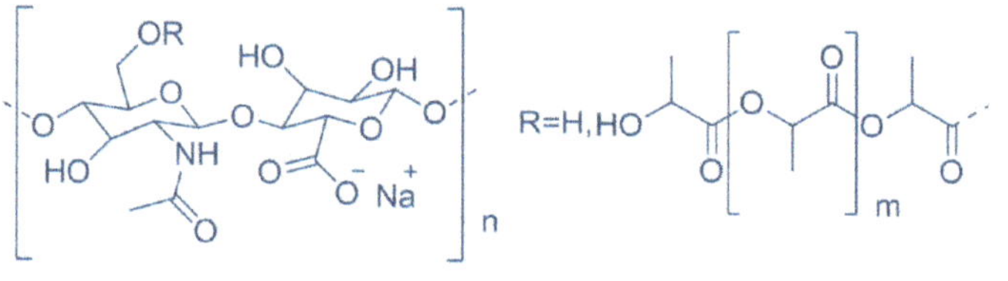

Figure 1. The molecular moieties (**a**) hyaluronic acid (HA) and (**b**) polylactic acid (PLA) forming the copolymer HA-PLA (b/a + b = 13% w/w).

In fact, this HA derivative is important for two main requisites to be reached. First, the stability of the hyaluronic acid chain against depolymerization, and secondly, the stability of the chemical junction between HA polymer and PLA moiety to confer to the polymer the desired physico-chemical characteristics. The sterilization procedure by ionizing through gamma ray irradiation induces a fragmentation of HA, as has been well reported [20]. The samples for the stability study have been synthesized by a new method property of Novagenit. The product has been named HA-PLA Novagenit

DAC® and obtained chemically in a sterile condition with an esterification of HA of about 13% (w/w) at the origin. The stability of the conjugate with sterile polylactic acid has been studied directly at the surface of titanium prosthesis as well as in the bulk region. For these samples, the physico-chemical characteristics, the stability of the copolymer, and its molecular weight dispersion have been studied by NMR spectroscopy [21] over time, which included a temperature cycle. The earlier proposed application of the diffusion-ordered spectroscopy (DOSY) NMR revealed it was particularly successful in this kind of study in order to distinguish among different molar diffusivities due to the different hydrodynamic radii that are often correlated with the different molecular weights [22–25]. In this field, very important advances have been made by DOSY NMR in general [26,27] for the separation of drugs in mixtures [28,29]. It is important that in similar studies addressed to the measure of precise molar diffusivity, should be performed usually with a gradient strength quite higher than that used in this study. Thus, the reported results are to be considered as apparent values that, notwithstanding, allowed us to monitor the macromolecular degradation and the extent of the hydrolysis of the conjugation. The characteristic of linear polymers (far from the usual spherical shape of macromolecules) led us to consider these results valid [30]. A discussion of the obtained results is detailed in a paragraph in the discussion section.

This study appears particularly important because of the diverse biological activities of the hyaluronan fragments [31]. In fact, depending on the polymer length, i.e., small, medium, and large fragments, small and medium ones have pro-angiogenic and anti-apoptotic properties which stimulate the synthesis of heat-shock proteins (HSP) as potent immunostimulants. Immunosuppressive and anti-angiogenic functions are mainly exerted by large polymers [12].

As reference standards with high molecular homogeneity, certified HAs at various molecular weights (MWs) have been used. This approach was already applied in other cases to reveal the presence of hyaluronic acid in mixtures [30]. In our case, this spectroscopic approach revealed to be more precise and sensitive than the results obtained by gel permeation chromatography reported below [32,33].

2. Materials and Methods

2.1. Materials

HA sodium salt HySilk reference standard with MW 13 kDa and 280 kDa were furnished by Giusto Favarelli (Via Medardo Rosso, 8, 20159 Milano, Italy), and HA sodium salt 50 kDa was purchased from Sigma-Aldrich (Milano, Italy). PLA "Purasorb" was purchased from Corbion Biomaterials Purac Biochem (Gorinchem, The Netherlands). HA-PLA in sterile and non-sterile form was prepared by Novagenit (Mezzolombardo-TN, Italy).

2.2. High Molecular Weight HA Sodium Salt

Sodium hyaluronate "high MW" with a value about MW ≥ 500 kDa for pharmaceutical and medical use was purchased from HTL (7 rue Alfred Kastler-ZI de l'Aumaillerie, 35133 Javené, France).

2.3. Synthesis of DAC®

DAC® copolymer was synthesized by esterification of HA with PLA according to the experimental protocol already registered by Novagenit [20]. In order to allow the grafting of PLA to HA, PLA was reacted with carbonyl di-imidazole to obtain the corresponding imidazole derivative (PLA-CI). Subsequently, the reaction between carboxyl-activated PLA and an organic soluble HA form (Hyaluronic acid tetrabutyl ammonium salt, HA-TBA) is carried out in N-methyl pyrrolidone (NMP), 48 h, 37 °C; finally, precipitation and ion exchange lead to the sodium salt form of HA-g-PLA. The final derivatization grade is PLA/(HA + PLA) = 13% w/w.

2.4. Stability Tests

Sample preparation and stability tests were performed according to UNI EN ISO 10993-9 [34], ISO 10993-13 [35] and ISO 13871:1995 [36] guidelines. The test was performed in triplicate, and was designed over 5 study times (from a few seconds to 15 days). For each study time, DAC® hydrogel was uniformly distributed over the roughened surface of the titanium disk. The system was topped with a second titanium disk, obtaining a "sandwich". Each double disk system was immersed in PBS solution at pH 7.2, each container was closed and placed at 37 °C, without stirring, to simulate the physiological condition of the medullary cavity. At each study time, for each sample, the double disk system was separated from its buffered solution, and disk samples and solutions were freeze-dried separately. For the initial study time, sudden disk removal, no sample is recovered from the buffer solution. For the final study time, 15 days, no sample is recovered from the interface of the two disks. A total of 24 samples are therefore recovered for the High Performance Liquid Chromatography (HPLC) and NMR study (Table 1).

Table 1. The 24 samples of HA-PLA (Novagenit DAC®) recovered for HPLC and NMR study both from gel on disk and in buffer solution as indicated (see text).

Time Days	Name	Quantity (mg)	Origin	Name	Quantity (mg)	Origin
0	T0D1	227.2		-	-	
0	T0D2	169.9		-	-	
0	T0D3	163.2		-	-	
1	T1D1	192.3		T1F1	709.3	
1	T1D2	212.3		T1F2	654.9	
1	T1D3	210.2		T1F3	668.7	
3	T3D1	126.2	Gel on disk	T3F1	727.8	Buffer solution
3	T3D2	134		T3F2	754.6	
3	T3D3	126.8		T3F3	733	
7	T7D1	7		T7F1	839.4	
7	T7D2	8		T7F2	818.4	
7	T7D3	7.2		T7F3	829.1	
15	T15D1	-		T15F1	893.1	
15	T15D2	-		T15F2	830.3	
15	T15D3	-		T15F3	848.3	

Sterile titanium disks 8.0 cm diameter and 0.4 cm thick were purchase by Adler Ortho (Via dell'Innovazione, 9, 20032 Cormano, Italy). Titanium disks 6.5 cm diameter and 0.4 cm thick were purchased by Adler and sterilized by gamma irradiation.

2.5. Degradation Method

The degradation method of a single batch of the original sample HA-PLA DAC® from Novagenit is reported in the proprietary patent.

Degradation was obtained by 1.0 mL of NaOH 0.2 M solution heated at 60 °C, then 30 mg of sample is added stirring at constant temperature for 30 min. The aqueous phase is extracted from dichloromethane to eliminate free PLA produced by hydrolysis.

2.6. Analytical Methods

The chromatographic experiments (HPLC-UV) were performed by Agilent instrument 1260 Series. 2 Waters UltraHydrogel columns in series (UH 500, 10 µm, 7.8 × 300 mm and UH250, 6 µm,

7.8 × 300 mm) were eluted with microfiltrated water milliQ PBS at pH 7.2, at 0.8 mL/min flow/rate. 20 µL injection was used. UV detector was set at 200 nm.

2.7. NMR Spectroscopy Measurements

^{1}H NMR were performed on Bruker Avance 300 MHz and Bruker Avance 400 MHz instruments. ^{1}H Diffusion ordered NMR (DOSY) experiments were run on Bruker Avance 700 MHz. Monodimensional NMR spectra were achieved by single-pulse NMR using the solvent suppression only in the case of samples in water solution by using a 90degree pulse and a repetition time as to allow the magnetization to relax completely. The spectra were registered in 8K data point and then transformed in 16K data point. Deuterated solvent for lock purpose was purchased from Sigma Aldrich (Milano, Italy): DMSO-d6, 99.9% D, or D$_2$O 99.9 atom % D. To estimate possible solvent effects on the spectrum of the sample other samples have been prepared in DMSO-d6–D$_2$O mixture (DMSO-d6:D$_2$O = 9:1) or (DMSO-d6:D$_2$O = 8:2) containing all 7.5 mg of sample.

The characteristic resonances of PLA and HA sodium salt were identified: for PLA 3H δ 1.4–1.6 CH3, 1H δ 5.0 CH and for HA sodium salt 3H δ 1.8 CH3(CO-NH); 2H δ 4.5–4.62.

In a diffusion ordered spectroscopy (DOSY) experiment, a series of NMR spectra is normally acquired in a spin or stimulated echo as a function of pulsed-field gradient amplitude, with the amplitude of each signal decaying at a rate determined by the diffusion coefficient. The ideal behavior for unrestricted diffusion is described by the Stejskal–Tanner equation [25].

The diffusion ordered spectroscopy (DOSY) experiments were performed by using the ledbpgppr2s pulse sequence of the Bruker library also in order to suppress the water signal at 4.7 ppm, when necessary. During the DOSY experiment 32 mono dimensional spectra were acquired with 64 scans in a linear increasing gradient varying from 5% to 95% with a pulsed gradient time δ of 70 ms and a diffusion time Δ of 2 ms. The spectra were then analyzed using the DOSY module implemented in Bruker software TOPSPIN 3.1 (Bruker Italy, Milano, Italy).

3. Results and Discussion

The samples were from the same batch of the original sample HA-PLA DAC® from Novagenit obtained as reported in the proprietary patent. The samples have been sterilized as indicated in the preparation protocol. The degree of degradation was obtained as reported in Materials and Methods.

3.1. NMR Spectroscopy Study

The NMR spectra of the HA sodium salt were obtained as reported in Materials and Methods. The observed chemical shifts were aligned with the literature data [8] and the integral of the signals (3H δ 2.08 CH3(CO-NH); 2H δ 4.5–4.62) showed the correct ratio of protons resonances. In total 16H (3 + 11 + 2). For PLA (3H δ 1.4–1.8 CH3;1H δ 5.2–5.4 CH, with a correct ratio between the resonance integrals 3H:1H). A typical spectrum of HA-PLA is reported in Figure 2.

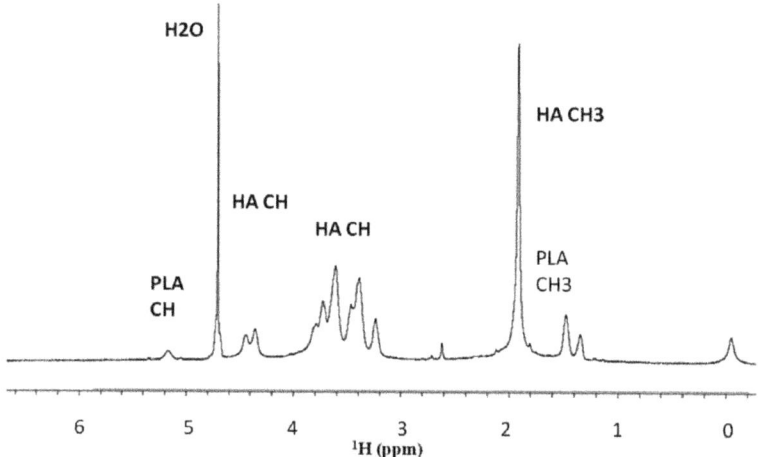

Figure 2. NMR spectrum for HA-PLA DAC® sterilized HA sodium salt–PLA at a concentration of 7.5 mg/mL in D_2O-PBS obtained as reported in Materials and Methods.

The samples of HA-PLA DAC® sterilized were preliminary studied in D_2O as to reduce the large water resonance dissolving at a 7.5 mg/mL concentration. The NMR resonance assignment was achieved by literature data distinguishing between HA and HA-Na salt. A mixture of D_2O-DMSO was also examined. In fact, in the NMR spectra, a marked dependence of the PLA resonances upon the different solvent has been observed. In the solvent mixtures DMSO-d6–D_2O (DMSO-d6:D_2O = 9:1) or (DMSO-d6:D_2O = 8:2) an increase in intensity of the PLA resonances with respect to the HA-Na resonances has been observed at the same concentration, with an increase of the intensity of the HA resonances upon the increase of the D_2O ratio (reported in Supplementary Materials, Figure S1).

During this study, the possibility of micellar aggregation in solvents was also considered. In fact, in the mixtures of water and DMSO at different ratios, it is easy to detect a probable formation of micelles. PLA is much more soluble in the non-aqueous system while HA strongly prefers the aqueous one. This effect has been already reported with copolymers like HA-PLA [17,33].

The stability of the presence of the conjugate between HA and PLA needed to be proved after sterilization. In fact, there was the possibility that the spectrum of Figure 2 would result as a simple additive overlap between the NMR spectrum of the HA and PLA. In our case, the DOSY experiment of HA-PLA after sterilization showed a unique diffusive front and, thus, confirmed the presence of a unique product (HA-PLA) chemically linked together. The observed diffusive front was 8.91×10^{-11} m^2 s^{-1}.

Moreover, DOSY experiments reported in Figure 3 for sterilized HA-PLA DAC® obtained as reported in Materials and Methods revealed that no differences are present in hydrodynamic volumes. This indicates there is only one population with the same diffusion coefficient and a unique molecular weight for the HA-PLA copolymer. Particularly the sharp shape of the peak on the left side of the DOSY representation indicates that nearly a unique species is present in the solution for the observed diffusive front. A weak diffusion is visible at 0 ppm probably due to a trace of the silyl moiety of DSS bound to the macromolecule, due to hydrophobicity of PLA moiety.

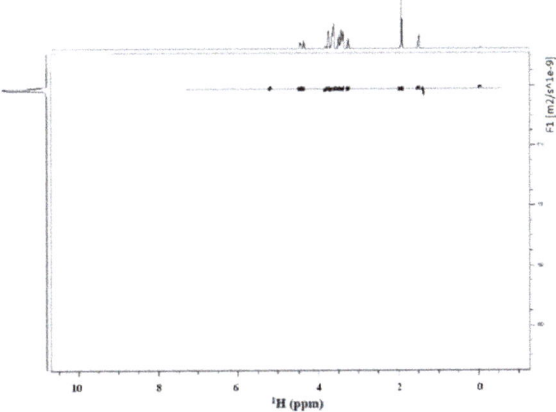

Figure 3. Diffusion ordered NMR (DOSY) experiment of the sample HA-PLA DAC® in D$_2$O-PBS after sterilization. The front of diffusion indicated a Diffusion coefficient D = 8.91 × 10^{-11} m^2 s^{-1}. A weak resonance is reported at 0 ppm probably due to the silyl moiety of DSS bound to the macromolecule.

Thus following the recommendations [29] of using standards with sufficient molecular homogeneity to calibrate the MW, a series of DOSY experiments were conducted on separate solutions of HA fragments with different MWs (13 kDa, 50 kDa, 208 kDa), to check the linearity of hydrodynamic radii and the MWs in hyaluronans. The results are reported in Figure 4 with a final log-log plot in Figure 5 reporting the good correlation found. An investigation was also performed to ascertain if an interference of macromolecules of different sizes in the measure of diffusion could occur. These results are reported in more detail in Supplementary Materials and in a dedicated section of the Discussion.

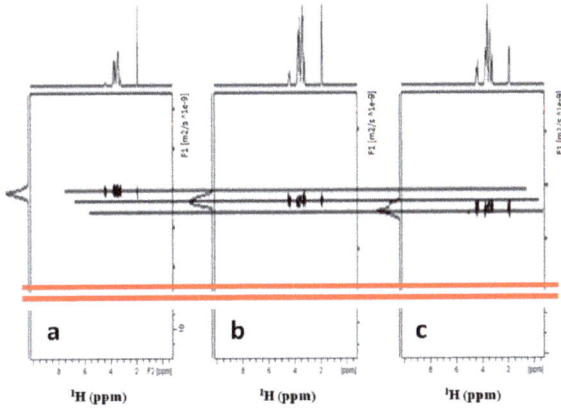

Figure 4. DOSY experiments of three different HA polymer standards in D2O with different MWs. (**a**) 13 kDa; (**b**) 50 kDa; (**c**) 208 kDa. To avoid confusion due to possible interferences, the experiments were run on single polymer samples. The diffusion coefficient obtained are used for the calibration curve in the log-log plot of the Figure 5.

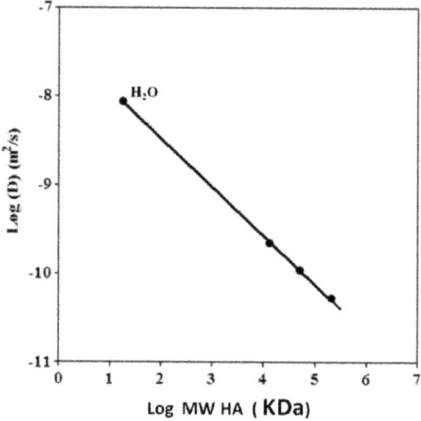

Figure 5. Calibration curve of the diffusion coefficient of fragments of HA with different MWs: 13 kDa; 50 kDa; 208 kDa. The diffusion of the trace of protonated water molecule (1% in D_2O) is also included with a correlation of 0.999.

The calibration curve of standards is reported in the following Figure 5, with a correlation coefficient of 0.999. The inspection of the results of the calibration curve indicated that the experiment is able to distinguish between fragments with MW lower than 13 kDa (from 5.0 to 13 kDa), in line with the expected range of MWs in the degradation experiments. Furthermore, the DOSY results included the resonance of protons of residual water, whose diffusion value displayed a good alignment with those of the fragments of HA observed. A more detailed discussion of the diffusion results is reported in the related paragraph in the Discussion section.

Moreover, a detailed study of the possible interferences between these types of polymers was carried out using a mix of the 13 kDa with the 50 kDa ones and of the 13 kDa with the 208 kDa standard samples. In addition, these results are reported in Supplementary Materials and in the related paragraph in the Discussion section.

3.2. Stability Studies of HA-PLA DAC® Sterilized

The stability studies have been performed by NMR and with DOSY NMR upon the time on the samples reported in Table 1. The NMR and DOSY of sample T15F1 are reported in Figure 6a–d and the NMR spectrum of T7F1 in D_2O are displayed in Figure 7.

In the spectrum in Figure 6a, the decrease of the resonance at 5.2 ppm and the rising of the resonances characteristic of the degradation of PLA into dimers and trimers of lactic acid are evident (Monomer: (CH3) δ 1.23; (CH) δ 4.03; Dimer: (CH3) δ 1.27–1.38; (CH) δ 4.19–4.92; Trimer: (CH3) δ 1.27–1.40–1.44; (CH) δ 4.20–4.98–5.10) [16]. Moreover, DOSY experiments showed evidently that different diffusive fronts due to different hydrodynamic radius are present (a) HA resonances without the resonances characteristic of PLA, (b) the resonances due to PLA moiety and the resonances due to oligomers of lactic acid.

In Figure 6b the left projection circled in red displays species with higher diffusivity than the (a) trace and, then with lower MW than the HA residual. These are also visible in the left projections of the DOSY spectrum, circled in red.

In Figure 6c,d the slices (a) and (b) of the DOSY in Figure 6b are reported and correspond to the spectral profiles of HA and PLA fragments. Furthermore, the diffusion coefficient of the HA polymer for this last study time appears close to that of the HA standard at 50 kDa.

The spectrum of the sample T7F1 in Figure 7 indicates that the ester bond is nearly completely hydrolyzed, and the PLA is completely transformed in lactic monomer, dimer, and trimer, for this study time.

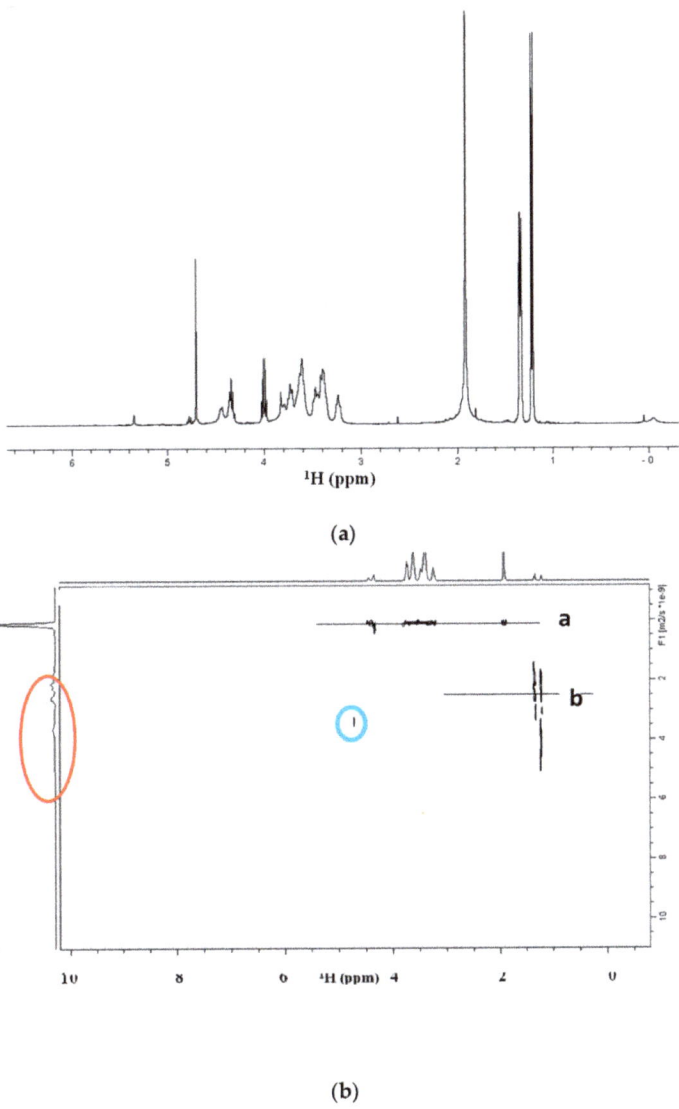

(a)

(b)

Figure 6. Cont.

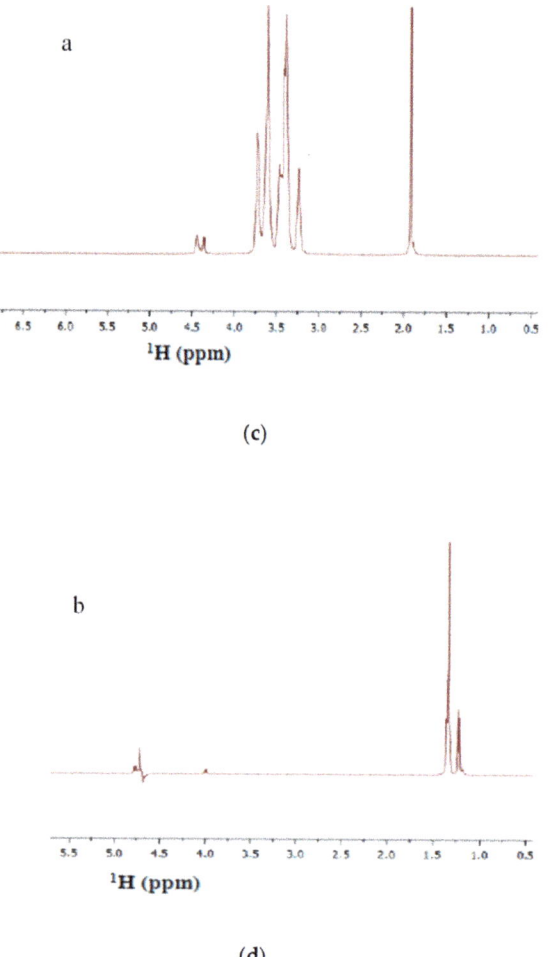

Figure 6. (a) NMR spectrum of the sample of T15F1 of HA-PLA DAC® sterilized and degraded as in Table 1 as reported in Materials and Methods. (b) DOSY spectrum of the sample T15F1. The DOSY indicates the presence of two diffusive fronts due to chemical species "a" and "b". Front "a" represents the residual HA polymer with a MW slightly decreased with respect to the result in Figure 3, showing D = 1.30×10^{-10} m² s⁻¹ corresponding to a MW of 37 kDa. Front "b" represents the fragments of PLA released by hydrolysis of the conjugated bridge. The region of the vertical profile circled in red shows the presence of the fragments with a higher diffusivity (3.9–2.50×10^{-9} m² s⁻¹). The region circled in blue is attributed to the diffusive front of the residual deuterium hydrogen oxide HDO at 4.76 ppm. For D₂O residual protons, the value of D found was 3.79×10^{-9} m² s⁻¹. (c) Slice of front "a" of the DOSY spectrum. (d) Slice of front "b" of the DOSY spectrum.

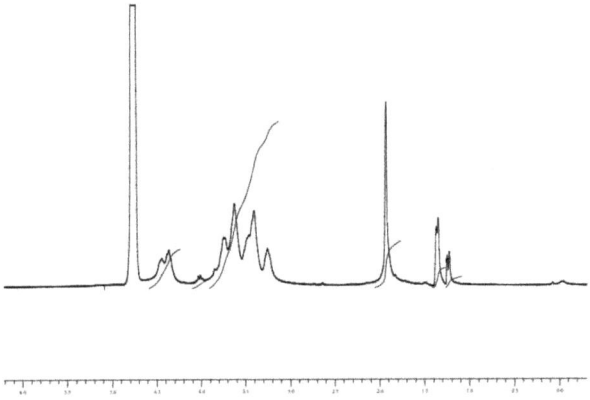

Figure 7. NMR spectrum of sample T7F1 in D2O.

3.3. Study by Gel Permeation Chromatography

After the analysis of the reference standards necessary to obtain the reference values, different batches of the sterile DAC® has been used at a fixed concentration of 1.0 mg/mL. The retention values (R.t.) found are reported in Table 2. The study required us to perform a separate experiment to describe correctly the hydrolysis behavior of the sterile DAC® recovered from the titanium disk surface (D samples) or, alternatively, from bulk exposed to biofluids (simplified as PBS buffer) (F samples). Thus, two series of samples have been studied D and F all in triplicate as listed in Table 1.

Table 2. Chromatographic retention values of samples of Table 1 examined as reported in Materials and Methods.

Name	Quantity (mg)	H_2O (mL)	Concentration (mg/mL)	Retention Time (min)
T0-D1	4.3	4.300	1.00	17.07
T0-D2	6	6.000	1.00	17.39
T0-D3	7.1	7.100	1.00	17.24
T1-D1	2.5	2.500	1.00	17.17
T1-D2	3.3	3.300	1.00	16.68
T1-D3	2.3	2.300	1.00	16.24
T1-F1	8.2	1.171	7.00	16.08
T1-F2	8	1.143	7.00	16.40
T1-F3	7.9	1.129	7.00	16.01
T3-D1	9	9.000	1.00	16.32
T3-D2	4.5	4.500	1.00	15.9
T3-D3	9.6	9.600	1.00	16.01
T3-F1	8	1.143	7.00	15.71
T3-F2	9	1.286	7.00	15.77
T3-F3	9.3	1.329	7.00	15.72
T7-D1	1.3	1.300	1.00	15.87
T7-D2	2.5	2.500	1.00	15.81

Table 2. *Cont.*

Name	Quantity (mg)	H$_2$O (mL)	Concentration (mg/mL)	Retention Time (min)
T7-D3	3	3.000	1.00	15.84
T7-F1	8.7	1.243	7.00	15.68
T7-F2	8.6	1.229	7.00	15.66
T7-F3	8.4	1.200	7.00	15.65
T15-F1	7.7	1.100	7.00	15.72
T15-F2	7.6	1.086	7.00	15.69
T15-F3	7.8	1.114	7.00	15.70

An important result is that the two D and F series show a similar behavior thus indicating that the proximity to the titanium disk surface it is not able to induce differences in the chemistry of the degradation.

The obtained results indicated that when the ester bond undergoes hydrolysis, the degraded samples lose progressively the amphiphilic character thus decreasing the micellar aggregation and decreasing the retention time significantly.

The chromatographic analysis results can be summarized as follows: the sterilized HA-PLA DAC® shows an apparent MW lower than that measured of the non-sterilized product; the retention time trend at the different study times suggests that hydrolysis occurs at the ester bond between HA and PLA.

In fact, the retention values (Table 2) have a clear decreasing trend upon the time of exposure to hydrolysis thus confirming what was reported in NMR study's section for HA-PLA DAC® with and without sterilization.

In summary, upon time the hydrolysis of the ester bond between HA and PLA increases thus decreasing the marked amphiphilic character of the copolymer and, then, decreasing the tendency to form micellar aggregation.

At the time T15 a single polymeric species is present identified as HA-Na with an apparent MW ≥ 50 kDa.

In Figure 8, the chromatograms of: HA-Na 13 kDa (blue), Ha-Na 50 kDa (red), and T15F1 (green) are reported, displaying a comparable apparent MW for the T15 sample with HA standard at 50 kDa. All the experiments were performed in triplicate as reported in Materials and Methods and the results showed an excellent reproducibility.

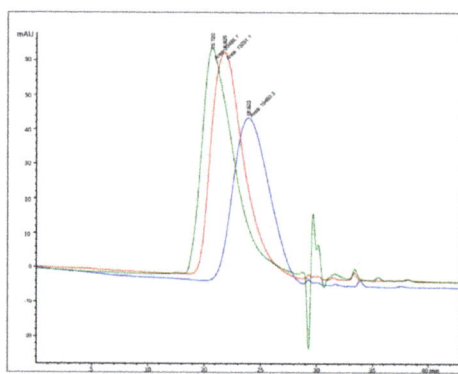

Figure 8. Overlay of the chromatograms of sample HA-Na 13 kDa (blue), Ha-Na 50 kDa (red), and T15F1 (green). In the horizontal axis the retention times in minutes are reported.

3.4. Considerations about Diffusion Data

Taking into consideration the limits of precision in the determination of the absolute value of diffusion coefficients by DOSY due to the available strength of the gradient of the 700 MHz instrument (6G), these results are an interesting example of the utility of this technique for the study of polymer fragmentation. In particular, the diffusion measure of the HA standards proved to discriminate properly the different MWs, considering that a good comparison can be performed for species possessing the same molecular characteristics. In fact, this is a recommendation formulated [29] for a good comparison of the molecular weights, where the use of similar structures eliminates systematic errors related to the use of standards with different chemical features.

In the diffusion measurements of DAC® in D_2O, the unique front shown in Figure 3 has a $D = 1.23 \times 10^{-9}$ m^2 s^{-1}. The values of the diffusive fronts in Figure 6b, are as follows: front (a) $D = 1.30 \times 10^{-10}$ m^2 s^{-1} and front (b) (fragment distribution) from 3.0 to 2.54×10^{-9} m^2 s^{-1}.

In the case of HA certified standard fragments the diffusion coefficients were: 13 kDa $D = 2.22 \times 10^{-10}$ m^2 s^{-1}; 50 kDa $D = 1.11 \times 10^{-10}$ m^2 s^{-1}; 208 kDa $D = 5.18 \times 10^{-11}$ m^2 s^{-1}.

Moreover, a detailed study of the possible interferences between these types of polymers was carried out. In fact, the possibility of entangling should be further investigated. These experiments consist of the DOSY measure of the mix of two standard fragments with different MWs. This was made by observing DOSY in the diffusion coefficient of the same standards using either a mix of the 13 kDa and the 208 kDa samples and the mix of the 50 kDa and 208 kDa. These experiments are reported in Figures S2 and S3.

In the first case (13 kDa and 208 kDa) the diffusion coefficients of the diffusive fronts were $D = 2.29 \times 10^{-10}$ m^2 s^{-1} for the 13 kDa fragments and 1.20×10^{-10} m^2 s^{-1} for the 50 kDa instead of the same fragments alone which were 2.22×10^{-10} m^2 s^{-1} and 1.11×10^{-10} m^2 s^{-1} respectively.

The other experiment was performed on a mix of the 13 kDa and the 208 kDa fragments. Results were for the 13 kDa front $D = 4.07 \times 10^{-10}$ m^2 s^{-1} and for the 208 kDa one $D = 4.09 \times 10^{-11}$ m^2 s^{-1}. Additionally, in this case, a difference is clearly visible between the diffusion D of the fragment alone reported above and the same in the mixture.

In fact, in the case of fragments alone for 13 kDa $D = 2.22 \times 10^{-10}$ m^2 s^{-1} and for 208 kDa $D = 5.18 \times 10^{-11}$ m^2 s^{-1}. The differences are not very large but are indicative that the presence of other macromolecules in the solution can affect the absolute value of the diffusion coefficient. This can be also due either to entangling between macromolecules or to an effect of the change in solution viscosity.

The diffusion of water in these systems (plot of Figure 3) was monitored. The D value of the trace of the deuterium hydrogen oxide (HDO) resonance was 1.86×10^{-9} m^2 s^{-1}.

On the other hand, in Figure 6b the diffusion front of a resonance at 4.76 ppm due to HDO shows a value of 3.79×10^{-9} m^2 s^{-1}. This is the value experimentally determined in the presence of HA-PLA macromolecules and the fragments upon fragmentation. Moreover, the D value of the resonance of HDO molecule in the mix of fragments reported above shows a variable value of D of HDO, in fact in the mix of 13 kDa and 208 kDa D was measured in 4.45×10^{-9} m^2 s^{-1}. On the basis of the literature diffusion of net water as a trace of the protonated water molecule, in practice HDO (1% of H_2O in D_2O) a value $D = 1.12 \times 10^{-9}$ m^2 s^{-1} is reported also with NMR [37–39] and with other physical techniques [40,41]. The variability of the diffusion of the HDO molecule is evidently strongly affected by the size and the dispersity of the macromolecules in the solution. It is also important to consider that HA contains a large number of hydroxyl groups able to alter the property of protons of HDO by the prototropic exchange.

In addition, the preliminary experiments included also the use of a High MW hyaluronic acid purchased as reported in Materials and Methods. The MW was with a value about MW ≥ 500 kDa.

The results are not suitable for publication due to the bad quality of the very noisy spectra but it is interesting that the diffusive profiles on the left of the DOSY spectra indicate the presence of three broad peaks with $D = 1.69 \times 10^{-11}$ m^2 s^{-1}, 2.51×10^{-11} m^2 s^{-1} and 4.68×10^{-11} m^2 s^{-1}. In the same experiment, the lactic acid molecule (100 Da) gave a value of 4.3×10^{-9} m^2 s^{-1}.

4. Conclusions

From the results of chromatographic analysis, HPLC-GPC-UV and NMR spectroscopy included DOSY NMR of sterilized HA-PLA DAC® it is possible to conclude that the sterilization process obtained with ionizing radiation induces a reduction of MW of HA-PLA DAC® as already reported by several authors.

Moreover, HA-PLA DAC® sterilized in physiological conditions (PBS), as in the proprietary Novagenit protocols for checking stability, shows the expected behavior obtaining a complete hydrolysis after about 15 days and the degradation into monomeric, dimeric, and trimeric species within the sensitivity of the techniques used.

The copolymer HA-PLA DAC® sterilized after 15 days in the PBS stability test shows an apparent MW of almost 50 kDa in both the techniques applied. The sterilized HA-PLA DAC® shows an amphiphilic character with micellar aggregation properties in the polar media as indicated by the results in agreement for all the applied techniques.

Finally, in our case, the diffusion studies by DOSY NMR spectroscopy confirm to be a very useful tool in investigating the behavior of chemically conjugated and bioconjugated compounds in pharmaceutical industry research.

Supplementary Materials: The following are available online at http://www.mdpi.com/2218-273X/10/11/1478/s1, Figure S1: Behaviour of HA-PLA conjugate DAC® after sterilization in different solvent mixture indicated, Figure S2: DOSY experiments conducted as reported in Materials and Methods on the mixture of two fragments of HA with different MWs, Figure S3: DOSY experiments conducted as reported in Materials and Methods on the mixture of two fragments of HA with different MWs.

Author Contributions: Conceptualization, A.T., T.G., D.B., M.P., C.M.; Methodology, T.G., R.N., W.M., F.B.; Validation, A.T., R.N., W.M.; Investigation, T.G., C.M., D.B., R.N., W.M., F.B., A.T., M.P.; Resources, A.T., D.B., T.G., M.P.; Data Curation, T.G., R.N., W.M., F.B.; Writing—Original Draft Preparation, C.M., A.T., T.G., M.P.; Writing—Review & Editing, A.T., T.G., D.B., M.P.; Visualization, T.G., C.M.; Supervision, A.T.; Project Administration, A.T., D.B., M.P.; Funding Acquisition, A.T., D.B.,M.P. All authors have read and agreed to the published version of the manuscript.

Funding: This research received no external funding.

Conflicts of Interest: The authors declare no conflict of interest.

References

1. Kowalski, P.S.; Bhattacharya, C.; Afewerki, S.; Langer, R. Smart Biomaterials: Recent Advances and Future Directions. *ACS Biomater. Sci. Eng.* **2018**, *4*, 3809–3817. [CrossRef]
2. Ulery, B.D.; Nair, L.S.; Laurencin, C.T. Biomedical applications of biodegradable polymers. *J. Polym. Sci. Part B Polym. Phys.* **2011**, *49*, 832–864. [CrossRef] [PubMed]
3. Sekhar Nair, N.R.; Nampoothiri, K.M.; Pandey, V.C.A. Current Developments in Biotechnology and Bioengineering: Production, Isolation and Purification of Industrial Products. *Biodegrad. Biopolym.* **2017**, *32*, 739–755.
4. Pitarresi, G.; Palumbo, F.S.; Calascibetta, F.; Fiorica, C.; Di Stefano, M.; Giammona, G. Medicated hydrogels of hyaluronic acid derivatives for use in orthopedic field. *Int. J. Pharm.* **2013**, *449*, 84–94. [CrossRef]
5. Fulzele, S.V.; Satturwar, P.M.; Dorle, A.K. Novel Biopolymers as Implant Matrix for the Deliveryof Ciprofloxacin: Biocompatibility, Degradation, and In Vitro Antibiotic Release. *J. Pharm. Sci.* **2007**, *96*, 132–144. [CrossRef]
6. Izawa, N.; Hanamizu, T.; Sone, T.; Chiba, K. Effects of fermentation conditions and soybean peptide supplementation on hyaluronic acid production by Streptococcus thermophilus strain YIT 2084 in milk. *J. Biosci. Bioeng.* **2010**, *109*, 356–360. [CrossRef]
7. Izawa, N.; Serata, M.; Sone, T.; Omasa, T.; Ohtake, H. Hyaluronic acid production by recombinant Streptococcus thermophilus. *J. Biosci. Bioeng.* **2011**, *111*, 665–670. [CrossRef]
8. Liu, L.; Du, G.; Chen, J.; Zhu, Y.; Wang, M.; Sun, J. Microbial production of low molecular weight hyaluronic acid by adding hydrogen peroxide and ascorbate in batch culture of Streptococcus zooepidemicus. *Bioresour. Technol.* **2009**, *100*, 362–367. [CrossRef]

9. Liu, L.; Wang, M.; Sun, J.; Du, G.; Chen, J. Application of a novel cavern volume controlled culture model to microbial hyaluronic acid production by batch culture of Streptococcus zooepidemicus. *Biochem. Eng. J.* **2010**, *48*, 141–147. [CrossRef]
10. Serban, M.A.; Scott, A.; Prestwich, G.D. Use of Hyaluronan-Derived Hydrogels for Three-Dimensional Cell Culture and Tumor Xenografts. *Curr. Protoc. Cell Biol.* **2008**, *40*, 10–14. [CrossRef]
11. Pouyani, T.; Prestwich, G.D. Functionalized Derivatives of Hyaluronic Acid Oligosaccharides: Drug Carriers and Novel Biomaterials. *Bioconjugate Chem.* **1994**, *5*, 339–347. [CrossRef] [PubMed]
12. Salwowska, N.M.; Bebenek, K.A.; Żądło, D.A.; Wcisło-Dziadecka, D.L. Physiochemical properties and application of hyaluronic acid: A systematic review. *J. Cosmet. Dermatol.* **2016**, *15*, 520–526. [CrossRef] [PubMed]
13. Schanté, C.E.; Zubera, C.; Herlinb, G.; Vandamme, T.F. Chemical modifications of hyaluronic acid for the synthesis of derivatives for a broad range of biomedical applications. *Carbohydr. Polym.* **2011**, *85*, 469–489. [CrossRef]
14. Arpicco, S.; Milla, P.; Stella, B.; Dosio, F. Hyaluronic Acid Conjugates as Vectors for the Active Targeting of Drugs, Genes and Nanocomposites in Cancer Treatment. *Molecules* **2014**, *19*, 3193–3230. [CrossRef]
15. Li, S.M.; Rashkov, I.; Espartero, J.L.; Manolova, N.; Vert, M. Synthesis, Characterization, and Hydrolytic Degradation of PLA/PEO/PLA Triblock Copolymers with Long Poly(L-lactic acid) Blocks. *Macromolecules* **1996**, *29*, 57–62. [CrossRef]
16. Wu, C.S.; Liao, H.T. A new biodegradable blends prepared from Polylactide and hyaluronic acid. *Polymer* **2005**, *46*, 10017–10026. [CrossRef]
17. Liu, C.C.; Chang, K.Y.; Wang, Y.-J. A novel biodegradable amphiphilic diblock copolymers based on poly(lactic acid) and hyaluronic acid as biomaterials for drug delivery. *J. Polym. Res.* **2009**, *17*, 459–469. [CrossRef]
18. Palumbo, F.S.; Pitarresi, G.; Mandracchia, D.; Tripodo, G.; Giammona, G. New graft copolymers of hyaluronic acid and polylactic acid: Synthesis and characterization. *Carbohydr. Polym.* **2006**, *66*, 379–385. [CrossRef]
19. Tripodo, G.; Trapani, A.; Torre, M.L.; Giammona, G.; Trapani, G.; Mandracchia, D. Hyaluronic acid and its derivatives in drug delivery and imaging: Recent advances and challenges. *Eur. J. Pharm. Biopharm.* **2015**, *97*, 400–416. [CrossRef]
20. Giammona, G.; Pitarresi, G.; Palumbo, F.; Romano, C.L.; Meani, E.; Cremascoli, E. Antibacterial Hydrogel and Use Thereof in Orthopedics. International Patent Application No. PCT/EP2010/051117, 5 August 2010.
21. Kvam, B.J.; Atzori, M.; Toffanin, R.; Paoletti, S.; Biviano, F. ^{1}H and ^{13}C NMR studies of solutions of hyaluronic acid esters and salts in methyl sulfoxide: Comparison of hydrogen-bond patterns and conformational behavior. *Carbohydr. Res.* **1992**, *230*, 1–13. [CrossRef]
22. Price, W.S. Pulsed-field gradient nuclear magnetic resonance as a tool for studying translational diffusion: Part 1. Basic theory. *Concepts Magn. Reson.* **1997**, *9*, 299–336. [CrossRef]
23. Price, W.S. Pulsed-field gradient nuclear magnetic resonance as a tool for studying translational diffusion: Part II. Experimental aspects. *Concepts Magn. Reson.* **1998**, *10*, 197–237. [CrossRef]
24. Stilbs, P. Fourier transform pulsed-gradient spin-echo studies of molecular diffusion. *Prog. Nucl. Magn. Reson. Spectrosc.* **1987**, *19*, 1–45. [CrossRef]
25. Stejskal, E.O.; Tanner, J.E. Spin Diffusion Measurements: Spin Echoes in the Presence of a Time-Dependent Field Gradient. *J. Chem. Phys.* **1965**, *42*, 288–292. [CrossRef]
26. Cohen, Y.; Avram, L.; Frish, L. Diffusion NMR spectroscopy. *Angew. Chem. Int. Ed.* **2005**, *44*, 520–554. [CrossRef] [PubMed]
27. Johnson, C. Diffusion ordered nuclear magnetic resonance spectroscopy: Principles and applications. *Prog. Nucl. Magn. Reson. Spectrosc.* **1999**, *34*, 203–256. [CrossRef]
28. Antalek, B. Spin Echo NMR for Chemical Mixture Analysis: How to Obtain OptimumResults. *Concepts Magn. Reson.* **2002**, *14*, 225–258. [CrossRef]
29. Bellomaria, A.; Nepravishta, R.; Mazzanti, U.; Marchetti, M.; Piccioli, P.; Paci, M. Determination of the presence of hyaluronic acid in preparations containing amino acids: The molecular weight characterization. *Eur. J. Pharm. Sci.* **2014**, *63*, 199–203. [CrossRef] [PubMed]
30. Zaccariaa, F.; Zuccacciaa, C.; Cipullo, R.; Macchionia, A. Extraction of Reliable Molecular Information from Diffusion NMR Spectroscopy: Hydrodynamic Volume or Molecular Mass? *Chem. Eur. J.* **2019**, *25*, 9930–9937. [CrossRef]

31. Stern, R.; Asari, A.A.; Sugahara, K.N. Hyaluronan fragments: An information-rich system. *Eur. J. Cell Biol.* **2006**, *85*, 699–715. [CrossRef]
32. Yeung, B.; Marecak, D. Molecular weight determination of hyaluronic acid by gel filtration chromatography coupled to matrix-assisted laser desorption ionization mass spectrometry. *J. Chromatogr. A* **1999**, *852*, 573–581. [CrossRef]
33. Pitarresi, G.; Saiano, F.; Cavallaro, G.; Mandracchia, D.; Palumbo, F. A new biodegradable and biocompatible hydrogel with polyaminoacid structure. *Int. J. Pharm.* **2007**, *335*, 130–137. [CrossRef] [PubMed]
34. International Organization for Standardization. *Biological Evaluation of Medical Devices–Part 9: Framework for Identification and Quantification of Potential Degradation Products*, 3rd ed.; ISO: Geneva, Switzerland, 2019.
35. International Organization for Standardization. *Biological Evaluation of Medical Devices–Part 13: Identification and Quantification of Degradation Products from Polymeric Medical Devices*, 2nd ed.; ISO: Geneva, Switzerland, 2010.
36. International Organization for Standardization/International Electrotechnical Commission. *Information Technology—Telecommunications and Information Exchange between Systems—Private Telecommunications Networks—Digital Channel Aggregation*, 1st ed.; ISO/IEC: Geneva, Switzerland, 1995.
37. Trappeniers, N.; Gerritsma, C.; Oosting, P. The self-diffusion coefficient of water, at 25 °C, by means of spin-echo technique. *Phys. Lett.* **1965**, *18*, 256–257. [CrossRef]
38. Murday, J.S. Self-Diffusion in Liquids: H_2O, D_2O, and Na. *J. Chem. Phys.* **1970**, *53*, 4724. [CrossRef]
39. Weingärtner, H.; Holz, M. 5 NMR studies of self-diffusion in liquids. *Annu. Rep. Prog. Chem., Sect. C Phys. Chem.* **2002**, *98*, 121–156. [CrossRef]
40. Tanaka, K. Self-diffusion Coefficients of Water in Pure Water and in Aqueous Solutioiis of Several Electrolytes with 18 O and 2 H as Tracers. *J. Chem. Soc. Faraday Trans.* **1978**, *74*, 1879–1881. [CrossRef]
41. Suárez-Iglesias, O.; Medina, I.; Sanz, M.D.L.Á.; Pizarro, C.; Bueno, J.L. Self-Diffusion in Molecular Fluids and Noble Gases: Available Data. *J. Chem. Eng. Data* **2015**, *60*, 2757–2817. [CrossRef]

Publisher's Note: MDPI stays neutral with regard to jurisdictional claims in published maps and institutional affiliations.

© 2020 by the authors. Licensee MDPI, Basel, Switzerland. This article is an open access article distributed under the terms and conditions of the Creative Commons Attribution (CC BY) license (http://creativecommons.org/licenses/by/4.0/).

Article

Misincorporation of Galactose by Chondroitin Synthase of *Escherichia coli* K4: From Traces to Synthesis of Chondbiuronan, a Novel Chondroitin-Like Polysaccharide

Mélanie Leroux [1,2], Julie Michaud [2], Eric Bayma [2], Sylvie Armand [2], Sophie Drouillard [2] and Bernard Priem [2,*]

1. HTL Biotechnology, 35133 Javene, France; mleroux@htlbiotech.com
2. CNRS, CERMAV, University Grenoble Alpes, 38000 Grenoble, France; a.julie.michaud@gmail.com (J.M.); eric.bayma@cermav.cnrs.fr (E.B.); Sylvie.Armand@cermav.cnrs.fr (S.A.); sophie.drouillard@cermav.cnrs.fr (S.D.)
* Correspondence: bernard.priem@cermav.cnrs.fr

Received: 25 November 2020; Accepted: 10 December 2020; Published: 12 December 2020

Abstract: Chondroitin synthase KfoC is a bifunctional enzyme which polymerizes the capsular chondroitin backbone of *Escherichia coli* K4, composed of repeated β3N-acetylgalactosamine (GalNAc)-β4-glucuronic acid (GlcA) units. Sugar donors UDP-GalNAc and UDP-GlcA are the natural precursors of bacterial chondroitin synthesis. We have expressed KfoC in a recombinant strain of *Escherichia coli* deprived of 4-epimerase activity, thus incapable of supplying UDP-GalNAc in the bacterial cytoplasm. The strain was also co-expressing mammal galactose β-glucuronyltransferase, providing glucuronyl-lactose from exogenously added lactose, serving as a primer of polymerization. We show by the mean of NMR analyses that in those conditions, KfoC incorporates galactose, forming a chondroitin-like polymer composed of the repeated β3-galactose (Gal)-β4-glucuronic acid units. We also show that when UDP-GlcNAc 4-epimerase KfoA, encoded by the K4-operon, was co-expressed and produced UDP-GalNAc, a small proportion of galactose was still incorporated into the growing chain of chondroitin.

Keywords: chondroitin; K4 chondroitin synthase; NDP-sugar misincorporation

1. Introduction

Bacterial chondroitin is a natural capsular component of some pathogenic strains such as the Gram negative *Escherichia coli* K4 [1], *Pasteurella multocida* type F [2,3], and *Avibacterium paragallinarum* [4]. Chondroitin belongs to the glycosaminoglycan (GAG) family which is found in animal extracellular matrices. Unlike in bacteria, animal chondroitin is normally sulfated and several types of chondroitin sulfate have been reported, depending of sulfation patterns [5]. There is a growing interest in the use of bacterial GAG synthases at the academic and industrial levels in order to produce high yield, structurally well-defined and animal contaminant free products. Synthesis can be achieved either chemo-enzymatically using purified enzymes, or through their expression in suitable, harmless, well known recombinant bacteria. For those purposes, it is critical to characterize and understand the behaviour of available enzymes.

Chondroitin synthase (CS) KfoC has been extensively studied and its tri-dimensional structure has been solved [6]. It is composed of a N-terminal domain A1 which catalyses GalNAc transfer, and a C-terminal domain A2 which catalyses GlcA transfer [7]. KfoC is being used in the engineering of bacterial strains to produce chondroitin at industrial scale [8,9]. Several studies addressed catalytic properties and specificity of KfoC regarding nucleotide-sugars and various acceptors [10–12]. However, studies have never been conducted regarding its behaviour in presence of UDP-galactose.

Here, we report the property of KfoC to incorporate galactose in the absence of UDP-GalNAc in vivo. The ability of some degrading enzymes to cut [β3Gal-β4GlcA]$_n$ polysaccharide we called "chondbiuronan" is also investigated.

2. Materials and Methods

2.1. Chemicals and Molecular Biology Supplies and Services

Chemicals, unless otherwise indicated, were purchased from Sigma-Aldrich Chimie (Saint-Quentin-Fallavier, France). Unsulfated chondroitin was from HTL Biotechnology (Javené, France). DNA polymerases, DNA modification, and restriction enzymes were purchased from Thermo-Fisher Scientific Inc. (Illkirch, France). Synthetic DNA was purchased from GenScript Biotech (Leiden, The Netherlands). Chondroitin AC lyase from *Flavobacterium heparinum* and hyaluronidase from bovine testis were purchased from Sigma-Aldrich. DNA sequencing was performed by Eurofins Genomics (Ebersberg, Germany GmbH). Recombinant DNA was isolated using miniprep columns (Qiagen Inc., Valencia, CA, USA). The *E. coli* strain Top10 was used for plasmid construction and strain K-12 for recombinant polysaccharide production in bioreactor.

2.2. Bacterial Strains and Recombinant Vectors

The *E. coli* DJ strain (i.e., strain DH1 *lacA lacZ wcaJ*), as well as plasmid constructions pBBR-glcATP-kfiD and pBS-kfoC have previously been used in our laboratory to produce heparosan and chondroitin [13,14]. DNA encoding for UDP-GlcNAc 4-epimerase of *E. coli* K4, KfoA [15] was cloned in Sac1-Sma1 sites of pBAD33 to obtain pBAD-kfoA. Engineered strains constructed in this study are summarized in Supporting information, Table S1.

2.3. Culture Media and Culture Conditions for the Production of Polysaccharides

Cultures were carried out in 0.5 L reactors (Infors HT Multifors, Bottmingen, Switzerland) containing 0.2 L of mineral culture medium, as previously described [13]. The high-cell density culture consisted of two phases: (1) an exponential-growth phase at 33 °C, which started with inoculation and lasted until exhaustion of the initially added glucose (17.5 g·L^{-1}), and (2) a 72 h fed-batch phase at 28 °C, with 90 mL of 50% glycerol feeding solution, provided at a flow rate of 1.375 mL·h^{-1}. isopropyl β-D-1-thiogalactopyranoside (IPTG) (0.2 mM) was added to the culture 3 h after the beginning of the glycerol feed. Lactose acceptor (750 mg) and arabinose (0.5% and 2% *w/v* when added) were put in the glycerol feed and supplied continuously. The selecting ampicillin, chloramphenicol, and tetracycline antibiotics were used at 100, 25, and 15 μg·mL^{-1}, respectively.

2.4. Extraction and Purification of Polysaccharides

After 72 h of fed-batch, the culture was centrifuged (10,000× *g*, 30 min) and the supernatant replaced by distilled water. The cells were then autoclaved at 105 °C for 10 min to be disrupted. The supernatant—designated as the "total extract"—was collected after centrifugation (10,000× *g*, 30 min). IR120 H+ Amberlite resin was used to lower its pH to 3.5 and precipitate proteins. This precipitate was then removed from the solution by centrifugation (10,000× *g*, 30 min). The pH was adjusted to 7 using NaOH, and 3 volumes of ice-cold 95% ethanol were added to the total extract to precipitate the polymer. Finally, the precipitated polysaccharide was isolated using centrifugation (10,000× *g*, 30 min, 4 °C). The pellet was then dissolved in distilled water. Pure chondroitin or chondbiuronan was obtained by anion-exchange chromatography on a MonoQ (HR16/10, GE Healthcare, Uppsala, Sweden) column run on an NGC medium-pressure chromatography system (Bio-Rad, Hercules, CA, USA). The column was washed with 20 mL of 10 mM Tris HCl pH 7.6 buffer, then a linear gradient of 50 mL of 0 to 1 M NaCl in the same buffer was applied. The flow rate was 2 mL·min^{-1} and fractions of 4 mL were collected. The presence of GAGs was monitored by UV detection at 210 nm and confirmed using a colorimetric assay [16]. The presence of contaminants was detected at 260 and 280 nm. Polysaccharide containing fractions were pooled and desalted on HighPrep 26/10 (GE Healthcare, Uppsala, Sweden) desalting column in water. The sample was finally lyophilized and stored at −20 °C for downstream analysis.

2.5. Enzymatic Degradation

Chondroitin and chondbiuronan (500 µg) were treated by chondroitin AC lyase from *Flavobacterium heparinum* (0.9 mg) in 500 µL of 50 mM Tris-HCl buffer pH 8 at 25 °C. The reactions were followed by monitoring the formed unsaturated reducing ends at 235 nm.

Chondroitin and chondbiuronan (500 µg) were incubated at 37 °C with hyaluronidase from bovine testis (350 µg) in 250 µL of 0.1 M sodium acetate buffer pH 5 containing 0.1 M NaCl. At regular time intervals, the amount of reducing sugars was determined using the ferricyanide method [17].

2.6. Carbohydrate Analyses

Chondroitin and chondbiuronan were quantified by colorimetric assay of uronic acid [16]. Average molecular weight and molecular weight distributions were determined using high-performance size-exclusion chromatography (HPSEC) (LC-20AD, Shimadzu, Marne La Vallée, France) with on-line multi-angle light scattering (MALS) (MiniDAWN TREOS, Wyatt Technology Corp. (Santa Barbara, CA, USA)) fitted with a K5 cell and a laser wavelength of 660 nm, a refractive index detector, and a viscometer. Columns (OHPAK SB-G guard column and OHPAK SB 806 HQ column (Shodex, Munich Germany) were eluted with 0.1 M NaNO$_3$ containing 0.03% NaN$_3$ at 0.5 mL·min^{-1}. Solvent and samples were filtered through 0.1 µm and 0.2 µm filter units (Merck-Millipore, Darmstadt, Germany), respectively. The dn/dc determined with a refractometer was of 0.1306 and 0.1293 for chondbiuronan and chondroitin, respectively. ^{13}C and ^{1}H-NMR spectra were recorded with a BRUKER Avance 400 spectrometer (Ettlingen, Germany) operating at a frequency of 100.618 MHz for ^{13}C and 400.13 MHz for ^{1}H. Samples were solubilized in D$_2$O and analysed at 353 K. Residual signal of the solvent was used as an internal standard: mono-deuterated water (HOD) at 4.25 ppm.

MALDI-TOF(-TOF)-MS: Matrix solution was prepared by dissolving 100 mg of 2,5-dihydroxybenzoic acid (DHB) in 1 mL of a 1:1 solution of water and acetonitrile then 20 µL of N,N-dimethylaniline (DMA) were added to the DHB matrix solution. The analyte (10 mg/mL in water) and DHB/DMA solutions were mixed (1:1) and spotted onto a polished steel MALDI target, then dried at ambient temperature. The MALDI mass analysis were performed in linear negative ion mode with an Autoflex Speed TOF/TOF spectrometer (BrukerDaltonics, Bremen-Lehe, Germany) equipped with a 355 nm Smartbeam II laser.

3. Results

3.1. Engineering of E. Coli to Produce Chondroitin/Chondbiuronan Lactose

Metabolic engineering of recombinant strains is shown in Figure 1.

The recombinant strain capable of polymerizing exogenously added lactose through the action of GAG synthases was previously described [14]. As a derivative of *E coli* DH1, the strain does not synthesize GalNAc, contrary to *E coli* K4, which expresses KfoA, an UDP-GlcNAc 4-epimerase.

Briefly, genes *lacZ* and *wcaJ* were knocked-out, resulting in the incapacity of the strain to degrade lactose (Lac), or to produce colanic acid which is a GlcA-containing polysaccharide [18]. In addition, recombinant mammal β1,3-glucuronyltransferase GlcAT-P and UDP-Glc dehydrogenase KfiD were expressed, thus allowing the production of glucuronyllactose (GlcA-Lac) upon Lac implementation [19,20]. Genes *kfiD* and *glcAT-P* were both expressed in low copy pBBR, while the gene *kfoC* encoding chondroitin synthase was expressed in high copy pBluescript.

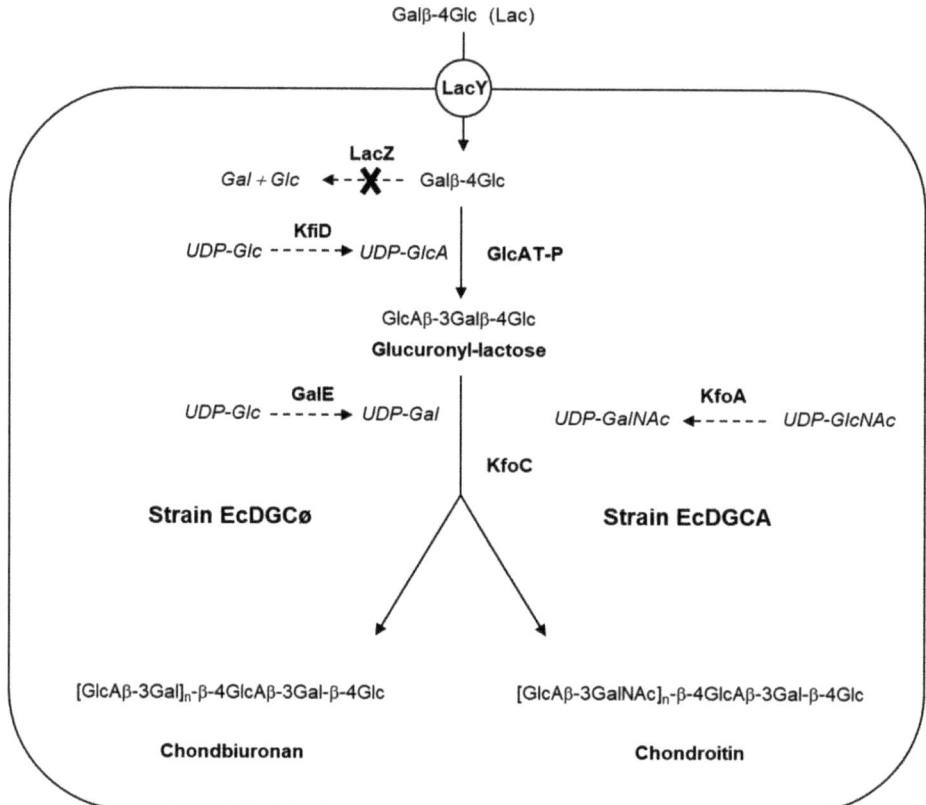

Figure 1. Synthesis of chondbiuronan and chondroitin in engineered *Escherichia coli*. Lactose (Lac) is taken up by lactose permease (LacY) and modified by the action of recombinant mouse β-1,3-glucuronyltransferase (GlcAT-P), leading to glucuronyl-lactose formation, UDP-GlcA being provided by the action of recombinant UDP-Glc dehydrogenase (KfiD) from the *E. coli* strain K5. Recombinant chondroitin synthase KfoC catalyses (i) synthesis of chondbiuronan in strain EcDGCø from UDP-GlcA and UDP-Gal constitutively produced by GalE or (ii) synthesis of chondroitin in strain EcDGCA from UDP-GlcA and UDP-GalNAc produced by recombinant epimerase KfoA. Genotypes: EcDGCø, strain DJ (pBS-kfoC, pBBR-kfiD); EcDGCA, strain DJ (pBS-kfoC, pBBR-kfiD, pBAD-kfoA).

The strain named "EcDGCø" expressing three recombinant genes *kfoC*, *glcAT-P*, and *kfiD* was constructed to examine the capacity of the enzyme KfoC to produce a polysaccharide in absence of UDP-GalNAc.

The strain "EcDGCA" was constructed in order to observe the incorporation of GalNAc while expressing UDP-GlcNAc 4-epimerase KfoA, providing UDP-GalNAc from UDP-GlcNAc epimerisation. Gene *kfoA* was cloned in medium copy pBAD33 carrying the tightly regulated promoter araBAD. Expression of the epimerase was induced by arabinose, independently of the other three recombinant genes induced by IPTG. It was thus possible to modulate the level of epimerase activity without changing the expression of the other recombinant genes. Strain EcDGCA was very similar to strain K-cho that we have described previously, capable of producing recombinant chondroitin from lactose and lactose-furyl [14].

Recombinant strains cultures were driven on minimal medium and cultivated for 72 h in fed-bach conditions (Figure 2).

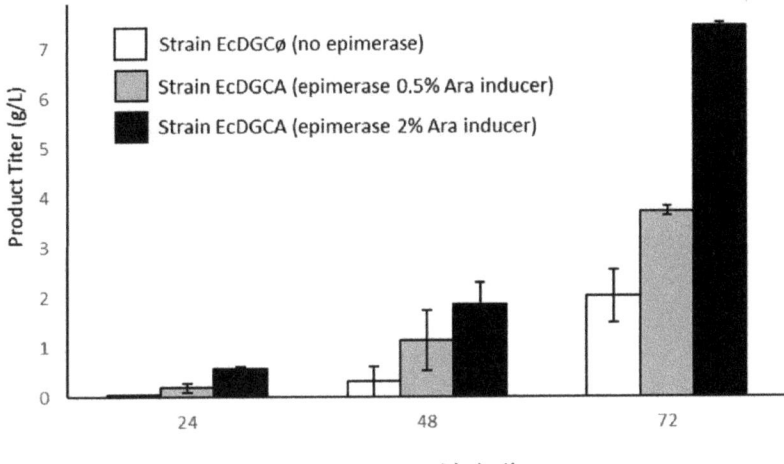

Figure 2. Production of recombinant polysaccharide during growth of engineered strains cultivated in bioreactor. Titer is calculated on the basis of the uronic acid assay. Error bars represent standard deviation of two independent cultures.

All cultures accumulated an ethanol-precipitated product intracellularly as the synthesis occurred from cytoplasmic lactose and *kps* genes involved in polysaccharide export were absent from K-12 strains [21]. Accordingly, strain EcDGCA accumulated several g·L^{-1} of chondroitin at the end of the culture. Interestingly enough, it could be observed that the rate of production increased with epimerase induction controlled by the addition of arabinose, suggesting that it was a limiting parameter of chondroitin synthesis. We can reasonably hypothesize that the epimerase induction level impacts the cytoplasmic concentration of UDP-GalNAc, probably limiting in the polymerization reaction. Strain EcDGCø culture lacking epimerase slightly accumulates a lower amount of unknown polysaccharide (compound 1). However, this compound was only noticeable after 2 days of glycerol feeding.

3.2. Structural Identification of Unknown Polymer Formed by KfoC in Absence of GalNAc

Compound **1** produced by strain EcDGCø as well as chondroitin produced by EcDGCA were purified and subjected to various analyses. The HPSEC-MALS analysis of compound **1** revealed a very low molecular weight, compared to chondroitin samples (Table 1).

Table 1. HPSEC-MALS analysis of chondbiuronan and chondroitin in this work. Error is the mean of two independent experiments.

Culture	Mn	Mw	Mw/Mn
EcDGCø + IPTG	6000 ± 1400	8000 ± 2000	1.34 ± 0.01
EcDGCA + IPTG + 0.5% Ara	13,700 ± 1400	24,000 ± 4800	1.79 ± 0.41
EcDGCA + IPTG + 2% Ara	18,600 ± 1800	30,400 ± 3000	1.63 ± 0.01

We then acquired the ^{1}H NMR (Figure 3A) and ^{13}C NMR (Figure 4A) spectra of compound **1**. They showed characteristic signals of glucuronic acid and galactose linked in β1-3 and β1-4 respectively. Indeed, this data proved that KfoC incorporated galactose in absence of UDP-GalNAc, enough to lead to the synthesis of a small chondroitin-like polysaccharide we named "chondbiuronan."

Considering the very low size of chondbiuronan, we decided to run a mass spectrometry analysis. MALDI-TOF-MS analysis in linear negative mode allowed to obtain several molecular ion clusters separated by 338.3 daltons which correspond to the [Gal-GlcA] motif (Figure 5).

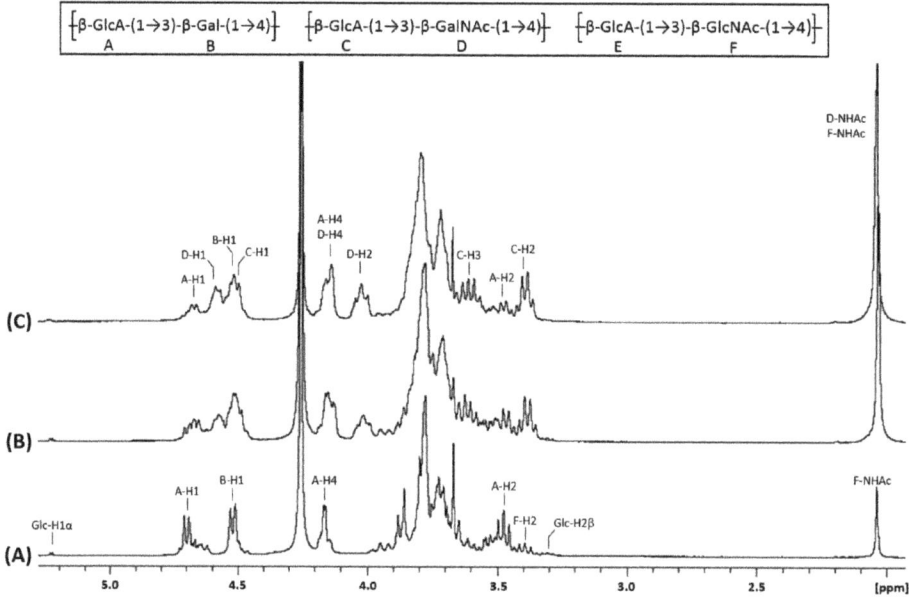

Figure 3. ^{1}H NMR spectra of chondbiuronan (**A**) and chondroitin obtained using 0.5% (**B**) and 2% (**C**) arabinose induction. Inset: chemical structure of the polysaccharides.

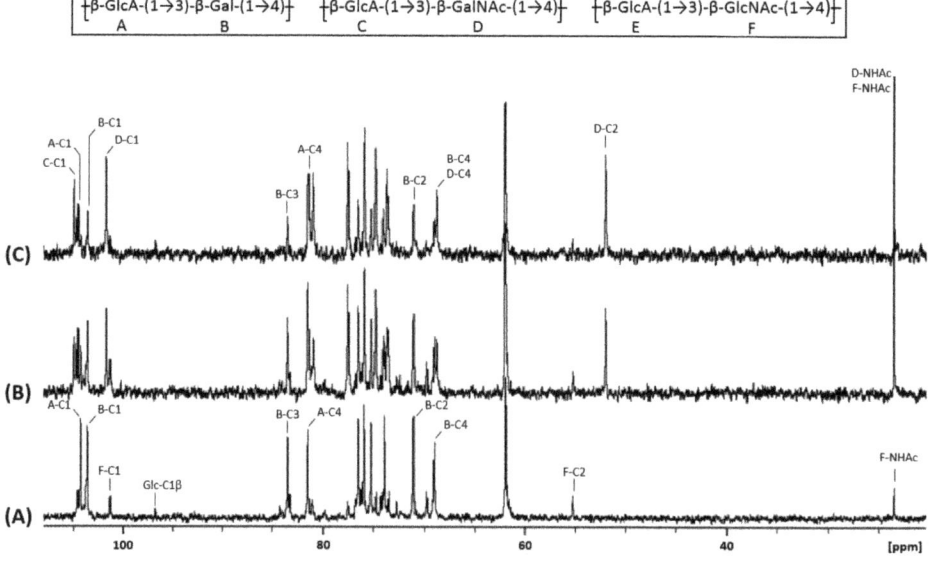

Figure 4. ^{13}C NMR spectra of chondbiuronan (**A**) and chondroitin obtained using 0.5% (**B**) and 2% (**C**) arabinose induction. Inset: chemical structure of the polysaccharides.

Although this analysis does not reflect the true size repartition of the sample as signal detection decreases with size, we clearly see that main molecular ions belong to odd-numbered polymers, indicating that they contain GlcA at non-reducing termini and lactosyl motif at the reducing termini. The fact that GlcA residues terminate elongating chains suggests that the glucuronyltransferase activity of KfoC was not limiting in the polymerization reaction, and confirms that the GalNAc-T activity was the limiting reaction instead, as suggested before.

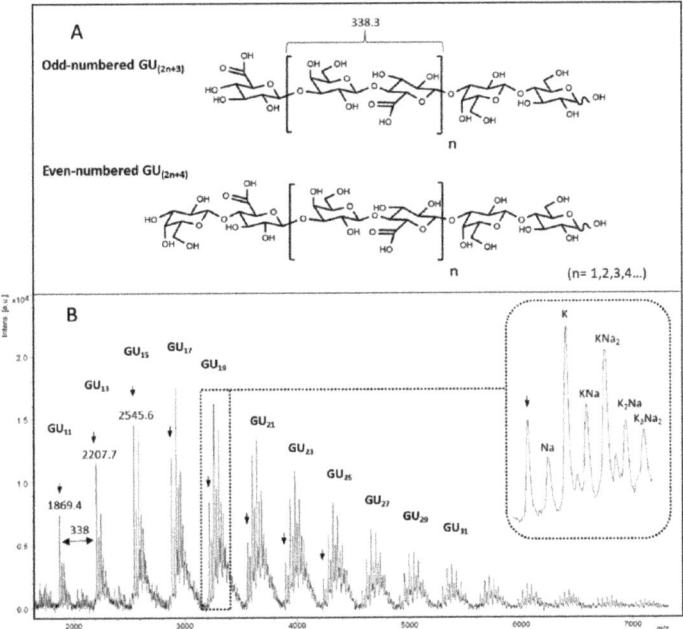

Figure 5. Possible structures of chondbiuronan depending on reducing termini (**A**) and negative-ion linear MALDI mass spectrum of compound **1** (**B**). Acidic [M-H]⁻ peaks are denoted by arrows. Na/H and K/H exchanges are observed increasingly with molecular weight (enlarged peaks).

3.3. Galactose Incorporation Remains in Recombinant Chondroitin

We compared the ^{1}H and ^{13}C NMR spectra of chondbiuronan (Figures 3A and 4A) to the ones of chondroitin produced by the strain EcDGCA at different induction levels of KfoA; 0.5% (Figures 3B and 4B) and 2% arabinose (Figures 3C and 4C). In addition to the signals found with chondbiuronan, they showed NMR signals that we attributed to N-Acetylgalactosamine (GalNAc) residues as well as glucuronic acid (GlcA) residues linked to GalNAc rather than Gal. The level of incorporated Gal could be calculated according to the ^{1}H NMR signals integration: the Gal:GalNAc ratio was of 4:6 and 2:8 at 0.5% and 2% arabinose, respectively, arguing for a competition between Gal and GalNAc incorporation, related to the level of induction of KfoA: the more UDP-GalNAc, the less galactose incorporation.

The chemical shifts of the corresponding protons and carbons were determined using heteronuclear single-quantum correlation experiments (HSQC, Supplementary Figure S1) and correlation spectroscopy (COSY, Supplementary Figure S3) and are reported in Table 2.

Table 2. ^{1}H and ^{13}C NMR chemical shifts (δ, ppm) of chondbiuronan (EcDGCø) and chondroitin (EcDGCA) obtained using 0.5% and 2% arabinose induction.

Sugar Residue		1	2	3	4	5	6 (6a,6b)	NHAc (CH3, CO)
Chondbiuronan EcDGCø								
A→4)-β–D-GlcA-(1→	^{1}H	4.70	3.47	3.65	3.78	3.86	no	
	^{13}C	104.24	73.91	75.25	81.43	76.51	175.78	
B→3)-β–D-Gal-(1→	^{1}H	4.52	3.71	3.79	4.16	3.72	3.79, 3.79	
	^{13}C	103.57	71.05	83.42	68.97	75.95	61.94	
Chondroitin EcDGCA								
A→4)–β–D-GlcA-(1→	^{1}H	4.68	3.47	3.65	3.78	3.85	no	
	^{13}C	104.53	74.01	75.25	81.43	76.53	175.78	
B→3)–β–D-Gal-(1→	^{1}H	4.52	3.71	3.79	4.16	3.72	3.79, 3.79	
	^{13}C	103.57	71.03	83.45	68.99	75.95	61.94	
C→4)–β–D-GlcA-(1→	^{1}H	4.51	3.39	3.61	3.78	3.72	no	
	^{13}C	104.99	73.65	74.80	80.93	77.52	175.11	
D→3)-β–D-GalNAc-(1→	^{1}H	4.58	4.02	3.82	4.14	3.72	3.69, 3.94	2.04
	^{13}C	101.71	52.01	81.36	68.76	75.88	62.00	23.47, 175.75

3.4. N-Acetylglucosamine: Another Misincorporation of KfoC

KfoC is known to incorporate GlcNAc from UDP-GlcNAc in vitro [7]. However, this incorporation has been reported to forbid further polymerization of the growing chain. The NMR experiments of our samples also show the presence of GlcNAc residues in the polysaccharide chains. The F-H3/F-C3 = 3.78/84.21 signals suggest that the third GlcNAc carbon is linked to another GlcA, which highlight the fact that in vivo, the chain elongation continues after the incorporation of a GlcNAc residue. Interestingly, the resulting motif corresponds to the hyaluronan structure. The percentage of GlcNAc incorporated (13%) in the polysaccharide chain was calculated from the ^{1}H NMR signals integration (Figure 3A) by comparing the integrals of anomeric protons and protons from the GlcNAc acetyl residue. The chemical shifts have been determined and are reported in the Supplementary Figure S2.

3.5. Enzymatic Susceptibility of Chondbiuronan

GAG can be enzymatically depolymerized either by eliminative cleavage with lyases (EC 4.2.2.-) resulting in disaccharides or oligosaccharides with Δ4,5-unsaturated uronic acid residues at their non-reducing end or by hydrolytic cleavage with hydrolases (EC 3.2.1.-).

Chondroitin AC lyase from *Flavobacterium heparinum* (EC 4.2.2.5) cleaves a variety of GAGs, including chondroitin sulfates A (chondroitin 4-sulfate) and C (chondroitin 6-sulfate), as well as the unsulfated chondroitin and hyaluronic acid [22]. Bovine testis hyaluronidase (EC 3.2.1.35) is an endo-β-N-acetyl-D-hexosaminidase that hydrolyses hyaluronan and to a lesser extent sulfated or unsulfated chondroitin [23].

Here, the capacity of chondbiuronan to be degraded by these enzymes was evaluated compared to unsulfated chondroitin. As shown in Figure 6, both enzymes could degrade chondbiuronan. Hydrolysis rates of chondbiuronan by chondroitin AC lyase and hyaluronidase were however decreased of about 10-fold and 30-fold respectively, when compared to chondroitin.

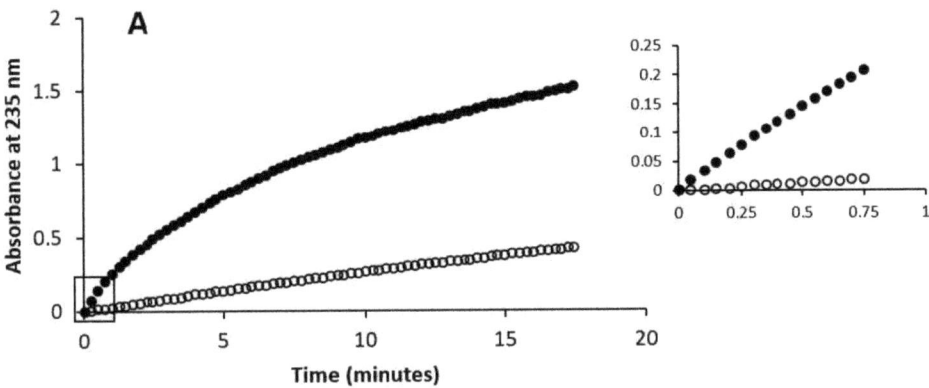

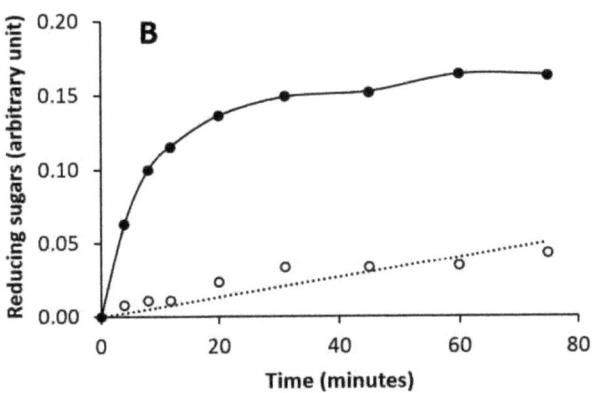

Figure 6. Enzymatic digestion of chondbiuronan and chondroitin by chondroitin AC lyase (**A**) and by hyaluronidase (**B**). The filled or hollow circles represent activity on chondroitin and chondbiuronan respectively. A_{235} is proportional to the concentration of unsaturated non reducing ends formed by the lyase action. Hyaluronidase activity was followed by measuring reducing sugars released from substrates with the ferricyanide method.

4. Discussion

Glycosyltransferases are assumed to be very specific enzymes and misincorporation of NDP-sugars is not common in normal circumstances. Nevertheless, the incorporation of non-natural analogues of N-acetamido sugars carrying various N-groups is a well-known approach for carbohydrate labelling or drug design [24,25]. To some extent, KfoC has been found to be able to take up UDP-GlcNAc as a donor, but no further elongation could be observed after that [7]. In this study, we showed that in vivo, elongation after the incorporation of GlcNAc was still possible, leading to a co-polymer composed of hyaluronic acid and chondroitin. To our knowledge, the ability of KfoC to use UDP-Gal as a donor has not been investigated, nor the presence of galactose among chondroitin backbone. The crystal structure of KfoC has been elucidated, but the binding of UDP-Gal in the binding pocket has not been investigated [26].

Here we showed that in vivo, in the absence of UDP-GalNAc, galactose was incorporated by KfoC leading to the synthesis of a chondroitin-like polymer. A similar phenomenon has already been observed with the heparosan synthase PmHS2. In vitro experiments show that PmHS2 incorporates glucose in absence of UDP-GlcNAc, resulting in a new heparosan-like polymer [-4-GlcAβ1-4-Glcα1-]n named hepbiuronan [26]. Following the same rationale, we named the new chondroitin-like polymer:

chondbiuronan. Interestingly, the PmHS enzyme used for in vitro hepbiuronic synthesis showed a 10-fold decrease in maximum velocity. Here, comparison of the chondbiuronan and chondroitin synthesis suggests a similar phenomenon occurring with KfoC.

Another work dealing with human α-1,3-GalNAc-T (GTA) involved in blood group reported that the enzyme could bind UDP-Gal but could not efficiently stabilize the complex in a catalytically active conformation [27]. Such a control based on the enzyme mobility does not seem to be very efficient in KfoC. This lack of selectivity may be due to the fact that in the context of recombinant strains, engineered metabolic pathways are different from what they are in wild type strains from which recombinant enzymes are issued, and this can result in enhancing aberrant enzymatic behaviour. In our work, only *kfoA* and *kfoC* genes belonging to the chondroitin operon of K4 have been expressed. It cannot be excluded that in normal circumstances, protein interactions occur, which play a role in enzymatic stability and specificity. Moreover, recombinant *kfoC* gene was cloned in high copy plasmid which supposedly differs from *E. coli* K4 wild type expression. Eventually, we pointed out the fact that galactose could be incorporated in the chondroitin backbone when overexpressing KfoC. The NMR study we have conducted should be of great help to investigate the presence of galactose in bacterial chondroitin produced in other biological systems.

We showed that chondbiuronan could be digested by chondroitin AC lyase and by hyaluronidase. Previously, it has been observed that hepbiuronan could be digested by the bacterial heparin lyase III [26]. These results suggest that the specificity for acting on the hexose-substituted GAGs is relaxed in both classes of bacterial enzymes which are widely used in the industry for quality control of GAGs.

5. Conclusions

The fact that chondroitin synthase KfoC could polymerize a GalNAc-free polysaccharide certainly is the most striking result in the present study. The very low molecular weight of the new polysaccharide specie we called chondbiuronan is quite an advantage for chemical coupling and possible use in material coating or drug delivery due to its chondroitin-like structural features and because it could be slowly eliminated by GAG degrading enzymes. The use of lactose derivatives carrying chemical groups such as propargyl, allyl, and furyl groups should be possible as it has been shown that those precursors could be efficiently taken up to produce GAG derivatives in vivo allowing conjugation by click-chemistry [14,28].

Supplementary Materials: The following are available online at http://www.mdpi.com/2218-273X/10/12/1667/s1, Figure S1: HSQC spectra of chondbiuronan (A) and chondroitin (B) obtained using 2% arabinose induction. Inset: chemical structure of the polysaccharides, Figure S2: HSQC spectrum of chondbiuronan showing the presence of Glc and GlcNAc incorporated in the polymer. Inset: chemical structure of the polysaccharides, Figure S3: COSY spectra of chondbiuronan (A) and chondroitin (B) obtained using 2% arabinose induction. Inset: chemical structure of the polysaccharides, Table S1: Genes, plasmids and *E. coli* strains used in this study.

Author Contributions: Conceptualization, M.L. and B.P.; strains, M.L. and B.P.; cultures and product extraction, M.L. and J.M.; SEC-MALS analyses, E.B.; NMR analyses, S.D.; enzymatic analyses, S.A.; original draft preparation, B.P.; supervision, B.P. All authors have read and agreed to the published version of the manuscript.

Funding: This research was supported by a CIFRE fellowship from ANRT (Association Nationale de la Recherche et de la Technologie), grant number 2016/0814; Labex ARCANE and CBH-EUR-GS (ANR-17-EURE-0003), Glyco@Alps (ANR-15-IDEX-02), and PolyNat Carnot Institut (ANR-16-CARN-0025-01).

Acknowledgments: We acknowledge Laure Fort from ICMG (FR 2607) for mass spectrometry analyses and the NMR platform for its technical support.

Conflicts of Interest: The authors declare no conflict of interest.

References

1. Rodriguez, M.L.; Jann, B.; Jann, K. Structure and serological characteristics of the capsular K4 antigen of *Escherichia coli* O5:K4:H4, a fructose-containing polysaccharide with a chondroitin backbone. *Eur. J. Biochem.* **1988**, *177*, 117–124. [CrossRef]
2. Rimler, R.B. Presumptive identification of *Pasteurella multocida* serogroup A, serogroup D and serogroup F by capsule depolymerization with mucopolysaccharidases. *Vet. Rec.* **1994**, *134*, 191–192. [CrossRef]

3. Rimler, R.B.; Register, K.B.; Magyar, T.; Ackermann, M.R. Influence of chondroitinase on indirect hemagglutination titers and phagocytosis of *Pasteurella multocida* serogroups A, D and F. *Vet. Microbiol.* **1995**, *47*, 287–294. [CrossRef]
4. Wu, J.-R.; Chen, P.-Y.; Shien, J.-H.; Shyu, C.-L.; Shieh, H.K.; Chang, F.; Chang, P.-C. Analysis of the biosynthesis genes and chemical components of the capsule of *Avibacterium paragallinarum*. *Vet. Microbiol.* **2010**, *145*, 90–99. [CrossRef]
5. Truppe, W.; Kresse, H. Uptake of Proteoglycans and Sulfated Glycosaminoglycans by Cultured Skin Fibroblasts. *JBIC J. Biol. Inorg. Chem.* **1978**, *85*, 351–356. [CrossRef]
6. Osawa, T.; Sugiura, N.; Shimada, H.; Hirooka, R.; Tsuji, A.; Shirakawa, T.; Fukuyama, K.; Kimura, M.; Kimata, K.; Kakuta, Y. Crystal structure of chondroitin polymerase from K4. *Biochem. Biophys. Res. Commun.* **2009**, *378*, 10–14. [CrossRef]
7. Ninomiya, T. Molecular Cloning and Characterization of Chondroitin Polymerase from *Escherichia coli* strain K4. *J. Biol. Chem.* **2002**, *277*, 21567–21575. [CrossRef]
8. Zanfardino, A.; Restaino, O.F.; Notomista, E.; Cimini, D.; Schiraldi, C.; De Rosa, M.; De Felice, M.; Varcamonti, M. Isolation of an *Escherichia coli* K4 kfoC mutant over-producing capsular chondroitin. *Microb. Cell Factories* **2010**, *9*, 34. [CrossRef]
9. Jin, P.; Zhang, L.; Yuan, P.; Kang, Z.; Du, G.; Chen, J. Efficient biosynthesis of polysaccharides chondroitin and heparosan by metabolically engineered *Bacillus subtilis*. *Carbohydr. Polym.* **2016**, *140*, 424–432. [CrossRef]
10. Sobhany, M.; Kakuta, Y.; Sugiura, N.; Kimata, K.; Negishi, M. The structural basis for a coordinated reaction catalysed by a bifunctional glycosyltransferase in chondroitin biosynthesis. *J. Biol. Chem.* **2012**, *287*, 36022–36028. [CrossRef]
11. Sugiura, N.; Shimokata, S.; Minamisawa, T.; Hirabayashi, J.; Kimata, K.; Watanabe, H. Sequential synthesis of chondroitin oligosaccharides by immobilized chondroitin polymerase mutants. *Glycoconj. J.* **2008**, *25*, 521–530. [CrossRef]
12. Xue, J.; Jin, L.; Zhang, X.; Wang, F.; Ling, P.; Sheng, J.-Z. Impact of donor binding on polymerization catalyzed by KfoC by regulating the affinity of enzyme for acceptor. *Biochim. Biophys. Acta (BBA) Gen. Subj.* **2016**, *1860*, 844–855. [CrossRef]
13. Barreteau, H.; Richard, E.; Drouillard, S.; Samain, E.; Priem, B. Production of intracellular heparosan and derived oligosaccharides by lyase expression in metabolically engineered *E. coli* K-12. *Carbohydr. Res.* **2012**, *360*, 19–24. [CrossRef]
14. Priem, B.; Peroux, J.; Colin-Morel, P.; Drouillard, S.; Fort, S. Chemo-bacterial synthesis of conjugatable glycosaminoglycans. *Carbohydr. Polym.* **2017**, *167*, 123–128. [CrossRef]
15. Zhu, H.-M.; Sun, B.; Li, Y.-J.; Meng, D.-H.; Ju-Zheng, S.; Wang, T.-T.; Wang, F.; Sheng, J.-Z. KfoA, the UDP-glucose-4-epimerase of *Escherichia coli* strain O5:K4:H4, shows preference for acetylated substrates. *Appl. Microbiol. Biotechnol.* **2017**, *102*, 751–761. [CrossRef]
16. Blumenkrantz, N.; Asboe-Hansen, G. New method for quantitative determination of uronic acids. *Anal. Biochem.* **1973**, *54*, 484–489. [CrossRef]
17. Kidby, D.; Davidson, D. A convenient ferricyanide estimation of reducing sugars in the nanomole range. *Anal. Biochem.* **1973**, *55*, 321–325. [CrossRef]
18. Stevenson, G.; Andrianopoulos, K.; Hobbs, M.; Reeves, P.R. Organization of the *Escherichia coli* K-12 gene cluster responsible for production of the extracellular polysaccharide colanic acid. *J. Bacteriol.* **1996**, *178*, 4885–4893. [CrossRef]
19. Yavuz, E.; Drouillard, S.; Samain, E.; Roberts, I.; Priem, B. Glucuronylation in *Escherichia coli* for the bacterial synthesis of the carbohydrate moiety of nonsulfated HNK-1. *Glycobiology* **2007**, *18*, 152–157. [CrossRef]
20. Bastide, L.; Priem, B.; Fort, S. Chemo-bacterial synthesis and immunoreactivity of a brain HNK-1 analogue. *Carbohydr. Res.* **2011**, *346*, 348–351. [CrossRef]
21. Studier, F.W.; Daegelen, P.; Lenski, R.E.; Maslov, S.; Kim, J.F. Understanding the differences between genome sequences of *Escherichia coli* B Strains REL606 and BL21(DE3) and comparison of the *E. coli* B and K-12 genomes. *J. Mol. Biol.* **2009**, *394*, 653–680. [CrossRef]
22. Rye, C.S.; Withers, S.G. Elucidation of the Mechanism of Polysaccharide Cleavage by Chondroitin AC Lyase from *Flavobacterium heparinum*. *J. Am. Chem. Soc.* **2002**, *124*, 9756–9767. [CrossRef]
23. Kakizaki, I.; Ibori, N.; Kojima, K.; Yamaguchi, M.; Endo, M. Mechanism for the hydrolysis of hyaluronan oligosaccharides by bovine testicular hyaluronidase. *FEBS J.* **2010**, *277*, 1776–1786. [CrossRef]

24. Wolf, S.; Warnecke, S.; Ehrit, J.; Freiberger, F.; Gerardy-Schahn, R.; Meier, C. Chemical synthesis and enzymatic testing of CMP-sialic acid derivatives. *ChemBioChem* **2012**, *13*, 2605–2615. [CrossRef]
25. Pouilly, S.; Bourgeaux, V.; Piller, F.; Piller, V. Evaluation of analogues of GalNAc as substrates for enzymes of the mammalian GalNAc salvage pathway. *ACS Chem. Biol.* **2012**, *7*, 753–760. [CrossRef]
26. Lane, R.S.; Ange, K.S.; Zolghadr, B.; Liu, X.; Schäffer, C.; Linhardt, R.J.; DeAngelis, P.L. Expanding glycosaminoglycan chemical space: Towards the creation of sulfated analogs, novel polymers and chimeric constructs. *Glycobiology* **2017**, *27*, 646–656. [CrossRef]
27. Gagnon, S.M.L.; Meloncelli, P.J.; Zheng, R.B.; Haji-Ghassemi, O.; Johal, A.R.; Borisova, S.N.; Lowary, T.L.; Evans, S.V. High resolution structures of the human ABO(H) blood group enzymes in complex with donor analogs reveal that the enzymes utilize multiple donor conformations to bind substrates in a stepwise manner. *J. Biol. Chem.* **2015**, *290*, 27040–27052. [CrossRef]
28. Fort, S.; Birikaki, L.; Dubois, M.-P.; Antoine, T.; Samain, E.; Driguez, H. Biosynthesis of conjugatable saccharidic moieties of GM2 and GM3 gangliosides by engineered *E. coli*. *Chem. Commun.* **2005**, *20*, 2558–2560. [CrossRef]

Publisher's Note: MDPI stays neutral with regard to jurisdictional claims in published maps and institutional affiliations.

© 2020 by the authors. Licensee MDPI, Basel, Switzerland. This article is an open access article distributed under the terms and conditions of the Creative Commons Attribution (CC BY) license (http://creativecommons.org/licenses/by/4.0/).

MDPI
St. Alban-Anlage 66
4052 Basel
Switzerland
Tel. +41 61 683 77 34
Fax +41 61 302 89 18
www.mdpi.com

Biomolecules Editorial Office
E-mail: biomolecules@mdpi.com
www.mdpi.com/journal/biomolecules

www.ingramcontent.com/pod-product-compliance
Lightning Source LLC
LaVergne TN
LVHW070217100526
838202LV00015B/2056